D1348871

Cosmology and Gravitation

Spin, Torsion, Rotation, and Supergravity

NATO ADVANCED STUDY INSTITUTES SERIES

A series of edited volumes comprising multifaceted studies of contemporary scientific issues by some of the best scientific minds in the world, assembled in cooperation with NATO Scientific Affairs Division.

Series B. Physics

Recent Volumes in this Series

Volume 52 – Physics of Nonlinear Transport in Semiconductors
edited by David K. Ferry, J. R. Barker, and C. Jacoboni

Volume 53 – Atomic and Molecular Processes in Controlled Thermonuclear Fusion
edited by M. R. C. McDowell and A. M. Ferendeci

Volume 54 – Quantum Flavordynamics, Quantum Chromodynamics, and Unified Theories
edited by K. T. Mahanthappa and James Randa

Volume 55 – Field Theoretical Methods in Particle Physics
edited by Werner Rühl

Volume 56 – Vibrational Spectroscopy of Molecular Liquids and Solids
edited by S. Bratos and R. M. Pick

Volume 57 – Quantum Dynamics of Molecules: The New Experimental Challenge to Theorists
edited by R. G. Woolley

Volume 58 – Cosmology and Gravitation: Spin, Torsion, Rotation, and Supergravity
edited by Peter G. Bergmann and Venzo De Sabbata

Volume 59 – Recent Developments in Gauge Theories
edited by G. 't Hooft, C. Itzykson, A. Jaffe, H. Lehmann, P. K. Mitter, I. M. Singer, and R. Stora

Volume 60 – Theoretical Aspects and New Developments in Magneto-Optics
Edited by Jozef T. Devreese

Volume 61 – Quarks and Leptons: *Cargèse 1979*
edited by Maurice Lévy, Jean-Louis Basdevant, David Speiser, Jacques Weyers, Raymond Gastmans, and Maurice Jacob

Volume 62 – Radiationless Processes
edited by Baldassare Di Bartolo

This series is published by an international board of publishers in conjunction with NATO Scientific Affairs Division

A Life Sciences
B Physics
Plenum Publishing Corporation
London and New York

C Mathematical and Physical Sciences
D. Reidel Publishing Company
Dordrecht, Boston and London

D Behavioral and Social Sciences
E Applied Sciences
Sijthoff & Noordhoff International Publishers
Alphen aan den Rijn, The Netherlands, and Germantown U.S.A.

Cosmology and Gravitation

Spin, Torsion, Rotation, and Supergravity

Edited by
Peter G. Bergmann
Syracuse University
Syracuse, New York

and
Venzo De Sabbata
University of Bologna
Bologna, Italy
and
University of Ferrara
Ferrara, Italy

PLENUM PRESS • NEW YORK AND LONDON
Published in cooperation with NATO Scientific Affairs Division

Library of Congress Cataloging in Publication Data

Nato Advanced Study Institute on Cosmology and Gravitation: Spin, Torsion, Rotation, and Supergravity, Bologna, 1979.
Cosmology and gravitation.

(Nato advanced study institutes series: Series B, Physics; v. 58)
"Proceedings of the NATO Advanced Study Institute on Cosmology and Gravitation: Spin, Torsion, Rotation, and Supergravity, held at the Ettore Majorana International Center for Scientific Culture, Bologna, Italy, May 6–18, 1979."
1. Gravitation–Congresses. 2. Gauge theories (Physics)–Congresses. 3. Supergravity–Congresses. 4. Torsion–Congresses. I. Bergmann, Peter Gabriel. II. De Sabbata, Venzo. III. North Atlantic Treaty Organization. Division of Scientific Affairs. IV. Title. V. Series.
QC178.N33 1979 530.1 80-23742
ISBN 0-306-40478-8

Proceedings of the NATO Advanced Study Institute on Cosmology and Gravitation:
Spin, Torsion, Rotation, and Supergravity, held at the Ettore Majorana International Center for Scientific Culture, Erice, Italy, May 6–18, 1979.

A Division of Plenum Publishing Corporation
227 West 17th Street, New York, N.Y. 10011

Printed in the United States of America

PREFACE

For the Sixth Course of the International School of Cosmology and Gravitation of the "Ettore Majorana" Centre for Scientific Culture we choose as the principal topics torsion and supergravity, because in our opinion it is one of the principal tasks of today's theoretical physics to attempt to link together the theory of elementary particles and general relativity. Our aim was to delineate the present status of the principal efforts directed toward this end, and to explore possible directions of work in the near future.

Efforts to incorporate spin as a dynamic variable into the foundations of the theory of gravitation were poineered by E. Cartan, whose contributions to this problem go back half a century. According to A. Trautman this so-called Einstein-Cartan theory is the simplest and most natural modification of Einstein's 1916 theory. F. Hehl has contributed a very detailed and comprehensive analysis of this topic, original view of non-Riemannian space-time.

Characteristic of Einstein-Cartan theories is the enrichment of Riemannian geometry by torsion, the non-symmetric part of the otherwise metric-compatible affine connection. Torsion has a impact on the theory of elementary particles. According to V. de Sabbata, weak interactions can be based on the Einstein-Cartan geometry, in that the Lagrangian describing weak interactions and torsion interaction possess analogous structures, leading to a unification of weak and gravitational forces.

Elementary particle physics, in its turn, suggests the importance of new types of symmetry, which lend themselves to gauging and which can serve as the point of departure for an effort towards a genuine, not merely formal unification effort, encompassing the various kinds of physical interaction. From that point of view unification might amount to a description of nature in terms of elementary particles and their mutual interactions. As the weak and the electromagnetic interaction appear to be aspects of a unitary gauge-symmetric field theory, it is natural to ask whether the strong interactions may not be brought into this picture and, more ambitiously, the gravitational interactions as well. This is in fact the center piece of this book, where there are to be found articles dealing with the current status of this quest.

The program of unification sketched out here presents particular difficulties for the inclusion of gravitations. Not only will one have to understand how gravitation is related to the other fundamental forces in nature, but there is at present no viable theory of gravitation consistent with the principles of quantum physics. In this respect "supergravity" may turn out to be a major step towards unification, with new ideas on both these problems. However, true unification is achieved only if the fields occurring in the theory cannot be decomposed invariantly under the group of symmetry operations.

The mathematical apparatus for supersymmetry and for supergravity, including graded Lie algebras, has been developed for our course in great detail by Y. Ne'eman and P. van Nieuvenhuizen. P. Bergman has shown that the rigidity of the space-time manifold with its world points as elementary constituents, is loosened already in standard general relativity if the theory is presented in terms of Dirac's Hamiltonian formalism.

In the context of supergauge theories the modern techniques of differential geometry, and especially of fiber bundles, have been emphasized by A. Trautman. Another kind of approach towards unification, involving complexification procedures, was presented by E. Newman and J. Goldberg, and also by R. Penrose, who talked about his development of twistor physics.

Individuals topics discussed included bimetric general relativity (N. Rosen), the motion of spinning particles (B. Średniawa), and a covariant formulation of quantum theory (E. Schumatzer). Last but far from least, R. Reasenberg presented recent results and current plans to measure gravitational effects to second post-Newtonian order with new imaginative procedures.

We hope that the readers of this book, which consists of the manuscripts contributed by our lecturers, will experience the same sense of exhiliration and of anticipation of things to come as those who were priviliged to attend the lectures and discussions in person.

Peter G. Bergmann

Venzo de Sabbata

CONTENTS

PART I: THEORIES WITH TORSION

Generalities on Geometric Theories of Gravitation. 1
A. Trautman

Four Lectures on Poincaré Gauge Field Theory 5
F.W. Hehl

The Macroscopic Limit of the Poincaré Gauge Field Theory of Gravitation. 63
J. Nitsch

QuasiClassical Limit of the Dirac Equation and the Equivalence Principle in the Riemann-Cartan Geometry 93
H. Rumpf

Contracted Bianchi Identities and Conservation Laws in Poincaré Gauge Theories of Gravity. 105
W. Szczyrba

The Gauge Symmetries of Gravitation. 117
M.A. Schweizer

The Motion of Test-Particles in Non-Riemannian Space-Time. 125
D.-E. Liebscher

Torsion and Strong Gravity in The Realm of Elementary Particles and Cosmological Physics 139
V. de Sabbata and M. Gasperini

PART III: SUPERSYMMETRIES, TWISTORS AND OTHER SYMMETRY GROUPS

The Fading World Point 173
P.G. Bergman

Superalgebras, Supergroups, and Geometric Gauging 177
Y. Ne'eman

Four Lectures at the 1979 Erice School on Spin, Torsion, Rotation, and Supergravity 227
P. van Nieuwenhuizen

Self Dual Fields 257
J.N. Goldberg

An Introduction to Complex Manifolds 275
E.T. Newman

A Brief Outline of Twistor Theory 287
R. Penrose

PART III: EXPERIMENTAL RELATIVITY AND OTHER TOPICS

Experimental Gravitation with Measurements Made from Within a Planetary System 317
R.D. Reasenberg

Tests of General Relativity at the Quantum Level 359
E. Fischbach

The Mass-Angular Momentum-Diagram of Astronomical Objects 375
P. Brosche

Bimetric General Relativity Theory 383
N. Rosen

Covariance and Quantum Physics-Need for a New Foundation of Quantum Theory? 407
E. Schmutzer

Relativistic Equations of Motion of "Spin Particles" 423
B. Średniawa

Angular Momentum of Isolated Systems in General Relativity 435
A. Ashtekar

Isometries and General Solutions of Non-Linear Equations 449
F.J. Chinea

On the Visual Geometry of Spinors and Twistors 457
H. Hellsten

Gravitation Photoproduction in Static Electromagnetic Fields and Some Astrophysical Applications 467
S.R. Valluri

APPENDIX

Invariant Deduction of the Gravitational Equations from the Principle of Hamilton 479
A. Palatini

On a Generalization of the Notion of Reimann Curvature and Spaces with Torsion 489
E. Cartan

Comments on the Paper by Elie Cartan: Sur une Generalisation de la Notion de Courbure de Riemann et les Espaces a Torsion 493
A. Trautman

INDEX 497

GENERALITIES ON GEOMETRIC THEORIES OF GRAVITATION

Andrzej Trautman

Instytut Fizyki Teoretycznej
Uniwersytet Warszawski
Hoza 69
00-681 Warszawa, Poland

All classical, local theories of spacetime and gravitation are based on a rather small number of assumptions about the geometry, the form of the field equations and the nature of the sources. The basic assumptions may be formulated in such a way as to allow an easy comparison between the theories. To achieve this, it is convenient to distinguish the 'kinematic' part of the assumptions, referring to the type of geometry, from the 'dynamic' part, which consists in specifying the form of the field equations.

The kinematics of essentially all theories is based on a four-dimensional differentiable manifold M as the model of spacetime. The manifold is endowed with at least two geometric structures: a connection and a metric structure. The connection is necessary to compare - and, in particular, to differentiate - objects such as vectors and tensors, needed to describe momenta, forces, field strengths, etc. In most cases a linear connection is used, but it is possible to develop all or parts of physics on the basis of other connections (affine, conformal). For example, to compare directions and to define straight (autoparallel) lines it suffices to consider a projective connection, defined as the equivalence class of linear connections whose coefficients $\Gamma^{\mu}_{\nu\rho}$ are related by

$$\bar{\Gamma}^{\mu}_{\nu\rho} = \Gamma^{\mu}_{\nu\rho} + \delta^{\mu}_{\nu}\lambda_{\rho} \quad .$$

A metric structure is needed to measure distances, time intervals, angles and relative velocities. A theory is relativistic if its metric structure is given by a metric tensor g of signature (+++-). A somewhat weaker metric structure is called conformal geometry: it is given by an equivalence class of metric tensors, two tensors g and $\bar{g}$ being considered as equivalent if and only if they differ by a point-dependent factor,

$$\bar{g} = e^{\sigma} g .$$

Conformal geometry is enough to write 'gauge equations' such as source-free Maxwell and Yang-Mills equations.

In most theories, the metric structure and the connection are assumed to be compatible. For example, a conformal geometry is compatible with a projective connection if the property of being a null direction is preserved under parallel transport. In a relativistic theory, the metric tensor is compatible with a linear connection iff the latter is metric, $\nabla_{\mu} g_{\nu\rho} = 0$.

From the kinematic point of view, the majority of viable theories falls into one of the following classes:
(i) Newtonian theories based on a linear connection Γ and a Galilean metric structure, given by a symmetric tensor $(h^{\mu\nu})$ of signature (+++0). A suitably normalized zero eigenform (τ_{μ}) of h is the 1-form of absolute simultaneity. In the standard theory, (τ_{μ}) is the gradient of absolute time.
(ii) Bimetric theories have, in addition to the metric tensor g , another symmetric tensor $(h_{\mu\nu})$ as a basic variable. Linear connection(s) are built from g (and/or h) by the Christoffel formula or its modifications. There are two main subclasses:

1. Linearized theories of gravity interpret h as the gravitational potential meaningful only up to transformations

$$h_{\mu\nu} \to h_{\mu\nu} + \nabla_{\mu} a_{\nu} + \nabla_{\nu} a_{\mu} ,$$

where (a_{μ}) is arbitrary and ∇ is defined by the Levi Civita connection of the metric g ;

2. Bimetric theories in the strict sense had been proposed by N. Rosen and recently considered also by A.A. Logunov. In these theories, matter variables couple to h only, but the Lagrangian of gravitation is allowed to depend on both g and h .

(iii) Riemann-Cartan theories assume a compatible pair (Γ, g) as determining the underlying geometry. The most important among them are:
1. Einstein's theory of 1915, based on Riemannian geometry, and
2. the Einstein-Cartan theory, which is a slight modification and generalization of Einstein's theory, obtained by allowing a non-zero torsion $Q^{\mu}{}_{\nu\rho}$.

Other possibilities are:

3. the Nordström theory, based on conformally flat Riemannian geometry;
4. theories with distant parallelism (teleparallelism); they are dual to Riemannian theories in the sense that they assume vanishing curvature, but $Q^{\mu}{}_{\nu\rho} \neq 0$.
They are also called tetrad theories (C. Møller) as they admit a family of preferred fields of tetrads, defined up to constant Lorentz rotations.

All theories listed under (ii) and (iii) are relativistic: the tangent spaces to the spacetime manifold in any of these theories have a geometry equivalent to that of Minkowski space.

The principle of equivalence played a heuristic role in arriving at Einstein's relativistic theory of gravitation. One is tempted to formulate it today in the following, rather sharp, way: in the vacuum, the geometry of spacetime defines in lowest differential order only one linear connection. This principle, if accepted, may be used to rule out many of the bimetric theories listed under (ii.2).

To develop a definite theory of gravitation it is necessary to
A. Specify its kinematics, i.e. the type of geometry;
B. Write the field equations of gravitation and the equations of motion of other types of matter;
C. Give a physical interpretation to the quantities occurring in, or derivable from, the geometrical model;
D. Study the consequences of the kinematic and dynamic aspects of the theory.
These general guidelines require comments which can be here only very brief. The field equations of gravitation usually follow the pattern of the Poisson equation

$$\Delta \varphi = 4\pi\rho$$

of the Newtonian theory: there is a "left-hand side" constructed from the potentials and a "right-hand side" describing the sources. In most cases, the left-hand side is obtained from a variational principle whereas the form of the sources follows from either
(a) a Lagrangian L depending on both gravitational and matter variables, or
(b) phenomenological considerations.
In the first case, the interaction of matter with gravitation is achieved by imposing a principle of minimal coupling which is a prescription how to go over from the special-relativistic Lagrangian to its general-relativistic counterpart without introducing explicitly the curvature tensor. The second approach is less satisfactory from the point of view of foundations, but more

suitable to astrophysical applications. In either case, it is necessary to specify which geometric elements of the theory have a dynamical significance, i.e. are determined from the field equations and which are 'absolute', independent of the particular physical situation. For example, the metric tensor is absolute in special relativity and dynamical in the general theory. Torsion is absolute - and zero - in Einstein's theory, but acquires a dynamical significance in generalized theories such as the Einstein-Cartan theory. The group of symmetries of a theory preserves the absolute elements.

It is important to remember that the physical interpretation of the mathematical notions occurring in a physical theory must be compatible with the equations of the theory. For example, it follows from Einstein's equations that the worldlines of 'dust particles' are geodesics; this determines the physical interpretation of the linear connection. When one goes over to a theory with torsion, it is not possible to "generalize" this result by postulating that test particles move along the autoparallels. It turns out that in the Einstein-Cartan theory spinless particles still move along the geodesics of the Riemannian connection. To measure torsion, one has to consider particles with spin.

Acknowledgments

This short article is based on the first of a series of four lectures I gave at Erice. It has been completed in July 1979 during my stay at the International Centre for Theoretical Physics, Trieste. I gratefully acknowledge financial support from the Norman Foundation, which made possible my visit to the Centre.

FOUR LECTURES ON POINCARÉ GAUGE FIELD THEORY*

Friedrich W. Hehl

Institute for Theoretical Physics
University of Cologne, W. Germany†
and
Center for Particle Theory and Center for Theoretical Physics, The University of Texas at Austin‡
Austin, Texas 78712

ABSTRACT

The Poincaré (inhomogeneous Lorentz) group underlies special relativity. In these lectures a consistent formalism is developed allowing an appropriate gauging of the Poincaré group. The physical laws are formulated in terms of points, orthonormal tetrad frames, and components of the matter fields with respect to these frames. The laws are postulated to be gauge invariant under local Poincaré transformations. This implies the existence of 4 translational gauge potentials $\underline{e}^{\alpha}$ ("gravitons") and 6 Lorentz gauge potentials $\underline{\Gamma}^{\alpha\beta}$ ("rotons") and the coupling of the momentum current and the spin current of matter to these potentials, respectively. In this way one is led to a Riemann-Cartan spacetime carrying torsion and curvature, richer in structure than the spacetime of general relativity. The Riemann-Cartan spacetime is controlled by the two general gauge field equations (3.44) and (3.45), in which material momentum and spin act as sources. The general framework of the

---

*Given at the 6th Course of the International School of Cosmology and Gravitation on "Spin, Torsion, Rotation, and Supergravity," held at Erice, Italy, May 1979.

†Permanent address.

‡Supported in part by DOE contract DE-AS05-76ER-3992 and by NSF grant PHY-7826592.

theory is summarized in a table in Section 3.6. - Options for picking a gauge field lagrangian are discussed (teleparallelism, ECSK). We propose a lagrangian quadratic in torsion and curvature governing the propagation of gravitons and rotons. A suppression of the rotons leads back to general relativity.

CONTENTS

Abstract

Lecture 1: General Background
1.1 Particle physics and gravity
1.2 Local validity of special relativity and quantum mechanics
1.3 Matter and gauge fields
1.4 Global inertial frames in the M_4 and action function of matter
1.5 Gauging the Poincaré group and gravity

Lecture 2: Geometry of Spacetime
2.1 Orthonormal tetrad frames and metric-compatible connection
2.2 Local P-transformation of the matter field
2.3 Commutation relations, torsion, and curvature
2.4 Local P-transformation of the gauge potentials
2.5 Closure of the local P-transformations
2.6 Local kinematical inertial frames
2.7 Riemann-Cartan spacetime seen anholonomically and holonomically
2.8 Global P-transformation and the M_4

Lecture 3: Coupling of Matter to Spacetime and the Two General Gauge Field Equations
3.1 Matter Lagrangian in a U_4
3.2 Noether identities of the matter Lagrangian: identification of currents and conservation laws
3.3 The degenerate case of macroscopic (scalar) matter
3.4 General gauge field Lagrangian and its Noether identities
3.5 Gauge field equations
3.6 The structure of Poincaré gauge field theory summarized

Lecture 4: Picking a Gauge Field Lagrangian
4.1 Hypothesis of quasi-linearity
4.2 Gravitons and rotons
4.3 Suppression of rotons I: teleparallelism
4.4 Suppression of rotons II: the ECSK-choice
4.5 Propagating gravitons and rotons
4.6 The Gordon-decomposition argument

Acknowledgments
Literature

LECTURE 1: GENERAL BACKGROUND

1.1 Particle Physics and Gravity

The recent development in particle physics seems to lead to the following overall picture: the fundamental constituents of matter are spin-one-half fermions, namely quarks and leptons, and their interactions are mediated by gauge bosons coupled to the appropriate conserved or partially conserved currents of the fermions. Strong, electromagnetic, and weak interactions can be understood in this way and the question arises, whether the gravitational interaction can be formulated in a similar manner, too. These lectures are dedicated to this problem.

General relativity is the most satisfactory gravitational theory so far. It applies to macroscopic tangible matter and to electromagnetic fields. The axiomatics of general relativity makes it clear that the notions of massive test particles* and of massless scalar "photons" underlie the riemannian picture of spacetime. Accordingly, test particles, devoid of any attribute other than mass-energy, trace the geodesics of the supposed riemannian geometry of spacetime. This highly successful conception of massive test particles and "photons" originated from classical particle mechanics and from the geometrical optics' limit of electrodynamics, respectively. It is indispensable in the general relativity theory of 1915 (GR).

Is it plausible to extrapolate riemannian geometry to microphysics? Or shouldn't we rather base the spacetime geometry on the supposedly more fundamental fermionic building blocks of matter?

1.2 Local Validity of Special Relativity and Quantum Mechanics

At least locally and in a suitable reference frame, special relativity and quantum mechanics withstood all experimental tests up to the highest energies available till now. Consequently we have to describe an isolated particle according to the rules of special relativity and quantum mechanics: Its state is associated with a unitary representation of the Poincaré (inhomogeneous Lorentz) group. It is characterized by its mass m and by its spin s. The universal applicability of the mass-spin classification scheme to all known particles establishes the Poincaré group as an unalterable element in any approach to spacetime physics.

Let us assume then at first the doctrine of special relativity. Spacetime is represented by a 4-dimensional differentiable manifold X_4 the points of which are labelled by coordinates x^i.

*To be more precise: massive, structureless, spherical symmetric, non-rotating, and neutral test particles....

On the X_4 a metric is given, and we require the vanishing of the riemannian curvature. Then we arrive at a Minkowski spacetime M_4. We introduce at each point an orthonormal frame of four vectors (tetrad frame)

$$\underline{e}_\alpha(x^k) = e^i_{\cdot\alpha}\partial_i \quad \text{with} \quad \underline{e}_\alpha \cdot \underline{e}_\beta = \eta_{\alpha\beta} = \text{diag.}(-+++) \quad . \qquad (1.1)$$

Here $\eta_{\alpha\beta}$ is the Minkowski metric.* We have the dual frame (co-frame) $\underline{e}^\alpha = e_i^{\cdot\alpha}dx^i$ and find $e_i^{\cdot\alpha}e^i_{\cdot\beta} = \delta^\alpha_\beta$.

In the framework of the Poincaré gauge field theory (PG) to be developed further down, the field of anholonomic tetrad frames $\underline{e}_\alpha(x^k)$ is to be considered an "irreducible" or primitive concept. We imagine spacetime to be populated with observers. Each observer is equipped with all the measuring apparatuses used in special relativity, in particular with a length and an orientation standard allowing him to measure spatial and temporal distances and relative orientations, respectively. Such local observers are represented by the tetrad field $\underline{e}_\alpha(x^k)$. Clearly this notion of "anholonomic observers" that lies at the foundations of the PG,† is alien to GR, as we saw above. It seems necessary, however, in order to accommodate, at least at a local level, the experimentally well established "Poincaré behavior" of matter, in particular its spinorial behavior.

1.3 Matter and Gauge Fields

After this general remark, let us come back to special relativity. In the M_4 the global Poincaré group with its 10 infinitesimal parameters (4 translations and 6 Lorentz-rotations) is the group of motions. Matter, as mentioned, is associated with unitary representations of the Poincaré group. The internal properties of matter, the flavors and colors, will be neglected in our presentation, since we are only concerned with its spacetime behavior. Accordingly, matter can be described by fields $\psi(x^k)$ which refer to the tetrad $\underline{e}_\alpha(x^k)$ and transform as Poincaré spinor-tensors,

---

*The anholonomic (tetrad or Lorentz) indices $\alpha,\beta,\gamma\cdots$ as well as the holonomic (coordinate or world) indices $i,j,k\cdots$ run from 0 to 3, respectively. For the notation and the conventions compare [1]. In the present article the object of anholonomity (1.5) is defined with a factor 2, however. GR = general relativity of 1915, PG = Poincaré gauge (field theory), P = Poincaré.

†During the Erice school I distributed Kerlick's translation of Cartan's original article.[2] It should be clear therefrom that it is Cartan who introduced this point of view.

respectively. Thereby, technically speaking, the $\psi(x^k)$ a priori carry only anholonomic spinor and tensor indices, which we'll suppress for convenience.

We will restrict ourselves to classical field theory, i.e. the fields $\psi(x^k)$ are unquantized c-number fields. Quantization will have to be postponed to later investigations.

The covariant derivative of a matter field reads

$$D_i\psi(x^k) = (\partial_i + \Gamma_i^{\cdot\alpha\beta} f_{\beta\alpha})\psi(x^k) \quad , \tag{1.2}$$

where the $f_{\alpha\beta}$ are the appropriate constant matrices of the Lorentz generators acting on $\psi(x^k)$. Their commutation relations are given by

$$[f_{\alpha\beta}, f_{\gamma\delta}] = \eta_{\gamma[\alpha} f_{\beta]\delta} - \eta_{\delta[\alpha} f_{\beta]\gamma} \quad . \tag{1.3}$$

The connection coefficients $\Gamma_i^{\cdot\alpha\beta}$, being referred to orthonormal tetrads on an M_4, can be expressed in terms of the object of anholonomity $\Omega_{ij}^{\cdot\cdot\alpha}$ according to

$$\Gamma_{\alpha\beta\gamma} : = e^{i}_{\cdot\alpha}\eta_{\beta\delta}\eta_{\gamma\varepsilon}\Gamma_i^{\cdot\delta\varepsilon} = -\frac{1}{2}\Omega_{\alpha\beta\gamma} + \frac{1}{2}\Omega_{\beta\gamma\alpha} - \frac{1}{2}\Omega_{\gamma\alpha\beta} \tag{1.4}$$

with

$$\Omega_{\alpha\beta\gamma} : = e^{i}_{\cdot\alpha} e^{j}_{\cdot\beta}\eta_{\gamma\delta}\Omega_{ij}^{\cdot\cdot\delta} \quad \text{and} \quad \Omega_{ij}^{\cdot\cdot\alpha} : = 2\partial_{[i} e_{j]}^{\cdot\,\alpha} \quad . \tag{1.5}$$

We can read off from (1.4) the antisymmetry of the connection coefficients,

$$\Gamma_i^{\cdot\alpha\beta} \equiv -\Gamma_i^{\cdot\beta\alpha} \quad , \tag{1.6}$$

i.e. neighboring tetrads are, apart from their relative displacement, only rotated with respect to each other. Furthermore we define $\partial_\alpha : = e^{i}_{\cdot\alpha}\partial_i$ and $D_\alpha : = e^{i}_{\cdot\alpha}D_i$. For the mathematics involved we refer mainly to ref. [3], see also [4].

By definition, a field possessing originally a holonomic index, cannot be a matter field. In particular, as it will turn out, gauge potentials like the gravitational potentials $e_i^{\cdot\alpha}$ and $\Gamma_i^{\cdot\alpha\beta}$ (see Section 2.4) or the electromagnetic potential A_i, emerge with holonomic indices as covariant vectors and do not represent matter fields.* The division of physical fields into matter fields

*Technically speaking gauge potentials are always one-forms with values in some Lie-algebra, see O'Raifeartaigh.[5]

$\psi(x^k)$ and gauge potentials like $e_i^{\cdot\alpha}$, $\Gamma_i^{\cdot\alpha\beta}$, A_i is natural and unavoidable in any gauge approach (other than supergravity). In our gauge-theoretical set-up, the gauge potentials and the associated fields will all be presented by holonomic totally antisymmetric covariant tensors (forms) or the corresponding antisymmetric contravariant tensor densities. Hence there is no need of a covariant derivative for holonomic indices and we require that the D_i acts only on anholonomic indices, i.e.

$$D_{[i}e_{j]}^{\cdot\alpha} = \partial_{[i}e_{j]}^{\cdot\alpha} + \Gamma_{[i|\gamma}^{\cdot\cdot\alpha}e_{|j]}^{\cdot\gamma} , \tag{1.7}$$

for example.†

We have seen that Poincaré matter is labelled by mass <u>and</u> spin. It is mainly this reason, why the description of matter by means of a field should be superior to a particle description: The spin behavior of matter can be better simulated in a field theoretical picture. Additionally, already in GR, and in any gauge approach to gravity, too, gravitation is represented by a field. Hence the coherence of the theoretical model to be developed would equally suggest a field-theoretical description of matter. After all, even in GR matter dust is represented hydrodynamically, i.e. field-theoretically. As a consequence, together with the notion of a particle, the notion of a path, so central in GR, will loose its fundamental meaning in a gauge approach to gravity. Operationally the linear connection will then have to be seen in a totally different context as compared to GR.‡ Only in a macroscopic limit will we recover the conventional path concept again.

1.4 Global Inertial Frames in the M_4 and Action Function of Matter

If we cover the M_4 with cartesian coordinates and orient all tetrads parallely to the corresponding coordinate lines, then we

†Our D_i-operator (see [1]) corresponds to the exterior covariant derivative of ref. [4].

‡In GR the holonomic connection $\tilde{\Gamma}_{ij}^{\cdot\cdot k}$ (the Christoffel) is expressible in terms of the metric and has, accordingly, no independent status. In the equation for the geodesics it represents a field strength acting on test particles. In PG it is the <u>an</u>holonomic $\Gamma_i^{\cdot\alpha\beta}$ which enters as a fundamental variable. It turns out to be the rotational gauge potential. For its measurement we need a Dirac spin, see Section 3.3.

find trivially for the tetrad coefficients

$$e_i^{\cdot\alpha} \stackrel{*}{=} \delta_i^\alpha \qquad (e^j_{\cdot\beta} \stackrel{*}{=} \delta^j_\beta) \quad , \tag{1.8}$$

i.e., in the M_4 we can build up global frames of reference, inertial ones, of course. With respect to these frames, the linear connection vanishes and we have for the corresponding connection coefficients

$$\Gamma_i^{\cdot\alpha\beta} \stackrel{*}{=} 0 \quad . \tag{1.9}$$

We will use these frames for the time being.

The lagrangian of the matter field will be assumed to be of first order $L = L[\eta_{ij}, \gamma^{i\cdots}, \psi(x), \partial_i\psi(x)]$. The action function reads

$$W_m = \int d^4x \; L(\eta_{ij}, \gamma^{i\cdots}, \psi, \partial_i\psi) \quad , \tag{1.10}$$

where γ^i denotes the Dirac matrices, e.g. The invariance of (1.10) under global Poincaré transformations yields momentum and angular momentum conservation, i.e., we find a conserved momentum current (energy-momentum tensor) and a conserved angular momentum current.

1.5 Gauging the Poincaré Group and Gravity

Now the gauge idea sets in. Global or rigid Poincaré invariance is of questionable value. From a field-theoretical point of view, as first pointed out by Weyl[6] and Yang and Mills,[7] and applied to gravity by Utiyama,[8] Sciama,[9] and Kibble,[10] it is unreasonable to execute at each point of spacetime the same rigid transformation. Moreover, what we know experimentally, is the existence of minkowskian metrics all over. How these metrics are oriented with respect to each other, is far less well known, or, in other words, local Poincaré invariance is really what is observed. Spacetime is composed of minkowskian granules, and we have to find out their relative displacements and orientations with respect to each other.

Consequently we substitute the (4 + 6) infinitesimal parameters of a Poincaré transformation by (4 + 6) spacetime-dependent functions and see what we can do in order to save the invariance of the action function under these extended, so-called local Poincaré transformations. (We have to introduce (4 + 6) compensating vectorial gauge potentials, see Lecture 2.)

This brings us back to gravity. According to the equivalence principle, there exists in GR in a freely falling coordinate frame the concept of the local validity of special relativity, too. Hence we see right away that gauging the Poincaré group must be related to gravitational theory. This is also evident from the fact that, by introducing local Poincaré invariance, the conservation of the momentum current is at disposition, inter alia. Nevertheless, the gauge-theoretical exploitation of the idea of a local Minkowski structure leads to a more general spacetime geometry, namely to a Riemann-Cartan or U_4 geometry, which seems to be at conflict with Einstein's result of a riemannian geometry. The difference arises because Einstein, in the course of heuristically deriving GR, treats material particles as described in holonomic coordinate systems, whereas we treat matter fields which are referred to anholonomic tetrads.

These lectures cover the basic features of the Poincaré gauge field theory ("Poincaré gauge," abbreviated PG). Our outlook is strictly <u>phenomenological</u>, hopefully in the best sense of the word. For a list of earlier work we refer to the review article.[1] The articles of Ne'eman,[11] Trautman,[12] and Hehl, Nitsch, and von der Heyde[13] in the Einstein Commemorative Volume together with informations from the lectures and seminars given here in Erice by Ne'eman,[14] Trautman,[15] Nitsch,[16] Rumpf,[17] W. Szczyrba,[18] Schweizer,[19] Yasskin,[20] and by ourselves, should give a fairly complete coverage of the subject. But one should also consult Tunyak,[21] who wrote a whole series of most interesting articles, the Festschrift for Ivanenko,[22] where earlier references of Ivanenko and associates can be traced back, and Zuo et al.[23]

LECTURE 2: GEOMETRY OF SPACETIME

We have now an option. We can either start with an M_4 and substitute the parameters in the P(oincaré)-transformation of the matter fields by spacetime dependent functions and work out how to compensate the invariance violating terms in the action function: this was carried through in ref. [1], where it was shown in detail how one arrives at a U_4 geometry with torsion and curvature. Or, following von der Heyde,[24,25] we can alternatively postulate a <u>local</u> P-structure everywhere on an X_4, derive therefrom in particular the transformation properties of the gauge potentials, and can subsequently recover the global P-invariance in the context of an M_4 as a special case. Both procedures lead to the same results. We shall follow here the latter one.

2.1 Orthonormal Tetrad Frames and Metric Compatible Connection

On an X_4 let there be given a sufficiently differentiable field of tetrad frames $\underline{e}_\alpha(x^k)$. Additionally, we assume the existence of a Minkowski metric $\eta_{\alpha\beta}$. Consequently, like in (1.1), we can choose the tetrad to be orthonormal, furthermore we can determine $e^i_{\cdot\alpha}$, $e_i^{\cdot\alpha}$, and $e := \det e_i^{\cdot\alpha}$. The relative position of an event with respect to the origin of a tetrad frame is given by $dx^\alpha = e_i^{\cdot\alpha}\, dx^i$ and the corresponding distance by $ds = (\eta_{\alpha\beta}dx^\alpha dx^\beta)^{1/2}$.

Let also be given a local standard of orientation. Then, starting from a tetrad frame $\underline{e}_\alpha(x^k)$, we are able to construct, at a point infinitesimally nearby, a parallelly oriented tetrad

$$\underline{e}''_\alpha(x^k + dx^k) = \underline{e}_\alpha(x^k) + dx^i \Gamma_{i\alpha}^{\cdot\cdot\beta}(x^k)\underline{e}_\beta(x^k) \quad , \tag{2.1}$$

provided the connection coefficients $\Gamma_{i\alpha}^{\cdot\cdot\beta}$ are given.

The metric, and thereby the length standard, are demanded to be defined globally, i.e. lengths and angles must stay the same under parallel transport:

$$\underline{e}''_\alpha(x^k + dx^k)\cdot\underline{e}''_\beta(x^k + dx^k) = \underline{e}_\alpha(x^k)\cdot\underline{e}_\beta(x^k) = \eta_{\alpha\beta} \quad . \tag{2.2}$$

Upon substitution of (2.1) into (2.2), we find a metric compatible connection*

$$\Gamma_i^{\cdot\alpha\beta} = -\Gamma_i^{\cdot\beta\alpha} \quad . \tag{2.3}$$

The (16 + 24) independent quantities $(e_i^{\cdot\alpha}, \Gamma_i^{\cdot\alpha\beta})$ will be the variables of our theory.† The anholonomic metric $\eta_{\alpha\beta}$ is a constant, the holonomic metric

*Observe that $\Gamma_i^{\cdot\alpha\beta}$ now represents an independent variable, it is no longer of the type as given in eq. (1.4). P-gauge invariance requires the existence of an independent rotational potential, see ref. [1].

†Instead of $e_i^{\cdot\alpha}$, we could also use $e^j_{\cdot\beta}$ as independent variable. This would complicate computations, however. We know from electrodynamics that the gauge potential is a covariant vector (one-form) as is the rotational potential $\Gamma_i^{\cdot\alpha\beta}$. Then $e_i^{\cdot\alpha}$, as a covariant vector, is expected to be more suitable as a gauge potential than $e^j_{\cdot\beta}$, and exactly this shows up in explicit calculations.

$$g_{ij} := e_i^{\cdot\alpha} e_j^{\cdot\beta} \eta_{\alpha\beta} \qquad (2.4)$$

a convenient abbreviation with no independent status.

The total arrangement of all tetrads with respect to each other in terms of their relative positions and relative orientations makes up the geometry of spacetime. Locally we can only recognize a special relativistic structure. If the global arrangement of the tetrads with respect to position and orientation is integrable, i.e. path-independent, then we have an M_4, otherwise a non-minkowskian spacetime, namely a U_4 or Riemann-Cartan spacetime.

2.2 Local P-Transformation of the Matter Field

We base our considerations on an active interpretation of the P-transformation. We imagine that the tetrad field and the coordinate system are kept fixed, whereas the matter field is "transported." A matter field $\psi(x^k)$, being translated from x^k to $x^k + \varepsilon^k$, where ε^k are the 4 infinitesimal parameters of translations and $\varepsilon^\gamma = e_k^{\cdot\gamma}\varepsilon^k$, has to keep its orientation fixed and, accordingly, the generator of translations is that of a parallel transport,* i.e. it is the covariant derivative operator

$$D_\gamma\psi(x) = e^i_{\cdot\gamma} D_i\psi(x) = e^i_{\cdot\gamma}(\partial_i + \Gamma_i^{\cdot\alpha\beta} f_{\beta\alpha})\psi(x) \quad . \qquad (2.5)$$

It acts only on anholonomic indices, see the analogous discussion in Section 1.3. This transformation of a translational type distinguishes the PG from gauge theories for internal symmetries, since the matter field is shifted to a different point in spacetime.

The Lorentz-rotations (6 infinitesimal parameters $\omega^{\alpha\beta}$) are of the standard, special relativistic type. Hence the local P-transformation Π of a field reads (see Figure 1):

$$(\Pi\psi)(x) = (1 - \varepsilon^\gamma(x) D_\gamma + \omega^{\alpha\beta}(x) f_{\beta\alpha})\psi(x) \quad . \qquad (2.6)$$

Here again, the $f_{\beta\alpha}$ are the matrices of the Lorentz generators obeying (1.3). Of course, setting up a gauge theory, the (4 + 6) infinitesimal "parameters" $(\varepsilon^\gamma, \omega^{\alpha\beta})$ are spacetime dependent functions. A matter field distribution $\psi(x)$, such is our postulate,

*For this reason, Ne'eman's title of his article[26] reads: "Gravity is the gauge theory of the parallel-transport modification of the Poincaré group."

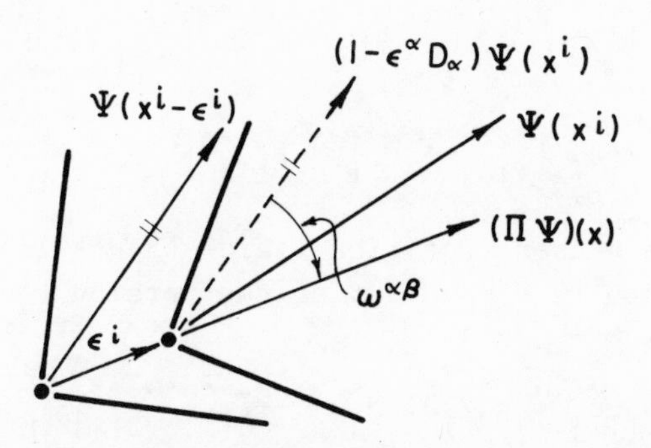

Fig. 1. An infinitesimal active local Poincaré transformation of a matter field: The field $\psi(x^i - \varepsilon^i)$ is first parallelly displaced over the infinitesimal vector $\varepsilon^i = e^i_{\cdot\gamma}\varepsilon^\gamma$, rotated by the angle $\omega^{\alpha\beta}$, and then compared with $\psi(x^i)$.

after the application of a local P-transformation Π, i.e. $\psi(x) \to (\Pi\psi)(x)$, is equivalent in all its measurable properties to the original distribution $\psi(x)$.

2.3 Commutation Relations, Torsion, and Curvature

The translation generators D_γ and the rotation generators $f_{\beta\alpha}$ fulfill commutation relations which we will derive now. The commutation relations for the $f_{\beta\alpha}$ with themselves are given by the special relativistic formula (1.3). For rotations and translations we start with the relation $f_{\alpha\beta}D_i = D_i f_{\alpha\beta}$, which is valid since D_i doesn't carry a tetrad index. Let us remind ourselves that the $f_{\alpha\beta}$, as operators, act on everything to their rights. Transvecting with $e^i_{\cdot\gamma}$, we find $e^i_{\cdot\gamma}f_{\alpha\beta}D_i = D_\gamma f_{\alpha\beta}$, i.e.,

$$[f_{\alpha\beta}, D_\gamma] = [f_{\alpha\beta}, e^i_{\cdot\gamma}]D_i = \eta_{\gamma[\alpha}D_{\beta]} \quad . \tag{2.7}$$

This formula is strictly analogous to its special relativistic pendant.

Finally, let us consider the translations under themselves. By explicit application of (2.5), we find

$$[D_i, D_j] = F_{ij}^{\cdot\cdot\alpha\beta} f_{\beta\alpha} \quad , \tag{2.8}$$

where

$$F_{ij\alpha}^{\cdot\cdot\cdot\beta} := 2\left(\partial_{[i}\Gamma_{j]\alpha}^{\cdot\ \cdot\beta} + \Gamma_{[i|\gamma}^{\cdot\ \cdot\beta}\Gamma_{|j]\alpha}^{\cdot\ \cdot\gamma}\right) \tag{2.9}$$

is the curvature tensor (rotation field strength). Now $D_i = e_i^{\cdot\alpha} D_\alpha$, substitute it into (2.8) and define the torsion tensor (translation field strength),

$$F_{ij}^{\cdot\cdot\alpha} := 2D_{[i}e_{j]}^{\cdot\ \alpha} = 2\left(\partial_{[i}e_{j]}^{\cdot\ \alpha} + \Gamma_{[i|\beta}^{\cdot\ \cdot\alpha}e_{|j]}^{\cdot\ \beta}\right) \quad . \tag{2.10}$$

Then finally, collecting all relevant commutation relations, we have

$$[D_\alpha, D_\beta] = -F_{\alpha\beta}^{\cdot\cdot\gamma} D_\gamma + F_{\alpha\beta}^{\cdot\cdot\gamma\delta} f_{\delta\gamma} \quad , \tag{2.11}$$

$$[f_{\alpha\beta}, D_\gamma] = \eta_{\gamma[\alpha} D_{\beta]} \quad , \tag{2.12}$$

$$[f_{\alpha\beta}, f_{\gamma\delta}] = \eta_{\gamma[\alpha} f_{\beta]\delta} - \eta_{\delta[\alpha} f_{\beta]\gamma} \quad . \tag{2.13}$$

For vanishing torsion and curvature we recover the commutation relations of global P-transformations. Local P-transformations, in contrast to the corresponding structures in gauge theories of internal symmetries, obey different commutation relations, in particular the algebra of the translations doesn't close in general. Observe, however, that it does close for vanishing curvature, i.e. in a spacetime with teleparallelism (see Section 2.7). In introducing our translation generators, we already stressed their unique features. This is now manifest in (2.11). Note also that the "mixing term" $2\Gamma_{[i|\beta}^{\cdot\ \cdot\alpha}e_{|j]}^{\cdot\ \beta}$ between translational and rotational potentials in the definition (2.10) of the translation field is due to the existence of orbital angular momentum.

In deriving (2.11), we find in torsion and curvature the tensors which covariantly characterize the possibly different arrangement of tetrads in comparison with that in special relativity. Torsion and curvature measure the non-minkowskian behavior of the tetrad arrangement. In (2.11) they relate to the translation and rotation generators, respectively. Consequently torsion represents the translation field strength and curvature the rotation field strength.

The only non-trivial Jacobi identity,

$$[D_\alpha,[D_\beta,D_\gamma]] + [D_\beta,[D_\gamma,D_\alpha]] + [D_\gamma,[D_\alpha,D_\beta]] = 0 \quad , \qquad (2.14)$$

leads, after substitution of (2.11), some algebra and using (2.44), to the two sets of Bianchi identities

$$D_{[i}F_{jk]}^{\cdot\cdot\ \beta} \equiv F_{[ijk]}^{\cdot\cdot\cdot\ \beta} \quad , \qquad (2.15)$$

$$D_{[i}F_{jk]\alpha}^{\cdot\cdot\ \cdot\beta} \equiv 0 \quad . \qquad (2.16)$$

2.4 Local P-Transformation of the Gauge Potentials

Let us now come back to our postulate of local P-invariance. A matter field distribution actively P-transformed, $\psi(x) \to (\Pi\psi)(x)$, should be equivalent to $\psi(x)$. How can it happen that a local observer doesn't see a difference in the field configuration after applying the P-transformation? The local P-transformation will induce not only a variation of $\psi(x)$, but also correct the values of the tetrad coefficients $e_i^{\cdot\alpha}$ and the connection coefficients $\Gamma_i^{\cdot\alpha\beta}$ such that a difference doesn't show up. In other words, the local P-transformation adjusts suitably the relative position and the relative orientation of the tetrads as determined by the corresponding coefficients $(e_i^{\cdot\alpha},\Gamma_i^{\cdot\alpha\beta})$. Thereby the P-transformation of the gauge potentials is a consequence of the local P-structure of spacetime.

Consider a matter field distribution $\psi(x)$, in particular its values at x^k and at a nearby point $x^k + \xi^k$. See Figure 2 where the matter field is symbolized by a vector. The relative position of $\psi(x+\xi)$ and $\psi(x)$ is determined by ξ^α, their relative orientation by $\zeta^{\alpha\beta} = \zeta^{[\alpha\beta]}$, the angle between $(1+\xi^\alpha D_\alpha)\psi(x)$ and $\psi(x)$. By a rotation $-\zeta^{\alpha\beta}$ of $(1+\xi^\alpha D_\alpha)\psi(x)$, we get $(\tilde{\Pi}\psi)(x) = (1+\xi^\alpha D_\alpha - \zeta^{\alpha\beta}f_{\beta\alpha})\times \psi(x)$, which is, of course, parallel to $\psi(x)$. The transformation $\tilde{\Pi}$ has the same structure as a P-transformation.*

Now P-transform $\psi(x)$ and $\psi(x+\xi)$. Then $(\tilde{\Pi}\psi)(x)$ must stay parallel to $\psi(x)$, i.e. it is required to transform as a P-spinor-tensor. Furthermore $(\Pi\psi)(x+\xi)$ and $(\Pi\psi)(x)$, the P-transforms of

*It is not a P-transformation, since we consider the untransformed matter field distribution.

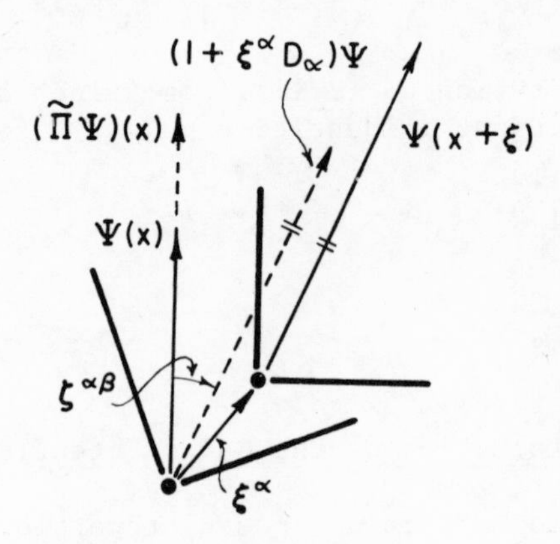

Fig. 2. A matter field distribution near x^k.

the fields $\psi(x+\xi)$ and $\psi(x)$, can be again related by a transformation of the type $\tilde{\Pi}$, i.e. there emerge new ξ^α and $\zeta^{\alpha\beta}$ which are the P-transforms of the old ones. Consequently, $D_\alpha\psi$ as well as $f_{\alpha\beta}\psi$ transform as P-spinor-tensors, i.e. if, according to (2.6) $\psi \to \Pi\psi$, then

$$D_\alpha\psi \to \Pi D_\alpha\psi \quad , \tag{2.17}$$

$$f_{\alpha\beta}\psi \to \Pi f_{\alpha\beta}\psi \quad . \tag{2.18}$$

This implies, via the commutation relations (2.11), that torsion $F_{\alpha\beta}^{\cdot\cdot\gamma}$ and curvature $F_{\alpha\beta}^{\cdot\cdot\gamma\delta}$ are also P-tensors, an information which will turn out to be useful in constructing invariant gauge field Lagrangians.

Now we have

$$\Pi D_\alpha\psi = (D_\alpha + \delta D_\alpha)(\psi + \delta\psi) = (D_\alpha + \delta D_\alpha)\Pi\psi \tag{2.19}$$

and, because Π deviates only infinitesimally from unity,

$$\delta D_\alpha = [\Pi, D_\alpha] \quad . \tag{2.20}$$

Analogously we find

$$\delta f_{\alpha\beta} = [\Pi, f_{\alpha\beta}] \quad . \tag{2.21}$$

Substitution of Π from (2.6) into (2.20), (2.21) and using the commutation relations (2.11)-(2.13), yields, respectively,

$$\begin{aligned}\delta D_\alpha &= -[\varepsilon^\gamma D_\gamma, D_\alpha] + [\omega^{\gamma\delta} f_{\delta\gamma}, D_\alpha] \\ &= -\left(-D_\alpha \varepsilon^\gamma + \omega_\alpha^{\cdot\gamma} - \varepsilon^\delta F_{\delta\alpha}^{\cdot\cdot\gamma}\right) D_\gamma + \left(-D_\alpha \omega_\varepsilon^{\cdot\gamma} - \varepsilon^\delta F_{\delta\alpha\varepsilon}^{\cdot\cdot\cdot\gamma}\right) f_\gamma^{\cdot\varepsilon} \quad ,\end{aligned} \tag{2.22}$$

$$\delta f_{\alpha\beta} = -[\varepsilon^\gamma D_\gamma, f_{\alpha\beta}] + [\omega^{\gamma\delta} f_{\delta\gamma}, f_{\alpha\beta}] = 0 \quad . \tag{2.23}$$

The coordinates are kept fixed during the active P-transformation. Then, using (2.5) and (2.23), we get

$$\delta D_i = \delta\left(\partial_i + \Gamma_i^{\cdot\alpha\beta} f_{\beta\alpha}\right) = \left(\delta\Gamma_i^{\cdot\alpha\beta}\right) f_{\beta\alpha} \tag{2.24}$$

or

$$\delta D_\alpha = \delta\left(e^i_{\cdot\alpha} D_i\right) = \left(\delta e^i_{\cdot\alpha}\right) D_i + e^i_{\cdot\alpha}\left(\delta\Gamma_i^{\cdot\alpha\beta}\right) f_{\beta\alpha} \quad . \tag{2.25}$$

A comparison with (2.22) and remembering $\delta(e^i_{\cdot\alpha} e_j^{\cdot\alpha}) = 0$, yields the desired relations*

$$\delta e_i^{\cdot\alpha} = -D_i \varepsilon^\alpha + \omega_\gamma^{\cdot\alpha} e_i^{\cdot\gamma} - \varepsilon^\gamma F_{\gamma i}^{\cdot\cdot\alpha} \quad , \tag{2.26}$$

$$\delta\Gamma_i^{\cdot\alpha\beta} = -D_i \omega^{\alpha\beta} - \varepsilon^\gamma F_{\gamma i}^{\cdot\cdot\alpha\beta} \quad . \tag{2.27}$$

From gauge theories of internal groups we are just not used to the non-local terms carrying the gauge field strengths; this is again an outflow of the specific behavior of the translations.

---

*It might be interesting to note that, in 3 dimensions, these relations represent essentially the two deformation tensors of a so-called Cosserat continuum as expressed in terms of their translation fields ε^α and rotation fields $\omega^{\alpha\beta}$. Those analogies suggested to us at first the existence of formulas of the type (2.26), (2.27). They were first proposed in ref. [1], cf. also refs. [27], [28].

Otherwise, the first terms on the right hand sides of (2.26), (2.27) namely $-D_i\varepsilon^\alpha$ and $-D_i\omega^{\alpha\beta}$, are standard. They express the non-homogeneous transformation behavior of the potentials under local P-gauge transformations, respectively, and it is because of this fact that the names translation and rotation gauge potential are justified. The term $(\omega_\gamma^{\cdot\alpha} e_i^{\cdot\gamma})$ in (2.26) shows that the tetrad, the translation potential, behaves as a vector under rotations. This leads us to expect, as indeed will turn out to be true, that $e_i^{\cdot\alpha}$, in contrast to the rotation potential $\Gamma_i^{\cdot\alpha\beta}$, should carry intrinsic spin.

Having a U_4 with torsion and curvature we know that besides local P-invariance, we have additionally invariance under general coordinate transformations. In fact, this coordinate invariance is also a consequence of our formalism (see [1]). Starting with a U_4, one can alternatively develop a "gauge" formalism with coordinate invariance and local Lorentz invariance as applied to tetrads as ingredients. The only difference is, however, that with our procedure we recover in the limiting case of an M_4 exactly the well-known global P-transformations of special relativity (see eq. (2.60)) including the corresponding conservation laws (see eqs. (3.12), (3.13)), whereas otherwise the global gauge limit in this sense is lost: We have only to remember that in special relativity energy-momentum conservation is not a consequence of coordinate invariance, but rather of invariance under translations.*

---

*Such an alternative formalism was presented by Dr. Schweizer in his highly interesting seminar talk.[19] His tetrad loses its position as a potential, since it transforms homogeneously under coordinate transformations. Furthermore, having got rid of the M_4-limit as discussed above, one has to be very careful about what to define as local Lorentz-invariance. Schweizer defines strong as well as weak local Lorentz-invariance. However, the former notion lacks geometrical significance altogether. Whereas we agree with Schweizer that there is nothing mysterious about the local P-gauge approach and that one can readily rewrite it in terms of coordinate and Lorentz-invariance (cf. [1], Sect. IV.C.3) and with the help of Lie-derivatives (cf. [29]), we hold that in our formulation there is a completely satisfactory place for the translation gauge (see also [30,26,31,32]). Hence we leave it to others "to gauge the translation group more attractively..."

2.5 Closure of the Local P-Transformations

Having discussed so far how the matter field $\psi(x)$ and the gauge potentials $e_i^{\cdot\alpha}$ and $\Gamma_i^{\cdot\alpha\beta}$ transform under local P-transformations, we would now like to show once more the intrinsic naturality and usefulness of the local P-formalism developed so far. We shall compute the commutator of two successive P-transformations $[\overset{2}{\Pi},\overset{1}{\Pi}]$. We will find out that it yields a third local P-transformation $\overset{3}{\Pi}$, the infinitesimal parameters $(\overset{3}{\varepsilon}{}^{\alpha},\overset{3}{\omega}{}^{\alpha\beta})$ of which depend in a suitable way on the parameters $(\overset{1}{\varepsilon}{}^{\alpha},\overset{1}{\omega}{}^{\alpha\beta})$ and $(\overset{2}{\varepsilon}{}^{\alpha},\overset{2}{\omega}{}^{\alpha\beta})$ of the two transformations $\overset{1}{\Pi}$ and $\overset{2}{\Pi}$, respectively.*

Take the connection as an example. We have

$$\overset{1}{\Gamma}{}_i^{\cdot\alpha\beta} := \overset{1}{\Pi}\Gamma_i^{\cdot\alpha\beta} = \Gamma_i^{\cdot\alpha\beta} - D_i\overset{1}{\omega}{}^{\alpha\beta} - \overset{1}{\varepsilon}{}^{\gamma} e_i^{\cdot\delta} F_{\gamma\delta}^{\cdot\cdot\alpha\beta} \quad . \tag{2.28}$$

In the last term we have purposely written the curvature in its totally anholonomic form. Then it is a P-tensor, as we saw in the last section. Applying now $\overset{2}{\Pi}$, we have to keep in mind that $\overset{2}{\Pi}$ acts with respect to the transformed tetrad coefficients $\overset{1}{e}{}_i^{\cdot\delta} := \overset{1}{\Pi}e_i^{\cdot\delta}$ as well as with respect to the transformed connection coefficients $\overset{1}{\Gamma}{}_i^{\cdot\alpha\beta}$. Consequently we have

$$\overset{21}{\Pi\Gamma}{}_i^{\cdot\alpha\beta} = \overset{21}{\Pi\Pi}\Gamma_i^{\cdot\alpha\beta} = \overset{1}{\Gamma}{}_i^{\cdot\alpha\beta} - \overset{1}{D}_i\overset{2}{\omega}{}^{\alpha\beta} - \overset{2}{\varepsilon}{}^{\gamma}\overset{1}{e}{}_i^{\cdot\delta}\overset{1}{F}{}_{\gamma\delta}^{\cdot\cdot\alpha\beta} \quad . \tag{2.29}$$

Now we will evaluate the different terms in (2.29). By differentiation and by use of (2.28) we find

$$\begin{aligned}\overset{1}{D}_i\overset{2}{\omega}{}^{\alpha\beta} &= D_i\overset{2}{\omega}{}^{\alpha\beta} + 2\left(\overset{1}{\delta\Gamma}{}_{i\mu}^{\cdot\cdot[\alpha|}\right)\overset{2}{\omega}{}^{\mu|\beta]} \\ &= D_i\overset{2}{\omega}{}^{\alpha\beta} - 2\left(D_i\overset{1}{\omega}{}_{\mu}^{\cdot[\alpha|}\right)\overset{2}{\omega}{}^{\mu|\beta]} - 2\overset{1}{\varepsilon}{}^{\gamma}F_{\gamma i\mu}^{\cdot\cdot\cdot[\alpha|}\overset{2}{\omega}{}^{\mu|\beta]} \quad .\end{aligned} \tag{2.30}$$

Let us turn to the last term in (2.29). If we apply (2.26), it reads in the appropriate order:

---

*The proof was first given by Nester.[29] Ne'eman and Takasugi[33] generalized it to supergravity including the ghost regime.

$$\overset{2}{\varepsilon}{}^{\gamma}\overset{1}{e}{}_i{}^{\cdot\delta}\overset{1}{F}{}_{\gamma\delta}{}^{\cdot\cdot\alpha\beta} = \overset{2}{\varepsilon}{}^{\gamma} e_i{}^{\cdot\delta}\overset{1}{F}{}_{\gamma\delta}{}^{\cdot\cdot\alpha\beta} + \overset{2}{\varepsilon}{}^{\gamma}\left(\overset{1}{\delta} e_i{}^{\cdot\delta}\right)F_{\gamma\delta}{}^{\cdot\cdot\alpha\beta}$$

$$= \overset{2}{\varepsilon}{}^{\gamma} e_i{}^{\cdot\delta}\overset{1}{F}{}_{\gamma\delta}{}^{\cdot\cdot\alpha\beta} + \overset{2}{\varepsilon}{}^{\gamma}\left(-D_i\overset{1}{\varepsilon}{}^{\delta} + \overset{1}{\omega}{}_{\nu}{}^{\cdot\delta} e_i{}^{\cdot\nu} - \overset{1}{\varepsilon}{}^{\nu} F_{\nu i}{}^{\cdot\cdot\delta}\right)F_{\gamma\delta}{}^{\cdot\cdot\alpha\beta} \quad . \tag{2.31}$$

In (2.31) there occurs the P-transform of the anholonomic curvature. Like any P-tensor, it transforms according to (2.6):

$$\overset{1}{F}{}_{\gamma\delta}{}^{\cdot\cdot\alpha\beta} = F_{\gamma\delta}{}^{\cdot\cdot\alpha\beta} - \overset{1}{\varepsilon}{}^{\mu} D_{\mu} F_{\gamma\delta}{}^{\cdot\cdot\alpha\beta} - 2\overset{1}{\omega}{}_{[\gamma|}{}^{\cdot\mu} F_{\mu|\delta]}{}^{\cdot\cdot\cdot\,\alpha\beta} + 2\overset{1}{\omega}{}_{\mu}{}^{\cdot[\alpha|} F_{\gamma\delta}{}^{\cdot\cdot\mu|\beta]} \, . \tag{2.32}$$

Now we substitute first (2.32) into (2.31). The resulting equation together with (2.28) and (2.30) are then substituted into (2.29). After some reordering we find:

$$\begin{aligned}\overset{21}{\Pi\Pi}\Gamma_i{}^{\cdot\alpha\beta} = {} & \Gamma_i{}^{\cdot\alpha\beta} - D_i\left(\overset{1}{\omega}{}^{\alpha\beta} + \overset{2}{\omega}{}^{\alpha\beta}\right) - F_{\gamma i}{}^{\cdot\cdot\alpha\beta}\left(\overset{1}{\varepsilon}{}^{\gamma} + \overset{2}{\varepsilon}{}^{\gamma}\right) + \\ & + F_{\gamma\delta}{}^{\cdot\cdot\alpha\beta} F_{\nu i}{}^{\cdot\cdot\delta}\overset{1}{\varepsilon}{}^{\nu}\overset{2}{\varepsilon}{}^{\gamma} + 2D_i\left(\overset{1}{\omega}{}_{\mu}{}^{\cdot[\alpha|}\right)\overset{2}{\omega}{}^{\mu|\beta]} + \\ & + 2F_{\gamma i\mu}{}^{\cdot\cdot\cdot[\alpha|}\left(\overset{1}{\varepsilon}{}^{\gamma}\overset{2}{\omega}{}^{\mu|\beta]} + \overset{2}{\varepsilon}{}^{\gamma}\overset{1}{\omega}{}^{\mu|\beta]}\right) + e_i{}^{\cdot\delta} F_{\mu\delta}{}^{\cdot\cdot\alpha\beta}\overset{2}{\varepsilon}{}^{\gamma}\omega_{\gamma}{}^{\cdot\mu} + \\ & + e_i{}^{\cdot\delta}\left(D_{\mu} F_{\gamma\delta}{}^{\cdot\cdot\alpha\beta}\right)\overset{2}{\varepsilon}{}^{\gamma}\overset{1}{\varepsilon}{}^{\mu} + F_{\gamma\delta}{}^{\cdot\cdot\alpha\beta}\overset{2}{\varepsilon}{}^{\gamma} D_i\overset{1}{\varepsilon}{}^{\delta} \quad .\end{aligned} \tag{2.33}$$

I am not happy myself with all those indices. Anybody is invited to look for a simpler proof. But the main thing is done. The right-hand-side of (2.33) depends only on the untransformed geometrical quantities and on the parameters. By exchanging the one's and the two's wherever they appear, we get the reversed order of the transformations:

$$\overset{12}{\Pi\Pi}\Gamma_i{}^{\cdot\alpha\beta} = (2.33) \quad \text{with} \quad \begin{matrix} 1 \to 2 \\ 2 \to 1 \end{matrix} \quad . \tag{2.34}$$

Subtracting (2.34) from (2.33) yields the commutator $[\overset{2}{\Pi},\overset{1}{\Pi}]\Gamma_i{}^{\cdot\alpha\beta}$.

After some heavy algebra and application of the 2nd Bianchi identity in its anholonomic form,

$$D_{[\alpha} F_{\beta\gamma]}{}^{\cdot\cdot\,\mu\nu} = F_{[\alpha\beta}{}^{\cdot\cdot\delta} F_{\gamma]\delta}{}^{\cdot\,\cdot\mu\nu} \quad , \tag{2.35}$$

we find indeed a transformation $\overset{3}{\Pi}\Gamma_i{}^{\cdot\alpha\beta}$ of the form (2.28) with the following parameters:

$$\overset{3}{\varepsilon}{}^{\alpha} = -\overset{2}{\omega}{}_{\gamma}{}^{\cdot\alpha}\overset{1}{\varepsilon}{}^{\gamma} + \overset{1}{\omega}{}_{\gamma}{}^{\cdot\alpha}\overset{2}{\varepsilon}{}^{\gamma} - \overset{2}{\varepsilon}{}^{\beta}\overset{1}{\varepsilon}{}^{\gamma}F_{\beta\gamma}{}^{\cdot\cdot\alpha} \quad , \tag{2.36}$$

$$\overset{3}{\omega}{}_{\alpha}{}^{\cdot\beta} = -\overset{2}{\omega}{}_{\alpha}{}^{\cdot\gamma}\overset{1}{\omega}{}_{\gamma}{}^{\cdot\beta} + \overset{1}{\omega}{}_{\alpha}{}^{\cdot\gamma}\overset{2}{\omega}{}_{\gamma}{}^{\cdot\beta} - \overset{2}{\varepsilon}{}^{\gamma}\overset{1}{\varepsilon}{}^{\delta}F_{\gamma\delta\alpha}{}^{\cdot\cdot\cdot\beta} \quad . \tag{2.37}$$

It is straightforward to show that the corresponding formulae for $[\overset{2}{\Pi},\overset{1}{\Pi}]$ as applied to the matter field and the tetrad lead to the same parameters (2.36), (2.37). These results are natural generalizations of the corresponding commutator in an M_4. One can take (2.36), (2.37) as the ultimate justification for attributing a fundamental significance to the notion of a local P-transformation.

2.6 Local Kinematical Inertial Frames

As we have seen in (1.8), (1.9), in an M_4 we can always trivialize the gauge potentials globally. Since spacetime looks minkowskian from a local point of view, it should be possible to trivialize the gauge potentials in a U_4 locally, i.e.,

$$\left\{\begin{aligned} &e_i^{\cdot\alpha}\left(x^k = \overset{}{\underset{0}{x}}{}^k\right) \overset{*}{=} \delta_i^{\alpha} \quad , \\ &\Gamma_i^{\cdot\alpha\beta}\left(x^k = \underset{0}{x}{}^k\right) \overset{*}{=} 0 \quad . \end{aligned}\right\} \tag{2.38}$$

The proof runs as follows: We rotate the tetrads according to $e'^{\cdot\alpha}_i = \Lambda_\beta^{\cdot\alpha}e_i^{\cdot\beta}$. This induces a transformation of the connection, namely the finite version of (2.27) for $\varepsilon^\gamma = 0$:

$$\Gamma'{}_{i\alpha}{}^{\cdot\cdot\beta} = \Lambda_\alpha^{\cdot\gamma}\Lambda_\delta^{\cdot\beta}\Gamma_{i\gamma}{}^{\cdot\cdot\delta} - \Lambda_\alpha^{\cdot\gamma}\partial_i\Lambda_\gamma^{\cdot\beta} \quad . \tag{2.39}$$

By a suitable choice of the rotation, we want these transformed connection coefficients to vanish. We put (2.39) provisionally equal to zero, solve for $\partial_i\Lambda_\beta^{\cdot\alpha}$, and find

$$\partial_i\Lambda_\beta^{\cdot\alpha} = \Lambda_\gamma^{\cdot\alpha}\Gamma_{i\beta}{}^{\cdot\cdot\gamma}\left(x^k = \underset{0}{x}{}^k\right) \quad . \tag{2.40}$$

For prescribed $\Gamma_{i\beta}{}^{\cdot\cdot\gamma}$ at $x^k = \underset{0}{x}{}^k$, we can always solve this first order linear differential equation, which concludes the first part of the proof. Then we adjust the holonomic coordinates. The connection $\Gamma_i^{\cdot\alpha\beta} \overset{*}{=} 0$, being a coordinate vector, stays zero, whereas the tetrad transforms as follows: $e'{}_i^{\cdot\alpha} = (\partial x^k/\partial x'^i)e_k^{\cdot\alpha}$. For a transformation of the type $x'^{i} = \delta^{i}_{\beta}\, e_{\ell}^{\cdot\beta} x^{\ell}$ + const we find indeed $e'^{\cdot\alpha}_{i} \overset{*}{=} \delta^{\alpha}_{i}$, q.e.d.*

What is the physical meaning of these trivial gauge frames existing all over spacetime? Evidently they represent in a Riemann-Cartan spacetime what was in Einstein's theory the freely falling non-rotating elevator. For these considerations it is vital, however, that from our gauge theoretical point of view the potentials $(e_{i}^{\cdot\alpha}, \Gamma_{i}^{\cdot\alpha\beta})$ are locally measurable, whereas torsion and curvature, as derivatives of the potentials, are only to be measured in a non-local way. For a local observer the world looks minkowskian. If he wants to determine, e.g., whether his world embodies torsion, he has to communicate with his neighbors thereby implying non-locality. This example shows that in the PG the question whether spacetime carries torsion or not (or curvature or not) is not a question one should ask one local observer.†

It is to be expected that non-local quantities like torsion and curvature, in analogy to Maxwell's theory and GR, are governed by field equations. In other words, whether, for instance, the world is riemannian or not, should in the framework of the PG not be imposed ad hoc but rather left as a question to dynamics.

We call the frames (2.38) "local kinematical inertial frames" in order to distinguish them from the local "dynamical" inertial frames in Einstein's GR. In the PG the notion of inertia refers to translation and rotation, or to mass <u>and</u> spin. A coordinate frame $\underline{\partial}_{i}$ of GR has to fulfill the differential constraint $\Omega_{ij}^{\cdot\cdot\alpha}(\underline{\partial}) \equiv 0$, see eq. (1.5). Hence a tetrad frame, which is unconstrained, can move more freely and is, as compared to the coordinate frame, a more local object. Accordingly, the notion of inertia in the PG is more local than that in GR. This is no surprise, since a test particle of GR carries a mass m which is a quantity won by integration over an extended energy-momentum distribution. The matter field $\psi(x)$, however, the object of consideration in the PG, is clearly a more localized being.

A natural extension of the Einstein equivalence principle to the PG would then be to postulate that in the frames (2.38) (these

*The proof was first given by von der Heyde,[34] see also Meyer.[35]

†Practically speaking, such non-local measurements may very well be made by one observer only. Remember that, in the context of GR, the Weber cylinder is also a non-local device for sensing curvature, i.e. the cylinder is too extended for an Einstein elevator.

are our new "elevators") special relativity is valid locally. Consequently the special-relativistic matter lagrangian L in (1.10) should in a U_4 be a lagrangian density $\mathcal{L}$ which couples to spacetime according to

$$\mathcal{L} = \mathcal{L}\left(\psi, \partial_i\psi, e_i^{\cdot\alpha}, \Gamma_i^{\cdot\alpha\beta}\right) \stackrel{*}{=} L(\psi, \partial_i\psi) \quad , \qquad (2.41)$$

i.e. in the local kinematical inertial frames everything looks special relativistic. Observe that derivatives of $(e_i^{\cdot\alpha}, \Gamma_i^{\cdot\alpha\beta})$ are excluded by our "local equivalence principle."* Strictly this discussion belongs into Lecture 3. But the geometry is so suggestive to physical applications that we cannot resist the temptation to present the local equivalence principle already in the context of spacetime geometry.

Naturally, as argued above, the local equivalence principle is not to be applied to directly observable objects like mass points, but rather to the more abstract notion of a lagrangian. In a field theory there seems to be no other reasonable option. And we have seen that the fermionic nature of the building blocks of matter require a field description, at least on a c-number level. Accordingly (2.41) appears to be the natural extension of Einstein's equivalence principle to the PG.

2.7 Riemann-Cartan Spacetime Seen Anholonomically and Holonomically

We have started our geometrical game with the $(e_i^{\cdot\alpha}, \Gamma_i^{\cdot\alpha\beta})$-set. We would now like to provide some machinery for translating this anholonomic formalism into the holonomic formalism commonly more known at least under relativists. Let us first collect some useful formulae for the anholonomic regime. The determinant $e := \det e_i^{\cdot\alpha}$ is a scalar density, furthermore, by some algebra we get $D_i e = \partial_i e$. If we apply the Leibniz rule and the definitions (2.5) and (2.10), we find successively ($F_\alpha := F_{\alpha\gamma}^{\cdot\cdot\gamma}$)

$$D_i(e e^i_{\cdot\alpha}) = eF_\alpha \quad , \qquad (2.42)$$

*This principle was formulated by von der Heyde,[34] see also von der Heyde and Hehl.[36] In his seminar Dr. Rumpf[17] has given a careful and beautiful analysis of the equivalence principle in a Riemann-Cartan spacetime. In particular the importance of his proof how to distinguish the macroscopically indistinguishable teleparallelism and Riemann spacetimes (see our Section 3.3) should be stressed.

$$2D_j\left(ee^{i}_{\bullet[\alpha}e^{j}_{\bullet\beta]}\right) = e\left(F_{\alpha\beta}^{\bullet\bullet i} + 2e^{i}_{\bullet[\alpha}F_{\beta]}\right) \quad , \tag{2.43}$$

$$D_{[\alpha}\left(e^{i}_{\bullet\beta}e^{j}_{\bullet\gamma]}\right) = e^{[i}_{\bullet[\alpha}F_{\beta\gamma]}^{\bullet\bullet\ j]} \quad . \tag{2.44}$$

The last formula was convenient for rewriting the 2nd Bianchi identity (2.16), which was first given in a completely anholonomic form. Eq. (2.43), defining the "modified torsion tensor" on its right hand side, will be used in the context of the field equations to be derived in Section 4.4.

Now, according to (2.1), $dx^i\Gamma_i^{\bullet\alpha\beta}\underline{e}_\beta$ is the relative rotation encountered by a tetrad $\underline{e}_\alpha$ in going from x^k to $x^k + dx^k$. From this we can calculate that the relative rotation of the respective coordinate frame $\underline{\partial}_j = e_j^{\bullet\alpha}\underline{e}_\alpha$ is $dx^i(\Gamma_{i\beta}^{\bullet\bullet\alpha}e_j^{\bullet\beta} + \partial_i e_j^{\bullet\alpha})\underline{e}_\alpha$. In a holonomic coordinate system, the parallel transport is thus given by

$$\nabla_i := \partial_i + \tilde{\Gamma}_{ij}^{\bullet\bullet k}h_k^{\bullet j} \quad , \tag{2.45}$$

where h represents the generator of coordinate transformations for tensor fields and

$$\tilde{\Gamma}_{ij}^{\bullet\bullet k} := e_j^{\bullet\alpha}e^{k}_{\bullet\beta}\Gamma_{i\alpha}^{\bullet\bullet\beta} + e^{k}_{\bullet\beta}\partial_i e_j^{\bullet\beta} \quad . \tag{2.46}$$

This relation translates the anholonomic into the holonomic connection. Observe that for a connection the conversion of holonomic to anholonomic indices and vice versa is markedly different from the simple transvection rule as applied to tensors. The holonomic components of the covariant derivative of a tensor A are given with respect to its anholonomic components by

$$\nabla_i A_{j\bullet\bullet}^{\ \ k\bullet\bullet} = e_j^{\bullet\alpha}\cdots e^{k}_{\bullet\beta}\cdots D_i A_{\alpha\bullet\bullet}^{\ \ \beta\bullet\bullet} \quad . \tag{2.47}$$

The concept of parallelism with respect to a coordinate frame, as defined in (2.46), is by construction locally identical with minkowskian parallelism, as is measured in a local tetrad. In a similar way, the local minkowskian length and angle measurements define the metric in a coordinate frame:

$$g_{ij}(x^k) := e_i^{\bullet\alpha}(x^k)e_j^{\bullet\beta}(x^k)\,\eta_{\alpha\beta} \quad . \tag{2.48}$$

From the antisymmetry $\Gamma_i^{\bullet\alpha\beta} = -\Gamma_i^{\bullet\beta\alpha}$ of the anholonomic connection and from (2.47) results again $D_i\eta_{\alpha\beta} = 0$, a relation which we used already earlier, and

$$\nabla_i g_{jk} = 0 \quad , \tag{2.49}$$

the so-called metric postulate of spacetime physics.

If we resolve (2.49) with respect to $\tilde{\Gamma}_{ij}^{\cdot\cdot k}$, we get

$$\tilde{\Gamma}_{ij}^{\cdot\cdot k} = \left\{ {k \atop ij} \right\} + \frac{1}{2} F_{ij}^{\cdot\cdot k} - \frac{1}{2} F_{j\cdot i}^{\cdot k} + \frac{1}{2} F^{k}_{\cdot ij} \tag{2.50}$$

and, taking (2.46) into account, the corresponding relation for the anholonomic connection:

$$\Gamma_{\alpha\beta\gamma} := e^{i}_{\cdot\alpha}\Gamma_{i\beta\gamma} = (-\Omega_{\alpha\beta\gamma} + \Omega_{\beta\gamma\alpha} - \Omega_{\gamma\alpha\beta} + F_{\alpha\beta\gamma} - F_{\beta\gamma\alpha} + F_{\gamma\alpha\beta})/2 \quad . \tag{2.51}$$

We have introduced here the Christoffel symbol $\left\{ {k \atop ij} \right\}$, the holonomic components of the torsion tensor,

$$F_{ij}^{\cdot\cdot k} = e^{k}_{\cdot\alpha} F_{ij}^{\cdot\cdot\alpha} = 2\tilde{\Gamma}_{[ij]}^{\cdot\cdot\ k} \quad , \tag{2.52}$$

and the object of anholonomity

$$\Omega_{\alpha\beta}^{\cdot\cdot\gamma} := e^{i}_{\cdot\alpha} e^{j}_{\cdot\beta} \Omega_{ij}^{\cdot\cdot\gamma} \ ; \qquad \Omega_{ij}^{\cdot\cdot\gamma} := 2\partial_{[i} e_{j]}^{\cdot\ \gamma} \quad . \tag{2.53}$$

Expressing the holonomic components of the curvature tensor in terms of $\tilde{\Gamma}_{ij}^{\cdot\cdot k}$ yields

$$F_{ijk}^{\cdot\cdot\cdot\ell} = 2\left(\partial_{[i} \tilde{\Gamma}_{j]k}^{\cdot\ \cdot\ell} + \tilde{\Gamma}_{[i|m}^{\cdot\ \cdot\ell} \tilde{\Gamma}_{|j]k}^{\cdot\ \cdot m} \right) \quad . \tag{2.54}$$

Finally, taking the antisymmetric part of (2.46) or using the definition of torsion (2.10), we get

$$F_{ij}^{\cdot\cdot\alpha} = \Omega_{ij}^{\cdot\cdot\alpha} - 2e_{[i}^{\cdot\beta}\Gamma_{j]\beta}^{\cdot\ \cdot\alpha} \quad , \tag{2.55}$$

a formula which will play a key role in discussing macroscopic gravity: the object of anholonomity mediates between torsion and the anholonomic connection.

Eq. (2.50) shows that instead of the potentials $(e_i^{\cdot\alpha}, \Gamma_i^{\cdot\alpha\beta})$ we can use holonomically the set $(g_{ij}, S_{ij}^{\cdot\cdot k})$, the geometry is always a Riemann-Cartan one, only the mode of description is different. In the anholonomic description $e_i^{\cdot\alpha}$ enters the definition of torsion $F_{ij}^{\cdot\cdot\alpha}$, hence we should not use $F_{ij}^{\cdot\cdot\alpha}$ as an independent variable in place of $\Gamma_i^{\cdot\alpha\beta}$. The anholonomic formalism is superior in a gauge

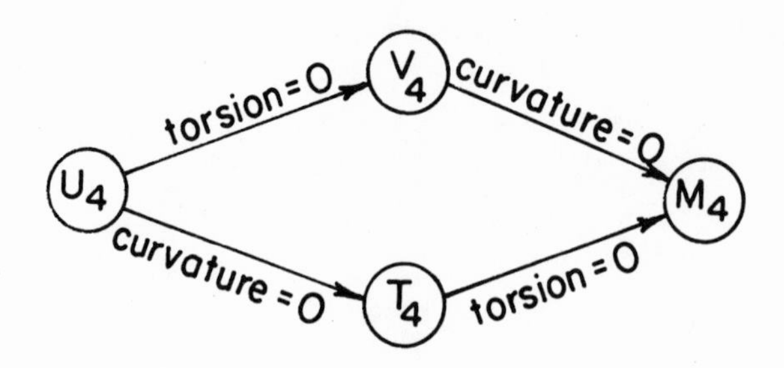

Fig. 3. The Riemann-Cartan spacetime U_4 and its limiting cases: T_4 = teleparallelism, V_4 = Riemann, M_4 = Minkowski.

approach, because $(e_i^{\cdot\alpha}, \Gamma_i^{\cdot\alpha\beta})$ are supposed to be directly measurable and have an interpretation as potentials and the corresponding transformation behavior, whereas the holonomic set $(g_{ij}, S_{ij}^{\cdot\cdot k})$ is a tensorial one.

If we put curvature to zero, we get a spacetime with teleparallelism T_4, a U_4 for vanishing torsion is a Riemann spacetime V_4, see Figure 3.

Perhaps surprisingly, as regards to their physical degrees of freedom, the T_4 and the V_4 are in some sense similar to each other. In a T_4 the curvature vanishes, i.e. the parallel transfer is integrable. Then we can always pick preferred tetrad frames such that the connection vanishes globally:

$$T_4 : \left(e_i^{\cdot\alpha}, \Gamma_i^{\cdot\alpha\beta} \overset{*}{=} 0 \right) \quad ; \quad \left(F_{ij}^{\cdot\cdot\alpha} \overset{*}{=} \Omega_{ij}^{\cdot\cdot\alpha}, F_{ij}^{\cdot\cdot\alpha\beta} \equiv 0 \right) \quad . \tag{2.56}$$

In these special anholonomic coordinates we are only left with the tetrad $e_i^{\cdot\alpha}$ as variable. In a V_4 the torsion vanishes, i.e., according to (2.51), the connection can be expressed exclusively in terms of the (orthonormal) tetrads:

$$V_4 : \left\{ e_i^{\cdot\alpha}, \Gamma_{\alpha\beta\gamma} = -\frac{1}{2}\,\Omega_{\alpha\beta\gamma} + \cdots \right\} ;$$

$$\left\{ F_{ij}^{\cdot\cdot\alpha} \equiv 0 \; , \; F_{ij\alpha}^{\cdot\cdot\cdot\beta} = \partial_i \Omega_{j\alpha}^{\cdot\cdot\beta} - \cdots \right\} \quad . \tag{2.57}$$

Again nothing but tetrads are left. In other words, in a T_4 as well as in a V_4 the gauge variables left over are the tetrad coefficients alone. We have to keep this in mind in discussing macroscopic gravity.

2.8 Global P-Transformation and the M_4

In an appropriate P-gauge approach one should recover the global P-transformation provided the U_4 degenerates to an M_4. In an M_4, because of (2.56), we can introduce a global tetrad system such that $\Gamma_i^{\cdot\alpha\beta} \overset{*}{=} 0$. Furthermore the torsion vanishes, $F_{ij}^{\cdot\cdot\alpha} \overset{*}{=} 2\partial_{[i} e_{j]}^{\cdot\alpha} = 0$, i.e. the orthonormal tetrads, in the system with $\Gamma_i^{\cdot\alpha\beta} \overset{*}{=} 0$, represent holonomic frames of the cartesian coordinates. This means that the conditions (2.38) are now valid on a <u>global,</u> level. If one only allows for P-transformations linking two such coordinate systems, then the potentials don't change under P-transformations and (2.26), (2.27) yield

$$\left\{ \begin{aligned} -\partial_i \varepsilon^\alpha + \omega_\gamma^{\cdot\alpha} \delta_i^{\cdot\gamma} &= 0 \\ -\partial_i \omega^{\alpha\beta} &= 0 \end{aligned} \right\} \tag{2.58}$$

or

$$\left\{ \begin{aligned} \varepsilon^\alpha &= \overset{0}{\varepsilon}{}^\alpha + \overset{0}{\omega}{}_\gamma^{\cdot\alpha} x^\gamma \\ \omega^{\alpha\beta} &= \overset{0}{\omega}{}^{\alpha\beta} \end{aligned} \right\} \tag{2.59}$$

with the constants $\overset{0}{\varepsilon}{}^\alpha$ and $\overset{0}{\omega}{}^{\alpha\beta}$. Substitution into (2.6) leads to the global P-transformation of the matter field

$$(\Pi\psi)(x) = [1 - \overset{0}{\varepsilon}{}^\alpha \partial_\alpha + \overset{0}{\omega}{}^{\alpha\beta}(x_{[\beta}\partial_{\alpha]} + f_{\beta\alpha})]\psi(x) \quad . \tag{2.60}$$

In terms of the "P-transformed" coordinates

$$x'^\gamma = x^\gamma - \overset{0}{\omega}{}_\alpha^{\cdot\gamma} x^\alpha - \overset{0}{\varepsilon} \quad , \tag{2.61}$$

it can be recast into the perhaps more familiar form

$$(\Pi\psi)(x') = \left(1 + \overset{0}{\omega}{}^{\alpha\beta} f_{\beta\alpha}\right)\psi(x) \quad . \tag{2.62}$$

Thus, in an M_4, the local P-transformation degenerates into the global P-transformation, as it is supposed to do.

We have found in this lecture that spacetime ought to be described by a Riemann-Cartan geometry, the geometrical gauge variables being the potentials $(e_i^{\cdot\alpha}, \Gamma_i^{\cdot\alpha\beta})$. The main results are, amongst other things, collected in the table of Section 3.6.

LECTURE 3: COUPLING OF MATTER TO SPACETIME AND THE TWO GENERAL GAUGE FIELD EQUATIONS

3.1 Matter Lagrangian in a U_4

In the last lecture we concentrated on working out the geometry of spacetime. But already in (2.41) we saw how to extend reasonably the Einstein equivalence principle to the "local equivalence principle" applicable to the PG. Hence we postulate the action function of matter as coupled to the geometry of spacetime to read

$$W_m = \int d^4x \mathcal{L}[\eta_{\alpha\beta}, \gamma^{\alpha}\cdots, \psi(x), \partial_i\psi(x), e_i^{\cdot\alpha}(x), \Gamma_i^{\cdot\alpha\beta}(x)] \quad . \tag{3.1}$$

Observe that the local Minkowski metric $\eta_{\alpha\beta}$, the Dirac matrices $\gamma^{\alpha}\cdots$ etc., since referred to the tetrads, maintain their special-relativistic values in (3.1).

In a local kinematical inertial frame (2.38), the potentials $(e_i^{\cdot\alpha}, \Gamma_i^{\cdot\alpha\beta})$ can be made trivial and, in the case of vanishing torsion and curvature, this can be done even globally, and then we fall back to the special-relativistic action function (1.10) we started with.

The lagrangian in (3.1) is of first order by assumption, i.e. only first derivatives of $\psi(x)$ enter. If we would allow for higher derivatives, we could not be sure of how to couple to $(e_i^{\cdot\alpha}, \Gamma_i^{\cdot\alpha\beta})$: the higher derivatives would presumably "feel" not only the potentials, but also the non-local quantities torsion and curvature. In such a case the gauge field strengths themselves would couple to $\psi(x)$ and thereby break the separation between matter and gauge field lagrangian. Then we would lose the special-relativistic limit seemingly necessary for executing a successful P-gauge approach.

Our postulate (3.1) applies to matter fields $\psi(x)$. These fields are anholonomic objects by definition. Gauge potentials of internal symmetries, like the electromagnetic potential $A_i(x)$, emerging as holonomic covariant vectors (one-forms), must not couple to $(e_i^{\cdot\alpha}, \Gamma_i^{\cdot\alpha\beta})$. Otherwise gauge invariance, in the case of A_i the U(1)-invariance, would be violated. This implies that gauge bosons other than the $(e_i^{\cdot\alpha}, \Gamma_i^{\cdot\alpha\beta})$-set, and in particular the photon field A_i, will be treated as P-scalars. Because of the natural division of physical fields into matter fields and gauge potentials (see Section 1.3), we cannot see any disharmony in exempting the internal gauge bosons from the coupling to the $(e_i^{\cdot\alpha}, \Gamma_i^{\cdot\alpha\beta})$-set.*

Besides the matter field $\psi(x)$, the (4 + 6) P-gauge potentials are new independent variables in (3.1). By means of the action principle we can derive the matter field equation. Varying (3.1) with respect to $\psi(x)$ yields†

$$\delta\mathcal{L}/\delta\psi = 0 \quad . \tag{3.2}$$

In our subsequent considerations we'll always assume that (3.2) is fulfilled.

3.2 Noether Identities of the Matter Lagrangian: Identification of Currents and Conservation Laws

The material action function (3.1) is a P-scalar by construction. Consequently it is invariant under active local P-transformations $\Pi(x)$. The next step will then consist in exploiting this invariance property of the action function according to the Noether procedure.

Let us call the field variables in (3.1) collectively $Q \in (\psi, e_i^{\cdot\alpha}, \Gamma_i^{\cdot\alpha\beta})$. Then the P-invariance demands

---

*There are opposing views, however, see Hojman, Rosenbaum, Ryan, and Shepley.[37,38] The tlaplon concept is a possibility to circumvent our arguments, even if not a very natural one, as it seems to us. According to Ni,[39] the tlaplon theories are excluded by experiment. See also Mukku and Sayed.[40]

†The variational derivative of a function $f = f(\psi, \partial_i\psi, \partial_i\partial_k\psi, \cdots)$ is defined by

$$\frac{\delta f}{\delta\psi} := \frac{\partial f}{\partial\psi} - \partial_i \frac{\partial f}{\partial\partial_i\psi} + \partial_i\partial_k \frac{\partial f}{\partial\partial_i\partial_k\psi} - + \cdots \quad .$$

$$\delta W_m := \int_{\Pi\Omega} d^4x\, \mathcal{L}(\Pi Q, \partial_i \Pi Q) - \int_{\Omega} d^4x\, \mathcal{L}(Q, \partial_i Q) \equiv 0 \quad , \tag{3.3}$$

where $\Pi\Omega$ is the volume translated by an amount $\varepsilon^{\alpha}(x)$. By the chain rule and by the Gauss law we calculate

$$\int_{\Omega} d^4x \left[\frac{\partial\mathcal{L}}{\partial Q}\,\delta Q + \frac{\partial\mathcal{L}}{\partial\partial_i Q}\,\delta\partial_i Q\right] + \int_{\partial\Omega} dA_i \varepsilon^i \mathcal{L}$$

$$= \int_{\Omega} d^4x \left[\frac{\delta\mathcal{L}}{\delta Q}\,\delta Q + \partial_i\left(\varepsilon^i \mathcal{L} + \frac{\partial\mathcal{L}}{\partial\partial_i Q}\,\delta Q\right)\right] \equiv 0 \quad . \tag{3.4}$$

Since this expression is valid for an arbitrary volume Ω, the integrand itself has to vanish. Furthermore we can substitute $\partial_i \to D_i$ in (3.4), since the expression in the parenthesis carries no anholonomic indices:

$$\frac{\delta\mathcal{L}}{\delta Q}\,\delta Q + D_i\left(\varepsilon^i \mathcal{L} + \frac{\partial\mathcal{L}}{\partial\partial_i Q}\,\delta Q\right) \equiv 0 \quad . \tag{3.5}$$

The identity (3.5) is valid quite generally for any lagrangian $\mathcal{L}(Q, \partial_i Q)$. We will need it later-on also for discussing the properties of the gauge field lagrangian.

Going back to (3.1), we find using (3.2),

$$\frac{\delta\mathcal{L}}{\delta e_i^{\cdot\alpha}}\,\delta e_i^{\cdot\alpha} + \frac{\delta\mathcal{L}}{\delta\Gamma_i^{\cdot\alpha\beta}}\,\delta\Gamma_i^{\cdot\alpha\beta} + D_i\left(\varepsilon^{\alpha} e_{\cdot\alpha}^{i}\mathcal{L} + \frac{\partial\mathcal{L}}{\partial\partial_i\psi}\,\delta\psi\right) \equiv 0 \quad . \tag{3.6}$$

Now we substitute the P-variations (2.26), (2.27), (2.6) of $(e_i^{\cdot\alpha}, \Gamma_i^{\cdot\alpha\beta})$ and of ψ, respectively, into (3.6):

$$\frac{\delta\mathcal{L}}{\delta e_i^{\cdot\alpha}}\left(-D_i\varepsilon^{\alpha} + \omega_{\gamma}^{\cdot\alpha} e_i^{\cdot\gamma} - \varepsilon^{\gamma} F_{\gamma i}^{\cdot\cdot\alpha}\right) + \frac{\delta\mathcal{L}}{\partial\Gamma_i^{\cdot\alpha\beta}}\left(-D_i\omega^{\alpha\beta} - \varepsilon^{\gamma} F_{\gamma i}^{\cdot\cdot\alpha\beta}\right) +$$

$$+ D_i\left[\varepsilon^{\alpha} e_{\cdot\alpha}^{i}\mathcal{L} + \frac{\partial\mathcal{L}}{\partial\partial_i\psi}\left(-\varepsilon^{\gamma} D_{\gamma} + \omega^{\alpha\beta} f_{\beta\alpha}\right)\psi\right] \equiv 0 \quad . \tag{3.7}$$

We differentiate the last bracket and order according to the independent quantities $D_i\varepsilon^{\alpha}$, $D_i\omega^{\alpha\beta}$, ε^{α}, $\omega^{\alpha\beta}$, the coefficients of which have to vanish separately. This yields the (10 + 40) identities

$$e\Sigma_{\alpha}^{\bullet i} := \frac{\delta \mathcal{L}}{\delta e_{i}^{\bullet\alpha}} \equiv e_{\bullet\alpha}^{i} \mathcal{L} - \frac{\partial \mathcal{L}}{\partial \partial_{i}\psi} D_{\alpha}\psi \quad , \tag{3.8}$$

$$e\tau_{\alpha\beta}^{\bullet\bullet i} := \frac{\delta \mathcal{L}}{\delta \Gamma_{i}^{\bullet\alpha\beta}} \equiv -\frac{\partial \mathcal{L}}{\partial \partial_{i}\psi} f_{\alpha\beta}\psi \quad , \tag{3.9}$$

$$D_{i}\left(e\Sigma_{\alpha}^{\bullet i}\right) \equiv F_{\alpha i}^{\bullet\bullet\beta} e\Sigma_{\beta}^{\bullet i} + F_{\alpha i}^{\bullet\bullet\beta\gamma} e\tau_{\beta\gamma}^{\bullet\bullet i} \quad , \tag{3.10}$$

$$D_{i}\left(e\tau_{\alpha\beta}^{\bullet\bullet i}\right) - e\Sigma_{[\alpha\beta]} \equiv 0 \quad . \tag{3.11}$$

Our considerations are valid for any U_4, especially for an M_4. In an M_4 in the global coordinates (1.8), (1.9), the equations (3.10), (3.11) degenerate to the special-relativistic momentum and angular momentum conservation laws, respectively:

$$\partial_{\gamma}\Sigma_{\alpha}^{\bullet\gamma} = 0 \quad , \tag{3.12}$$

$$\partial_{\gamma}\tau_{\alpha\beta}^{\bullet\bullet\gamma} - \Sigma_{[\alpha\beta]} = 0 \quad . \tag{3.13}$$

Consequently $\Sigma_{\alpha}^{\bullet i}$ is the canonical momentum current, linked via (3.8) to the translational potential $e_{i}^{\bullet\alpha}$, and $\tau_{\alpha\beta}^{\bullet\bullet i}$ is the canonical spin current, linked via (3.9) to the rotational potential $\Gamma_{i}^{\bullet\alpha\beta}$. Moreover, (3.10), (3.11) are recognized as the momentum and angular momentum conservation laws in a U_4.

It comes to no surprise that in a U_4 the volume-force densities of the Lorentz type $F_{\alpha i}^{\bullet\bullet\beta}(e\Sigma_{\beta}^{\bullet i})$ and $F_{\alpha i}^{\bullet\bullet\beta\gamma}(e\tau_{\beta\gamma}^{\bullet\bullet i})$, respectively, appear on the right hand side of the momentum conservation law. The analog of the latter force is known in GR as the Matthisson force acting on a spinning particle,* and because of the similar couplings of translations and rotations, the force $F_{\alpha i}^{\bullet\bullet\beta}(e\Sigma_{\beta}^{\bullet i})$ is to be expected, too. The left hand side of (3.10) contains second derivatives of the matter field $\psi(x)$. It is because of this "non-locality" that the local equivalence principle doesn't apply on this level. Hence the volume forces just discussed, do not violate the local equivalence principle.

*In [41] we compared in some detail the standing of the Matthisson force in GR with that in the U_4-framework. Clearly the Matthisson force emerges much more natural in the PG.

One further observation in the context of the Noether identities (3.8), (3.9) is of importance. Because of

$$-\frac{\partial\mathcal{L}}{\partial\partial_i\psi} f_{\alpha\beta}\psi = \frac{\partial\mathcal{L}}{\partial\partial_k\psi}\frac{\partial(D_k\psi)}{\partial\Gamma_i^{\bullet\alpha\beta}} \stackrel{!}{=} \frac{\partial\mathcal{L}}{\partial\Gamma_i^{\bullet\alpha\beta}} \quad , \tag{3.14}$$

the connection must only show up in the lagrangian in terms of $D_i\psi$. Similarly, $e_i^{\bullet\alpha}$ can only enter as $e = \det e_i^{\bullet\alpha}$ and in transvecting $D_i\psi$ according to the substitution (cf. (1.10))

$$L(\psi,\partial_i\psi) \to eL(\psi,e^{i}_{\bullet\alpha}D_i\psi) = \mathcal{L}(\psi,D_\alpha\psi,e) \quad , \tag{3.15}$$

because then

$$\frac{\partial\mathcal{L}}{\partial e_i^{\bullet\alpha}} = \frac{\partial\mathcal{L}}{\partial e}\frac{\partial e}{\partial e_i^{\bullet\alpha}} + \frac{\partial\mathcal{L}}{\partial D_\beta\psi}\frac{\partial(D_\beta\psi)}{\partial e_i^{\bullet\alpha}} = e\Sigma_\alpha^{\bullet i} \quad , \tag{3.16}$$

q.e.d. Therefore the so-called minimal coupling (3.15) is a consequence of (2.41), (3.1) and of local P-invariance. It is derived from the local equivalence principle.*

The main results of the kinematical considerations of Sections 3.1 and 3.2 are again collected in the table of Section 3.6.

3.3 The Degenerate Case of Macroscopic (Scalar) Matter

From GR we know that the equations of motion for a test particle moving in a given field are derived by integrating the momentum conservation law. This will be similar in the PG. However, one has to take into account the angular momentum conservation law additionally.

A test particle in GR, as a macroscopic body, will consist of many elementary particles. Hence in order to derive its properties from those of the elementary particles, one has, in the sense of a statistical description, to average over the ensemble of particles constituting the test body.

Mass is of a monopole type and adds up, whereas spin is of a dipole type and normally tends to be averaged out (unless some force aligns the spins like in ferromagnets or in certain superfluids).

*There have been several attempts to develop non-minimal coupling procedures, see Cho,[42] for instance.

Accordingly, macroscopic matter, and in particular the test particles of GR, will carry a finite energy-momentum whereas the spin is averaged out, i.e. $\langle \tau_{\alpha\beta}^{\cdot\cdot i} \rangle \sim 0$. Consequently, because of the macroscopic analog of (3.11), the macroscopic energy-momentum tensor $eT_{\alpha}^{\cdot i} = \langle e\Sigma_{\alpha}^{\cdot i} \rangle$ turns out to be symmetric, as we are used to it in GR. What effect will this averaging have on the momentum conservation law (3.10)? Provided the curvature doesn't depend algebraically on the spin, the Matthisson force is averaged out and we expect the macroscopic analog of (3.10) to look like

$$D_i(eT_{\alpha}^{\cdot i}) \sim \langle F_{\alpha i}^{\cdot\cdot\beta} \rangle eT_{\beta}^{\cdot i} \quad . \tag{3.17}$$

These arguments are, of course, not rigorous. But we feel justified in modelling macroscopic matter by a scalar, i.e. a spinless field $\Phi(x)$. And for a scalar field $\Phi(x)$ our derivations will become rigorous. We lose thereby the information that macroscopic matter basically is built up from fermions and should keep this in mind in case we run into difficulties.

Let us then consider a U_4-spacetime with only scalar matter present.* The lagrangian of the scalar field $\Phi(x)$ reads

$$\mathcal{L} = eL\left(\Phi, e^{i}_{\cdot\alpha}\partial_i\Phi\right) = \mathcal{L}(\Phi, \partial_{\alpha}\Phi, e) \quad . \tag{3.18}$$

If we denote the momentum current by $\sigma_{\alpha}^{\cdot i}$, we find via (3.9), (3.11) $\tau_{\alpha\beta}^{\cdot\cdot i} = 0$ and $\sigma_{[\alpha\beta]} = 0$. Thus (3.10) yields

$$D_i\left(e\sigma_{\alpha}^{\cdot i}\right) - F_{\alpha i}^{\cdot\cdot\beta} e\sigma_{\beta}^{\cdot i} = 0 \quad , \tag{3.19}$$

and only one type of volume-force density is left.

Because of the symmetry of $\sigma_{\alpha\beta}$, the term $\sim\Gamma_{\alpha\gamma}^{\cdot\cdot\beta}\sigma_{\beta}^{\cdot\gamma}$ contained in the volume force density vanishes identically and we find

$$\partial_i\left(e\sigma_{\alpha}^{\cdot i}\right) - \Omega_{\alpha i}^{\cdot\cdot\beta} e\sigma_{\beta}^{\cdot i} = 0 \quad , \tag{3.20}$$

or, after some algebra,

$$\partial_i\left(e\sigma_{k}^{\cdot i}\right) - \left\{{\ell \atop ik}\right\} e\sigma_{\ell}^{\cdot i} = \overset{\{\}}{\nabla}_i\left(e\sigma_{k}^{\cdot i}\right) = 0 \quad , \tag{3.21}$$

*Compare for these considerations always the lecture of Nitsch[16] and the diploma thesis of Meyer[35] and references given there.

i.e. the rotational potential $\Gamma_i^{\cdot\alpha\beta}$ drops out from the momentum law of a scalar field altogether. The covariant derivative in (3.21) is understood with respect to the Christoffel symbol. We stress that the volume force of (3.19) is no longer manifest, it got "absorbed." Consequently a scalar field $\Phi(x)$ is not sensitive to the connection $\Gamma_i^{\cdot\alpha\beta}$. It is perhaps remarkable that this property of $\Phi(x)$ is a result of the Noether identities as applied to an arbitrary scalar matter lagrangian, i.e. we need no information about the gauge field part of the lagrangian in order to arrive at (3.20) and (3.21), respectively. It is a universal property of any scalar matter field embedded in a general U_4.

For the Maxwell potential A_i, which is treated in the PG as a scalar (-valued one-form), all these considerations apply mutatis mutandi. It should be understood, however, that spinning matter, say Dirac matter, couples to $\Gamma_i^{\cdot\alpha\beta}$, and in this case there is no ambiguity left as to whether we live in a V_4, a T_4 or in a general U_4. Therefore a Dirac electron can be used as a probe for measuring the rotational potential* $\Gamma_i^{\cdot\alpha\beta}$.

Let us conclude with some plausibility considerations: In our model universe filled only with scalar matter, $\Phi(x)$ does not feel $\Gamma_i^{\cdot\alpha\beta}$, as we saw. Hence one should expect that it doesn't produce it either, or, in other words, the "scalar" universe should obey a teleparallelism geometry T_4 with the rigid $F_{ij}^{\cdot\cdot\alpha\beta} = 0$ constraint, since then, according to (2.56), we could make $\Gamma_i^{\cdot\alpha\beta}$ vanish globally. Because of the equivalence of (3.19), (3.20), and (3.21), scalar matter would move along geodesics of the attached V_4, nevertheless. If one took care that the field equations of the T_4 were appropriately chosen, one could produce a T_4-theory which is, for scalar matter, indistinguishable from GR.

To similar conclusions leads the following argument: Suppose there existed only scalar matter. Then there is no point in gauging the rotations since $\Phi(x)$ is insensitive to it. Repeating all considerations of Lecture 2, yields immediately a T_4 as the spacetime appropriate for a translational gauge theory, in consistence with the arguments as given above.

---

*The equations of motion of a Dirac electron in a U_4, and in particular its precession in such a spacetime, were studied in detail by Rumpf.[17] For earlier references see [1]. Recent work includes Hojman,[43] Balachandran et al.,[44] and the extensive studies of Yasskin and Stoeger.[20,45,46]

Summing up: scalar (macroscopic) matter is uncoupled from the rotational potential $\Gamma_i^{\cdot\alpha\beta}$, as proven in (3.20), and is expected to span a T_4-spacetime.

3.4 General Gauge Field Lagrangian and its Noether Identities

In order to build up the total action function of matter plus field, one has to add to the matter lagrangian in (3.1) a gauge field lagrangian V representing the effect of the free gauge field. We will assume, in analogy to the matter lagrangian, that the gauge field lagrangian is of first order in the gauge potentials:

$$V = V\left(\kappa_1,\kappa_2\cdots,\eta_{\alpha\beta},e_i^{\cdot\alpha},\Gamma_i^{\cdot\alpha\beta},\partial_k e_i^{\cdot\alpha},\partial_k\Gamma_i^{\cdot\alpha\beta}\right) \quad . \tag{3.22}$$

The quantities $\kappa_1,\kappa_2\cdots$ denote some universal coupling constants to be specified later and for parity reasons we assume that V must not depend on the Levi-Cività symbol $\varepsilon^{\alpha\beta\gamma\delta}$. Then the gauge field equations, in analogy to Maxwell's theory, will turn out to be of second order in the potentials in general.

Applying the Noether identity (3.5) to (3.22) yields*

$$\frac{\delta V}{\delta e_i^{\cdot\alpha}}\,\delta e_i^{\cdot\alpha} + \frac{\delta V}{\delta\Gamma_i^{\cdot\alpha\beta}}\,\delta\Gamma_i^{\cdot\alpha\beta} + D_j\left(\varepsilon^j V + \mathcal{H}_\alpha^{\cdot ij}\delta e_i^{\cdot\alpha} + \mathcal{H}_{\alpha\beta}^{\cdot\cdot ij}\delta\Gamma_i^{\cdot\alpha\beta}\right) \equiv 0, \tag{3.23}$$

where we have introduced the field momenta†

$$\mathcal{H}_\alpha^{\cdot ij} := \frac{\partial V}{\partial\partial_j e_i^{\cdot\alpha}} \quad , \qquad \mathcal{H}_{\alpha\beta}^{\cdot\cdot ij} := \frac{\partial V}{\partial\partial_j\Gamma_i^{\cdot\alpha\beta}} \quad , \tag{3.24}$$

---

*The identities of this section and of Section 3.2 can be also found in the paper of W. Szczyrba.[18]

†In spite of current practice in theoretical physics, it should be stressed that even in microphysical vacuum electrodynamics it is advisable to introduce the "induction" tensor density $\mathcal{H}^{ij}$ as an independent concept amenable to direct operational interpretation (see Post[47]). One is in good company then (Maxwell). For a "practical" application of such ideas see the discussion preceding eq. (4.55). Recent work of Rund[48] seems to indicate that also in Yang-Mills theories such a distinction between induction and field could be useful.

canonically conjugated to the potentials $e_i^{\bullet\alpha}$ and $\Gamma_i^{\bullet\alpha\beta}$, respectively. Substitute (2.26), (2.27) into (3.23) and get

$$\frac{\delta V}{\delta e_i^{\bullet\alpha}}\left(-D_i\varepsilon^\alpha+\omega_\gamma^{\bullet\alpha}e_i^{\bullet\gamma}-\varepsilon^\gamma F_{\gamma i}^{\bullet\bullet\alpha}\right)+\frac{\delta V}{\delta\Gamma_i^{\bullet\alpha\beta}}\left(-D_i\omega^{\alpha\beta}-\varepsilon^\gamma F_{\gamma i}^{\bullet\bullet\alpha\beta}\right)+$$
$$+D_j\left[e^j_{\bullet\alpha}\varepsilon^\alpha V+\mathcal{H}_\alpha^{\bullet ij}\left(-D_i\varepsilon^\alpha+\omega_\gamma^{\bullet\alpha}e_i^{\bullet\gamma}-\varepsilon^\gamma F_{\gamma i}^{\bullet\bullet\alpha}\right)+\right. \tag{3.25}$$
$$\left.+\mathcal{H}_{\alpha\beta}^{\bullet\bullet ij}\left(-D_i\omega^{\alpha\beta}-\varepsilon^\gamma F_{\gamma i}^{\bullet\bullet\alpha\beta}\right)\right]\equiv 0 \quad .$$

Again we have to differentiate the bracket. This time, however, we find second derivatives of the translation and rotation parameters, namely, collecting these terms,

$$-\mathcal{H}_\alpha^{\bullet ij}D_jD_i\varepsilon^\alpha-\mathcal{H}_{\alpha\beta}^{\bullet\bullet ij}D_jD_i\omega^{\alpha\beta}=$$
$$-\mathcal{H}_\alpha^{\bullet ij}D_{(i}D_{j)}\varepsilon^\alpha-\mathcal{H}_{\alpha\beta}^{\bullet\bullet ij}D_{(i}D_{j)}\omega^{\alpha\beta}+ \tag{3.26}$$
$$+\frac{1}{2}\mathcal{H}_\alpha^{\bullet ij}F_{ij\gamma}^{\bullet\bullet\bullet\alpha}\varepsilon^\gamma+\mathcal{H}_{\alpha\beta}^{\bullet\bullet ij}F_{ij\gamma}^{\bullet\bullet\bullet\alpha}\omega^{\gamma\beta} \quad ,$$

where we have used (2.8). Since there are no other second derivative terms in (3.25) but the ones which show up in (3.26), the coefficients of $D_{(i}D_{j)}\varepsilon^\alpha$ and $D_{(i}D_{j)}\omega^{\alpha\beta}$ in (3.26) have to vanish identically, i.e.

$$\mathcal{H}_\alpha^{\bullet(ij)}\equiv 0 \quad , \qquad \mathcal{H}_{\alpha\beta}^{\bullet\bullet(ij)}\equiv 0 \tag{3.27}$$

or

$$\frac{\partial V}{\partial\partial_{(j}e_{i)}^{\bullet\ \alpha}}\equiv 0 \quad , \qquad \frac{\partial V}{\partial\partial_{(j}\Gamma_{i)}^{\bullet\ \alpha\beta}}\equiv 0 \quad . \tag{3.28}$$

Accordingly, the derivatives $\partial_j e_i^{\bullet\alpha}$ and $\partial_j\Gamma_i^{\bullet\alpha\beta}$ can only enter V in the form $\partial_{[j}e_{i]}^{\bullet\ \alpha}$ and $\partial_{[j}\Gamma_{i]}^{\bullet\ \alpha\beta}$, i.e. in the form present in torsion and curvature. Algebraically one cannot construct out of $\Gamma_i^{\bullet\alpha\beta}$ a tensor piece for V because of (2.38). Hence, using (3.24), we have

$$\mathcal{H}_\alpha^{\bullet ij}=2\,\frac{\partial V}{\partial F_{ji}^{\bullet\bullet\alpha}} \quad , \qquad \mathcal{H}_{\alpha\beta}^{\bullet\bullet ij}=2\,\frac{\partial V}{\partial F_{ji}^{\bullet\bullet\alpha\beta}} \tag{3.29}$$

or

$$V = V\left(\kappa_1,\kappa_2\cdots,\eta_{\alpha\beta},e_i^{\cdot\alpha},F_{ij}^{\cdot\cdot\alpha},F_{ij}^{\cdot\cdot\alpha\beta}\right) \quad . \tag{3.30}$$

Eq. (3.29) shows that $(\mathcal{H}_\alpha^{\cdot ij},\mathcal{H}_{\alpha\beta}^{\cdot\cdot ij})$ are both tensor densities, a fact which was not obvious in their definition (3.24).

After (2.18) we saw already that $(F_{\alpha\beta}^{\cdot\cdot\gamma},F_{\alpha\beta}^{\cdot\cdot\gamma\delta})$ are P-tensors. Consequently (3.30) can be simplified and the most general first order gauge field lagrangian reads

$$V = eV\left(\kappa_1,\kappa_2\cdots,\eta_{\alpha\beta},F_{\alpha\beta}^{\cdot\cdot\gamma},F_{\alpha\beta}^{\cdot\cdot\gamma\delta}\right) \quad . \tag{3.31}$$

It is remarkable that the potentials don't appear explicitly in V.

Let us now collect the coefficients of the $(D_i\varepsilon^\alpha,D_i\omega^{\alpha\beta})$-terms in (3.25). The calculation yields

$$\frac{\delta V}{\delta e_i^{\cdot\alpha}} \equiv -D_j\mathcal{H}_\alpha^{\cdot ij} + \varepsilon_\alpha^{\cdot i} \tag{3.32}$$

and

$$\frac{\delta V}{\delta \Gamma_i^{\cdot\alpha\beta}} \equiv -D_j\mathcal{H}_{\alpha\beta}^{\cdot\cdot ij} + \varepsilon_{\alpha\beta}^{\cdot\cdot i} \tag{3.33}$$

with

$$\varepsilon_\alpha^{\cdot i} := e_{\cdot\alpha}^{i}V - F_{\alpha j}^{\cdot\cdot\gamma}\mathcal{H}_\gamma^{\cdot ji} - F_{\alpha j}^{\cdot\cdot\gamma\delta}\mathcal{H}_{\gamma\delta}^{\cdot\cdot ji} \tag{3.34}$$

and

$$\varepsilon_{\alpha\beta}^{\cdot\cdot i} := \mathcal{H}_{[\beta\alpha]}^{\cdot\cdot\ i} \quad , \tag{3.35}$$

respectively. Of course, the quantities (3.34), (3.35) are well-behaved tensor densities.

The interpretation of (3.35) is obvious. Because of (3.24) we find

$$\varepsilon_{\alpha\beta}^{\cdot\cdot i} = e_{k[\alpha}\mathcal{H}_{\beta]}^{\cdot\ ki} = \frac{\partial V}{\partial\partial_i e_k^{\cdot[\beta}}\, e_{|k|\alpha]} = \frac{\partial V}{\partial\partial_i e_k^{\cdot\gamma}}\, f_{\beta\alpha}e_k^{\cdot\gamma} \quad . \tag{3.36}$$

A comparison with (3.9) shows that (3.36) represents the canonical spin current of the translational gauge potential $e_i^{\cdot\alpha}$. Analogously,

(3.34) has the structure and the dimension of a canonical momentum current of both potentials $(e_i^{\bullet\alpha}, \Gamma_i^{\bullet\alpha\beta})$, as is evidenced by a comparison with (3.8).

We don't need the identities resulting from the $(\varepsilon^\alpha, \omega^{\alpha\beta})$-terms, since they will become trivial consequences of the field equations (3.44), (3.45) as substituted into the conservation laws (3.10), (3.11).

3.5 Gauge Field Equations

The total action function of the interacting matter and gauge fields reads

$$W = \int d^4x \left[\mathcal{L}\left(\eta_{\alpha\beta}, \gamma^\alpha \cdots, \psi, \partial_i\psi, e_i^{\bullet\alpha}, \Gamma_i^{\bullet\alpha\beta}\right) + \right.$$
$$\left. + V\left(\kappa_1, \kappa_2 \cdots, \eta_{\alpha\beta}, e_i^{\bullet\alpha}, \Gamma_i^{\bullet\alpha\beta}, \partial_k e_i^{\bullet\alpha}, \partial_k \Gamma_i^{\bullet\alpha\beta}\right)\right] \quad . \tag{3.37}$$

Local P-invariance implies two things, inter alia: It yields, as applied to $\mathcal{L}$, the minimal coupling prescription (3.15), and it leads, as applied to V, to the gauge field lagrangian (3.31) as the most general one allowed. Consequently we have

$$W = \int d^4x \; e\left[L\left(\eta_{\alpha\beta}, \gamma^\alpha \cdots, \psi, D_\alpha\psi\right) + \right.$$
$$\left. + V\left(\kappa_1, \kappa_2 \cdots, \eta_{\alpha\beta}, F_{\alpha\beta}^{\bullet\bullet\gamma}, F_{\alpha\beta}^{\bullet\bullet\gamma\delta}\right)\right] \quad . \tag{3.38}$$

The independent variables are*

$$\psi = \text{matter field} \;, \qquad (e_i^{\bullet\alpha}, \Gamma_i^{\bullet\alpha\beta}) = \text{gauge potentials} \;. \tag{3.39}$$

The action principle requires

$$\delta_{\psi, e, \Gamma} W = 0 \quad . \tag{3.40}$$

We find successively

---

*Independent variation of tetrad and connection is usually and mistakenly called "Palatini variation." In order to give people who only cite Palatini's famous paper a chance to really read it, an English translation is provided in this volume on my suggestion. Sure enough, this service will not change habits.

$$\frac{\delta\mathcal{L}}{\delta\psi} = 0 \quad , \tag{3.41}$$

see (3.2), and

$$-\frac{\delta V}{\delta e_i^{\cdot\alpha}} = \frac{\delta\mathcal{L}}{\delta e_i^{\cdot\alpha}} \quad , \tag{3.42}$$

$$-\frac{\delta V}{\delta \Gamma_i^{\cdot\alpha\beta}} = \frac{\delta\mathcal{L}}{\delta \Gamma_i^{\cdot\alpha\beta}} \quad . \tag{3.43}$$

By (3.8), (3.9) the right hand sides of (3.42), (3.43) are identified as material momentum and spin currents, respectively. Their left hand sides can be rewritten using the tensorial decompositions (3.32), (3.33). Therefore we get the following two gauge field equations:

$$\boxed{D_j\mathcal{H}_\alpha^{\cdot ij} - \varepsilon_\alpha^{\cdot i} = e\Sigma_\alpha^{\cdot i} \quad ,} \tag{3.44}$$

$$\boxed{D_j\mathcal{H}_{\alpha\beta}^{\cdot\cdot ij} - \varepsilon_{\alpha\beta}^{\cdot\cdot i} = e\tau_{\alpha\beta}^{\cdot\cdot i} \quad .} \tag{3.45}$$

We call these equations 1st (or translational) field equation and 2nd (or rotational) field equation, respectively, and we remind ourselves of the following formulae relevant for the field equations, see (3.29), (3.34), (3.35):

$$\mathcal{H}_\alpha^{\cdot ij} = 2\,\frac{\partial V}{\partial F_{ji}^{\cdot\cdot\alpha}} \quad ; \quad \mathcal{H}_{\alpha\beta}^{\cdot\cdot ij} = 2\,\frac{\partial V}{\partial F_{ji}^{\cdot\cdot\alpha\beta}} \quad , \tag{3.46}$$

$$\varepsilon_\alpha^{\cdot i} = e_{\cdot\alpha}^{i}V - F_{\alpha j}^{\cdot\cdot\gamma}\mathcal{H}_\gamma^{\cdot ji} - F_{\alpha j}^{\cdot\cdot\gamma\delta}\mathcal{H}_{\gamma\delta}^{\cdot\cdot ji} \quad , \tag{3.47}$$

$$\varepsilon_{\alpha\beta}^{\cdot\cdot i} = \mathcal{H}_{[\beta\alpha]}^{\cdot\cdot\ i} \quad . \tag{3.48}$$

In general, without specifying a definite field lagrangian V, the field momenta $\mathcal{H}_\alpha^{\cdot ij}$ and $\mathcal{H}_{\alpha\beta}^{\cdot\cdot ij}$ are of first order in their corresponding potentials $e_i^{\cdot\alpha}$ and $\Gamma_i^{\cdot\alpha\beta}$, i.e. (3.44) and (3.45) are generally second order field equations for $e_i^{\cdot\alpha}$ and $\Gamma_i^{\cdot\alpha\beta}$, respectively:

$$\left\{\begin{array}{ll} \partial\partial e + \cdots \sim \Sigma & , \\ \partial\partial\Gamma + \cdots \sim \tau & . \end{array}\right\} \tag{3.49}$$

Furthermore the currents $\Sigma_{\alpha}^{\cdot i}$ and $\tau_{\alpha\beta}^{\cdot\cdot i}$ couple to the potentials $e_{i}^{\cdot\alpha}$ and $\Gamma_{i}^{\cdot\alpha\beta}$, respectively. Clearly then the field equations are of the Yang-Mills type (cf. eq. (12a) in [7]), as expected in faithfully executing the gauge idea.

There is the fundamental difference, however. The universality of the P-group induces the existence of the tensorial currents $(\varepsilon_{\alpha}^{\cdot i}, \varepsilon_{\alpha\beta}^{\cdot\cdot i})$ of the gauge potentials themselves.* In other words, in the PG it is not only the material currents $(e\Sigma_{\alpha}^{\cdot i}, e\tau_{\alpha\beta}^{\cdot\cdot i})$ which produce the fields, but rather the sum of the material and the gauge currents $(e\Sigma_{\alpha}^{\cdot i} + \varepsilon_{\alpha}^{\cdot i},\ e\tau_{\alpha\beta}^{\cdot\cdot i} + \varepsilon_{\alpha\beta}^{\cdot\cdot i})$, this sum being a tensor density again.

Taking into account (3.36) and the discussion following it, it is obvious that $\varepsilon_{\alpha}^{\cdot i}$ is the <u>momentum current</u> (energy-momentum density) and $\varepsilon_{\alpha\beta}^{\cdot\cdot i}$ the <u>spin current</u> (spin angular momentum density) <u>of the gauge fields</u>. Whereas both gauge potentials carry momentum, as is evident from (3.47), only the translational potential $e_{i}^{\cdot\alpha}$ gives rise to a tensorial spin, as one would expect, according to (2.26), (2.27), from the behavior of the $(e_{i}^{\cdot\alpha}, \Gamma_{i}^{\cdot\alpha\beta})$-set under local rotations.

Hence the rotational potential as a quasi-intrinsic gauge potential has vanishing dynamical spin in the sense of the PG, and this fact goes well together with the vanishing dynamical spin of the Maxwell field A_i.

Within the objective of finding a genuine gauge field theory of the P-group, the structure put forward so far, materializing in particular in the two general field equations† (3.44), (3.45), seems

*As a by-product of our investigations, we found the energy-momentum tensor $\varepsilon_{\alpha}^{\cdot i}$ of the gravitational field. Hence the tetrad (or rather T_4-) people who searched for this quantity for quite a long time, were not all that wrong, as will become clear from the lecture of Nitsch.[16] As we will see in Section 4.4, for a V linear in curvature, $\varepsilon_{\alpha}^{\cdot i}$ turns out to be just the Einstein tensor of the U_4, for a quadratic lagrangian $\varepsilon_{\alpha}^{\cdot i}$ is of the type of a contracted Bel-Robinson tensor (see [49]) or, to speak in electrodynamical terms, of the type of Minkowski's energy-momentum tensor.

†In words we could summarize the structure of the field equations as follows: gauge covariant divergence of field momentum = (gauge + material) current. In a pure Yang-Mills theory, just omit the phrase "gauge +."

to us final. The picking of a suitable field lagrangian, which is the last step in establishing a physical theory, is where the real disputation sets in (see Lecture 4).

To our knowledge there exist only the following objections against the PG:

- The description of matter in terms of classical fields is illegitimate. This is a valid objection which can be met by quantizing the theory.
- The PG is too special, it needs grading, i.e. instead of a PG we should rather have a graded PG. Then we end up with supergravity in a space with supertorsion and supercurvature (see [14,50,51]). Up to now, it is not clear whether this step is really compulsory.
- The PG is too special, it needs the extension to the gauge theory of the 4-dimensional real affine group GA(4,R) (metric-affine theory). There are in fact a couple of independent indications pointing in this direction. Therefore we developed a tensorial [52] and a spinorial [53-55] version of such a theory. What is basically happening is that the 2nd field equation (3.45) loses its antisymmetry in $\alpha\beta$, i.e. new intrinsic material currents (dilation plus shear) which, together with spin, constitute the hypermomentum current, couple to the newly emerging gauge potential $\Gamma_{i}{}^{\cdot(\alpha\beta)}$.
- The PG is too special, it needs the extension to a GA(4,R)-gauge and it needs grading as well ([56] and refs. therein). May be.

Any of these objections, however, doesn't make a thorough investigation into the PG futile, it is rather a prerequisite for a better understanding of the extended frameworks.

3.6 The Structure of Poincaré Gauge Field Theory Summarized

		translation	rotation	phase change U(1)
GEOMETRY	infinitesimal generator	D_α 4	$f_{\alpha\beta}$ 6	q 1 charge
	gauge potential	$e_i^{\cdot\alpha}$ tetrad	$\Gamma_i^{\cdot\alpha\beta}$ connection	A_i
	gauge field strength	$F_{ij}^{\cdot\cdot\alpha}$ torsion	$F_{ij}^{\cdot\cdot\alpha\beta}$ curvature	F_{ij}
	Bianchi identity	$D_{[i}F_{jk]}^{\cdot\cdot\alpha} \equiv F_{[ijk]}^{\cdot\cdot\cdot\alpha}$	$D_{[i}F_{jk]}^{\cdot\cdot\alpha\beta} \equiv 0$	$\partial_{[i}F_{jk]} \equiv 0$
KINEMATICS	material current	$\hat{\Sigma}_\alpha^{\cdot i} = \delta\mathfrak{L}/\delta e_i^{\cdot\alpha}$	$\hat{\tau}_{\alpha\beta}^{\cdot\cdot i} = \delta\mathfrak{L}/\delta\Gamma_i^{\cdot\alpha\beta}$	$\mathcal{J}^i = \delta\mathfrak{L}/\delta A_i$
	conservation law	$D_i\hat{\Sigma}_\alpha^{\cdot i} = F_{\alpha i}^{\cdot\cdot\beta}\hat{\Sigma}_\beta^{\cdot i} + F_{\alpha i}^{\cdot\cdot\beta\gamma}\hat{\tau}_{\beta\gamma}^{\cdot\cdot i}$	$D_i\hat{\tau}_{\alpha\beta}^{\cdot\cdot i} - \hat{\Sigma}_{[\alpha\beta]} = 0$	$\partial_i\mathcal{J}^i = 0$
DYNAMICS	field momentum	$\mathcal{H}_\alpha^{\cdot ij} = 2\partial V/\partial F_{ji}^{\cdot\cdot\alpha}$	$\mathcal{H}_{\alpha\beta}^{\cdot\cdot ij} = 2\partial V/\partial F_{ji}^{\cdot\cdot\alpha\beta}$	$\mathcal{H}^{ij} = 2\partial V_{Max}/\partial F_{ji}$
	field equation	$D_j\mathcal{H}_\alpha^{\cdot ij} - \varepsilon_\alpha^{\cdot i} = \hat{\Sigma}_\alpha^{\cdot i}$ first (translational)	$D_j\mathcal{H}_{\alpha\beta}^{\cdot\cdot ij} - \varepsilon_{\alpha\beta}^{\cdot\cdot i} = \hat{\tau}_{\alpha\beta}^{\cdot\cdot i}$ second (rotational)	$\partial_j\mathcal{H}^{ij} = \mathcal{J}^i$ inhom. Maxwell
	ECSK choice	$\mathcal{H}_\alpha^{\cdot ij} \equiv 0$	$\mathcal{H}_{\alpha\beta}^{\cdot\cdot ij} = e e^i_{\cdot[\alpha} e^i_{\cdot\beta]}/\ell^2$	for vacuum
	our choice	$\mathcal{H}_\alpha^{\cdot ij} = e\left(F^{ij}_{\cdot\cdot\alpha} + 2e^{[i}_{\cdot\alpha}F^{j]\gamma}_{\cdot\cdot\gamma}\right)/\ell^2$	$\mathcal{H}_{\alpha\beta}^{\cdot\cdot ij} = eF^{ij}_{\cdot\cdot\alpha\beta}/\kappa$	$\mathcal{H}^{ij} = \hat{\mu}F^{ij}$

(The caret "^" means that the quantity be multiplied by $e = \det e_i^{\cdot\alpha}$. The momentum current of the field is denoted by $\varepsilon_\alpha^{\cdot i}$, see (3.47), the spin current by $\varepsilon_{\alpha\beta}^{\cdot\cdot i} := \mathcal{H}_{[\beta\alpha]}^{\cdot\cdot i}$.) The Riemann-Cartan geometry is dictated by a proper application of the gauge idea to the Poincaré group. To require at that stage a constraint like $F_{ij}^{\cdot\cdot\alpha\beta} = 0$ (teleparallelism) or $F_{ij}^{\cdot\cdot\alpha} = 0$ (riemannian geometry) would seem without foundation. Provided one takes a first-order lagrangian of the type $\mathfrak{L}(\psi,\partial\psi,e,\Gamma) + V(e,\Gamma,\partial e,\partial\Gamma)$, the coupling of matter to geometry and the two gauge field equations are inequivocally fixed and we find $V = eV(F_{\alpha\beta}^{\cdot\cdot\gamma}, F_{\alpha\beta}^{\cdot\cdot\gamma\delta})$.

LECTURE 4: PICKING A GAUGE FIELD LAGRANGIAN

Let us now try to find a suitable gauge field lagrangian in order to make out of our PG-framework a realistic physical theory.

4.1 Hypothesis of Quasi-Linearity

The leading terms of our field equations are

$$\partial_i \mathcal{H}_\alpha{}^{\bullet ij} \sim \Sigma_\alpha{}^{\bullet i} \quad , \tag{4.1}$$

$$\partial_j \mathcal{H}_{\alpha\beta}{}^{\bullet\bullet ij} \sim \tau_{\alpha\beta}{}^{\bullet\bullet i} \quad . \tag{4.2}$$

In general the field momenta will depend on the same variables as V:

$$\mathcal{H}_\alpha{}^{\bullet ij} = \mathcal{H}_\alpha{}^{\bullet ij}\left(\kappa_1, \kappa_2 \cdots, \eta, e, \Gamma, \partial e, \partial\Gamma\right) \quad , \tag{4.3}$$

$$\mathcal{H}_{\alpha\beta}{}^{\bullet\bullet ij} = \mathcal{H}_{\alpha\beta}{}^{\bullet\bullet ij}\left(\kappa_1, \kappa_2 \cdots, \eta, e, \Gamma, \partial e, \partial\Gamma\right) \quad . \tag{4.4}$$

The translational momentum, being a third-rank tensor, cannot depend on the derivatives of the connection $(\partial_\ell \Gamma_k{}^{\bullet\gamma\delta})$, because this expression is of even rank. Furthermore, algebraic expressions of $\Gamma_k{}^{\bullet\gamma\delta}$ never make up a tensor. Hence we have

$$\mathcal{H}_\alpha{}^{\bullet ij} = \mathcal{H}_\alpha{}^{\bullet ij}\left(\kappa_1, \kappa_2 \cdots, \eta_{\gamma\delta}, e_k{}^{\bullet\gamma}, F_{k\ell}{}^{\bullet\bullet\gamma}\right) \quad . \tag{4.5}$$

Similarly we find for the rotational momentum

$$\mathcal{H}_{\alpha\beta}{}^{\bullet\bullet ij} = \mathcal{H}_{\alpha\beta}{}^{\bullet\bullet ij}\left(\kappa_1, \kappa_2 \cdots, \eta_{\gamma\delta}, e_k{}^{\bullet\gamma}, F_{k\ell}{}^{\bullet\bullet\gamma\delta}, \left[F_{k\ell}{}^{\bullet\bullet\gamma}\right]^2\right) \quad . \tag{4.6}$$

This time, both curvature and torsion are allowed, the torsion must appear at least as a square, however.

In order to narrow down the possible choices of a gauge field lagrangian, we will assume quasi-linearity of the field equations as a working hypothesis. This means that the second derivatives in (4.1), (4.2) must only occur linearly. To our knowledge, any successful field theory developed so far in physics obeys this principle.* As a consequence the derivatives of $(e_k{}^{\bullet\gamma}, \Gamma_k{}^{\bullet\gamma\delta})$ in

---

*Instead of the quasi-linearity hypothesis one would prefer having theorems of the Lovelock type available, see Aldersley[57] and references given there.

(4.5), (4.6) can only occur linearly or, in other words, (4.5) and (4.6) are linear in torsion and curvature, respectively. For the translational momentum we have

$$\mathcal{H}_{\alpha}^{\bullet ij} \sim g\left(\kappa_1,\kappa_2\cdots,\eta_{\gamma\delta},e_k^{\bullet\gamma}\right) + \mathrm{lin}_t\left(\kappa_1,\kappa_2\cdots,\eta_{\gamma\delta},e_k^{\bullet\gamma},F_{k\ell}^{\bullet\bullet\gamma}\right) \quad . \tag{4.7}$$

Out of $\eta_{\gamma\delta}$ and $e_k^{\bullet\gamma}$ we cannot construct a third rank tensor, i.e. g has to vanish. Accordingly we find

$$\mathcal{H}_{\alpha}^{\bullet ij} \sim \mathrm{lin}_t\left(\kappa_1,\kappa_2\cdots,\eta_{\gamma\delta},e_k^{\bullet\gamma},F_{k\ell}^{\bullet\bullet\gamma}\right) \quad , \tag{4.8}$$

$$\mathcal{H}_{\alpha\beta}^{\bullet\bullet ij} \sim h\left(\kappa_1,\kappa_2\cdots,\eta_{\gamma\delta},e_k^{\bullet\gamma}\right) + \mathrm{lin}_c\left(\kappa_1,\kappa_2\cdots,\eta_{\gamma\delta},e_k^{\bullet\gamma},F_{k\ell}^{\bullet\bullet\gamma\delta}\right) \quad , \tag{4.9}$$

where lin_t and lin_c denote tensor densities being linear and homogeneous in torsion and curvature, respectively. The possibility of having the curvature-independent term h in (4.9) is again a feature particular to the PG.

We note that the hypothesis of quasi-linearity constrains the choice of the "constitutive laws" (4.8), (4.9) and of the corresponding gauge field lagrangian appreciably. The lagrangian V, as a result of (3.46) and of (4.8), (4.9), is at most quadratic in torsion and curvature (compare (3.31)):

$$V \sim e(\mathrm{const} + \mathrm{torsion}^2 + \mathrm{curvature} + \mathrm{curvature}^2) \quad . \tag{4.10}$$

By the definition of the momenta, we recognize the following correspondence between (4.10) and (4.8), (4.9):

$$\mathrm{const} \quad \to \mathcal{H}_{\alpha}^{\bullet ij} \equiv 0 \;, \quad \mathcal{H}_{\alpha\beta}^{\bullet\bullet ij} \equiv 0 \quad , \tag{4.11}$$

$$\mathrm{torsion}^2 \quad \to \mathrm{lin}_t \quad , \tag{4.12}$$

$$\mathrm{curvature} \quad \to h \quad , \tag{4.13}$$

$$\mathrm{curvature}^2 \to \mathrm{lin}_c \quad . \tag{4.14}$$

The correspondence (4.11) can be easily understood. For vanishing field momenta we find $(-\mathrm{const})e^{i}_{\bullet\alpha} = \Sigma_{\alpha}^{\bullet i}$ and $\tau_{\alpha\beta}^{\bullet\bullet i} = 0$. Clearly then this term in V is of the type of a cosmological-constant-term in GR, i.e. $V = e \times \mathrm{const}$ doesn't make up an own theory, it can only supplement another lagrangian.

In regard to (4.13) we remind ourselves that in a U_4 we have the identities

$$F_{ij}^{\cdot\cdot\alpha\beta} \equiv F_{[ij]}^{\cdot\cdot\ \alpha\beta} \equiv F_{ij}^{\cdot\cdot[\alpha\beta]} \quad . \tag{4.15}$$

Hence, apart from a sign difference, there is only one way to contract the curvature tensor to a scalar:

$$F := e^{i}_{\cdot\beta} e^{j}_{\cdot\alpha} F_{ij}^{\cdot\cdot\alpha\beta} = e^{i}_{\cdot[\beta} e^{j}_{\cdot\alpha]} F_{ij}^{\cdot\cdot\alpha\beta} \quad . \tag{4.16}$$

Hence the term linear in curvature in (4.10) is just proportional to the curvature scalar F.

Since we chose units such that $\hbar = c = 1$, the dimension of V has to be $(\text{length})^{-4}$:

$$[V] = \ell^{-4} \quad . \tag{4.17}$$

Furthermore we have

$$\left[e_{i}^{\cdot\alpha}\right] = 1 \,, \qquad \left[\Gamma_{i}^{\cdot\alpha\beta}\right] = \ell^{-1} \quad , \tag{4.18}$$

and

$$\left[F_{ij}^{\cdot\cdot\alpha}\right] = \ell^{-1} \,, \quad \left[F_{ij}^{\cdot\cdot\alpha\beta}\right] = \ell^{-2} \quad . \tag{4.19}$$

Accordingly a more definitive form of (4.10) reads

$$V \sim e\left[\frac{1}{L_0^4} + \frac{1}{L_1^2}\,(\text{torsion})^2 + \frac{1}{L_2^2}\,(\text{curv.scalar}) + \frac{1}{\kappa}\,(\text{curvature})^2\right] \tag{4.20}$$

with $[L_0] = [L_1] = [L_2] = \ell$, $[\kappa] = 1$. Of course, any number of additional dimensionless coupling constants are allowed in (4.20). If we put $L_0 = \mu L$, $L_1 = L$, $L_2 = \chi^{1/2} L$, then (4.20) gets slightly rewritten and we have

$$V \sim e\left[\frac{1}{(\mu L)^4} + \frac{1}{L^2}\,\{(\text{torsion})^2 + \frac{1}{\chi}\,(\text{curv.scalar})\} + \frac{1}{\kappa}\,(\text{curvature})^2\right] \tag{4.21}$$

with

$$[L] = \ell \,, \qquad [\mu] = [\chi] = [\kappa] = 1 \quad . \tag{4.22}$$

Therefore (4.8), (4.9) finally read

$$e^{-1}\mathcal{H}_{\alpha}^{\bullet ij} \sim L^{-2}\ \mathrm{lin}_t\left(d_1, d_2\cdots, \eta_{\gamma\delta}, e_k^{\bullet\gamma}, F_{k\ell}^{\bullet\bullet\gamma}\right) \quad , \tag{4.23}$$

$$e^{-1}\mathcal{H}_{\alpha\beta}^{\bullet\bullet ij} \sim \chi^{-1}L^{-2}e^{i}_{\bullet[\alpha}e^{j}_{\bullet\beta]} + \kappa^{-1}\ \mathrm{lin}_c\left(f_1, f_2\cdots, \eta_{\gamma\delta}, e_k^{\bullet\gamma}, F_{k\ell}^{\bullet\bullet\gamma\delta}\right), \tag{4.24}$$

with

$$[d_1] = [d_2] = \cdots = [f_1] = [f_2] = \cdots = 1 \quad . \tag{4.25}$$

Provided we don't only keep the curvature-square piece in the lagrangian (4.21) alone, which leads to a non-viable theory,* we need one fundamental length, three primary dimensionless constants μ (scaling the cosmological constant), χ (fixing the relative weight between torsion-square and curvature scalar), κ (a measure of the "roton" coupling), and a number of secondary dimensionless constants d_1, $d_2\cdots$ and f_1, $f_2\cdots$.

We have discussed in this section, how powerful the quasi-linearity hypothesis really is. It doesn't leave too much of a choice for the gauge field lagrangian.

4.2 Gravitons and Rotons

We shall take the quasi-linearity for granted. Then, according to (4.23), (4.24), the leading derivatives of the field momenta in general are

$$\mathcal{H}_{\alpha}^{\bullet ij} \sim L^{-2}\partial e \quad , \tag{4.26}$$

$$\mathcal{H}_{\alpha\beta}^{\bullet\bullet ij} \sim \kappa^{-1}\partial\Gamma \quad . \tag{4.27}$$

Substitute (4.26), (4.27) into the field equations (3.44), (3.45) and get the scheme:

$$\left.\begin{aligned} \partial\partial e + \cdots &\sim L^2\Sigma \quad , \\ \partial\partial\Gamma + \cdots &\sim \kappa\tau \quad . \end{aligned}\right\} \tag{4.28}$$

Let us just for visualization tentatively take the simplest toy theory possible for describing such a behavior, patterned after Maxwell's theory,

$$\mathcal{H}_{\alpha}^{\bullet ij} \sim L^{-2}F^{ij}_{\bullet\bullet\alpha} \quad , \tag{4.29}$$

*...to the Stephenson-Kilmister-Yang ansatz, see [58] and references given there.

$$\mathcal{H}_{\alpha\beta}^{\cdot\cdot ij} \sim \kappa^{-1} F^{ij}_{\cdot\cdot\alpha\beta} \quad , \tag{4.30}$$

i.e.

$$V \sim L^{-2} F^{ij}_{\cdot\cdot\alpha} F^{\cdot\cdot\alpha}_{ij} + \kappa^{-1} F^{ij}_{\cdot\cdot\alpha\beta} F^{\cdot\cdot\alpha\beta}_{ij} \quad . \tag{4.31}$$

Then we have for both potentials "kinetic energy" - terms in (4.31). Observe that the index positions in (4.29), (4.30) are chosen in such a way that in (4.31) only pure squares appear of each torsion or curvature component, respectively. This is really the simplest choice.

Clearly then, the PG-framework in its general form allows for two types of propagating gauge bosons: gravitons $e_i^{\cdot\alpha}$ ("weak Einstein gravity") and rotons $\Gamma_i^{\cdot\alpha\beta}$("strong Yang-Mills gravity").* These and only these two types of interactions are allowed and emerge quite naturally from our phenomenological analysis of the P-group. <u>We postulate that both types of P-gauge bosons exist in nature.</u>[58]

From our experience with GR we know that the fundamental length L of (4.22) in the PG has to be identified with the Planck length ℓ (K = relativistic gravitational constant)

$$L = \ell = (K)^{1/2} \approx 10^{-32} \text{ cm} \quad , \tag{4.32}$$

whereas we have no information so far on the magnitude of the dimensionless constant κ coupling the rotons to material spin.

Gravitons, as we are assured by GR exist, but the rotons need experimental verification. As gauge particles of the Lorentz (rotation-) group SO(3,1), they have much in common with, say, SU(2)-gauge bosons. They can be understood as arising from a quasi-internal symmetry SO(3,1). It is tempting then to relate the rotons to strong interaction properties of matter. However, it is not clear up to now, how one could manage to exempt the leptons from roton interactions. One should also keep in mind

---

*The f-g-theory of gravity of Zumino, Isham, Salam, and Strathdee (for the references see [58]) appears more fabricated as compared to the PG. The term "strong gravity" we borrowed from these authors. There should be no danger that our rotons be mixed up with those of liquid helium. Previously we called them "tordions" [36,1], see also Hamamoto [59], but this gives the wrong impression as if the rotons were directly related to torsion. With the translation potential $e_i^{\cdot\alpha}$ there is associated a set of 4 vector-bosons of dynamical spin 1. It would be most appropriate to call these quanta "translatons" since the graviton is really a spin-2 object.

that the rotons, because of the close link between $e_i^{\cdot\alpha}$ and $\Gamma_i^{\cdot\alpha\beta}$, have specific properties not shared by SU(2)-gauge bosons, their propagation equation, the second field equation, carries an effective mass-term $\sim\kappa\ell^{-2}\Gamma$, for example.*

Before studying the roton properties in detail, one has to come up with a definitive field lagrangian.

4.3 Suppression of Rotons I: Teleparallelism

Since the rotons haven't been observed so far, one could try to suppress them. Let us look how such a mechanism works.

By inspection of (4.1), (4.2) and of (4.23), (4.24) we recognize that the second derivatives of $\Gamma_k^{\cdot\gamma\delta}$ enter the 2nd field equation by means of the rotational momentum (4.24), or rather by means of the lin_c-term of it. Hence there exist two possibilities of getting rid of the rotons: drop $\mathcal{H}_{\cdot\alpha\beta}^{\cdot\cdot ij}$ altogether or drop only its lin_c-piece. We'll explore the first possibility in this section, the second one in Section 4.4.

Let us put

$$\mathcal{H}_{\alpha\beta}^{\cdot\cdot ij} \equiv 0 \quad . \tag{4.33}$$

Then the field equations (3.44), (3.45) read

$$\left\{\begin{aligned} D_j\mathcal{H}_\alpha^{\cdot ij} - e_{\cdot\alpha}^{i}V + F_{\alpha j}^{\cdot\cdot\gamma}\mathcal{H}_\gamma^{\cdot ji} &= e\Sigma_\alpha^{\cdot i} \quad , &(4.34)\\ \mathcal{H}_{[\alpha\beta]}^{\cdot\cdot\ i} &= e\tau_{\alpha\beta}^{\cdot\cdot i} \quad . &(4.35)\end{aligned}\right\}\ \text{(inconsistent)}$$

For vanishing material spin $\tau_{\alpha\beta}^{\cdot\cdot i} = 0$, however, the tetrad spin $\mathcal{H}_{\alpha\beta}^{\cdot\cdot i}$ would vanish, too, and therefore force the tetrads out of business. We were left with the term $-e_{\cdot\alpha}^{i}V = e\Sigma_\alpha^{\cdot i}$ related to the cosmological constant, see (4.11). Hence this recipe is not successful.

But it is obvious that eq. (4.34) is of the desired type, because it is a second-order field equation in $e_k^{\cdot\gamma}$. We know from (4.23) that $\mathcal{H}_\alpha^{\cdot ij}$ can only be linear and homogeneous in $F_{k\ell}^{\cdot\cdot\gamma}$.

*Some attempts to relate torsion to weak interaction, were recently criticized by DeSabbata and Gasperini.[60]

Additionally one can accommodate the Planck length in the constitutive law (4.23). Consequently, the most general linear relation

$$\overset{T}{\mathcal{H}}_{\alpha}^{\cdot\, ij} = e\left(d_1 F^{ji}_{\cdot\cdot\alpha} + d_2 F_{\alpha}^{\cdot\,[ij]} + d_3 e^{[i}_{\cdot\alpha} F^{j]\gamma}_{\cdot\ \cdot\gamma}\right)/\ell^2 \quad , \tag{4.36}$$

as substituted in (4.34), would be just an einsteinian type of field equation provided the constants d_1, d_2, and d_3 were appropriated fixed.

Our goal was to suppress the rotons. In (2.56) and (2.57) we saw that we can get rid of an independent connection in a T_4 as well as in a V_4. In view of (4.36) the choice of a V_4 would kill the whole lagrangian. Therefore we have to turn to a T_4 and we postulate the lagrangian

$$\overset{T}{V} = e\left(d_1 F_{ijk}F^{ijk} + d_2 F_{kji}F^{ijk} + d_3 F^{\cdot\cdot k}_{ik} F^{i\ell}_{\cdot\cdot\ell} + \Lambda_{\alpha\beta}^{\cdot\cdot ij} F_{ij}^{\cdot\cdot\alpha\beta}\right)/4\ell^2 \,, \tag{4.37}$$

with $\lambda_{\alpha\beta}^{\cdot\cdot ij}$ as a lagrangian multiplier, i.e. we have imposed onto (4.36) and onto our U_4-spacetime the additional requirement of vanishing curvature. The translational momentum $\overset{T}{\mathcal{H}}_{\alpha}^{\cdot\, ij}$ is still given by (4.36), but the rotational momentum, against our original intention, surfaces again:

$$\mathcal{H}_{\alpha\beta}^{\cdot\cdot ij} = e\Lambda_{\alpha\beta}^{\cdot\cdot ij}/4\ell^2 \quad . \tag{4.38}$$

To insist on a vanishing $\mathcal{H}_{\alpha\beta}^{\cdot\cdot ij}$ turns out to be not possible in the end, but the result of our insistence is the interesting teleparallelism lagrangian (4.37).

The field equations of (4.37) read (see [35])

$$D_j \overset{T}{\mathcal{H}}_{\alpha}^{\cdot\, ij} - e^{i}_{\cdot\alpha}\overset{T}{V} + F_{\alpha j}^{\cdot\cdot\gamma}\overset{T}{\mathcal{H}}_{\gamma}^{\ ji} = e\Sigma_{\alpha}^{\cdot\, i} \quad , \tag{4.39}$$

$$D_j\left(e\Lambda_{\alpha\beta}^{\cdot\cdot ij}/2\ell^2\right) - \overset{T}{\mathcal{H}}_{[\beta\alpha]}^{\cdot\cdot\ i} = e\tau_{\alpha\beta}^{\cdot\cdot i} \quad , \tag{4.40}$$

$$F_{ij}^{\cdot\cdot\alpha\beta} = 0 \quad . \tag{4.41}$$

One should compare these equations with the inconsistent set (4.34), (4.35). Observe that in (4.39) the term in $\overset{T}{V}$ carrying the lagrangian multiplier vanishes on substituting (4.41), i.e. from the point of view of the 1st field equation, its value is irrelevant.

We imposed a T_4-constraint onto spacetime by the lagrangian multiplier term in (4.37). In a T_4 the $\Gamma_i^{\cdot\alpha\beta}$ are made trivial. Therefore, for consistency, we cannot allow spinning matter (other than as test particles) in such a T_4; spin is coupled to the $\Gamma_i^{\cdot\alpha\beta}$, after all. Accordingly, we take macroscopic matter with vanishing spin $\tau_{\alpha\beta}^{\cdot\cdot i} \equiv 0$. Then (4.40) is of no further interest, the multiplier just balances the tetrad spin. We end up with the "tetrad field equation" in a teleparallelism spacetime T_4

$$D_j \overset{T}{\mathcal{H}}_{\alpha}^{\cdot ij} - \frac{1}{4} e^{i}_{\cdot\alpha}\left(F_{k\ell}^{\cdot\cdot\gamma}\overset{T}{\mathcal{H}}_{\gamma}^{\cdot \ell k}\right) + F_{\alpha j}^{\cdot\cdot\gamma}\overset{T}{\mathcal{H}}_{\gamma}^{\cdot ji} = e\sigma_{\alpha}^{\cdot i} \quad , \tag{4.42}$$

$$F_{ij}^{\cdot\cdot\alpha\beta} = 0 \quad , \tag{4.43}$$

with $\overset{T}{\mathcal{H}}_{\alpha}^{\cdot ij}$ as given by (4.36).

As shown in teleparallelism theory,* there exists a one-parameter family of teleparallelism lagrangians all leading to the Schwarzschild solution including the Birkhoff theorem and all in coincidence with GR up to 4th post-newtonian order:

$$d_1 = -\frac{1}{2}(\lambda + 1) \ , \quad d_2 = \lambda - 1 \ , \quad d_3 = 2 \quad . \tag{4.44}$$

We saw already in Section 3.3 that the equations of motion for macroscopic matter in a T_4 coincide with those in GR. For all practical purposes this whole class of teleparallelism theories (4.44) is indistinguishable from GR. The choice $\lambda = 0$ leads to a locally rotation-invariant theory which is exactly equivalent to GR.

Let us sum up: In suppressing rotons we found a class of viable teleparallelism theories for macroscopic gravity (4.42), (4.43) with (4.36), (4.44). According to (4.37), they derive from a torsion-square lagrangian supplemented by a multiplier term in order to enforce a T_4-spacetime. The condition $\lambda = 0$ yields a theory indistinguishable from GR.

4.4 Suppression of Rotons II: The ECSK-Choice

This route is somewhat smoother and instead of finding a T_4, we are finally led to a V_4. As we have seen, there exists the

---

*Nitsch[16] and references given there, compare also Hayashi and Shirafuji,[61] Liebscher,[62] Meyer,[35] Møller,[63] and Nitsch and Hehl.[64]

option of only dropping the curvature piece of (4.24). This leads to the ansatz (we put $\chi = 1$):

$$\hat{\mathcal{H}}_{\alpha\beta}^{\cdot\cdot ij} = e e^{i}_{\cdot[\alpha} e^{j}_{\cdot\beta]} / \ell^2 \quad . \tag{4.45}$$

A look at the field equations (3.44), (3.45) convinces us that we are not in need of a non-vanishing translational momentum now:

$$\hat{\mathcal{H}}_{\alpha}^{\cdot ij} \equiv 0 \quad . \tag{4.46}$$

With (4.46) the field equations reduce to

$$F_{\alpha j}^{\cdot\cdot\gamma\delta} \hat{\mathcal{H}}_{\gamma\delta}^{\cdot\cdot ji} - e^{i}_{\cdot\alpha} \hat{V} = e \Sigma_{\alpha}^{\cdot i} \quad , \tag{4.47}$$

$$D_j \hat{\mathcal{H}}_{\alpha\beta}^{\cdot\cdot ij} = e \tau_{\alpha\beta}^{\cdot\cdot i} \quad . \tag{4.48}$$

Substitution of (4.45) and using (2.43) yields

$$F_{\gamma\alpha}^{\cdot\cdot i\gamma} - \frac{1}{2} e^{i}_{\cdot\alpha} F_{\gamma\delta}^{\cdot\cdot\delta\gamma} = \ell^2 \Sigma_{\alpha}^{\cdot i} \quad , \tag{4.49}$$

$$\frac{1}{2} F_{\alpha\beta}^{\cdot\cdot i} + e^{i}_{\cdot[\alpha} F_{\beta]\gamma}^{\cdot\ \cdot\gamma} = \ell^2 \tau_{\alpha\beta}^{\cdot\cdot i} \quad . \tag{4.50}$$

These are the field equations of the Einstein-Cartan-Sciama-Kibble (ECSK)-theory of gravity* derivable from the lagrangian

$$\hat{V} = \frac{1}{2} F_{ji}^{\cdot\cdot\alpha\beta} \hat{\mathcal{H}}_{\alpha\beta}^{\cdot\cdot ij} = e e^{i}_{\cdot[\alpha} e^{j}_{\cdot\beta]} F_{ji}^{\cdot\cdot\alpha\beta} / 2\ell^2 = F / 2\ell^2 \quad . \tag{4.51}$$

The ECSK-theory has a small additional contact interaction as

---

*Sciama, who was the first to derive the field equations (4.49), (4.50), judges this theory from today's point of view as follows (private communication): "The idea that spin gives rise to torsion should not be regarded as an ad hoc modification of general relativity. On the contrary, it has a deep group theoretical and geometric basis. If history had been reversed and the spin of the electron discovered before 1915, I have little doubt that Einstein would have wanted to include torsion in his original formulation of general relativity. On the other hand, the numerical differences which arise are normally very small, so that the advantages of including torsion are entirely theoretical."

compared to GR.* For vanishing matter spin we recover GR. Hence in this framework we got rid of an independent $\Gamma_i^{\bullet\alpha\beta}$ in a V_4 spacetime in consistency with (2.57).

Observe that something strange happened in (4.49), (4.50): The rotation field strength $F_{ij}^{\bullet\bullet\alpha\beta}$ is controlled by the translation current (momentum), the translation field strength $F_{ij}^{\bullet\bullet\alpha}$ by the rotation current (spin). It is like putting "Chang's cap on Li's head."[23]

Linked with this intertwining of translation and rotation is a fact which originally led Cartan to consider such a type of theory: In the ECSK-theory the contracted Bianchi identities (2.15), (2.16) are, upon substitution of the field equations (4.49), (4.50), identical to the conservation laws (3.10), (3.11), for details see [1]. From a gauge-theoretical point of view this is a pure coincidence. Probably this fact is a distinguishing feature of the ECSK-theory as compared to other theories in the PG-framework.

Consequently a second and perhaps more satisfactory procedure for suppressing rotons consists in picking a field lagrangian proportional to the U_4-curvature scalar.

4.5 Propagating Gravitons and Rotons

After so much suppression it is time to liberate the rotons. How could we achieve this goal? By just giving them enough kinetic energy $\sim[(\partial e)^2 + (\partial\Gamma)^2]$ in order to enable them to get away.

Let us take recourse to our toy theory (4.29), (4.30), (4.31). The lagrangian (4.31) carries kinetic energy of both potentials, and the gravitational constant, or rather the Planck length, appears, too. But the game with the teleparallelism theories made us wiser. The first term on the right hand side of (4.31) would be inconsistent with macroscopic gravity, as can be seen from (4.37) with (4.44). We know nothing about the curvature-square term, hence we don't touch it and stick with the simplest choice. Consequently the ansatz

$$\mathcal{H}_\alpha^{\bullet ij} = e\left[-\frac{1}{2}(\lambda+1)F^{ji}_{\bullet\bullet\alpha} + (\lambda-1)F_\alpha^{\bullet[ij]} + 2e^{[i}_{\bullet\alpha}F^{j]\gamma}_{\bullet\,\bullet\gamma}\right]/\ell^2 \quad (4.52)$$

$$= \langle e\ell^{-2}\eta_{\alpha\gamma}g^{k[i}g^{j]\ell}\rangle\left[-\frac{1}{2}(\lambda+1)F_{\ell k}^{\bullet\bullet\gamma} + (\lambda-1)F^\gamma_{\bullet k\ell} + 2e_k^{\bullet\gamma}F_{\ell\delta}^{\bullet\bullet\delta}\right],$$

*A certain correspondence between the ECSK-theory and GR was beautifully worked out by Nester.[65] For a recent analysis of the ECSK-theory see Stelle and West.[66,67]

$$\mathcal{H}_{\alpha\beta}^{\cdot\cdot ij} = eF^{ij}_{\cdot\cdot\alpha\beta}/\kappa \quad , \tag{4.53}$$

or, by Euler's theorem for homogeneous functions, the corresponding field lagrangian,

$$\begin{aligned} V &= \frac{1}{4}\left[F_{ji}^{\cdot\cdot\alpha}\mathcal{H}_{\alpha}^{\cdot ij} + F_{ji}^{\cdot\cdot\alpha\beta}\mathcal{H}_{\alpha\beta}^{\cdot\cdot ij}\right] \\ &= (e/4\ell^2)\left[-\frac{1}{2}(\lambda+1)F_{ij}^{\cdot\cdot\alpha}F^{ij}_{\cdot\cdot\alpha} + (\lambda-1)F_{ij}^{\cdot\cdot\alpha}F_{\alpha}^{\cdot ji} + 2F^{i\beta}_{\cdot\cdot\beta}F_{i\gamma}^{\cdot\cdot\gamma}\right] + \\ &\quad + (e/4\kappa)\left[-F_{ij}^{\cdot\cdot\alpha\beta}F^{ij}_{\cdot\cdot\alpha\beta}\right] \quad , \end{aligned} \tag{4.54}$$

would appear to be the simplest choice which encompasses macroscopic gravity in some limit.

Consider the "constitutive assumption" (4.52). In analogy with Maxwell's theory one would like to have the metric appearing only in the "translational permeability" $\langle\ \rangle$ and not in the $(\lambda - 1)$-term in the bracket: $F^{\gamma}_{\cdot k\ell} = g_{\ell m}g^{no}e^{m}_{\cdot\delta}e_{n}^{\cdot\gamma}F_{ok}^{\cdot\cdot\delta}$. For harmony one would then like to cancel this term by putting $\lambda = 1$. This is our choice. Substitute $\lambda = 1$ into (4.52) and use (2.43). Then our translational momentum can be put into a very neat form:

$$\begin{aligned} \mathcal{H}_{\alpha}^{\cdot ij} &= e\left(F^{ij}_{\cdot\cdot\alpha} + 2e^{[i}_{\cdot\alpha}F^{j]\gamma}_{\cdot\ \cdot\gamma}\right)/\ell^2 \\ &= 2e^{i\mu}e^{j\nu}e_{m\alpha}D_n\left(ee^{m}_{\cdot[\mu}e^{n}_{\cdot\nu]}\right)/\ell^2 \quad . \end{aligned} \tag{4.55}$$

There is another choice which is distinguished by some property. This is the choice à la Einstein $\lambda = 0$, since $\lambda = 0$ leads, if one enforces a T_4, to a locally rotation-invariant theory, as was remarked on in Section 4.3.* Apart from these two possibilities, there doesn't exist to our knowledge any other preferred choice of λ.

---

*After the proposal[24,58] of the lagrangian with $\lambda = 1$, Rumpf[68] worked out an analysis of the lagrangian in terms of differential forms and formulated a set of guiding principles yielding the $(\lambda = 1)$-choice. It was also clear from his work that this choice is the most natural one obeying (4.44) from a gauge-theoretical point of view. Recently Wallner,[69] in a most interesting paper, advocated the use of $\lambda = 0$.

Before we substantiate our choice by some deeper-lying ideas, let us look back to our streamlined toy theory (4.54) cum (4.52), (4.53) and compare it with the general quasi-linear structure (4.21) cum (4.23), (4.24):

Since we want propagating rotons, the curvature-square term in (4.21) is indispensable, i.e. κ is finite. We could accommodate more dimensionless constants by looking for the most general linear expression lin_c in (4.24). Furthermore, if we neglect the cosmological term, then there is only to decide of how to put gravitons into the theory, either by means of the torsion-square term or by means of the curvature scalar (or with both together). Now, curvature is already taken by the roton interaction in the quadratic curvature-term, i.e. the rotons should be suppressed in the limit of vanishing curvature (then the rotons' kinetic energy is zero). In other words, curvature is no longer at our disposal and the torsion-square piece has to play the role of the gravitons' kinetic energy. And we know from teleparallelism that it can do so. Since we don't need the curvature scalar any longer, we drop it and put $\chi = \infty$, even if that is not necessarily implied by the arguments given. It seems consistent with this picture that theories in a U_4 with (curvature scalar) + $(\text{curvature})^2$ don't seem to have a decent newtonian limit.*

Collecting all these arguments, we see that the gauge field lagrangian (4.54) has a very plausible structure both from the point of view of allowing rotons to propagate and of being consistent with macroscopic gravity in an enforced T_4-limit.

4.6 The Gordon Decomposition Argument

The strongest argument in favor of the choice with $\lambda = 1$ [24, 58,13,25] comes from other quarters, however. Take the lagrangian $\mathcal{L}$ of a Dirac field $\psi(x)$ and couple it minimally to a U_4 according to the prescription (3.15). Compute the generalized Dirac equation according to (3.2) and the momentum and spin currents according to (3.8) and (3.9), respectively.[77] We find

$$\Sigma_{\alpha}^{\cdot i} = \frac{i}{2}\,\bar{\psi}\gamma^{i}D_{\alpha}\psi + \text{h.c.} \quad , \tag{4.56}$$

$$\tau_{\alpha\beta}^{\cdot\cdot i} = \frac{i}{2}\,\bar{\psi}\gamma^{i}f_{\alpha\beta}\psi + \text{h.c.} \quad . \tag{4.57}$$

*Theories of this type have been investigated by Anandan,[70,71] Fairchild,[72] Mansouri and Chang,[73] Neville,[74,75] Ramaswamy and Yasskin,[76] Tunyak,[21] and others.

(i = imaginary unit, h.c. = hermitian conjugate, $\overline{\psi}$ = Dirac adjoint).

Execute a Gordon decomposition of both currents and find (we will give here the results for an M_4, they can be readily generalized to a U_4):

$$\Sigma_{\alpha}^{\bullet i} = \overset{\text{conv}}{\Sigma_{\alpha}^{\bullet i}} + \partial_j \left(M^{ij}_{\bullet\bullet\alpha} + 2e^{[i}{}_{\bullet\alpha} M^{j]\beta}_{\bullet\ \bullet\beta} \right) \quad , \tag{4.58}$$

$$\tau_{\alpha\beta}^{\bullet\bullet i} = \overset{\text{conv}}{\tau_{\alpha\beta}^{\bullet\bullet i}} + \partial_j M^{ij}_{\bullet\bullet\alpha\beta} + M^{i}_{\bullet[\alpha\beta]} + e^{i}_{\bullet[\beta} M^{\bullet\ \bullet\gamma}_{\alpha]\gamma} \quad . \tag{4.59}$$

The convective currents are of the usual Schrödinger type

$$\overset{\text{conv}}{\Sigma_{\alpha}^{\bullet i}} := \frac{1}{2m} (\partial^i \overline{\psi}) \partial_\alpha \psi + \text{h.c.} + \delta^i_\alpha \overset{\text{conv}}{\mathcal{L}} \quad , \tag{4.60}$$

$$\overset{\text{conv}}{\tau_{\alpha\beta}^{\bullet\bullet i}} := \frac{1}{2m} (\partial^i \overline{\psi}) f_{\alpha\beta} \psi + \text{h.c.} \quad , \tag{4.61}$$

with

$$\overset{\text{conv}}{\mathcal{L}} := -\frac{1}{2m} [(\partial_\alpha \overline{\psi}) \partial^\alpha \psi - m^2 \overline{\psi} \psi] \quad . \tag{4.62}$$

In analogy with the Dirac-Maxwell theory (i.e. with Dirac plus U(1)-gauge, whereas we have Dirac plus PG) in (4.58), (4.59) there emerge the translational and the rotational <u>gravitational moments</u> of the electron field

$$M^{ij}_{\bullet\bullet\alpha} := \frac{1}{m} \overline{\psi} f^{ji} \partial_\alpha \psi \tag{4.63}$$

and

$$M^{ij}_{\bullet\bullet\alpha\beta} := \frac{1}{m} \overline{\psi} f^{ji} f_{\alpha\beta} \psi \quad , \tag{4.64}$$

respectively. We stress that these two new expressions for the gravitational moments of the electron are measurable in principle. Hence there is a way to decide, whether it makes physical sense to Gordon-decompose the momentum and the spin currents of the electron field.

If we introduce the polarization currents

$$\overset{\text{pol}}{\Sigma_{\alpha}^{\bullet i}} := \Sigma_{\alpha}^{\bullet i} - \overset{\text{conv}}{\Sigma_{\alpha}^{\bullet i}} \quad , \quad \overset{\text{pol}}{\tau_{\alpha\beta}^{\bullet\bullet i}} := \tau_{\alpha\beta}^{\bullet\bullet i} - \overset{\text{conv}}{\tau_{\alpha\beta}^{\bullet\bullet i}} \quad , \tag{4.65}$$

then we have finally

$$\overset{\text{pol}}{\Sigma}{}_{\alpha}^{\cdot i} = \partial_j \left(M^{ij}_{\cdot\cdot\alpha} + 2e^{[i}_{\cdot\alpha} M^{j]\gamma}_{\cdot\cdot\gamma} \right) , \tag{4.66}$$

$$\overset{\text{pol}}{\tau}{}_{\alpha\beta}^{\cdot\cdot i} = \partial_j M^{ij}_{\cdot\cdot\alpha\beta} + M^{i}_{\cdot[\alpha\beta]} + e^{i}_{\cdot[\beta} M^{\cdot\,\cdot\gamma}_{\alpha]\gamma} . \tag{4.67}$$

Observe that the last two terms in (4.67) are required by angular momentum conservation (3.13).

Up to now we just carried through some special-relativistic Dirac algebra. Let us now turn to our field equations (3.44), (3.45) and linearize them:

$$\Sigma_{\alpha}^{\cdot i} \sim \partial_j \mathcal{H}_{\alpha}^{\cdot ij} , \tag{4.68}$$

$$\tau_{\alpha\beta}^{\cdot\cdot i} \sim \partial_j \mathcal{H}_{\alpha\beta}^{\cdot\cdot ij} + \mathcal{H}_{[\alpha\beta]}^{\cdot\cdot\ i} . \tag{4.69}$$

A comparison with (4.66), (4.67) will reveal immediately a similarity in structure. Read off the working hypothesis: The translational and rotational <u>field strengths</u> couple in an analogous way to the canonical currents, as the respective gravitational <u>moments</u> to the polarization currents. Consequently we get

$$\mathcal{H}_{\alpha}^{\cdot ij} \sim F^{ij}_{\cdot\cdot\alpha} + 2e^{[i}_{\cdot\alpha} F^{j]\gamma}_{\cdot\cdot\gamma} , \tag{4.70}$$

$$\mathcal{H}_{\alpha\beta}^{\cdot\cdot ij} \sim F^{ij}_{\cdot\cdot\alpha\beta} . \tag{4.71}$$

Properly adjusting the dimensions, leads straightaway to the field lagrangian of our choice

$$V = (e/4\ell^2)\left(-F^{ij}_{\cdot\cdot\alpha} F^{\cdot\cdot\alpha}_{ij} + 2F^{i\gamma}_{\cdot\cdot\gamma} F^{\cdot\cdot\delta}_{i\delta} \right) + (e/4\kappa)\left(-F^{\cdot\cdot\cdot\beta}_{ij\alpha} F^{ij\alpha}_{\cdot\cdot\cdot\beta} \right) \tag{4.72}$$

without all the involved reasonings used earlier.*

*According to Wallner,[69] one can reformulate the Gordon decomposition such that one arrives at the choice $\lambda = 0$. We are not able to understand his arguments, however.

The most remarkable achievement of the working hypothesis as formulated above is the following: Without having ever talked about gravity, one arrives at a lagrangian which, in the enforced T_4-limit, yields the Schwarzschild solution, the newtonian approximation, etc. Some consequences of (4.72), like a "confinement" potential in a weak field approximation, have been worked out already.[58,13] But since we are running out of space and time, we shall discuss these matters elsewhere.

ACKNOWLEDGMENTS

For many enlightening and helpful discussions on the PG I am most grateful to Yuval Ne'eman as well as to Jürgen Nitsch. I'd like to thank P. G. Bergmann and V. DeSabbata for inviting me to give these lectures in Erice and I highly appreciate the support of Y. Ne'eman and E. C. G. Sudarshan (CPT) and of J. A. Wheeler (CTP) of The University of Texas at Austin, where these notes were written up.

LITERATURE

1. F. W. Hehl, P. von der Heyde, G. D. Kerlick, and J. M. Nester, Rev. Mod. Phys. 48:393 (1976).
2. E. Cartan, C. R. Acad. Sci. (Paris) 174:593 (1922). A translation was provided by G. D. Kerlick, see his Ph.D. Thesis, Princeton University (1975).
3. J. A. Schouten, Ricci Calculus, 2nd ed., Springer, Berlin (1954).
4. Y. Choquet-Bruhat, C. DeWitt-Morette, and M. Dillard-Bleick, "Analysis, Manifolds and Physics," North-Holland, Amsterdam (1977).
5. L. O'Raifeartaigh, Rep. Progr. Phys. 42:159 (1979).
6. H. Weyl, Z. Phys. 56:330 (1929).
7. C. N. Yang and R. L. Mills, Phys. Rev. 96:191 (1954).
8. R. Utiyama, Phys. Rev. 101:1597 (1956).
9. D. W. Sciama, Recent Developments in General Relativity (Festschrift for Infeld), Pergamon, Oxford (1962), p.415.
10. T. W. B. Kibble, J. Math. Phys. 2:212 (1961).
11. Y. Ne'eman, "Gravity, Groups, and Gauges," Einstein Commemorative Volume (A. Held et al., eds.), Plenum Press, New York, to be published (1980).
12. A. Trautman, "Fiber Bundles, Gauge Fields, and Gravitation," ibid.
13. F. W. Hehl, J. Nitsch, and P. von der Heyde, "Gravitation and the Poincaré Gauge Field Theory with Quadratic Lagrangian," ibid.
14. Y. Ne'eman, this volume.

15. A. Trautman, this volume.
16. J. Nitsch, this volume.
17. H. Rumpf, this volume.
18. W. Szczyrba, this volume.
19. M. A. Schweizer, this volume.
20. P. B. Yasskin, see his Ph.D. Thesis, University of Maryland (1979).
21. V. N. Tunyak, Sov. Phys. J. (Izv. VUZ. Fiz., USSR) 18:74,77 (1975); 19:599 (1976); 20:538,1337,1537 (1977).
22. Current Problems in Theoret. Physics (Festschrift for Ivanenko), Moscow Univ. (1976).
23. Z. Zuo, P. Huang, Y. Zhang, G. Li, Y. An, S. Chen, Y. Wu, Z. He, L. Zhang, and Guo, Scientia Sinica 22:628 (1979).
24. P. von der Heyde, Phys. Lett. 58A:141 (1976); Z. Naturf. 31a:1725 (1976).
25. P. von der Heyde, talk given at the "9th Texas Symposium on Relativistic Astrophysics," during the Workshop "Gauge Theories of Gravity," held at Munich in December 1978 (unpublished), and unpublished draft.
26. Y. Ne'eman, Lecture Notes in Mathematics (Springer) No. 676, p.189 (1978).
27. R. Jackiw, Phys. Rev. Lett. 41:1635 (1978).
28. F. Mansouri and C. Schaer, Phys. Lett. 83B:327 (1979).
29. J. M. Nester, Ph.D. Thesis, University of Maryland (1977).
30. R. J. Petti, Gen. Rel. Gravitation J. 7:869 (1976).
31. Y. Ne'eman and T. Regge, Riv. Nuovo Cimento 1, No. 5, 1 (1978).
32. G. Cognola, R. Soldati, M. Toller, L. Vanzo, and S. Zerbini, Preprint Univ. Trento (1979).
33. Y. Ne'eman and E. Takasugi, Preprint University of Texas at Austin (1979).
34. P. von der Heyde, Lett. Nuovo Cim. 14:250 (1975).
35. H. Meyer, Diploma Thesis, University of Cologne (1979).
36. P. von der Heyde and F. W. Hehl, Proc. of the First Marcel Grossmann Meeting on General Relativity, held at Trieste in 1975 (R. Ruffini, ed.) North-Holland, Amsterdam (1977), p.255.
37. S. Hojman, M. Rosenbaum, M. P. Ryan, and L. C. Shepley, Phys. Rev. D17:3141 (1978).
38. S. Hojman, M. Rosenbaum, and M. P. Ryan, Phys. Rev. D19:430 (1979).
39. W.-T. Ni, Phys. Rev. D19:2260 (1979).
40. C. Mukku and W. A. Sayed, Phys. Lett. 82B:382 (1979).
41. F. W. Hehl, Rep. on Math. Phys. (Toruń) 9:55 (1976).
42. Y. M. Cho, J. Phys. A11:2385 (1978).
43. S. Hojman, Phys. Rev. 18:2741 (1978).
44. A. P. Balachandran, G. Marmo, B. S. Skagerstam, and A. Stern, CERN-preprint TH.2740 (1979).
45. W. R. Stoeger and P. B. Yasskin, Gen. Rel. Gravitation J. (to be published).

46. P. B. Yasskin and W. R. Stoeger, Preprint Harvard University (1979).
47. E. J. Post, "Formal Structure of Electromagnetics," North-Holland, Amsterdam (1962).
48. H. Rund, J. Math. Phys. 20:1392 (1979).
49. C. W. Misner, K. S. Thorne, and J. A. Wheeler, "Gravitation," Freeman, San Francisco (1973).
50. P. van Nieuwenhuizen, this volume.
51. J. Wess, talk given at the Einstein Symposium Berlin, preprint Univ. Karlsruhe (1979).
52. F. W. Hehl, G. D. Kerlick, and P. von der Heyde, Z. Naturforsch. 31a:111,524,823 (1976); Phys. Lett. 63B:446 (1976).
53. F. W. Hehl, E. A. Lord, and Y. Ne'eman, Phys. Lett. 71B:432 (1977); Phys. Rev. D17:428 (1978).
54. E. A. Lord, Phys. Lett. 65A:1 (1978).
55. Y. Ne'eman and Dj. Šijački, Ann. Phys. (N.Y.) 120:292 (1979).
56. Y. Ne'eman, in "Jerusalem Einstein Centennial Symposium," (Y. Ne'eman, ed.), Addison-Wesley, Reading, to be published (1980).
57. S. J. Aldersley, Gen. Rel. Gravitation J. 8:397 (1977).
58. F. W. Hehl, Y. Ne'eman, J. Nitsch, and P. von der Heyde, Phys. Lett. 78B:102 (1978).
59. S. Hamamoto, Progr. Theor. Physics 61:326 (1979).
60. V. DeSabbata and M. Gasperini, Lett. Nuovo Cimento 21:328 (1978).
61. K. Hayashi and T. Shirafuji, Phys. Rev. D19:3524 (1979).
62. D.-E. Liebscher, this volume.
63. C. Møller, Mat. Fys. Medd. Dan. Vid. Selsk. 39, No. 13 (1978).
64. J. Nitsch and F. W. Hehl, Preprint Univ. Cologne (1979).
65. J. M. Nester, Phys. Rev. D16:2395 (1977).
66. K. S. Stelle and P. C. West, J. Phys. A12:L205 (1979).
67. K. S. Stelle and P. C. West, Preprint Imperial College London (1979).
68. H. Rumpf, Z. Naturforsch. 33a:1224 (1978).
69. R. P. Wallner, Preprint Univ. Vienna (1979); Gen. Rel. Gravitation J. (to be published).
70. J. Anandan, Preprint Univ. of Maryland (1978); Nuovo Cimento (to be published).
71. J. S. Anandan, in "Quantum Theory and Gravitation," Proc. New Orleans Conf., May 1979 (A. R. Marlow, ed.), Academic Press, New York (to be published).
72. E. E. Fairchild, Phys. Rev. D16:2438 (1977).
73. F. Mansouri and L. N. Chang, Phys. Rev. D13:3192 (1976).
74. D. E. Neville, Phys. Rev. D18:3535 (1978).
75. D. E. Neville, Preprint Temple Univ. (1979).
76. S. Ramaswamy and P. B. Yasskin, Phys. Rev. D19:2264 (1979).
77. F. W. Hehl and B. K. Datta, J. Math. Phys. 12:1334 (1971).

THE MACROSCOPIC LIMIT OF THE POINCARÉ GAUGE FIELD THEORY OF GRAVITATION

Jürgen Nitsch

Institut for Theoretical Physics of the
University of Cologne
D-5000 Cologne 41, W. Germany

ABSTRACT

In the framework of the recently[1] formulated Poincaré gauge field theory of gravitation we study the limit of macroscopic spinless matter modelled by a scalar matter field Φ. We define local active Poincaré transformations for scalar matter and the corresponding invariance. We derive a Yang-Mills-type tetrad field equation and assume the gauge field Lagrangian to be quadratic in torsion. In order to find a definitive "viable" field Lagrangian, we investigate the weak field approximation. We discuss groups of motions in a U_4 with vanishing curvature ($F_{ij\alpha}^{\cdot\cdot\cdot\beta} = 0$) and prove, among other things, that (i) all spherical symmetric solutions of the Poincaré gauge field theory for macroscopic matter (translational gauge field theory) are also solutions in Einstein's theory and vice versa; (ii) in fifth order of the post-Newtonian approximation there is a deviation from Einstein's theory. This deviation shows up in the spin precession of a Dirac test particle moving in the gravitational field of a rotating body.

INTRODUCTION

Einstein's theory of general relativity (GR) from 1915 gives a well-founded and consistent description of all known gravitational phenomena, and the specific experimental tests of the theory are performed with a very high degree of accuracy. The underlying principles of GR are of such generality and simplicity that strong arguments are needed to formulate a theory of gravitation for macroscopic matter from a different point of view. The occurrence of essential singularities in GR, however, may be interpreted as a sign that the domain of applicability of the theory has been exceeded.

In order to circumvent the problem of singularities, Møller[2] constructed a tetrad theory of gravitation, retaining at the same time all satisfactory features of GR. In a recent work it was shown by Meyer[3] that Møller's tetrad theory is completely equivalent to the translational gauge field theory of gravitation[1,4]. The translational gauge field theory, in turn, can be understood as a limiting case of the Poincaré gauge field theory of gravitation[1]. We add here some reasons favoring an alternative theory of gravitation for macroscopic matter which have their roots in the microphysical area of the foundation of the Poincaré gauge field theory[5].

The spin aspect of elementary matter, in particular, comes to mind. We believe that the "existence" of spinors in nature necessarily enforces us to describe the spinor dynamics and the implied physical consequences with the aid of anholonomic frames (tetrads). As soon as we, as a matter of principle, relate physical phenomena to anholonomic measuring tetrads, we are inevitably led, as an immediate consequence of an "equivalence principle", to an independent asymmetric linear connection $\Gamma_{i\alpha}^{\cdot\cdot\beta}$.

In Section II we establish a gravitational theory for macroscopic matter based on the assumption of the anholonomicity of the local reference frames. We introduce the basic kinematical conceptions in terms of the 4 translational potentials $e_i^{\cdot\alpha}(x)$ and of the 6 integrable rotational potentials $\Gamma_{i\alpha}^{\cdot\cdot\beta}(x)$ describing the relative orientations of the reference frames.

Active local Poincaré transformations of scalar matter are defined with respect to unchanged frames and coordinates. Scalar matter distributions are said to be physical equivalent iff they can be transformed into each other by a suitable local Poincaré transformation; thereby the relative positions of the field amplitudes are kept fixed. Since we only deal with scalar matter it does not make any sense to speak about relative orientation of neighboring field amplitudes. Scalar matter is not appropriate for the introduction of an independent orientation standard.

The "equivalence property" of matter distributions is used to derive the gauge transformations of the potentials. We briefly discuss the structure of the local Poincaré transformations and notice its non-local character which becomes evident by the appearance of the torsion field.

Then we turn to the dynamics of scalar matter. The invariance of the material action under local Poincaré transformations leads to eight identities. The first four identities lead to the identification of the dynamical current with the canonical energy-momentum current. The second four identities yield the energy-momentum conservation law.

The gravitational field equation results from the variation of the action. In addition to the material source $\sigma_{\alpha}{}^{\cdot i}$ the gravitational field itself also contributes a tensorial part to the source. We write down the most general first-order field Lagrangians involving the generalized Weitzenböck invariants with three unknown dimensionless constants.

In Section III we limit the arbitrariness in the choice of the field Lagrangian by requiring the theory to give the same experimental results as GR for solar system tests, i.e. our theory coincides with GR in linear approximation. This requirement fixes two of the three free constants, and we are left with a one parameter class of possible Lagrangians.

We study the transformation behavior of the tetrad field in the weak field limit and show that the linearized field equation (III.8) possesses a "restricted" local Lorentz invariance. It is this invariance which allows us to impose certain gauge conditions on the symmetric (deDonder condition) as well as on the antisymmetric ("Lorentz" condition) part of the tetrad field. Then, in the weak field approximation, the tetrads turn out to be "symmetric", and the "metric content" of our theory is the same as in GR.

To make the deviation from Einstein's field equation completely evident, we recast the tetrad field equation in a quasi-Einsteinian form, splitting off the Einstein tensor of the Riemannian piece of the connection.

Section IV deals with groups of affine motions and exact solutions of the general field equation. Infinitesimal affine motions are defined in a U_4 with vanishing curvature, i.e. in a space with absolute parallelism, and the fundamental equations for a group of affine motions are developed. The results are applied to the problem of determining the form of the tetrad field for a spherical-symmetric spacetime. We find spherical-symmetric solutions of the field equation and emphasize the validity of Birkhoff's theorem.

In Section V we perform a post Newtonian expansion and develop the tetrad field up to fifth order in (v/c). We compute the deviation from Einstein's theory in this order. This deviation is due to the appearance (in fifth order) of the antisymmetric part of the tetrad field which differs from the Einsteinian contribution of the third order by the ratio of the Schwarzschild radius to the radius of the rotating body under consideration.

Finally, in the last section, Section VI, we apply our results of Section V to the spin of a Dirac test particle, moving in a stationary, axial-symmetric gravitational field of a spherical rigidly rotating body. In the spin precession of electrons there occurs a non-Einsteinian contribution which might indirectly become measurable in the polarization of the emission spectra of electrons of a dilute plasma in the vicinity of presumably existing black holes.

Acknowledgements

I am indebted to Prof. Dr. F.W. Hehl for introducing me to research concerened with the Poincaré gauge field theory of gravitation and for many helpful and elucidating discussions. The support of Prof. Dr. P. Mittelstaedt, his interest in this work, expressed by discussions and critical remarks, is gratefully acknowledged.

Conventions

Our conventions are taken from Reference 1. The Latin alphabet i,j,k,... refers to holonomic indices which in general run from 0 to 3. The Greek alphabet $\alpha,\beta,\gamma,\ldots$ refers to anholonomic Lorentz indices which generally run from 0 to 3. The sign := means "is defined to be", $\overset{*}{=}$ means "equal to, in a suitable coordinate system", $e := \det(e_i^{\cdot\alpha})$; $\hbar = c = 1$; l(Planck length) $= \sqrt{k}$; $k = 8\pi G$; G = Newton's gravitational constant. Square brackets denote skewsymmetrization, e.g. $x_{[\alpha\beta]} = (1/2!)(x_{\alpha\beta} - x_{\beta\alpha})$, while round brackets indicate symmetrization, e.g. $x_{(\alpha\beta)} = (1/2!)(x_{\alpha\beta} + x_{\beta\alpha})$. Other notation and abbreviations are taken from Schouten.[15]

POINCARÉ GAUGE KINEMATICS AND DYNAMICS FOR SCALAR MATTER DISTRIBUTIONS

Consider a spacetime U_4 filled with scalar matter only. In this case it is justifiable to choose the curvature of the U_4 equal to zero ($\varkappa \to 0$)[+]

$$F_{ij\alpha}^{\cdot\cdot\cdot\beta}(U_4) = 0, \tag{II.1}$$

i.e., we have a Riemann-Cartan spacetime with a flat, metric-compatible connection. For this situation we would like to define, following the ideas of Hehl[5] and von der Heyde[7], local active Poincaré transformations for a given scalar matter field $\Phi(x)$ and to derive the corresponding gravitational field equation in a manifest locally Lorentz invariant form. Since we formulate local Poincaré invariance exclusively in terms of special relativistic kinematics, it is reasonable to use the following special relativistic notions:[7]

[+] $\varkappa$, the strong coupling constant, is of microphysical origin and refers to the spin aspect of matter. Cf. also Section 9 in Ref. 1.

(i) A Local Orthonormal Tetrad Field $\underline{e}_\alpha(x)$ and their duals $\underline{e}^\alpha(x)$. With a local coordinate system (x^i) on U_4 we have the relation

$$\underline{e}_\alpha(x) = e^i_{\cdot\alpha}(x)\, \underline{\partial}_i \;, \quad \underline{e}^\alpha(x) = e_i^{\cdot\alpha}(x)\, \underline{dx}^i \;. \tag{II.2}$$

The infinitesimal distance ds between the events x and x + dx is then given by

$$ds = (\eta_{\alpha\beta} dx^\alpha dx^\beta)^{1/2} \;, \quad dx^\alpha = e_i^{\cdot\alpha} dx^i \;, \tag{II.3}$$

and is obviously independent of the special choice of the coordinate system (x^i). The constant, diagonal Minkowski metric $\eta_{\alpha\beta}$ has the signature (-1,+1,+1,+1).

(ii) A Local Standard of Relative Orientation.
At each $x \in U_4$ we uniquely construct for each direction dx a tetrad $\underline{e}^{\parallel}_\alpha(x,dx)$ which is parallel (in the special relativistic sense) to the tetrad $\underline{e}_\alpha(x + dx)$:

$$\underline{e}^{\parallel}_\alpha(x,dx) = (\delta^\beta_\alpha + dx^i \Gamma_{i\alpha}^{\;\;\;\beta})\underline{e}_\beta(x) \;. \tag{II.4}$$

In special relativity a parallel transport does not change lengths and angles. Therefore we have

$$(\underline{e}^{\parallel}_\alpha \cdot \underline{e}^{\parallel}_\beta)(x,dx) = (\underline{e}_\alpha \cdot \underline{e}_\beta)(x + dx) = \eta_{\alpha\beta} \tag{II.5}$$

which implies

$$\Gamma_{i(\alpha\beta)}(x) = 0 \;. \tag{II.6}$$

The position standard, defined in (i), and the orientation standard, defined in (ii), are independent of each other. The 4 x 4 = 16 tetrad functions $e_i^{\cdot\alpha}(x)$ and the 4 x 6 = 24 connection functions $\Gamma_{i\alpha}^{\;\;\;\beta}(x)$ are the fundamental geometrical quantities of our theory. Because of eq. (II.1), however, the linear connection $\underline{\Gamma}$ is integrable. Hence, the number of independent connection functions will drastically be reduced (cf. eq. (II.50)).

The local active Poincaré transformation

$$\Pi : \Phi(x) \quad \rightarrow \quad (\Pi\Phi)\;(x) \tag{II.7}$$

of the matter field leaves the coordinates and frames fixed and is defined by

$$(\Pi\Phi)(x) := \Phi(x) + \delta\Phi(x) \,, \quad \delta\Phi(x) := -\mathcal{L}_\varepsilon \Phi(x) = -\varepsilon^\gamma(x)\, D_\gamma \Phi\,(x). \tag{II.8}$$

Although a local Poincaré transformation involves (6 + 4) infinitesimal functions $\omega^{[\alpha\beta]}(x) = \omega^{\alpha\beta}(x)$ and $\varepsilon^{\gamma}(x)$, only the infinitesimal translations $\varepsilon^{\gamma}(x)$ occur explicitely in eq. (II.8). The infinitesimal rotations $\omega^{\alpha\beta}(x)$ do not appear. The Lorentz generators belonging to Φ are equal to zero. Thus the amplitude $\Phi''(x) =: (1 - dx^i D_i)\Phi(x)$ parallel to $\Phi(x - dx)$ and generated by $D_i = \partial_i + {}^i\Gamma_{i\alpha}{}^{\beta} f_{\beta}^{\cdot\alpha}$ is not distinguishable from $(\Pi\Phi)(x)$; i.e. the scalar matter distribution is not sensitive to local rotations.

We demand that the matter amplitude $(\Pi\Phi)$, together with Φ, is compatible with the corresponding matter dynamics, since the local Poincaré transformation should lead to measurable equivalent matter distributions. In order to guarantee such an equivalence of Φ and $(\Pi\Phi)$ the potentials $e_i^{\cdot\alpha}(x)$ and $\Gamma_{i\alpha}{}^{\beta}(x)$ have to transform in a suitable way. This gauge transformation of the potentials can be determined by purely kinematical methods, without knowing any matter dynamics. "Measurably equivalent" of the fields $\Phi(x)$ and $(\Pi\Phi)(x)$ means that we translate and rotate the field amplitudes in an infinitesimal extended region of spacetime keeping their relative positions and orientations fixed; thereby the matter piece under consideration behaves like a rigid body. We have to keep in mind, however, that we deal with scalar matter which is not appropriate for determining relative orientations.

We conclude that, together with $\Phi(x)$ and $(\Pi\Phi)(x)$, the fields $D_\alpha\Phi(x)$ and $\Pi[(D_\alpha\Phi)](x)$ also have to be Poincaré tensors. This result uniquely leads to the transformation law of the potentials. We can write

$$\Pi(D_\alpha\Phi) = (D_\alpha + \delta D_\alpha)(\Pi\Phi) \tag{II.9}$$

with

$$\delta D_\alpha := [\Pi, D_\alpha] . \tag{II.10}$$

Using the formulae

$$D_\alpha = e^{i}_{\cdot\alpha} D_i = e^{i}_{\cdot\alpha}(\partial_i + \Gamma_{i\beta}{}^{\gamma} f_{\gamma}^{\cdot\beta}) , \tag{II.11}$$

$$\Pi := 1 - \varepsilon^{\gamma} D_\gamma , \tag{II.12}$$

and the commutation relation[5]

$$[D_\alpha, D_\beta] = -e^{i}_{\cdot\alpha} e^{j}_{\cdot\beta} F_{ij}^{\cdot\cdot\gamma} D_\gamma , \tag{II.13}$$

we obtain for eq. (II.10):

$$[\Pi, D_\alpha] = (\varepsilon^{\delta} F_{\delta\alpha}^{\cdot\cdot\gamma} + D_\alpha \varepsilon^{\gamma}) D_\gamma . \tag{II.14}$$

The (translational) field strength is given by

$$F_{ij}^{\cdot\cdot\gamma} := 2D_{[i}e_{j]}^{\cdot\gamma} = 2(\partial_{[i}e_{j]}^{\cdot\gamma} + \Gamma_{[i}^{\ \alpha\gamma} e_{j]}^{\cdot\beta}\eta_{\alpha\beta}) \tag{II.15}$$

and turns out to be the torsion tensor. On the other hand we have

$$\delta D_i = \delta\Gamma_{i\alpha}^{\ \ \beta} f_{\beta}^{\cdot\alpha} , \tag{II.16}$$

and, together with eq. (II.11), we find:

$$e_j^{\cdot\alpha}\delta D_\alpha = -e_{\cdot\alpha}^{i}\delta e_j^{\cdot\alpha} D_i + \delta\Gamma_{j\alpha}^{\cdot\cdot\beta} f_{\beta}^{\cdot\alpha} . \tag{II.17}$$

A comparison of eqs. (II.14) and (II.17) finally yields

$$\delta e_i^{\cdot\alpha} = -(D_i\boldsymbol{\varepsilon}^\alpha + \boldsymbol{\varepsilon}^\gamma F_{\gamma i}^{\ \ \alpha}) , \tag{II.18}$$

and

$$\delta\Gamma_{i\alpha}^{\ \ \beta} = 0 . \tag{II.19}$$

It has independently been shown by Nester[8] and Schweizer[9] that eq. (II.18) has a very simple geometrical meaning: Consider an orthonormal Lorentz frame Θ^α and its expansion in a coordinate frame

$$\Theta^\alpha = e_i^{\cdot\alpha}\, dx^i . \tag{II.20}$$

Then the Lie transport of Θ^α along the flow lines of the infinitesimal vector field $\boldsymbol{\varepsilon}^\alpha$ can be expressed by

$$-\pounds_{\boldsymbol{\varepsilon}}\Theta^\alpha = \delta e_i^{\cdot\alpha}\, dx^i . \tag{II.21}$$

Before we finish the kinematics we would like to add some useful relations concerning the structure of the local Poincaré transformation. The Jacobi identity reads:

$$[[D_\alpha, D_\beta], D_\gamma] + \text{cycl.} \equiv 0 . \tag{II.22}$$

Apply eq. (II.13) twice to this equation and get

$$D_{[\alpha}F_{\beta\gamma]}^{\cdot\cdot\ \delta} \equiv F_{[\alpha\beta}^{\cdot\cdot\ \tau} F_{\gamma]\tau}^{\cdot\ \ \delta} . \tag{II.23}$$

By means of the relation

$$D_{[\alpha}e_{\cdot\beta}^{i}\, e^j_{\ \gamma]} = e^{[i}_{\cdot[\alpha}F_{\beta\gamma]}^{\cdot\cdot\ j]} \tag{II.24}$$

it is easy to write eq. (II.23) in the holonomic form:

$$D_{[i}F_{jk]}^{\cdot\cdot\,\alpha} \equiv 0 \quad . \tag{II.25}$$

Now we turn to the dynamics starting from a total action function, given by

$$W(V,\mathcal{L},\mathfrak{V}) = \int_V dV\,[\mathcal{L}(\Phi,\partial\Phi,e) + \mathfrak{V}(e,\partial e,\Gamma) \quad , \tag{II.26}$$

where we take Φ and $e_i^{\cdot\alpha}$ to be the independent variables+). V denotes a finite 4-dimensional volume, $\mathcal{L}(\Phi,\partial\Phi,e)$ is the matter- and $\mathfrak{V}(e,\partial e,\Gamma)$ the field-Lagrangian. Physical equivalence of Φ and $\Pi\Phi$ implies that

$$\delta W_m := W(\tilde{V},\Pi\mathcal{L}) - W(V,\mathcal{L}) \equiv 0 \quad . \tag{II.27}$$

$\tilde{V}$ is the volume occupied by the matter distribution $\Pi\Phi$. An analogous equation holds for $W(V,\mathfrak{V})$. We assume

$$\delta\mathcal{L}/\delta\Phi = 0 \quad , \tag{II.28}$$

$$\partial\mathcal{L}/\partial(\partial_i e_j^{\cdot\alpha}) = 0 \tag{II.29}$$

and perform the variation (II.27). Using eqs. (II.18) and (II.19), we obtain the following identities as the coefficients of the independent quantities $D_i\varepsilon^\alpha$ and ε^α, respectively:

$$\delta\mathcal{L}/\delta e_i^{\cdot\alpha} \equiv -(\partial\mathcal{L}/\partial(\partial_i\Phi))D_\alpha\Phi + e_\alpha^{\cdot i}\mathcal{L} = :e\sigma_\alpha^{\cdot i}, \tag{II.30}$$

$$D_i(e\sigma_\alpha^{\cdot i}) \equiv eF_{\alpha i}^{\quad\beta}\sigma_\beta^{\cdot i} \quad . \tag{II.31}$$

Here, the canonical energy-momentum tensor $\sigma_\alpha^{\cdot i}$ turns out to be symmetric, since only the metric defining part of the tetrad enters the minimally coupled matter Lagrangian. Eq. (II.31) represents a generalization of the special realtivistic conservation law

+) Of course, we have also to vary the potential $\Gamma_{i\alpha}^{\quad\beta}$. This variation can be performed in a U_4 with a field Lagrangian quadratic in torsion and curvature, but under the constraint of vanishing U_4-curvature. Therefore we introduce Lagrange multipliers $\lambda_{ij}^{\cdot\cdot\alpha\beta}$ in front of an additional linear curvature term according to $\mathfrak{V}(U_4) \to \mathfrak{V}(U_4;F_{ij\alpha}^{\quad\beta} = 0) = \mathfrak{V}(U_4) + (e/4l^2)\lambda_{\alpha\beta}^{\cdot\cdot ij}F_{ij}^{\cdot\cdot\alpha\beta}$. Alternatively, it is possible to get $F_{ij\alpha}^{\quad\beta} = 0$ in the limit $\varkappa \to 0$ from the second field equation of the "full" Poincaré gauge field theory (cf. Reference 5). Thus we essentially have only to deal with the field equation which arises from the variation with respect to the translational potential $e_i^{\cdot\alpha}$.

$$\partial_i \sigma_{\alpha}^{\cdot\, i} = 0 \quad . \tag{II.32}$$

Eq. (II.31) shows that the material momentum is not conserved because of the presence of the gauge field $F_{ij}^{\cdot\cdot\,\alpha}$. This field exerts an exterior force on matter.

It is worthwhile to point out the fact that eq. (II.31) can be rewritten and eventually results in (cf. References 10, 3)

$$\partial_k (e\sigma_l^{\cdot\, k}) = \left\{ {m \atop kl} \right\} e\sigma_m^{\cdot\, k} \quad , \tag{II.33}$$

the usual energy-momentum theorem of standard general relativity. Thus, we conclude that spinless matter behaves like macroscopic matter in Einstein's theory. Both eq. (II.31) and eq. (II.33) lead to the geodesic motion of structureless test particles. Therefore, Einstein's equation of motion is not at all a characteristic feature of Einstein's theory of gravitation. And by no means does the geodesic motion of macroscopic matter enforce a Riemannian geometry.

The variation of the field action $W(V,\mathfrak{V})$, performed in close analogy to that of the matter action, results in a first order Yang-Mills type field eqaution

$$D_j \mathcal{H}_{\alpha}^{\cdot\, ij} = e\sigma_{\alpha}^{\cdot\, i} + \varepsilon_{\alpha}^{\cdot\, i} \quad . \tag{II.34}$$

To derive this result we have applied Hamilton's principle $\delta(\mathcal{L}+\mathfrak{V})/\delta e_i^{\cdot\,\alpha} = 0$ and have introduced the gravitational field momentum

$$\mathcal{H}_{\alpha}^{\cdot\, ij} := \partial \mathfrak{V} / \partial(\partial_j e_i^{\cdot\,\alpha}) \quad . \tag{II.35}$$

The energy-momentum tensor density of the tetrad field

$$\varepsilon_{\alpha}^{\cdot\, i} := e^{\cdot\, i}_{\alpha} \mathfrak{V} - F_{\alpha j}^{\cdot\cdot\,\gamma} \mathcal{H}_{\gamma}^{\cdot\, ji} \tag{II.36}$$

contributes in a natural way to the material source.

On account of the equation

$$[D_i, D_j] = 0 \tag{II.37}$$

the field momentum $\mathcal{H}_{\alpha}^{\cdot\, ij}$ which occurs as coefficient in front of the (independent) quantity $D_i D_j \varepsilon^{\alpha}$

$$\mathcal{H}_{\alpha}^{\cdot\, ij} D_i D_j \varepsilon^{\alpha} \equiv 0 \tag{II.38}$$

turns out to be antisymmetric in the indices ij, i.e.

$$\mathcal{H}_{\alpha}^{\cdot\,(ij)} \equiv 0 \quad . \tag{II.39}$$

This equation indicates that the field Lagrangian $\mathfrak{V}(e, \partial e, \Gamma)$ contains the derivatives of the potentials only in the form $\partial_{[i} e_{j]}{}^{\alpha}$. Therefore the gravitational field itself enters the field Lagrangian, since it is not possible to derive tensorial quantities from the differences involving the potentials $\Gamma_{i\alpha}{}^{\beta}$. For this reason we have

$$\mathfrak{H}_{\alpha}^{\cdot\, ij} = 2\, \partial \mathfrak{V} / \partial F_{ji}{}^{\alpha} , \tag{II.40}$$

$$\mathfrak{V} = \mathfrak{V}(\eta_{\alpha\beta}, e_i{}^{\alpha}, F_{ij}{}^{\alpha}) \quad . \tag{II.41}$$

Assuming proportionality $\mathfrak{H} \sim F$ between the field momentum $\mathfrak{H}$ and the field strength F, the field Lagrangian becomes quadratic in torsion and reads:

$$\mathfrak{V} = \sum_{i=1}^{3} d_i I_i / l^2 \quad . \tag{II.42}$$

l is the Planck length, d_1, d_2, d_3 are unknown dimensionless constants, and I_1, I_2, I_3 are the three (generalized) Weitzenböck invariants:

$$I_1 = eF_{ijk}F^{ijk} ; \; I_2 = eF_{kji}F^{ijk} ; \; I_3 = eF_{ij}{}^{j}F^{ik}{}_{k} \quad . \tag{II.43}$$

The tetrad field equation (II.34) and this field Lagrangian are manifest locally Lorentz invariant+ (see also Reference 3), irrespective of whether the connection is integrable or not.

The condition (II.1) is known[11] to be a necessary and sufficient condition in order that there exists an anholonomic coordinate system such that the $\Gamma_{i\alpha}{}^{\beta}$ vanish:

$$\Gamma_{i\alpha}{}^{\beta}(x) \overset{*}{=} 0 \quad . \tag{II.44}$$

In this special system the holonomic representation of the (non-symmetric) connection is expressed in terms of linearly independent parallel vector fields $\underline{\tilde{e}}_{\alpha} = \tilde{e}^{i}_{\cdot\alpha}\, \underline{\partial}_i$:

$$\tilde{\Gamma}_{ij}{}^{k} \overset{*}{=} \tilde{e}^{k}_{\cdot\alpha}\, \partial_i \tilde{e}_{j}^{\cdot\alpha} \quad . \tag{II.45}$$

Consequently, in these coordinates, the covariant derivative operator $\underline{D}$ agrees with the usual gradient $\underline{\partial}$

$$D_i \overset{*}{=} \partial_i \quad , \tag{II.46}$$

+) We would like to emphasize that the local Lorentz invariance of the field Lagrangian (II.42) by no means predicts the "Einstein choice" of $\mathfrak{V}$, i.e. $d_1 = -1/2$; $d_2 = -1$; $d_3 = 2$. This result clearly contradicts that of Cho[12].

and the field eq. (II.34) reduces to the form:

$$\partial_j \mathfrak{H}_{\alpha}^{\cdot ij} \overset{*}{=} e\sigma_{\alpha}^{\cdot i} + \varepsilon_{\alpha}^{\cdot i} \quad , \tag{II.47}$$

with the torsion tensor

$$F_{ij}^{\cdot\cdot\alpha} \overset{*}{=} 2\Omega_{ij}^{\cdot\cdot\alpha} := 2\partial_{[i} e_{j]}^{\alpha} \quad . \tag{II.48}$$

Perform a Lorentz rotation $\Lambda^{\alpha}_{\cdot\beta}(x)$ of the tetrad field $\tilde{e}_i^{\beta}(x)$

$$e_i^{\cdot\alpha}(x) = \Lambda^{\alpha}_{\cdot\beta}(x)\tilde{e}_i^{\cdot\beta} \tag{II.49}$$

and evaluate the anholonomic connection with respect to the basis $\underline{e}_{\alpha}$:

$$\Gamma_{i\alpha}^{\beta} = \Lambda_{\gamma}^{\cdot\beta}\partial_i \Lambda^{\gamma}_{\alpha} \quad . \tag{II.50}$$

This connection contains at most six independent functions of space-time. In a straightforward calculation it can be proved that the curvature tensor

$$F_{ij\alpha}^{\cdot\cdot\cdot\beta} := 2(\partial_{[i}\Gamma_{j]\alpha}^{\beta} + \Gamma_{[i|\rho|}^{\beta}\Gamma_{j]\alpha}^{\rho}), \tag{II.51}$$

corresponding to the connection (II.50), vanishes identically.

In the following sections all investigations and calculations will take place in the anholonomic coordinate system indicated by the equations (II.44) and (II.45). However, we will omit the star over the equal sign and the tilde over the tetrad field.

WEAK-FIELD APPROXIMATION AND THE CHOICE OF THE FIELD LAGRANGIAN

Without any doubt, every new gravitational theory has to agree with all present day experiments. This important requirement does not only concern those experiments which occur in the Newtonian limit but also gravitational tests in the solar system, measuring the non-Newtonian aspects of the gravitational field. It is this reason, as we shall demonstrate, that will lead to a "reduction" of the class of the general field Lagrangians (II.42).

We start the weak field approximation of the tetrad field equation (II.47), expanding the gravitational field variables $e_i^{\cdot\alpha}(x)$ in a post-Newtonian approximation up to third order in (v/c), i.e.

$$e_{\cdot\alpha}^{i} = \delta_{\alpha}^{i} - (1/2)\,{}^{(2,3)}h_{\cdot\alpha}^{i} \tag{III.1}$$

with

$$|\overset{(2,3)}{h}{}^{i}_{\cdot\alpha}| << 1 \text{ and } v \sim O(\epsilon) \quad . \tag{III.2}$$

The superscript numbers $\overset{(n,m)}{h}$ refer to the expansion of the tetrads to $O(\epsilon^{n,m})$. In the following we will drop the superscript numbers. Then, neglecting terms of order $O(\epsilon^{n})$, for $n \geq 4$, we are left with

$$\partial_j \mathcal{H}_{\alpha}^{\cdot ij} = e\sigma_{\alpha}^{\cdot i} \quad , \tag{III.3}$$

$$\mathcal{H}_{\alpha}^{\cdot ij} = (2e/l^2)(d_1 \Omega^{ji}_{\cdot\cdot\alpha} + d_2 \Omega_{\alpha}^{\cdot [ij]} + d_3 e^{[i}_{\cdot\alpha} \Omega^{j]\sigma}_{\cdot\cdot\sigma}), \tag{III.4}$$

and

$$e_i^{\cdot\alpha} = \delta^{\alpha}_{i} + (1/2)h_i^{\cdot\alpha} \quad , \quad \Omega_{ij}^{\cdot\cdot\alpha} = (1/2)\partial_{[i}h_{j]}^{\cdot\alpha} \quad . \tag{III.5}$$

Decompose the tetrad field $h_i^{\cdot\alpha}$ into its symmetric

$$h_{(ij)} =: \gamma_{ij} + (1/2)\eta_{ij}h_k^{\cdot k} \tag{III.6}$$

and antisymmetric part

$$h_{[ij]} =: a_{ij} \quad , \tag{III.7}$$

insert expression (III.4) for $\mathcal{H}_{\alpha}^{\cdot ij}$ into eq. (III.3), use eqs. (III.5)-(III.7) and eventually find the result:

$$\begin{aligned}
&((1/2)d_1 + (1/4)d_2)\Box\gamma_{\alpha\beta} - (1/2)((1/2)d_1 + (1/4)d_2 + (1/4)d_3)\eta_{\alpha\beta}\Box\gamma + \\
&+ (1/2)((1/2)d_1 + (1/4)d_2 + (1/4)d_3)\partial_{\alpha}\partial_{\beta}\gamma + (-(1/2)d_1 + (1/4)d_2)\Box\mathbf{a}_{\alpha\beta} \\
&- ((1/2)d_1 + (1/4)d_2)\partial^{\gamma}\partial_{\beta}\gamma_{\alpha\gamma} + (1/4)d_3(\partial^{\gamma}\partial_{\alpha}\gamma_{\beta\gamma} - \eta_{\alpha\beta}\partial^{\gamma}\partial_{\sigma}\gamma^{\sigma}_{\cdot\gamma}) + \\
&+ ((1/2)d_1 - (1/4)d_2)\partial^{\gamma}\partial_{\beta}\mathbf{a}_{\alpha\gamma} + ((1/2)d_2 + (1/4)d_3)\partial^{\gamma}\partial_{\alpha}\mathbf{a}_{\beta\gamma} = l^2\sigma_{\alpha\beta} \; .
\end{aligned} \tag{III.8}$$

The symbol $\Box$ denotes the usual d'Alembert operator; $\Box := \partial^i\partial_i$.

We want to choose the constants d_i such that our theory gives the same results as GR in the linear approximation of weak fields. To get rid of the derivatives of the trace terms in eq. (III.8), the constants d_i have to fulfill the condition:

$$d_1 + (1/2)d_2 + (1/2)d_3 = 0 \quad . \tag{III.9}$$

Then the linearized field equation (III.8) becomes invariant under the substitutions

$$\gamma_{\alpha\beta} \rightarrow \gamma_{\alpha\beta} + \partial_{(\alpha}\xi_{\beta)} - (1/2)\eta_{\alpha\beta}\partial^{\gamma}\xi_{\gamma} \quad , \tag{III.10}$$

and

$$a_{\alpha\beta} \rightarrow a_{\alpha\beta} + \partial_{[\alpha}\eta_{\beta]} \quad , \tag{III.11}$$

for any arbitrary vector fields ξ_i and η_i. Equations (III.10) and (III.11) allow us to impose the deDonder and the "Lorentz" conditions, respectively,

$$\partial^{\gamma}\gamma_{\alpha\gamma} = 0 \quad , \tag{III.12}$$

$$\partial^{\gamma}a_{\alpha\gamma} = 0 \quad . \tag{III.13}$$

We require the symmetric part of eq. (III.8), in the gauge (III.12), to coincide with Einstein's linearized field equation

$$-(1/2)\Box\gamma_{\alpha\beta} = l^2\sigma_{\alpha\beta} \quad . \tag{III.14}$$

We obtain, besides the condition (III.9), the value for the constant d_3:

$$d_3 = 2 \quad . \tag{III.15}$$

For the three constants d_i $(i = 1,2,3)$ we now have two relations:

$$d_1 = -(1 + (1/2)d_2) \quad \text{and} \quad d_3 = 2. \tag{III.16}$$

The antisymmetric part of the field equation (III.8) reads[+]

$$(1/2)(1 + d_2)\Box a_{\alpha\beta} = 0 \quad . \tag{III.17}$$

It is worth noting that in Einstein's theory $d_2 = -1$ and therefore $a_{\alpha\beta}$ remains undetermined. However, since Einstein's theory is locally Lorentz invariant, one can - with the aid of a local Lorentz rotation - always adjust the tetrad field such that in the linearized theory it becomes symmetric[13]. In our case we obtain, under suitable boundary conditions, as unique solution of eq. (III.17):

+) Hayashi and Shirafuji[14] have introduced a teleparallelism theory in which they also describe spinor matter. On the right hand side of their equation, corresponding to our (III.17), there occurs the antisymmetric part of the energy-momentum tensor for spin-1/2 particles. A Weitzenböck space-time, however, is not the appropriate arena to describe spinning matter, too. This can satisfactorally be done only in a U_4 (with non-vanishing curvature).

$$a_{\alpha\beta} = 0 \quad . \tag{III.18}$$

Thus, in the weak field approximation, our theory has the same "metric content" as GR.

With the choice (III.16) we have a one-parameter class of admissible field Lagrangians:

$$\mathfrak{V} = \mathfrak{V}_E - (3e/2l^2)\lambda\Omega_{[ijk]}\Omega^{[ijk]} \tag{III.19}$$

with

$$\mathfrak{V}_E := ((-1/2)I_1 - I_2 + 2I_3)/l^2 \tag{III.20}$$

and

$$\lambda := 1 + d_2 \quad . \tag{III.21}$$

$\mathfrak{V}_E$ denotes the effective, locally Lorentz invariant Einstein Lagrangian. The constant λ is not determined by solar system experiments. Rather, the determination of λ can only be performed in the microphysical realm,where fundamental spin-(1/2) particles interact with each other, or at least in those physical domains, where we can observe the gravitational interaction between the orbital angular momentum of a massive rotating body and the elementary spin of a Dirac particle (cf. Section VI). Hehl[5] and von der Heyde[10] have given cogent arguments for the choice $\lambda = 1$, executing Gordon decompositions of the currents Σ and τ. We join this choice of λ and shall use $\lambda = 1$ throughout our present work. Sometimes we shall write λ instead of 1, to indicate the consequences of different choices of λ.

To gain some more insight into the meaning of the gauge transformations (III.10),(III.11) and to study the invariance behavior of the linearized field eq. (III.8) for our choice of the field Lagrangian, i.e.

$$d_1 = -1, \quad d_2 = 0, \quad d_3 = 2 \quad , \tag{III.22}$$

we investigate the transformation behavior of the tetrad field in the weak field limit. Based on the fact that the tetrad functions $e_i^{\cdot\alpha}(x)$ carry two different indices (i.e., the holonomic vector index i and the anholonomic Lorentz index α) we have to take into account a different transformation behavior of $e_i^{\cdot\alpha}(x)$ with respect to their indices. A Lorentz rotation $\Lambda_\alpha^{\cdot\beta}(x)$ only affects the anholonomic index α, whereas a coordinate transformation $(x^i \rightarrow \tilde{x}^i)$, refers to the index i. Performing simultaneously a Lorentz rotation as well as a coordinate transformation, the tetrad functions transform by virtue of the law:

$$\tilde{e}_{i}^{\cdot\alpha} = \Lambda_{\beta}^{\cdot\alpha}(\partial x^{j}/\partial \tilde{x}^{i})e_{j}^{\cdot\beta} \quad , \tag{III.23}$$

$$e_{l}^{\cdot\gamma} = \Lambda_{\cdot\alpha}^{\gamma}(\partial \tilde{x}^{k}/\partial x^{l})\tilde{e}_{k}^{\cdot\alpha} \quad . \tag{III.24}$$

For the last equation we have used the orthogonality relation

$$\Lambda_{\alpha}^{\cdot\beta}\Lambda_{\cdot\gamma}^{\alpha} = \delta_{\gamma}^{\beta} \tag{III.25}$$

of the Lorentz rotation.

Now, in the weak field approximation, we consider a coordinate transformation of the form

$$\tilde{x}^{i} = x^{i} + \xi^{i}(x) \tag{III.26}$$

and an infinitesimal Lorentz rotation

$$\Lambda_{\alpha}^{\cdot\beta} = \delta_{\alpha}^{\beta} + \omega_{\alpha}^{\cdot\beta} \quad , \tag{III.27}$$

where $\xi^{i}(x)$ and $\omega_{\alpha}^{\cdot\beta}(x)$ are of the order $h_{i}^{\cdot\alpha}$, and the rotation matrix $\omega_{\alpha\beta}$ is antisymmetric:

$$\omega_{\alpha\beta} = -\omega_{\beta\alpha} \quad . \tag{III.28}$$

Then we have:

$$e_{l}^{\cdot\gamma} = (\delta_{\alpha}^{\gamma} + \omega_{\cdot\alpha}^{\gamma})(\delta_{l}^{k} + (\partial\xi^{k}/\partial x^{l}))\tilde{e}_{k}^{\cdot\alpha} \quad . \tag{III.29}$$

Expanding $\tilde{e}_{k}^{\cdot\alpha}$

$$\tilde{e}_{k}^{\cdot\alpha} = \delta_{k}^{\alpha} + (1/2)\tilde{h}_{k}^{\cdot\alpha} \quad , \tag{III.30}$$

and retaining only terms of first order we find:

$$\tilde{h}_{l}^{\cdot\gamma} = h_{l}^{\cdot\gamma} - 2(\partial\xi^{\gamma}/\partial x^{l}) + 2\omega_{l}^{\cdot\gamma} \quad . \tag{III.31}$$

From this relation we obtain for the symmetric and antisymmetric parts of $h_{l}^{\cdot\gamma}$, respectively:

$$h_{(l\gamma)} = \tilde{h}_{(l\gamma)} + 2\xi_{(\gamma,l)} \quad , \tag{III.32}$$

$$h_{[l\gamma]} = \tilde{h}_{[l\gamma]} + 2\xi_{[\gamma,l]} + 2\omega_{\gamma l} \quad . \tag{III.33}$$

Define, in analogy to eqs. (III.6), (III.7),

$$\tilde{\gamma}_{lk} := \tilde{h}_{(lk)} - (1/2)\tilde{h}_{i}^{\cdot i} \quad , \tag{III.34}$$

$$\tilde{a}_{lk} := \tilde{h}_{[lk]} \tag{III.35}$$

and rewrite eqs. (III.32), (III.33) as follows:

$$\tilde{\gamma}_{lk} = \gamma_{lk} - 2\xi_{(l,k)} + \eta_{lk}\xi^{i}_{,i} \quad , \tag{III.36}$$

$$\tilde{a}_{lk} = a_{lk} + 2(\omega_{lk} + \xi_{[l,k]}) \quad . \tag{III.37}$$

Observe that the symmetric part of the field eq. (III.8), for the parameter choice (III.22),

$$-(1/2)\Box\gamma_{ij} + \partial_{(i}\partial^{k}\gamma_{j)k} - (1/2\eta_{ij}\partial^{k}\partial^{l}\gamma_{kl} = l^{2}\sigma_{ij} \tag{III.38}$$

is invariant under the transformations (III.36) and (III.37). The antisymmetric part of eq. (III.8)

$$(3/2)\partial^{k}\partial_{[k}a_{ij]} = 0 \quad , \tag{III.39}$$

however, is invariant only with respect to the transformation (III.36) and is not invariant under (III.37), since in general $\partial_{[\alpha}\omega_{\beta\gamma]} \neq 0$. For the special choice of the infinitesimal rotation matrix

$$\omega_{\alpha\beta} = \partial_{[\alpha}a_{\beta]} + \partial_{[\alpha}b_{\beta]} + \partial_{[\alpha}c_{\beta]} + \ldots , \tag{III.40}$$

where a_{α} , b_{α} , c_{α}, ... are arbitrary vector fields, eq. (III.39) remains invariant under the transformation (III.37).

In the new coordinates and basis we can, as usual, impose the de Donder gauge on $\tilde{\gamma}_{lk}$ with the vector field ξ^{k} satisfying the equation

$$\partial^{k}\gamma_{lk} = \Box\xi_{l} \quad . \tag{III.41}$$

In eq. (III.40) we choose $a_{k} = \xi_{k}$ and use the invariance of (III.39) under

$$\tilde{a}_{lk} = a_{lk} + 2(\partial_{[l}b_{k]} + \partial_{[l}c_{k]} + \ldots) \tag{III.42}$$

to impose the "Lorentz" condition (III.13) on $\tilde{a}_{lk}$. In this way we recover again our former eqs. (III.14) and (III.17,18) for the quantities $\tilde{\gamma}_{lk}$ and $\tilde{a}_{lk}$. But, in addition, we have elucidated an important point: even in anholonomic coordinate systems, indicated by eq.

(II.44), our linearized gravitational theory turns out to possess (restricted; cf. eq. (III.40)) local Lorentz invariance. Hence, we interpret the substitution (III.11) as a special infinitesimal local Lorentz rotation.

Having fixed the constants d_i, we turn our consideration to the field equation (II.47). In order to compare our theory with Einstein's, we would like to recast (II.47) into a quasi-Einsteinian form. For this reason we establish the relation between the field momenta $\overset{E}{\mathfrak{H}}{}_\alpha{}^{ij}$ and $\overset{T}{\mathfrak{H}}{}_\alpha{}^{ij}$ corresponding to "Einstein's choice"

$$d_1 = -1/2, \quad d_2 = -1, \quad d_3 = 2 \tag{III.43}$$

and to the choice (III.22):

$$\overset{T}{\mathfrak{H}}{}_\alpha{}^{ij} = \overset{E}{\mathfrak{H}}{}_\alpha{}^{ij} + (3e/l^2)\eta_{\alpha\beta}\, e_k{}^\beta \Omega^{[ijk]} \quad . \tag{III.44}$$

Furthermore we obtain:

$$\mathfrak{V}_{T,E} = (1/2)\Omega_{ji}{}^{\alpha}\, \overset{T,E}{\mathfrak{H}}{}_\alpha{}^{ij} \quad . \tag{III.45}$$

By virtue of the last two equations we derive from (II.47) the more familiar forms:

$$G_{\alpha\beta}(V_4) + \lambda\tilde{G}_{\alpha\beta} = l^2\sigma_{\alpha\beta} \quad , \tag{III.46}$$

$$\lambda A_{\alpha\beta} = 0 \quad . \tag{III.46'}$$

The (symmetric) Einstein tensor $G_{\alpha\beta}(V_4)$ is defined by

$$G_{\alpha\beta}(V_4) := (l^2/e)e_i{}^\sigma \eta_{\sigma\beta}(\partial_j \overset{E}{\mathfrak{H}}{}_\alpha{}^{ij} + 2\Omega_{\alpha j}{}^{\gamma}\overset{E}{\mathfrak{H}}{}_\gamma{}^{ji} - e_\alpha{}^i \mathfrak{V}_E) \tag{III.47}$$

and the symmetric and antisymmetric tensors $\tilde{G}_{\alpha\beta}$ and $A_{\alpha\beta}$, respectively, are given by:

$$\tilde{G}_{\alpha\beta} := 3\Omega^{\mu\nu}{}_{\beta}\Omega_{[\mu\nu\alpha]} - 6\Omega_{\alpha}{}^{\tau\gamma}\Omega_{[\beta\tau\gamma]} + (3/2)\eta_{\alpha\beta}\Omega_{[\mu\nu\sigma]}\Omega^{[\mu\nu\sigma]} \quad , \tag{III.48}$$

$$A_{\alpha\beta} := (3/e)\,\partial_j(e\Omega^{[j}{}_{\alpha\beta]}) = 3\partial_\mu(\eta^{\mu\gamma}\Omega_{[\alpha\beta\gamma]}) + 6\Omega^{\gamma\tau}{}_{\tau}\Omega_{[\alpha\beta\gamma]} \,. \tag{III.49}$$

$G_{\alpha\beta}(V_4)$ only depends on the metric

$$g_{ij} := e_i{}^\alpha e_j{}^\beta \eta_{\alpha\beta} \tag{III.50}$$

and is known to be invariant under local Lorentz rotations. The tensor $A_{\alpha\beta}$ can be expressed with the aid of the field momentum $\overset{T}{\mathfrak{H}}{}_\alpha{}^{ij}$:

$$A_{\alpha\beta} = (l^2/e)\partial_j \overset{T}{\mathfrak{H}}{}_{[\alpha\beta]}{}^{j} = (1/e)(l^2\partial_j \overset{E}{\mathfrak{H}}{}_{[\alpha\beta]}{}^{j} + 3\partial_j(e\Omega^{[j}{}_{\alpha\beta]})) \,. \tag{III.51}$$

Comparing eqs. (III.49) and (III.51) we read off the important result

$$\partial_j \overset{E}{\mathcal{H}}_{[\alpha\beta]}{}^{\cdot\cdot j} \equiv 0$$

which is a special feature of the "Einstein choice" (III.43).

GROUPS OF AFFINE MOTIONS AND EXACT SOLUTIONS OF THE TETRAD FIELD EQUATION

Since we consider a U_4 spacetime with vanishing curvature, we can investigate groups of motions in two different ways: firstly, we think of a U_4, subjected to the subsidiary condition of vanishing curvature, and know that an infinitesimal deformation[15]

$$x^i \to x^i + v^i dt \quad , \qquad \text{(IV.1)}$$

describes a motion if

$$\underset{v}{\mathcal{L}} g_{ij} = 0 \quad ; \quad \underset{v}{\mathcal{L}} F_{ij}{}^k = 0 \qquad \text{(IV.2)}$$

$\underset{v}{\mathcal{L}}$: = Lie derivative with respect to the field v^i. $v^i(x)$ is a contravariant vector field and dt is an infinitesimal.

Secondly, we may think of a 4 dimensional spacetime L_4 (cf. Ref. 16) provided with a linear integrable connection $\Gamma_{ij}^{\cdot\cdot k}$ (the metric can be understood as a derived quantity). In such an L_4 with absolute parallelism we can study groups of affine motions. An infinitesimal affine motion in L_4 is characterized by[15]:

$$\underset{v}{\mathcal{L}} \Gamma_{ij}^{\cdot\cdot k} = 0 \quad . \qquad \text{(IV.3)}$$

In the present section we refer to the second possibility to study groups of motions. An L_4 is said to possess absolute parallelism or teleparallelism if for any two points P and Q of a certain region of the spacetime the parallel displacement of any quantity from P to Q along a curve joining P and Q gives a result at Q which does not depend on the choice of the curve. In such a spacetime we fix a point x_o^i and consider 4 linearly independent contravariant vectors $e^j_{\cdot\alpha}(x_o^i)$ at this point. Since the parallel displacement is independent of the curve along which the vectors $e^j_{\cdot\alpha}(x_o^i)$ are displaced from x_o^i to an arbitrary point x^i of the spacetime, we obtain fields of vectors $e^j_{\cdot\alpha}(x^i)$. For this tetrad field holds:

$$\nabla_i e^j_{\cdot\alpha} := \partial_i e^j_{\cdot\alpha} + \Gamma_{ik}^{\cdot\cdot j} e^k_{\cdot\alpha} = 0 \quad , \qquad \text{(IV.4)}$$

from which we get (cf. eq. (II.45))

$$\Gamma_{ij}^{\cdot\cdot k} = e^{k}_{\cdot\alpha}\, \partial_i e^{\cdot\alpha}_{j} \quad , \tag{IV.5}$$

where the dual field $e^{\cdot\alpha}_{j}$ is defined by

$$e^{\cdot\alpha}_{j}\, e^{j}_{\cdot\beta} = \delta^{\alpha}_{\beta} \quad . \tag{IV.6}$$

A case of particular interest arises when we consider the absolute parallelism defining vector fields $e^{i}_{\cdot\alpha}(x)$ as orthonormal ones.

By means of eqs. (IV.3) and (IV.4) the following theorem can be proved:

$$\underset{v}{\mathcal{L}} e^{i}_{\cdot\alpha} = c_{\alpha}^{\cdot\beta}\, e^{i}_{\cdot\beta} \quad ; \quad c_{\alpha}^{\cdot\beta} = \text{constants} \ , \tag{IV.7}$$

i.e., in order that an L_4 with absolute parallelism admits an infinitesimal affine motion, it is necessary and sufficient that the Lie derivatives of 4 linearly independent absolutely parallel contravariant vectors be linear combinations of these vectors with constant coefficients.

We next consider the integrability conditions of $\underset{v}{\mathcal{L}}\Gamma_{ij}^{\cdot\cdot k} = 0$. Eq. (IV.3) can be written as[15]

$$0 = \underset{v}{\mathcal{L}}\Gamma_{ij}^{\ \ k} = \nabla_i \nabla_j v^k + 2\nabla_i \Omega_{lj}^{\ \ k} v^l =: \nabla_i v_j^{\cdot k} , \tag{IV.8}$$

$$\nabla_j v^k = v_j^{\ k} - 2\Omega_{lj}^{\ \ k} v^l \tag{IV.9}$$

with the unknowns v^k and $v_j^{\cdot k}$ and the first integrability condition:

$$\underset{v}{\mathcal{L}}\Omega_{ij}^{\cdot\cdot k} = 0 \quad . \tag{IV.10}$$

As for an affine motion the covariant differentiation and the Lie derivation with respect to (IV.1) are commutative, the second and following integrability conditions are found by covariant differentiation of (IV.10):

$$\underset{v}{\mathcal{L}}\nabla_{i_m} \dots {}_{i_1} \Omega_{jk}^{\ \ l} = 0 \quad ; \ m = 0,1,2,3 \ \dots \ . \tag{IV.11}$$

If there exists a positive integer N such that the (N+1)st set of eqs. (IV.10), (IV.11) algebraically depends on the sets for m = 0,1,2, ..., N, then all other sets for m > N depend on the foregoing ones. In order to find solutions for eqs. (IV.8), (IV.9) it has to be ensured that at some point x^i_o there exist values v^i_o, $v_{oi}^{\cdot j}$ satisfying the equations (IV.11). Then these values can be taken as initial values at x^i_o leading to a unique solution of the differential eqs. (IV.8), (IV.9). The number of independent sets of values v^i_o, $v_{oi}^{\ j}$ satisfying (IV.11) is finite. Therefore the set of one-parameter groups of transformations generated by the v^i depend on a finite number of parameters. Having two solutions u^i and v^i, i.e.

$$\underset{u}{\mathcal{L}}\Gamma_{ij}^{\ \ k} = 0 \quad ; \quad \underset{v}{\mathcal{L}}\Gamma_{ij}^{\ \ k} = 0 \tag{IV.12}$$

then also

$$w^i := \underset{u}{\mathcal{L}} v^i \tag{IV.13}$$

is a solution:

$$\underset{w}{\mathcal{L}}\Gamma_{ij}^{\ \ k} = 0 \quad . \tag{IV.14}$$

Hence we have the result that the one-parameter groups form a group, too (cf. Schouten[15]).

Of course, instead of studying the integrability conditions of eq. (IV.3) we could have derived those for the fundamental eq. (IV.7). This has been done in extenso by Robertson[17].

We now apply the above results to the problem of determining the form of a spherical symmetric spacetime, i.e. a spacetime suitable for describing the gravitational field of a spherically symmetric body. We impose O(3)-spherical symmetry on the tetrad field and set up the O(3)-corresponding generators to solve eq. (IV.3) or eq. (IV.7). As solution we find (cf. also Ref.[17]) the most general spherical symmetric tetrad field which, in addition, carries spacetime reflection symmetry and consists of only two independent tetrad functions:

$$e_{0}^{\cdot 0} = F(r,t) \quad ; \quad e_{a}^{\cdot\mu} = B(r,t)\,\delta_{a}^{\ \mu} \quad (\mu, a = 1,2,3). \tag{IV.15}$$

The former possibility to study groups of motions in a U_4, indicated by eq. (IV.2), has been applied by Baekler[18] and Yasskin[19] to the spherical symmetric case. Under the additional condition of spacetime reflection symmetry they found, as usual, two independent metric functions and four functions for the torsion. Taking into account the restriction of eq. (II.1) Baekler[20] has explicitely shown that, for the static case, the Schwarzschild solution is a vacuum solution of eq. (II.34) and that all four torsion functions are completely determined by the two metric functions. Because of eq. (IV.15) the same results should also hold for the time dependent situation.

In conclusion of this section we turn back to the field equations (III.46) and (III.46') and try to find exact solutions. It becomes immediately evident that the vanishing of the "axial" vector piece $\Omega_{[\alpha\beta\gamma]}$ of the torsion tensor is a sufficient condition for the coincidence of our theory with Einstein's theory:

$$\Omega_{[\alpha\beta\gamma]} = 0 \curvearrowright \text{"}U_4(F_{ij\alpha}^{\cdot\cdot\cdot\beta} = 0) - \text{theory} = \text{Einstein theory"}. \tag{IV.16}$$

Consider a diagonal metric

$$g_{ij} = g_{ii)}\delta_{ij} \tag{IV.17}$$

and suppose that the functions $g_{ii)}(x)$ are solutions of Einstein's field equation (III.46, for $\lambda = 0$). Then we can construct the standard tetrad field[2] corresponding to that metric by

$$e^{i}_{\cdot\alpha} = (1/\sqrt{g_{ii)}})\delta^{i}_{\cdot\alpha} \quad . \tag{IV.18}$$

For this tetrad field the "axial vector" $\Omega_{[\alpha\beta\gamma]}$ vanishes identically. Thus, the field defined by (IV.18) also becomes an exact solution of eqs. (III.46) and (III.46'). As a result we find that all spherical symmetric solutions in Einstein's theory are also solutions of the Poincaré gauge theory for macroscopic matter. Thereby we recover the Schwarzschild and the Friedman solutions, for instance.

On the other hand we have the validity of eq. (IV.15). For this tetrad field we again obtain

$$\Omega_{[\alpha\beta\gamma]} \equiv 0 \quad , \tag{IV.19}$$

and the associated metric components are those of the spherical symmetric in isotropic coordinates. Consequently the reverse theorem is proved: spherical symmetric solutions of our theory are also, via the construction (IV.18), solutions of Einstein's theory. This result in turn implies the validity of Birkhoff's theorem. And, in addition, the meaning of the "axial vector" piece of the torsion tensor, $\Omega_{[\alpha\beta\gamma]}$, becomes clear: it represents a "measure" for the deviation from spherical symmetry.

POST-NEWTONIAN APPROXIMATION

In connection to our investigations of Section III we extend the post-Newtonian approach of the field eqs. (III.46), (III.46') up to the order $O(\epsilon^5)$. It is our aim to find a solution of these equations which makes our theory distinguishable from Einstein's theory. In the foregoing section we have recognized that this can only happen if the "axial vector" $\Omega_{[\alpha\beta\gamma]}$ does not vanish. As we shall see below, $\Omega_{[\alpha\beta\gamma]}$ "occurs" for the first time in fifth order in (v/c).

We stay in the post-Newtonian scheme and expand the tetrad field, in analogy to the post-Newtonian expansion of the metric field in Einstein's theory[21], as follows:

$$e_{i}^{\cdot\alpha} = \delta_{i}^{\alpha} + (1/2)(\overset{(2,3)}{h}{}_{i}^{\cdot\alpha} + \overset{(4,5)}{k}{}_{i}^{\cdot\alpha}) \quad . \tag{V.1}$$

From this equation we obtain the inverse tetrad field $e^{\cdot i}_{\alpha}$ via the definition (IV.6). Solve eq. (IV.6) for the Ansatz (V.1) and find

$$e^{\cdot j}_{\alpha} = \delta^{j}_{\alpha} - (1/2)(\overset{(2,3)}{h}{}^{\cdot j}_{\alpha} + \overset{(4,5)}{k}{}^{\cdot j}_{\alpha}) + (1/4)\overset{(2,3)}{h}{}^{\cdot\tau}_{\alpha}\overset{(2,3)}{h}{}^{\cdot j}_{\tau} . \tag{V.2}$$

From now on we will drop in general the superscript numbers. The conversion of the indices from holonomic to anholonomic ones (and vice versa) is performed with the Kronecker symbol δ^{α}_{i}.

We compute the "axial vector" piece of the torsion tensor

$$\Omega_{[\alpha\beta\gamma]} = e^{i}_{[\alpha} e^{j}_{\beta} \eta_{\gamma]\delta} \Omega_{ij}^{\cdot\cdot\delta} \tag{V.3}$$

with

$$\Omega_{ij}^{\cdot\cdot\delta} = \partial_{[i} e_{j]}^{\cdot\delta} = (1/2)(\partial_{[i} h_{j]}^{\cdot\delta} + \partial_{[i} k_{j]}^{\cdot\delta}) \tag{V.4}$$

by means of the expansions (V.1) and (V.2). After a lengthy but straightforward calculation we get

$$\Omega_{[\alpha\beta\gamma]} = (1/2)(\partial_{[\alpha} h_{\beta\gamma]} + (1/2) h_{[\alpha}^{\cdot j} \partial_{\beta} h_{|j|\gamma]} -$$
$$- (1/2) h_{[\alpha}^{\;\; i} \partial_{|i|} h_{\beta\gamma]} + \partial_{[\alpha} k_{\beta\gamma]}) \quad . \tag{V.5}$$

Use the decomposition (III.6) and (III.7) of the tetrad field $h_{i}^{\cdot\alpha}$ into its symmetric and antisymmetric part, respectively, define

$$\tilde{a}_{\alpha\beta} := k_{[\alpha\beta]} \tag{V.6}$$

and rewrite eq. (V.5), taking into account the gauges (III.12), (III.13), and the corresponding solution (III.18):

$$\Omega_{[\alpha\beta\gamma]} = (1/2)((1/2)\gamma^{\cdot\tau}_{[\alpha} \partial_{\beta} \gamma_{\gamma]\tau} + \partial_{[\alpha} \tilde{a}_{\beta\gamma]}) \quad . \tag{V.7}$$

In the following we will proceed in two steps. First we want to show that $\Omega_{[\alpha\beta\gamma]}$ vanishes up to fourth order of $(v/c)^{+}$. For

---

$^{+}$ This result was first derived by Schweizer and Straumann[22] and by Smalley[23]. Our proof, presented here, is somewhat different from the proof of these authors and fits quite well into the second step: the calculation of $\Omega_{[\alpha\beta\gamma]}$ up to the order of $O(\epsilon^{5})$.

this purpose we have to know the second order solution of Einstein's linearized field equation. Since (cf. Ref. 21)

$$\gamma_{00} = O(\epsilon^2) \tag{V.8}$$

is the only non-vanishing component of this order, we immediately find

$$\overset{(4)}{\Omega}_{[\alpha\beta\gamma]} = (1/2)\,\partial_{[\alpha}\overset{(4)}{\tilde{a}}_{\beta\gamma]} \quad . \tag{V.9}$$

Introduce (V.9) into the field equation (III.49). This yields

$$\overset{(4)}{A}_{\alpha\beta} = (3/2)\,\partial^{\gamma}\partial_{[\alpha}\overset{(4)}{\tilde{a}}_{\beta\gamma]} = 0 \quad . \tag{V.10}$$

The invariance of this equation under the substitution

$$\overset{(4)}{\tilde{a}}_{ij} \longrightarrow \overset{(4)}{\tilde{a}}_{ij} + \partial_{[i}\overset{(4)}{\eta}_{j]} \quad , \tag{V.11}$$

where $\overset{(4)}{\eta}_i$ is an arbitrary vector field of the same order as $\overset{(4)}{\tilde{a}}_{ij}$, allows us to impose the "Lorentz" condition

$$\partial^j \overset{(4)}{\tilde{a}}_{ij} = 0 \quad . \tag{V.12}$$

Then the solution becomes $\overset{(4)}{\tilde{a}}_{ij} = 0$, i.e. up to the order $O(\epsilon^4)$ the tetrads turn out to be "symmetric". Therefore, up to this order, there is no distinction between Einstein's theory and the Poincaré gauge field theory for macroscopic tangible matter ($\varkappa \to 0$; $F_{ij\alpha}{}^{\beta}(U_4) = 0$). This result also implies[24] that the velocities of light and of gravitational waves coincide.

In the second step we perform the post-Newtonian approximation of $\Omega_{[\alpha\beta\gamma]}$ in fifth order. Insert eq. (V.7) into the antisymmetric part of the tetrad field equation (III.46') and find in the gauge

$$\partial^j \tilde{a}_{ij} = 0 \tag{V.13}$$

the expression

$$\overset{(5)}{A}_{\alpha\beta} = (1/2)(\Box\,\tilde{a}_{\alpha\beta} + \tilde{L}_{\alpha\beta}) = 0 \tag{V.14}$$

with

$$\tilde{L}_{\alpha\beta} := -(1/2)\gamma^{\cdot\tau}_{[\alpha}\Box\gamma_{\beta]\tau} + (1/2)(\partial^{\gamma}\gamma^{\cdot\tau}_{[\alpha})(\partial_{\beta]}\gamma_{\gamma\tau}) +$$

$$+ (1/2)\, \gamma^{\cdot\tau}_{\gamma}\, \partial^{\gamma} \partial_{[\alpha} \gamma_{\beta]\tau} \quad . \tag{V.15}$$

The integration of eq. (V.14) by standard methods yields

$$\tilde{a}_{\alpha\beta}(\underline{r},t) = (1/4\pi) \int d^3x \tilde{L}_{\alpha\beta}(x^b, t-R)/R \quad , \tag{V.16}$$

where the quantity R is defined by

$$R := \sum_{b=1}^{3} (r^b - x^b)^2 \quad . \tag{V.17}$$

Observe that the symmetric γ-field serves as source for the antisymmetric $\tilde{a}$-field. Therefore the six tetrad functions $\tilde{a}_{\alpha\beta}$ do not have an "independent status". They are determined by the ten symmetric "metric" functions $\gamma_{\alpha\beta}$, which in turn are the solutions of Einstein's linearized field equation. This is not surprising, since the energy-momentum tensor of the scalar matter distribution is symmetric.

We summarize the results of this section by indicating explicitely the "order of deviation" from Einstein's theory in the field equations (III.46) and (III.46'):

$$G_{\alpha\beta}(V_4) + \overset{(7)}{O} + \ldots \qquad = l^2 \sigma_{\alpha\beta} \quad , \tag{V.18}$$

$$3\partial^{\gamma} \overset{(5)}{\Omega}_{[\alpha\beta\gamma]} + 3\partial^{\gamma} \overset{(6)}{\Omega}_{[\alpha\beta\gamma]} + \overset{(7)}{O} + \ldots = 0 \quad . \tag{V.19}$$

Because of eq. (V.18), the metric coincides with Einstein's up to the order $O(\epsilon^6)$, whereas by virtue of eq. (V.19) deviations from Einstein occur in the order $O(\epsilon^5)$.

THE SPIN MOTION OF A DIRAC TEST PARTICLE IN AN AXIAL SYMMETRIC STATIONARY GRAVITATIONAL FIELD

Some effort has been spent in deriving the propagation equation of a test spin in a Riemann-Cartan space-time. Hehl[25] and Trautman[26] have derived this equation, starting from the conservation laws for the canonical energy-momentum tensor and for the angular momentum tensor. In addition, they have assumed special simplified forms for the energy-momentum tensor and for the spin tensor. The method of deriving the spin motion used by Yasskin[19] and by Stoeger and Yasskin[27] is very similar to that introduced by Mathisson[28], Papapetrou[29] and Dixon[30] in a Riemannian space-time.

Recently, by means of purely algebraic methods applied to the Dirac equation, Rumpf[6] gave a very satisfactory derivation of the spin motion in a U_4-theory. He showed that in the classical limit the spin is not parallely transported with respect to the Riemann-Cartan (U_4) connection, rather its motion was given most transparently in the geodesic Fermi frame (parallel transport with respect to the Riemannian connection) and was expressed in the limit of a teleparallelism geometry by

$$\frac{d}{ds}\underline{w} = -3 \overset{(5)}{\underline{\Omega}} \times \underline{w} \quad , \tag{VI.1}$$

where w^a denotes the spin vector of the particle, and $\overset{(5)}{\underline{\Omega}}$ is defined by

$$\overset{(5)}{\underline{\Omega}} := (\overset{(5)}{\Omega}_{[023]}, \overset{(5)}{\Omega}_{[031]}, \overset{(5)}{\Omega}_{[012]}) \quad . \tag{VI.2}$$

The fourth component $\Omega_{[123]}$ of the "axial vector" $\Omega_{[\alpha\beta\gamma]}$ of the torsion tensor is of the order $O(\epsilon^6)$ and will be ignored henceforth. As in Section V, the superscript number $\overset{(n)}{\underline{\Omega}}$ refers to the order $O(\epsilon^n)$ of the approximation.

Equation (VI.1) represents a basis on which it becomes possible to distinguish experimentally between Einstein's theory and the "macroscopic" Poincaré gauge field theory. The appearance of $\overset{(5)}{\underline{\Omega}}$ in eq. (VI.1) contributes an additional, non-Einsteinian amount to the spin precession. To decide whether this contribution is sufficiently large to be detected in a certain experiment with present-day technology, we need to give an order of magnitude estimate for $\overset{(5)}{\underline{\Omega}}$.

Consider a stationary, axial symmetric body (e.g. the Earth or a neutron star, etc.) with radius R and mass M, rotating at a slow and constant rate $\underline{\omega}$. The rotation of the body produces off-diagonal terms γ_{0b} of the metric. Hence we obtain as the only non-vanishing components of $\gamma_{\alpha\beta}$ (see Ref. 21):

$$\gamma_{00}(\underline{r}) = 4\int d^3r' \frac{\rho(\underline{r}')}{|\underline{r} - \underline{r}'|} = O(\epsilon^2) \quad , \tag{VI.3}$$

$$\underline{\gamma}(\underline{r}) := (\gamma_{0b}(\underline{r})) = -4\int d^3r' \, \frac{\rho(\underline{r}')\underline{v}(\underline{r}')}{|r - r'|} = O(\epsilon^3) \tag{VI.4}$$

For the stationary case under consideration the solution of eq. (V.14) reads

$$\underline{\tilde{a}}(\underline{r}) = (1/4\pi)\int d^3r' \frac{\underline{\tilde{L}}(\underline{r}')}{|\underline{r} - \underline{r}'|} \tag{VI.5}$$

where we have introduced the quantities

$$\underline{\tilde{a}} := (\tilde{a}_{01}, \tilde{a}_{02}, \tilde{a}_{03}) \quad ; \quad \underline{\tilde{L}} := (\tilde{L}_{01}, \tilde{L}_{02}, \tilde{L}_{03}) \tag{VI.6}$$

For $\underline{\tilde{L}}$ we obtain

$$\underline{\tilde{L}} = (1/4)(\gamma_{00}\Delta\underline{\gamma} - \underline{\gamma}\Delta\gamma_{00}) - (1/4)\underline{\nabla}\overset{\curvearrowright}{(\underline{\gamma}\cdot\underline{\nabla})}\gamma_{00} + $$
$$+ (1/4)(\underline{\gamma}\cdot\underline{\nabla})\underline{\nabla}\gamma_{00} \tag{VI.7}$$

The other components of $\tilde{L}_{\alpha\beta}$ turn out to be of the order $O(\epsilon^6)$. Now we are able to write $\overset{(5)}{\underline{\Omega}}$ in a very concise form:

$$\overset{(5)}{\underline{\Omega}} = -(1/6)(\underline{\nabla} \times \underline{\tilde{a}}) + (1/24)(-\gamma_{00}(\underline{\nabla} \times \underline{\gamma}) - \underline{\gamma} \times \underline{\nabla}\gamma_{00}). \tag{VI.8}$$

The operator $\underline{\nabla}$ denotes the usual gradient.

It is well known that in GR the term $(\underline{\nabla} \times \underline{\gamma})$ gives rise to the "dragging of inertial frames". This Thirring-Lense effect also occurs in eq. (VI.8), but it is modified by the factor $\gamma_{00} \sim (M/R)$. In suitable units, all terms on the r.h.s. of eq. (VI.8) are proportional to the square of the gravitational constant.

In order to facilitate our further calculation of $\overset{(5)}{\underline{\Omega}}$, we assume a spherical rigidly rotating body with constant density, i.e.

$$\underline{v} = \underline{\omega} \times \underline{r} \qquad \text{and} \qquad \rho(\underline{r}) = \rho_0 \quad . \tag{VI.9}$$

Then we find for the angular momentum $\underline{J}$ of the rotating sphere

$$\underline{J} = (2/5)MR^2\underline{\omega} \tag{VI.10}$$

Using the "parameters" M and $\underline{J}$ of the matter distribution, we compute $\gamma_{00}(\underline{r})$ and $\underline{\gamma}(\underline{r})$ for the two distinct domains: $r \leq R$ and $r \geq R$. The results are:

$$\gamma_{00}^{\text{int.}}(\underline{r}) = (4M/R)((-1/2)(r^2/R^2) + (3/2)) \ , \ (r \leq R) \ ; \tag{VI.11}$$

$$\gamma_{00}^{\text{ext.}}(\underline{r}) = 4M/r \ , \qquad (r \geq R) \quad ; \tag{VI.12}$$

$$\underline{\gamma}^{\text{int.}}(\underline{r}) = (\underline{J} \times \underline{r})((3r^2/R^2)-5)/R^3, \qquad (r \leq R) \quad ; \tag{VI.13}$$

$$\underline{\gamma}^{\text{ext.}}(\underline{r}) = -2(\underline{J} \times \underline{r})/r^3 \quad , \quad (r \gtrsim R) \quad . \tag{VI.14}$$

To get an explicit formula for $\underline{\tilde{a}}(\underline{r})$ in the outside region ($r \gtrsim R$) of the matter distribution, we insert eqs. (VI.11) - (VI.14) into eq. (VI.7). This yields an expression for $\underline{\underline{\tilde{L}}}(\underline{r})$ which in turn is inserted into eq. (VI.5). We obtain:

$$\underline{\tilde{a}}(\underline{r}) = (M/R)(\underline{J} \times \underline{r})((24/7) - (R/r))/r^3 \quad , \quad (r \gtrsim R) \quad . \tag{VI.15}$$

Comparing eqs. (VI.14) and (VI.15), we observe that the ratio

$$|\underline{\tilde{a}}(\underline{r})|/|\underline{\gamma}^{\text{ext.}}(\underline{r})| \sim (M/R) = (1/2)(\text{Schwarzschild radius/radius}) . \tag{VI.16}$$

However, for r-values outside the body, the main contribution to $\underline{\Omega}^{(5)}$ originates from the antisymmetric part of the tetrad field, i.e. from

$$-(1/6)(\underline{\nabla} \times \underline{\tilde{a}}) = -\, 4M(-\underline{J} + 3\underline{e}_0(\underline{J}\cdot\underline{e}_0))/7Rr^3 \; +$$
$$+ \; (M/r)(-\underline{J} + 2\underline{e}_0(\underline{J}\cdot\underline{e}_0))/r^3, \quad (r \gtrsim R), \tag{VI.17}$$

whereas the symmetric part of the tetrad field gives an over-all contribution of

$$(2/3)(M/r)\underline{e}_0(\underline{J}\cdot\underline{e}_0)/r^3 \quad , \quad (r \gtrsim R) \tag{VI.18}$$

to $\underline{\Omega}^{(5)}$. The unit vector $\underline{e}_0$ points in the direction of $\underline{r}$. We add the equations (VI.17) and (VI.18) to the final result for $\underline{\Omega}^{(5)}$:

$$\underline{\Omega}^{(5)} = -(4/7)(M/R)(-\underline{J} + 3\underline{e}_0(\underline{J}\cdot\underline{e}_0))/r^3 \; +$$
$$+ \; (1/3)(M/r)(-\underline{J} + 4\underline{e}_0(\underline{J}\cdot\underline{e}_0))/r^3 \quad , \quad (r \gtrsim R) \quad . \tag{VI.19}$$

On the surface ($r = R$) of the rotating body both terms on the r.h.s. of eq. (VI.19) are of equal size, and we obtain:

$$\underline{\Omega}^{(5)}(\underline{R}) = -\,(5/21)(M/R)(-\underline{J} + (8/5)\underline{e}_0(\underline{J}\cdot\underline{e}_0))/R^3 \quad . \tag{VI.20}$$

Unfortunately, on the surface of the Earth, for example, the absolute amount of $\underline{\Omega}^{(5)}$ becomes

$$|\underline{\Omega}^{(5)}|_{R_\oplus} \sim 10^{-24}\,\text{s}^{-1} \quad , \tag{VI.21}$$

which is hopelessly beyond present-day experimental detectability (cf. Ref. 31). Although the polarization of the emission spectra of electrons in the vicinity of neutron stars is influenced by the spin precession, there is no chance to measure such an effect due to the large magnetic fields of the neutron stars. This effect may, however become observable with electrons of a very dilute plasma in the neighborhood of presumably existing black holes.

REFERENCES

1. F.W. Hehl, Y. Ne'eman, J. Nitsch, and P. Von der Heyde, Phys. Lett. 78B:102 (1978).
   F.W. Hehl, J. Nitsch, and P. Von der Heyde, Gravitation and the Poincaré Gauge Field Theory with Quadratic Lagrangian, in: Einstein Volume 1, A. Held (ed.), Plenum Press (to appear 1979/80).
2. C. Møller, Kgl. Dan. Vid. Selsk. Mat. Fys. Medd. 39:Nr. 13 (1978)
3. H. Meyer, Diploma Thesis, Universität Köln (1979).
4. J. Nitsch and F.W. Hehl, Translational Gauge Theory of Gravity: Post-Newtonian Approximation and Spin Precession, University of Cologne, preprint (1979).
5. F.W. Hehl, Lecture given at the 6th Course of the International School of Cosmology and Gravitation on "Spin, Torsion, Rotation, and Supergravity", held at Erice, Italy, May 1979; and article in this volume.
6. H. Rumpf, Quasiclassical Limit of the Dirac Equation and the Equivalence Principle in the Riemann-Cartan Geometry: article in this volume.
7. P. von der Heyde, Phys. Lett. 58A:141 (1976).
8. J.M. Nester, Canonical Formalism and the ECSK Theory, Ph.D. Thesis, University of Maryland (1977).
9. M. Schweizer, The Gauge Symmetries of Gravitation: article in this volume.
10. P. von der Heyde, Z. Naturf. 31a:1725 (1976).
11. L.P. Eisenhart, Non-Riemannian Geometry: American Mathematical Society, Volume VIII (1927).
12. Y.M. Cho, Phys. Rev. D14:2521 (1976).
13. F.I. Fedorov, Current Problems in Gravitational Theory; Proc. 2nd Sov. Conf. on Grav., Tiflis, April 1965; Univ. Press, Tiflis 67
14. K. Hayashi and T. Shirafuji, Phys. Rev. D 19:3524 (1979).
15. A. Schouten, "Ricci Calculus," 2nd ed., Springer, Berlin (1954).
16. K. Yano, "The Theory of the Lie Derivatives and its Applications" North-Holland Publishing Co., Amsterdam (1957).
17. H.P. Robertson, Annals of Math. 33:496 (1932).
18. P. Baekler and P.B. Yasskin, All Torsion-free Spherical Vacuum Solutions of the Quadratic Poincaré Gauge Field Theory, University of Cologne, paper in preparation (1979).
19. P.B. Yasskin, Metric-Connection Theories of Gravity, Ph.D. Dissertation, University of Maryland (1979).
20. P. Baekler, private communication (1979).

21. C.W. Misner, K.S. Thorne, and J.A. Wheeler, "Gravitation", W.H. Freeman and Company, San Francisco (1973).
22. M. Schweizer and N. Straumann, Phys. Lett. 71A:493 (1979).
23. L.L. Smalley, Post-Newtonian Approximation of the Poincaré Gauge Theory of Gravitation, preprint 1979 (to appear in Phys. Rev. D).
24. D.-E. Liebscher, Lecture given at the 6th Course of the International School of Cosmology and Gravitation on "Spin, Torsion, Rotation, and Supergravity" held at Erice, Italy, May 1979.
25. F.W. Hehl, Phys. Lett. 36A:225 (1971).
26. A. Trautman, Bull. Acad. Pol. Sci., Ser. Sci. Math. Astron. Phys. 20:895 (1972).
27. W.R. Stoeger and P.B. Yasskin, Can a Macroscopic Gyroscope Feel Torsion?, preprint (1979).
28. M. Mathisson, Acta Phys. Pol. 6:163 (1937).
29. A. Papapetrou, Proc. Roy. Soc. Lond. A209:248 (1951).
30. W.G. Dixon, Proceedings of the International School of Physics "Enrico Fermi", Course LXVII: Isolated Gravitating Systems in General Relativity, J. Ehlers (ed.) (1979).
31. E. Fishbach, Tests of General Relativity at the Quantum Level, article in this volume.

QUASICLASSICAL LIMIT OF THE DIRAC EQUATION AND THE EQUIVALENCE PRINCIPLE IN THE RIEMANN-CARTAN GEOMETRY

Helmut Rumpf **

Institute for Theoretical Physics

University of Cologne

Abstract

A natural extension of the weak equivalence principle to the Riemann-Cartan geometry is proposed and then shown to be violated by the classical equations of translational and spin motion derived from the Dirac equation. This classical limit is obtained by a purely algebraic method. We point out that it is possible to distinguish experimentally between macroscopically equivalent Riemannian and teleparallelism geometries by measuring the precession of spinning particles.

** Present Address: Institute for Theoretical Physics, University of Berne, Sidlerstrasse 5, CH-3012 Berne, Switzerland

1. INTRODUCTION

The possibility that the equivalence principle (EP) underlying Einstein's General Theory of Relativity (GR) is violated by the electron has been considered by several authors (see e.g. Peres (1978) and the references cited therein). In this context it has always been understood that the validity of the EP is guaranteed by the minimal coupling of the Dirac equation to the space-time geometry. Although this statement turns out to be true (approximately) in the Riemannian case, it is not trivial. For minimal coupling is a rather abstract concept, whereas the EP makes predictions about directly observable particle properties such as ionization tracks and polarization. If the geometry of space-time is of a more general type, it might very well be that minimal coupling of the Dirac field eventually translates into an EP-violating gravitational coupling of Dirac particles. Starting from a natural generalization of the ordinary EP, I want to point out in this seminar that this is just what happens in the Poincaré gauge theory of gravitation, which was the subject of the course given by Professor Hehl in this school. By the effect discussed here a new possibility of distinguishing experimentally between this theory and GR is offered.

2. ON THE FORMULATION OF THE EQUIVALENCE PRINCIPLE IN A RIEMANN-CARTAN SPACE-TIME

Since we shall treat Dirac particles as test particles, only the kinematical content of the Poincaré gauge theory will be needed, i.e. the fact that the space-time geometry is of the Riemann-Cartan (U_4) type. This geometry attributes new degrees of freedom to the gravitational field, which are present in a general metric-compatible connection $\Gamma^{a}_{\mu b}$ (anholonomic indices will be denoted by Roman letters). We stress that because of these additional degrees of freedom the formulation of the weak EP in a U_4 becomes a non-trivial task: Besides mass, there exists now a second type of passive gravitational charge, namely spin. So we might expect that the generalized weak EP for test bodies will read as follows:

inertial mass (spin) = passive gravitational mass (spin) (W)

Note, however, that in contrast to "inertial spin", which is defined by non-gravitational forces (e.g. via the Larmor frequency if the particle carries a magnetic moment), the notion of "passive gravitational spin" has yet to be defined. In fact most of this section of my seminar will be devoted to this issue.

First, however, I want to discuss the strong EP, whose formulation can be taken over from GR without change:

> For any fundamental field equation of physics there exists for any space-time point P a local inertial system of reference (LIS) such that the equation acquires its special-relativistic form at P in the LIS. (S)

Note that it is not required that for all fields gravity is "transformed away" in the same LIS. As a matter of fact, even in GR two different types of LIS are necessary: Riemannian normal coordinate systems (RNCS) and (for spinning matter) the anholonomic coordinate systems in which the Ricci rotation coefficients vanish. Does the property (S) hold in a U_4? In order to show that it does not I want to discuss a notion of LIS that was introduced by von der Heyde (1975).

> LIS-Proposal 1: Define the LIS at P by a "trivial gauge" frame $\Sigma_0(P)$, for which $e_a^{\mu} = \delta_a^{\mu}$, $\Gamma_{\mu b}^{a} = 0$ at P.

The existence of a $\Sigma_0(P)$ is a consequence of Poincaré gauge invariance. Every $\Sigma_0(P)$ is distinguished by the following property: If the Lagrangian L_{U_4} of some matter field ψ does not contain higher than first derivatives of ψ and if ψ is minimally coupled to the U_4 background, i. e. if

$$L_{U_4}(\psi, e_a\psi) = L_{SR}(\psi, \overset{\Gamma}{\nabla}_a\psi) \tag{1}$$

where L_{SR} is the special-relativistic Lagrangian, then

$$L_{U_4} = L_{SR} \text{ at P in any } \Sigma_0(P) \tag{2}$$

Obviously, $\Sigma_0(P)$ is not an appropriate LIS as defined by the EP (S), since eq. (2) does not imply that the equations of motion become special-relativistic at P. Of course, one could try to replace definition (S) of the strong EP by the weaker requirement of eq. (2), as was proposed by von der Heyde. But then any gauge field interaction would fulfil an EP, because in any gauge field theory there exists an analog of $\Sigma_0(P)$!

In contrast to eq. (2), def. (S) requires a "dynamical" LIS. For the scalar, Maxwell, and other tensor fields such an LIS always exists: Since these fields couple only to the metric, any RNCS is appropriate. The problem comes in with the Dirac equation

$$(i\gamma^{\mu}\overset{\Gamma}{\nabla}_{\mu} + \frac{1}{2} K_{\rho\mu}{}^{\rho}\gamma^{\mu})\psi = 0 \tag{3}$$

where

$$K_{\lambda\mu\cdot}{}^{\nu} := \left\{ {\nu \atop \lambda\mu} \right\} - \Gamma_{\lambda\mu}^{\nu} \tag{4}$$

is the "contortion" tensor. First we should stress that eq. (3) is the unique equation derived from a minimally coupled Lagrangian. However, because of the appearance of the trace of the contortion tensor, neither $\Sigma_0(P)$ (which is appropriate in GR because the anholonomic connection appears in the covariant derivative), nor any other frame of reference can define an LIS as required by def. (S). Hence the strong EP as defined by (S) is violated in the Poincaré gauge theory.

Still one might hope that the weak EP is valid. In order to prepare a precise definition of it, we now turn to an analysis of the equations of motion for the most common classical model of a spinning particle, which is due to Mathisson and Papapetrou. In this model the spin of the classical particle is described by an antisymmetric tensor $s^{\mu\nu}$ obeyin the condition

$$s^{\mu\nu}u_{\nu} = 0 \tag{5}$$

where u^{ν} is the 4-velocity.

We consider first the translational motion of the particle. In a U_4 this motion is governed by the momentum transport equation (Hehl 1971, Trautman 1972)

$$\frac{dp^{\mu}}{ds} = -\left\{ {\mu \atop \nu\lambda} \right\} u^{\lambda}p^{\nu} - g^{\mu\rho} K_{\rho\sigma\tau} u^{\sigma}p^{\tau} + R^{\mu}{}_{\cdot\nu\rho\sigma}(\Gamma)\, s^{\rho\sigma}u^{\nu} \tag{6}$$

From this we see that in an RNCS the special-relativistic law of momentum transport holds approximately (i.e. if the contortion and curvature tensors are suitably bounded). Note that in order to obtain the worldline of the particle a relation between momentum and velocity is needed and that such a relation will in general not yield geodesic motion even in Minkowski space. However, what can be concluded from eq. (6) is that the motion of the center of mass of the particle is approximately geodesic.

For the motion of spin in a U_4 the following equation was obtained by Adamowicz and Trautman (1975):

$$\frac{Ds^{\mu}}{ds} + u^{\mu}\frac{Du^{\nu}}{ds}s_{\nu} = 0 \tag{7}$$

$$s_{\lambda} := \frac{1}{2}\varepsilon_{\lambda\mu\nu\rho}s^{\mu\nu}u^{\rho} \tag{8}$$

The absolute derivative D/ds has to be taken with respect to the U_4 connection Γ. Eq. (7) is equivalent to the covariantized spin transport equation of Mathisson

$$\frac{Ds^{\mu\nu}}{ds} = p^{\mu}u^{\nu} - p^{\nu}u^{\mu} \qquad (9)$$

and the assumption

$$p^{\mu}u_{\mu} = m \qquad (10)$$

(Eqs. (5), (8), (9) imply the momentum-velocity relation

$$p^{\mu} = mu^{\mu} - s^{\mu\nu}\frac{Du_{\nu}}{ds} \qquad (11)$$

Because of eq. (5), eq. (7) amounts to Fermi propagation of the spin vector with respect to Γ. Eq. (7) becomes even more transparent physically, if we introduce a Fermi frame transported parallely along the geodesic world line of the center of mass of the particle with respect to the Levi-Civita connection. We shall refer to this frame as a geodesic Fermi frame for short. In this frame eq. (7) reduces to

$$\frac{d\vec{s}}{dt} = \vec{K} \times \vec{s} \qquad (12)$$

$$K_1 := K_{023} \text{ etc.} \qquad (13)$$

Thus, as compared with macroscopic gyroscopes, the spinning particles in this model undergo an additional <u>universal spin precession</u> of frequency $\vec{K}$. As noted by Adamowicz and Trautman, it is now possible to give meaning to the notion of "passive gravitational spin" by rewriting eq. (12) in the form

$$\frac{d\vec{s}_{in}}{dt} = \vec{K} \times \vec{s}_{grav} \qquad (14)$$

$$\vec{s}_{grav} = \vec{s}_{in} \qquad (15)$$

With this clarification we are now able to propose a more refined version of the weak EP as compared to (W).

<u>LIS-Proposal 2</u>: The LIS's for translational (spin) motion of a test particle are of the RNCS (Σ_0) type.

Observe that in Σ_0 due to the vanishing of Γ eq. (7) becomes the special-relativistic law of Thomas precession. The latter originates from the fact that a freely falling particle is accelerated with respect to Σ_0.

The classical particle model considered has thus led us to the conclusion that both the "mass aspect" and the "spin aspect" of gravity can be transformed away (approximately), though not simultaneously. But has that model anything to do with reality? In particular, will electrons obey the EP just proposed or do they possess an "anomalous gravitational moment" as we could term any deviation from the Fermi propagation law (7)? The answer to this question has to be deduced from the Dirac equation.

3. QUASICLASSICAL LIMIT OF THE DIRAC EQUATION

In order to derive classical equations of motion from quantum theory one usually starts from a WKB expansion of a wavefunction. It was by this method that Pauli (1932) concluded that the translational motion of an electron in an electromagnetic field is unaffected by its spin. Rubinow and Keller (1963) were the first to derive also a classical equation for the spin motion from the Dirac theory. This equation is identical with the so-called BMT equation (Bargmann, Michel, Telegdi 1959)

$$\frac{ds^{\mu}}{ds} = \frac{e}{m} F^{\mu\nu} s_{\nu} \tag{16}$$

There exists, however, a different method of obtaining the classical limit, which does not use partial differential equations, but is purely algebraic. A somewhat sketchy outline of this method can be found in the textbook of Corben (1968). I will present it in some detail here.

Consider first the simple case of a free Dirac particle. Its C*-algebra of observables is generated by elements x^a, p_b, γ^c (with the usual physical interpretation) obeying the (anti-)commutation relations

$$[x^a, p_b] = i\hbar\delta^a_b \tag{17}$$

$$\{\gamma^a, \gamma^b\} = 2\eta^{ab} \tag{18}$$

$$[x^a, \gamma^b] = [p_c, \gamma^d] = 0 \tag{19}$$

The *-operation of this algebra will be denoted by a bar, since it corresponds to the familiar Dirac adjoint:

$$\bar{x}^a = x^a, \quad \bar{p}_b = p_b, \quad \bar{\gamma}^c = \gamma^c \tag{20}$$

The dynamics of the particle is determined by the Heisenberg equation for the evolution of an observable O in the proper time s

$$\dot{O} = \frac{dO}{ds} = \frac{i}{\hbar}[O, H] \tag{21}$$

$$H = \gamma^a p_a \tag{22}$$

In the usual representation H is the Dirac mass operator. Next we list the equations of motion for some relevant observables:

$$\dot{x}^a = \gamma^a \tag{23}$$

$$\dot{p}^a = 0 \tag{24}$$

$$\ddot{x}^a = \dot{\gamma}^a = \frac{2}{\hbar}\sigma^{ab}p_b \tag{25}$$

$$\sigma^{ab} := \frac{i}{2}[\gamma^a, \gamma^b] \tag{26}$$

The tensor σ^{ab} is proportional to the spin tensor

$$s^{ab} = \frac{\hbar}{2}\sigma^{ab} \tag{27}$$

We note in passing that the equations

$$\{s^{ab}, \gamma_b\} = 0 \tag{28}$$

$$\dot{s}^{ab} = p^a\gamma^b - p^b\gamma^a \tag{29}$$

are exact quantum counterparts of eqs. (5) and (9) of the classical model. In analogy to eq. (8) we introduce the polarization vector

$$w^a := \frac{1}{2}\varepsilon^{abcd}\{\sigma_{bc}, \gamma_d\} = \gamma_5\gamma^a \tag{30}$$

where

$$\gamma_5 := i\gamma^0\gamma^1\gamma^2\gamma^3 \tag{31}$$

and obtain

$$\{w^a, \gamma_a\} = 0 \tag{32}$$

$$\sigma^{ab} = -\frac{1}{2}\varepsilon^{abcd}\gamma_c w_d \tag{33}$$

$$\dot{w}^a = \frac{i}{\hbar}\gamma_5 p^a \tag{34}$$

Eqs. (28) and (29) imply the momentum-velocity relation

$$p^a = -\frac{1}{2}\{s^{ab}, \dot{\gamma}_b\} \tag{35}$$

which should be compared with eq. (11). It seems extremely difficult to solve the equations for the worldline $x^a(s)$ and the evolution of the spin vector $w^a(s)$ even in the classical limit, where all observables commute. But this is not even desirable. For we know that particle states are always eigenstates $|m\rangle$ of H,

$$H|m\rangle = m|m\rangle \tag{36}$$

and therefore only the expectation values of observables in these eigenstates are physically interesting. Fortunately it is much easier to evaluate the evolution of these expectation values than that of the observables themselves. The clue is contained in the following observation. If O is replaced by

$$O' := \frac{1}{2m}\{O,\ H\} \tag{37}$$

all the physical expectation values remain the same:

$$\langle m|O|m\rangle = \langle m|O'|m\rangle = \langle m|O''|m\rangle = \ldots \tag{38}$$

On the other hand it is possible to derive simple equations of motion if the replacement $O \rightarrow O'$ is allowed! It follows from eqs. (21), (37) that

$$\frac{dO'}{ds} = \left(\frac{dO}{ds}\right)' = \frac{i}{\hbar}\left[O,\ \frac{H^2}{2m}\right] \tag{39}$$

Therefore, if we accept the postulate that in the operator equations of motion, from which the classical limit is to be derived, only equal numbers of "$\dot{}$" and "'" operations should be applied to observables, then the introduction of the "'" operation amounts to a formal redefinition of the dynamics of the particle through the replacement of H by $H^2/2m$. As will be shown below, this postulate yields results which are in perfect agreement with experiment. Moreover it is also involved implicitly in the WKB type approaches to the classical limit, all of which make use of the iterated Dirac equation. The fact that mass eigenstates cannot be localized and the examples given below indicate that physically the "'" operation has to be interpreted as an averaging over the "Zitterbewegung" of the Dirac particle.

Before we apply this formalism to the U_4 background, let us test its meaningfulness in the free and electromagnetic case:

a) Free Particle:

$$\gamma'^a = p^a/m \tag{40}$$

$$\sigma'^{ab} = \frac{1}{m}\varepsilon^{abcd} p_c w_d \Rightarrow (\sigma^{ab} p_b)' = 0 \tag{41}$$

$$w'^a = \frac{1}{2m}\varepsilon^{abcd} p_b \sigma_{cd} \Rightarrow (w^a p_a)' = 0 \tag{42}$$

We recognize that w'^a is the Pauli-Lubanski vector.

b) External Electromagnetic Field:

$$\ddot{x}'^{a} = \frac{e}{m}F^{a}{}_{b}\dot{x}'^{b} + \frac{e\hbar}{4m}\sigma^{bc}\eta^{ad}F_{bc,d} \tag{43}$$

$$\dot{w}'^{a} = \frac{e}{m}F^{a}{}_{c}w^{c} \tag{44}$$

The equation of translational motion (43) contains, in contrast to Pauli's (1932) result, a gradient term, from which the gyromagnetic ratio g = 2 of the electron can be deduced. We consider this result to be a strong argument in favor of the formal redefinition of the dynamics that we postulated (cf. in this connection also de Broglie's (1952) criticism of Pauli's result). Eq. (44) becomes in the classical limit the BMT equation (16).

We are now prepared to derive our main result, the equations of motion for the electron in a Riemann-Cartan space-time. The mass operator is given by

$$H = \frac{1}{2}\gamma^{a}(e^{\mu}_{a}p_{\mu} + \bar{p}_{\mu}e^{\mu}_{a}) - \frac{\hbar}{8}\Gamma_{bcd}\{\gamma^{b}, \sigma^{cd}\} \tag{45}$$

$$\bar{p}_{\mu} = \frac{1}{e}p_{\mu}e \tag{46}$$

$$e: = \det(e^{\mu}_{a}) \tag{47}$$

Eq. (45) can be deduced from the well-known classical action of the minimally coupled Dirac field ψ (obeying eq. (3))

$$S[\psi] = -\langle\psi|(\hat{H} - m)\psi\rangle \tag{48}$$

$$\langle\psi_1|\psi_2\rangle: = \int d^4x\, e\,\bar{\psi}_1\psi_2 \tag{49}$$

$$\hat{H} = \frac{i\hbar}{2}(\gamma^{\mu}\vec{\nabla}_{\mu} - \overleftarrow{\nabla}_{\mu}\gamma^{\mu}) \tag{50}$$

$\hat{H}$ is just the usual representation of H, and (46) is a consequence of (49). We also note that eq. (50) automatically yields H self-adjoint in every order $\hbar$.

The following exact operator equations of motion are relevant in the classical limit:

$$\ddot{x}'^{\mu} = -\left\{{\mu \atop \lambda\nu}\right\}\dot{x}'^{\mu}\dot{x}'^{\nu} + O(\hbar) + O(\hbar^2) \tag{51}$$

$$\dot{w}'^{a} = -\overset{*}{\Gamma}{}^{a}_{\mu b}\dot{x}'^{\mu}w^{b} + O(\hbar) \tag{52}$$

$$\overset{*}{\Gamma}{}^{\lambda}_{\mu\nu}: = \Gamma^{\lambda}_{\mu\nu} + 2S_{\nu\cdot\mu}{}^{\lambda} = \left\{{\lambda \atop \mu\nu}\right\} + 3g^{\lambda\rho}S_{[\mu\nu\rho]} \tag{53}$$

$$S_{\lambda\mu}{}^{\nu}: = \Gamma^{\lambda}_{[\mu\nu]} \tag{54}$$

The order $\hbar$ and $\hbar^2$ terms are not covariant and should be dropped in the classical limit. Lawrence (1970) derived eq. (51) in the special case of linearized GR and interpreted the $O(\hbar)$ term as a Papapetrou term. But in the general case only the derivative part of the curvature tensor associated with the connection (53) appears.

In the classical limit electrons move on Riemannian geodesics. However, for the spin motion we do not get the Fermi transport law (7) of the classical particle model. Rather, the spin of electrons is transported parallely with respect to the new connection $\overset{*}{\Gamma}$. Note that due to the antisymmetry of the torsion tensor (54) $\overset{*}{\Gamma}$ is metric-compatible.

The non-Riemannian contribution to eq. (52) becomes most transparent in the geodesic Fermi frame introduced in Sec. 2. In this frame we have

$$\dot{\vec{w}}' = 3\,\vec{S} \times \vec{w} \tag{55}$$

$$S^{\alpha} := -\frac{1}{3!}\,\varepsilon^{\alpha}{}_{\beta\gamma\delta}\,S^{\beta\gamma\delta} \tag{56}$$

If the torsion tensor is totally antisymmetric, the spin vector precesses around the same direction as in the classical model, but with a frequency three times as great. Incidentally, the precession law (55) can be obtained from exact solutions of the Dirac equation in the special case of a constant totally antisymmetric torsion field in a metrically flat U_4 (Rumpf 1979).

While the coupling of the electron spin to the space-time geometry is not as trivial as anticipated in LIS-Proposal 2, it seems still possible that spin motion is universal. It must be left to further investigations whether eqs. (52) and (53) apply also to particles of higher spin.

4. EXPERIMENTAL DISTINCTION BETWEEN RIEMANNIAN AND TELEPARALLELISM GEOMETRIES

In the Poincaré gauge theory of Hehl et al. (1978a, 1978b) the Newtonian and Schwarzschild limits are obtained in a teleparallelism geometry rather than a Riemannian one (for further details cf. the seminar of Dr. Nitsch). It has been pointed out by Schweizer and Straumann (1979), however, that this difference is not likely to be detectable by the solar system experiments proposed so far. Thus the measurement of the spin precession of elementary particles could become an experimentum crucis: Whereas in the Riemannian case spin is parallely transported along geodesics, in the teleparallelism case an "anomalous" precession occurs as soon as the torsion axial vector S^{α} (56) is different from zero.

What is this axial vector like in the case of an isolated source of the gravitational field? As shown by Nitsch, this vector vanishes in the case of spherical symmetry. Since so far no axialsymmetric solution of the field equations of Hehl et al. has been found, I can present only a guess for $\vec{S}$ based on dimensional arguments:

$$\vec{S} = \frac{G}{c^2 r^3}\left[\alpha\vec{L} + \beta(\vec{L}\cdot\vec{r}_0)\vec{r}_0\right] \tag{57}$$

$\vec{L}$ is the angular momentum of the source, and α, β are dimensionless constants. If $\alpha, \beta \sim 1$, we get on the surface of the earth

$$|\vec{S}| \sim 10^{-15} s^{-1}$$

which is far beyond present experimental detectability. So far a polarization accuracy of $\sim 10^{-6}$ rad has been achieved in the measurement of the electric dipole moment of the neutron, and the best neutron storage times are of the order of seconds (cf. the seminar of Prof. Fishbach). Still it is conceivable that in the vicinity of neutron stars emission spectra are influenced by the spin precession to such an extent as to render the effect observable indirectly.

NOTE ADDED

After the completion of this work we learned that eqs. (51) and (52) have been derived also by Hayashi and Shirafuji (1979) by a different method in the special case of teleparallelism.

REFERENCES

Adamowicz, W. and Trautman, A., 1975, Bull. Acad. Pol. Sci., Ser. Sci. Math. Astr. Phys. 23, 339

Bargmann, V., Michel, L., Telegdi, V. L., 1959, Phys. Rev. Lett. 2, 435

de Broglie, L., 1952, "La Theorie des Particules de Spin 1/2", Gauthier-Villars, Paris, pp. 132, 128

Corben, H. C., 1968, "Classical and Quantum Theory of Spinning Particles"

Hayashi, K., and Shirafuji, 1979, "New General Relativity", Univ. of Tokyo preprint

Hehl, F. W., 1971, Phys. Lett. 36 A, 225

Hehl, F.W., Ne'eman, Y., Nitsch, J., von der Heyde, P., 1978a, Phys. Lett. 78 B, 102

Hehl, F. W., Nitsch, J., von der Heyde, P., 1978b, "Gravitation and Poincaré Gauge Field Theory with Quadratic Lagrangian", to appear in "The Einstein Memorial Volume" ed. by A. Held, Plenum Press

Lawrence, J., 1970, Ann. Phys. (N. Y.) 58, 47

Pauli, W., 1932, Helvetia Phys. Acta 5, 1979

Peres, A., 1978, Phys. Rev. D 18, 2739
Rubinow, S. I., and Keller, J. B., 1963, Phys. Rev. 131, 2789
Rumpf, H., 1979, "Creation of Dirac Particles in General Relativity with Torsion and Electromagnetism III: Matter Production in a Model of Torsion", Gen. Rel. Grav. 10
Schweizer, M., and Straumann, N., 1979, "Poincaré Gauge Theory of Gravitation and the Binary Pulsar 1913 + 16", Zürich Univ. preprint
Trautman, A., 1972, Bull. Acad. Pol. Sci. Math. Astr. Phys. 20, 895
von der Heyde, P., 1975, Lett. Nuovo Cim. 14, 250

CONTRACTED BIANCHI IDENTITIES AND CONSERVATION LAWS IN POINCARÉ GAUGE THEORIES OF GRAVITY

Wiktor Szczyrba

Institute for Theoretical Physics
University of Cologne
5000 Cologne, Germany *

ABSTRACT: For Poincaré gauge theories of gravity we derive differential identities of the Belinfante-Rosenfeld and of the Bianchi type. We construct two conserved Noether 3-forms which are related to energy -momentum and angular momentum, respectively.

1. INTRODUCTION

It is known that for any (pseudo)riemannian metric $(g_{\mu\nu})$ on space time M and the corresponding Einstein tensor

$$G_{\mu\nu} = R_{\mu\nu} - \frac{1}{2}g_{\mu\nu}R \qquad (1.1)$$

we have

$$\nabla_{\mu}G^{\mu}{}_{\nu} = 0 \qquad (1.2)$$

This result known as "the contracted Bianchi identities" is very important in classical general relativity especially for the initial value formulation of the theory cf. [1,2,4,5,15,18].

In the recent last few years a growing interest in gauge formulations of theories of gravity can be observed cf. [3,7-10,14,22,23,24]. There are different interesting directions of research in such approaches to gravity. One of them is the initial value formulation and the question, how to pose correctly the initial value problem for gauge theories of gravity. A comparision with general relativity tells us that we have first to generalize (1.2) for such theories. In the present paper we prove the contracted Bianchi identities for a local Poincaré gauge theory of gravity with the presence of an arbitrary (tensor) matter field. The first order gravitational and matter field lagrangians are not specified, we require only that they have to be invariant with respect to the action of the gauge

*Permanent address: Institute of Mathematics, Polish Academy of Sciences, 00-950 Warsaw, Poland.

group. It turns out that these invariance properties give rise to an elegant generalization of (1.2)

$$\nabla_{\mu}\left(\frac{\delta\tilde{L}}{\delta e^{A}_{\mu}}\right) = 0 \qquad (1.3)$$

where $\tilde{L}$ is the lagrangian density of the system.

The second problem we deal with is a construction of Noether quantities corresponding to the action of the local Poincaré group. The invariance properties with respect to the action of the diffeomorphism group of space time (active coordinate transformations in M) give raise to a conserved quantity E^{λ} and the invariance properties with respect to the action of the local Lorentz group give rise to a conserved quantity J^{λ}. We prove that the conservation laws

$$\nabla_{\lambda}E^{\lambda} = 0 \quad ; \quad \nabla_{\lambda}J^{\lambda} = 0 \qquad (1.4)$$

are equivalent to the field equations.

It can be shown that the quantities E^{λ}, J^{λ} are generators of the dynamics of the system. This problem and its connection with the Bianchi identities (1.3) will be discussed in the next paper.

The main results presented in this paper are original. Some results in this direction have already been found earlier, however. The contracted Bianchi identities and their applications for the metric affine theory of gravity (a non gauge approach) with the Einstein gravitational lagrangian were given in [19] . For the Einstein -Cartan theory an equivalent result was proved by Trautman [22] (but no applications were given in that paper).

The quantity E^{λ} was defined recently by Kijowski [12] for a generalized Einstein theory (symmetric, non metric compatible connection) and by the present author [21] for the metric affine theory with the Einstein gravitational lagrangian. These constructions generalize the formula given by Komar,[13].

The differential identities following from the invariance properties of lagrangians we get by means of differential-geometric methods based on the notions of multisymplectic manifolds and the Hamilton-Cartan forms. These identities have been obtained recently by Hehl, von der Hayde et al,[7-10] for a general case and by Trautman[22] and Szczyrba [19] in special cases.

The results of the paper were obtained during my stay at the Institute for Theoretical Physics, University of Cologne (spring 1979) I would like to thank Professor F.W.Hehl for stimulating my interest in gauge theories of gravity, many fruitful discussions and critical remarks. Special thanks are due to the Humboldt Foundation for the award of the fellowship.

2.NOTATION

Geometry of space time M is described by means of a field of tetrads (of covectors) $(\underline{e}^{A})^{3}_{A=0}$ and a linear connection $\underline{\Gamma}$ on M. Let (x^{λ}) be a local coordinate system on M, then the covectors

$(\underline{e}^A)_{A=0}^3$ and their duals $(\underline{e}_A)_{A=0}^3$ are

$$\underline{e}^A = e^A_\mu dx^\mu \quad ; \quad \underline{e}_A = e^\mu_A \frac{\partial}{\partial x^\mu} \tag{2.1}$$

and the matrices $[e^A_\mu]$, $[e^\mu_A]$ satisfy

$$e^A_\mu e^\nu_A = \delta^\nu_\mu \quad ; \quad e^A_\mu e^\mu_B = \delta^A_B \tag{2.2}$$

Let (Γ^A_{CB}) be anholonomic components of the connection $\underline{\Gamma}$ cf.[17] refered to the tetrad $(\underline{e}^A)_{A=0}^3$. We define mixed components

$$\Gamma^A_{\lambda B} = e^C_\lambda \Gamma^A_{CB} \tag{2.3}$$

A matter tensor field $\underline{\phi}$ can be described either by means of its anholonomic components $\phi^{A_1 \ldots A_k}_{B_1 \ldots B_s}$ (components with respect to the bases $\underline{e}^A$, $\underline{e}_A$) or by means of its holonomic components $\phi^{\alpha_1 \cdots \alpha_k}_{\beta_1 \cdots \beta_s}$ (components with respect to the bases dx^μ , $\partial/\partial x^\mu$).

We have the following relation

$$\phi^{A_1 \ldots A_k}_{B_1 \ldots B_s} = e^{A_1}_{\mu_1} \ldots e^{A_k}_{\mu_k} e^{\nu_1}_{B_1} \ldots e^{\nu_s}_{B_s} \phi^{\mu_1 \cdots \mu_k}_{\nu_1 \cdots \nu_s} \tag{2.4}$$

If $(\Gamma^\lambda_{\mu\nu})$ are holonomic components of the connection $\underline{\Gamma}$ then

$$\Gamma^\lambda_{\mu\nu} = \Gamma^A_{\mu B} e^\lambda_A e^B_\nu + e^\lambda_C \partial_\mu e^C_\nu \tag{2.5}$$

For technical reasons we use also mixed components of a tensor field e.g. $\phi^{A\mu}$, ϕ^A_ν , $\phi^{\mu A}_B$ etc....

The covariant derivative of such a quantity, e.g. , is given by

$$D_\lambda \phi^{A\mu}_{B\nu} = \partial_\lambda \phi^{A\mu}_{B\nu} + \Gamma^A_{\lambda C} \phi^{C\mu}_{B\nu} - \Gamma^E_{\lambda B} \phi^{A\mu}_{E\nu} + \Gamma^\mu_{\lambda\varepsilon} \phi^{A\varepsilon}_{B\nu} - \Gamma^\tau_{\lambda\nu} \phi^{A\mu}_{B\tau} \tag{2.6}$$

i.e. anholonomic indices are to be differentiated by means of $\Gamma^A_{\lambda B}$ and holonomic indices by means of $\Gamma^\varepsilon_{\lambda\tau}$.

Let η_{AB} be the diagonal, constant Minkowski metric with the signature (-1,+1,+1,+1) and η^{AB} be the inverse matrix. The scalar product between covectors and vectors, respectively is

$$\underline{e}^A \cdot \underline{e}^B = \eta^{AB} \quad ; \quad \underline{e}_A \cdot \underline{e}_B = \eta_{AB} \tag{2.7}$$

Therefore tetrads are always (Minkowski) orthonormal. A field of tetrads on M defines a metric $\underline{g}$ on M whose anholonomic components are η_{AB} and whose holonomic components are

$$g_{\mu\nu} = e^A_\mu e^B_\nu \eta_{AB} \tag{2.8}$$

Let

$$\Gamma^{AB}_\lambda = \Gamma^A_{\lambda C} \eta^{CB} \tag{2.9}$$

We assume that

$$\Gamma^{AB}_\lambda = - \Gamma^{BA}_\lambda \tag{2.10}$$

This condition yields

$$D_\lambda \eta_{AB} = 0 \; ; \; D_\lambda g_{\mu\nu} = 0 \tag{2.11}$$

i.e. the connection $\underline{\Gamma}$ is compatible with the metric $\underline{g}$.

The Riemann tensor $\underline{R} = (R^{\lambda}_{\ \tau\mu\nu}) = (R^{A}_{\ B\mu\nu})$ and the torsion tensor $\underline{Q} = (Q^{\lambda}_{\ \mu\nu}) = (Q^{A}_{\ \mu\nu})$ are given by

$$R^{A}_{\ B\mu\nu} = \partial_{\mu}\Gamma^{A}_{\ \nu B} - \partial_{\nu}\Gamma^{A}_{\ \mu B} + \Gamma^{C}_{\ \nu B}\Gamma^{A}_{\ \mu C} - \Gamma^{C}_{\ \mu B}\Gamma^{A}_{\ \nu C}$$

$$Q^{A}_{\ \mu\nu} = \partial_{\mu}e^{A}_{\ \nu} - \partial_{\nu}e^{A}_{\ \mu} + \Gamma^{A}_{\ \mu C}e^{C}_{\ \nu} - \Gamma^{A}_{\ \nu C}e^{C}_{\ \mu}$$

$$R^{A}_{\ B\mu\nu} = e^{A}_{\ \alpha}e^{\beta}_{\ B}R^{\alpha}_{\ \beta\mu\nu} = e^{A}_{\ \alpha}e^{\beta}_{\ B}(\partial_{\mu}\Gamma^{\alpha}_{\ \nu\beta} - \partial_{\nu}\Gamma^{\alpha}_{\ \mu\beta} + \Gamma^{\tau}_{\ \nu\beta}\Gamma^{\alpha}_{\ \mu\tau} - \Gamma^{\tau}_{\ \mu\beta}\Gamma^{\alpha}_{\ \nu\tau})$$

$$Q^{A}_{\ \mu\nu} = e^{A}_{\ \lambda}Q^{\lambda}_{\ \mu\nu} = e^{A}_{\ \lambda}(\Gamma^{\lambda}_{\ \mu\nu} - \Gamma^{\lambda}_{\ \nu\mu}) \tag{2.12}$$

The metric condition (2.11) implies

$$r^{\varepsilon}_{\ \lambda\sigma}g^{\sigma\tau} + r^{\tau}_{\ \lambda\sigma}g^{\sigma\varepsilon} = 0 \tag{2.13}$$

where

$$r^{\varepsilon}_{\ \lambda\sigma} = \Gamma^{\varepsilon}_{\ \lambda\sigma} - \{^{\ \varepsilon}_{\lambda\sigma}\} \tag{2.14}$$

and $\{^{\ \varepsilon}_{\lambda\sigma}\}$ is the riemannian connection corresponding to $g_{\mu\nu}$.
We have also

$$r^{\varepsilon}_{\ \varepsilon\sigma} = Q^{\varepsilon}_{\ \varepsilon\sigma} \quad ; \quad r^{\varepsilon}_{\ \sigma\varepsilon} = 0 \tag{2.15}$$

3. FIELD EQUATIONS AND DIFFERENTIAL IDENTITIES

A physical system under consideration is described by a field of tetrads $(e^{A}_{\ \lambda})$, a metric compatible connection $\Gamma^{AB}_{\ \ \lambda}$ and a tensor field $\phi^{\{A\}}$ on M. The field equations are derived from a lagrangian L which splits into the sum

$$L = L_{gr} + L_{mat} \tag{3.1}$$

The independent variables are $e^{A}_{\ \lambda}$, $\Gamma^{AB}_{\ \ \lambda}$, $\phi^{\{A\}}$ and their first partial derivatives. The lagrangians depend on their variables in the following way

$$L_{gr} = L_{gr}(e^{A}_{\ \lambda}, \partial_{\tau}e^{A}_{\ \lambda}, \Gamma^{AB}_{\ \ \lambda}, \partial_{\tau}\Gamma^{AB}_{\ \ \lambda}) \tag{3.2}$$

$$L_{mat} = L_{mat}(e^{A}_{\ \lambda}, \Gamma^{AB}_{\ \ \lambda}, \phi^{\{A\}}, \partial_{\tau}\phi^{\{A\}})$$

If $e = \det[e^{A}_{\ \lambda}] = (-\det g_{\mu\nu})^{\frac{1}{2}}$ is the volume element on M then

$$\tilde{L}_{gr} = eL_{gr} \quad ; \quad \tilde{L}_{mat} = eL_{mat} \quad ; \quad \tilde{L} = eL \tag{3.3}$$

denote the corresponding densities.

The field equations are the Euler-Lagrange equations for $\tilde{L}$

$$(\text{Eq.I})^{\lambda}_{\ A} = (1/e)(\delta\tilde{L}/\delta e^{A}_{\ \lambda}) = 0 \tag{3.4}$$

$$(\text{Eq.II})^{\lambda}_{\ AB} = (1/e)(\delta\tilde{L}/\delta\Gamma^{AB}_{\ \ \lambda}) = 0 \tag{3.5}$$

$$(\text{Eq.mat})_{\{A\}} = (1/e)(\delta\tilde{L}/\delta\phi^{\{A\}}) = 0 \tag{3.6}$$

Let $\underline{L} = [L^A{}_B]$ be a Lorentz matrix, i.e. $\underline{L}$ preserves the metric

$$\eta_{AB} L^A{}_E L^B{}_F = \eta_{EF} \qquad (3.7)$$

The rotation of the tetrad at $\underline{x} \in M$ performed by means of $\underline{L}$ does not affect the metric structure of M, we assume that this rotation preserves also the tensor field $\underline{\phi}$. This means, if

$$'e^A{}_\mu = \overset{-1}{L}{}^A{}_B e^B{}_\mu \; ; \quad 'e^\nu{}_A = L^B{}_A e^\nu{}_B \qquad (3.8)$$

then

$$'\phi^{A_1 \ldots A_k}{}_{B_1 \ldots B_s} = \overset{-1}{L}{}^{A_1}{}_{C_1} \ldots \overset{-1}{L}{}^{A_k}{}_{C_k} L^{E_1}{}_{B_1} \ldots L^{E_s}{}_{B_s} \phi^{C_1 \ldots C_k}{}_{E_1 \ldots E_s} \qquad (3.9)$$

Let $M \ni \underline{x} \to \underline{L}(\underline{x}) = [L^A{}_B(\underline{x})]$ be a field of Lorentz matrices (an element of the local Lorentz group). To get a correct formula for the transformed covariant derivative, we have to take the following formula for the transformed connection

$$'\Gamma^A{}_{\lambda B} = \overset{-1}{L}{}^A{}_C \Gamma^C{}_{\lambda D} L^D{}_B - (\partial_\lambda \overset{-1}{L}{}^A{}_C) L^C{}_B \qquad (3.10)$$

The transformation rules for partial derivatives of tetrad, connection and matter field can be derived from (3.8-3.10).

Remark: One can easily check that the holonomic components of the connection and the matter field do not change under the transformations(3.8-3.10). It means that the geometry of the system is invariant with respect to local Lorentz rotations.

An active coordinate transformation in M (an element of DiffM) can be in a natural way extended on the set of our variational variables. In Section 5 we give the rigorous definition of this action and prove the following results

Proposition 1

If the gravitational lagrangian L_{gr} is invariant with respect to local Lorentz transformations and the action of Diff M and if we denote

$$P^{\lambda\tau}{}_{AB} = \partial L_{gr}/\partial(\partial_\lambda \Gamma^{AB}{}_\tau) \; ; \quad U^{\lambda\tau}{}_A = \partial L_{gr}/\partial(\partial_\lambda e^A{}_\tau) \qquad (3.11)$$

then we have the following set of differential identities

$$P^{\lambda\tau}{}_{AB} = - P^{\tau\lambda}{}_{AB} \; ; \quad U^{\lambda\tau}{}_A = - U^{\tau\lambda}{}_A \qquad (3.12)$$

$$\Gamma^{AB}{}_\tau (\partial L_{gr}/\partial \Gamma^{AB}{}_\lambda) + (\partial L_{gr}/\partial e^A{}_\lambda) e^A{}_\tau + P^{\varepsilon\lambda}{}_{AB} \partial_\varepsilon \Gamma^{AB}{}_\tau + P^{\lambda\varepsilon}{}_{AB} \partial_\tau \Gamma^{AB}{}_\varepsilon + U^{\varepsilon\lambda}{}_A \partial_\varepsilon e^A{}_\tau + U^{\lambda\varepsilon}{}_A \partial_\tau e^A{}_\varepsilon = 0 \qquad (3.13)$$

$$\partial L_{gr}/\partial \Gamma^{AB}{}_\lambda + \frac{1}{2}(U^{\lambda C}{}_B \eta_{CA} - U^{\lambda C}{}_A \eta_{CB}) + P^{\tau\lambda}{}_{EA} \Gamma^E{}_{\tau B} - P^{\tau\lambda}{}_{EB} \Gamma^E{}_{\tau A} = 0 \qquad (3.14)$$

$$\frac{1}{2}(P^{\tau\lambda}{}_{EB} R^E{}_{A\tau\lambda} - P^{\tau\lambda}{}_{EA} R^E{}_{B\tau\lambda}) - \frac{1}{2}((\partial L_{gr}/\partial e^A{}_\lambda) e^E{}_\lambda \eta_{EB} - (\partial L_{gr}/\partial e^B{}_\lambda) e^E{}_\lambda \eta_{EA}) +$$

$$- \frac{1}{4}(Q^E{}_{\tau\varepsilon} U^{\tau\varepsilon}{}_A \eta_{EB} - Q^E{}_{\tau\varepsilon} U^{\tau\varepsilon}{}_B \eta_{EA}) + \frac{1}{2}(\Gamma^E{}_{\tau A} U^{\tau F}{}_E \eta_{FB} - \Gamma^E{}_{\tau B} U^{\tau F}{}_E \eta_{FA}) = 0 \qquad (3.15)$$

From (3.12),(3.14) follows

Corollary 1

The gravitational lagrangian can be expressed in terms of curvature, torsion and tetrad, i.e.

$$L_{gr}(e^{A}_{\mu}, \partial_{\lambda} e^{A}_{\mu}, \Gamma^{AB}_{\mu}, \partial_{\lambda}\Gamma^{AB}_{\mu}) = L_{1gr}(e^{A}_{\mu}, R^{A}_{B\mu\nu}, Q^{A}_{\mu\nu}) \tag{3.16}$$

Using (3.13),(3.14) and (3.16) we have

Corollary 2

$$L_{gr} = L_{2gr}(R^{A}_{BEF}, Q^{A}_{EF}) \tag{3.17}$$

Proposition 2

If the matter field lagrangian is invariant with respect to local Lorentz transformations and with respect to the action of Diff M and if we denote

$$\begin{aligned} p^{\lambda}_{\{A\}} &= \partial L_{mat}/\partial(\partial_{\lambda}\phi^{\{A\}}) \\ s^{\lambda}_{AB} &= \partial L_{mat}/\partial \Gamma^{AB}_{\lambda} \end{aligned} \tag{3.18}$$

then

$$e^{A}_{\tau}(\partial L_{mat}/\partial e^{A}_{\lambda}) + p^{\lambda}_{\{A\}}\partial_{\tau}\phi^{\{A\}} + s^{\lambda}_{AB}\Gamma^{AB}_{\tau} = 0 \tag{3.19}$$

$$s^{\lambda}_{AB} = \frac{1}{2}(v^{\lambda}_{AB} - v^{\lambda}_{BA}) \tag{3.20}$$

where

$$\begin{aligned} v^{\lambda\ B}_{\ A} = {} & p^{\lambda D_1 \ldots D_s}_{AC_2 \ldots C_k}\,\phi^{BC_2 \ldots C_k}_{D_1 \ldots D_s} + \ldots + p^{\lambda D_1 \ldots D_s}_{C_1 \ldots A}\,\phi^{C_1 \ldots B}_{D_1 \ldots D_s} + \\ & - p^{\lambda BD_2 \ldots D_s}_{C_1 \ldots C_k}\,\phi^{C_1 \ldots C_k}_{AD_2 \ldots D_s} - \ldots - p^{\lambda D_1 \ldots B}_{C_1 \ldots C_k}\,\phi^{C_1 \ldots C_k}_{D_1 \ldots A} \end{aligned} \tag{3.21}$$

From (3.20),(3.21) follows

Corollary 3

$$L_{mat}(e^{A}_{\mu}, \Gamma^{AB}_{\mu}, \phi^{\{A\}}, \partial_{\lambda}\phi^{\{A\}}) = L_{1mat}(e^{A}_{\mu}, \phi^{\{A\}}, D_{\lambda}\phi^{\{A\}}) \tag{3.22}$$

i.e. the matter field lagrangian can be expressed as a function of the matter field, its covariant derivatives and the tetrad.

Remark: In the literature s^{λ}_{AB} is called the spin tensor of the field ϕ cf. [7-10, 22].

Let

$${}_{cm}T^{\alpha}_{\ \beta} = \delta^{\alpha}_{\beta} L_{mat} - p^{\alpha}_{\{A\}} D_{\beta}\phi^{\{A\}} \tag{3.23}$$

be the canonical energy-momentum tensor of the matter field. We have from (3.19),(3.20) and (3.21)

$${}_{cm}T^{\alpha}_{\ \beta} = \partial(eL_{mat})/\partial e^{A}_{\alpha}\cdot e^{A}_{\beta}\frac{1}{e} = \partial(eL_{1mat})/\partial e^{A}_{\alpha}\cdot e^{A}_{\beta}\frac{1}{e} \tag{3.24}$$

Proposition 3

If the matter field lagrangian is invariant with respect to local Lorentz transformations and if the matter field equations (3.6) are satisfied, then

$$D_{\tau}v^{\tau}_{[AB]} - Q^{\varepsilon}_{\varepsilon\tau}v^{\tau}_{[AB]} - {}_{cm}T_{[AB]} = 0 \tag{3.25}$$

By analogy with (3.23) we define the canonical energy momentum tensor of the gravitational field

$$ {}_{cg}T^{\tau}{}_{\beta} = \delta^{\tau}_{\beta} L_{gr} - P^{\tau\mu}{}_{CD} R^{CD}{}_{\beta\mu} - U^{\tau\mu}{}_{C} Q^{C}{}_{\beta\mu} \qquad (3.26) $$

combining (3.13)-(3.15) we get

$$ {}_{cg}T_{[AB]} + \frac{1}{2}(P^{\tau\lambda}{}_{EB} R^{E}{}_{A\tau\lambda} - P^{\tau\lambda}{}_{EA} R^{E}{}_{B\tau\lambda}) + $$

$$ - \frac{1}{4}(Q^{E}{}_{\tau\varepsilon} U^{\tau\varepsilon}{}_{A} \eta_{EB} - Q^{E}{}_{\tau\varepsilon} U^{\tau\varepsilon}{}_{B} \eta_{EA}) = 0 \qquad (3.27) $$

$$ {}_{cg}T^{\tau}{}_{\beta} = \frac{1}{e}(\partial(eL_{gr})/\partial e^{A}_{\tau}) e^{A}_{\beta} + U^{\tau\varepsilon}{}_{C} \Gamma^{C}{}_{\varepsilon A} e^{A}_{\beta} = \frac{1}{e}(\partial(eL_{Igr})/\partial e^{A}_{\tau}) e^{A}_{\beta} \qquad (3.28) $$

Remark: The differential identities of the Belinfante-Rosenfeld type presented in Propositions 1-3 coincide with those given by Hehl et al. [7-10].

4. THE CONTRACTED BIANCHI IDENTITIES

In the previous Sections we have used the covariant derivative D_μ defined by means of the connection $\underline{\Gamma}$. However, because space time is equipped with a metric tensor $\underline{g}$ we have also the natural riemannian connection $\{^{\lambda}_{\mu\nu}\}$ corresponding to $g_{\mu\nu}$ and the riemannian covariant derivative ∇_μ. It turns out that some formulas look simpler if we use ∇ instead of D, especially those which are related to integral formulas (conservation laws cf. Section 6).

Now we prove the following

Theorem 1

If the equations (3.5) and (3.6) are satisfied, then

$$ \nabla_{\mu}(\mathrm{Eq.I})^{\mu}_{A} = 0 \qquad (4.1) $$

The proof is based on the following Lemmas

Lemma 1

If the matter field equations (3.6) are satisfied, then

$$ \nabla_{\mu}({}_{cm}T^{\mu}{}_{\nu}) = s^{\alpha\beta}{}_{\tau} R^{\tau}{}_{\beta\nu\alpha} + r^{\omega}{}_{\nu\mu} g_{\omega\lambda} (D_{\tau}(s^{\tau\lambda}{}_{\sigma} g^{\sigma\mu}) - Q^{\varepsilon}{}_{\varepsilon\tau} s^{\tau\lambda}{}_{\sigma} g^{\sigma\mu}) \qquad (4.2) $$

Lemma 1 follows from (2.13),(3.20),(3.25),(3.23).

Let be

$$ (\mathrm{Eq.GI})^{\lambda}_{A} = (1/e)(\delta \tilde{L}_{gr} / \delta e^{A}_{\lambda}) \qquad (4.3) $$

$$ (\mathrm{Eq.GII})^{\lambda}_{AB} = (1/e)(\delta \tilde{L}_{gr} / \delta \Gamma^{AB}_{\lambda}) \qquad (4.4) $$

We have

$$ (\mathrm{Eq.GI})^{\lambda}_{A} = {}_{cg}T^{\lambda}{}_{A} + (- D_{\tau}U^{\tau\lambda}{}_{A} + Q^{\tau}{}_{\tau\varepsilon} U^{\varepsilon\lambda}{}_{A} + \frac{1}{2}Q^{\lambda}{}_{\tau\varepsilon} U^{\tau\varepsilon}{}_{A}) \qquad (4.5) $$

$$ (\mathrm{Eq.GII})^{\lambda}_{AB} = \frac{1}{2}(U^{\lambda C}{}_{A}\eta_{CB} - U^{\lambda C}{}_{B}\eta_{CA}) + $$

$$ + (- D_{\tau}P^{\tau\lambda}{}_{AB} + Q^{\tau}_{\tau\varepsilon}P^{\varepsilon\lambda}{}_{AB} + \frac{1}{2}Q^{\lambda}_{\tau\varepsilon}P^{\tau\varepsilon}{}_{AB}) \qquad (4.6) $$

$$(\mathrm{Eq.I})^{\lambda}_{A} = (\mathrm{Eq.GI})^{\lambda}_{A} + {}_{cm}T^{\lambda}_{A} \tag{4.7}$$

$$(\mathrm{Eq.II})^{\lambda}_{AB} = (\mathrm{Eq.GII})^{\lambda}_{AB} + s^{\lambda}_{AB} \tag{4.8}$$

We get from (3.13),(314),(3.26),(3.27)

Lemma 2

$$\nabla_{\tau}({}_{cg}T^{\tau}{}_{\beta}) = r^{\alpha}_{\beta\lambda}({}_{cg}T^{\lambda}{}_{\alpha}) + (\mathrm{Eq.GII})^{\lambda}_{AB}R^{AB}{}_{\beta\lambda} + \frac{1}{2}U^{\tau\mu}{}_{\varepsilon}R^{\varepsilon}{}_{\beta\tau\mu} + \left(- D_{\tau}U^{\tau\lambda}{}_{\alpha} + Q^{\tau}_{\tau\varepsilon}U^{\varepsilon\lambda}{}_{\alpha} + \frac{1}{2}Q^{\lambda}_{\tau\varepsilon}U^{\tau\varepsilon}{}_{\alpha} \right) Q^{\alpha}_{\beta\lambda} \tag{4.9}$$

Using (2.13),(3.27),(4.5) and (4.9) we have

Lemma 3

$$\nabla_{\lambda}(\mathrm{Eq.GI})^{\lambda}_{\alpha} = (\mathrm{Eq.GII})^{\lambda}_{\varepsilon\omega}g^{\omega\tau}R^{\varepsilon}{}_{\tau\alpha\lambda} + r^{\beta}_{\alpha\omega}g_{\beta\sigma}\left(D_{\varepsilon}(\mathrm{Eq.GII})^{\varepsilon}_{\mu\nu}g^{\mu\omega}g^{\nu\sigma} - Q^{\tau}_{\tau\varepsilon}(\mathrm{Eq.GII})^{\varepsilon}_{\mu\nu}g^{\mu\omega}g^{\nu\sigma} \right) \tag{4.10}$$

We see that (4.1) follows from (4.2),(4.7) and (4.10).

From the results of this Section we are able to get more information concerning the divergences of the left sides of gravitational field equations. From (3.25),(3.20),(3.27),(4.7),(4.8) and (4.10) we have

$$\nabla_{\lambda}(\mathrm{Eq.I})^{\lambda}_{\alpha} = (\mathrm{Eq.II})^{\lambda}_{\varepsilon\tau}R^{\varepsilon\tau}{}_{\alpha\lambda} + \frac{1}{2}r^{\sigma}_{\alpha\tau}g^{\tau\omega}\left((\mathrm{Eq.I})^{\beta}_{\sigma}g_{\beta\omega} - (\mathrm{Eq.I})^{\beta}_{\omega}g_{\beta\sigma}\right) \tag{4.11}$$

$$\nabla_{\lambda}(\mathrm{Eq.II})^{\lambda}_{\nu\mu} - r^{\alpha}_{\lambda\nu}(\mathrm{Eq.II})^{\lambda}_{\alpha\mu} + r^{\alpha}_{\lambda\mu}(\mathrm{Eq.II})^{\lambda}_{\alpha\nu} = \frac{1}{2}\left((\mathrm{Eq.I})^{\alpha}_{\mu}g_{\alpha\nu} - (\mathrm{Eq.I})^{\alpha}_{\nu}g_{\alpha\mu} \right) \tag{4.12}$$

Theorem 2

If the matter field equations (3.6) are satisfied, then the divergence of the left sides of the gravitational field equations (3.4),(3.5) are linear combinations of the left sides of the gravitational field equations.

Remark: For the Einstein-Cartan theory this result was proved earlier by Trautman ,[22] .

In the next paper we will show that Theorems 1 and 2 allow us to split the system (3.5),(3.6) into dynamical equations and constraints and that such a splitting is important for the initial value formulation of the theory (cf. [19,20]).

5. THE MULTISYMPLECTIC BUNDLE AND THE ACTION OF THE GAUGE GROUP

In Section 3 we discussed how tetrads, tensor field components and connection components change under local Lorentz rotations. In the present Section we will show that the local Lorentz group and the group Diff M act in a natural way in some bundle $\mathcal{P}$. The bundle $\tau : \mathcal{P} \to M$ describes all dynamical variables of the theory and their first partial derivatives. According to the general procedure to define $\mathcal{P}$ one has to specify local coordinates in fibres and their transformation properties with respect to a change of local coordinates in M.

Local coordinates in the fibre over $\underline{x} \in M$ are

$$(e^A_\lambda ; \Gamma^{AB}_\lambda ; e^A_{\lambda',\tau} ; \Gamma^{AB}_{\lambda',\tau} ; \phi^{\{A\}} ; \phi^{\{A\}}_{',\tau}) \tag{5.1}$$

where e^A_λ Γ^{AB}_λ and $\phi^{\{A\}}_{',\tau}$ transform like covectors under transformations of local coordinates in M; $\phi^{\{A\}}$ transform like scalars and $e^A_{\lambda',\tau}$ $\Gamma^{AB}_{\lambda',\tau}$ transform like partial derivatives of covectors.

The following formulas define two differential 4-forms on $\mathcal{P}$ (the Hamilton-Cartan forms) cf. [6, 11]

$$\begin{aligned}\Theta_{gr} &= eU^{\tau\mu}_A dx^o \wedge \ldots \wedge de^A_\mu \wedge \ldots \wedge dx^3 + \\ &+ eP^{\tau\mu}_{AB} dx^o \wedge \ldots \wedge d\Gamma^{AB}_\mu \wedge \ldots \wedge dx^3 + \\ &- e(U^{\tau\mu}_A e^A_{\mu',\tau} + P^{\tau\mu}_{AB}\Gamma^{AB}_{\mu',\tau} - L_{gr})dx^o \wedge \ldots \wedge dx^3\end{aligned} \tag{5.2}$$

$$\begin{aligned}\Theta_{mat} &= ep^\tau_{\{A\}} dx^o \wedge \ldots \wedge d\phi^{\{A\}} \wedge \ldots \wedge dx^3 + \\ &- e(p^\tau_{\{A\}}\phi^{\{A\}}_{',\tau} - L_{mat})dx^o \wedge \ldots \wedge dx^3\end{aligned} \tag{5.3}$$

Let locLor be the set of space time depending Lorentz matrices

$$M \ni \underline{x} \to [L^A_B(\underline{x})] \in SO(1,3)$$

This set has a natural group structure and we call it the local Lorentz group. We have the action of locLor in $\mathcal{P}$

$$\begin{aligned}A_{Lor}([L^A_B(\cdot)])(x^\lambda;e^A_\mu;\Gamma^{AB}_\mu;e^A_{\mu',\tau};\Gamma^{AB}_{\mu',\tau};\phi^{\{A\}};\phi^{\{A\}}_{',\tau}) &= \\ = (x^\lambda;'e^A_\mu;'\Gamma^{AB}_\mu;'e^A_{\mu',\tau};'\Gamma^{AB}_{\mu',\tau};'\phi^{\{A\}};'\phi^{\{A\}}_{',\tau})\end{aligned} \tag{5.4}$$

where the transformed variables are given by equations (3.8-10) and their derivatives , respectively.

The action of Diff M in the bundle $\mathcal{P}$ is given in the following way cf. [19]. If $\Psi \in$ Diff M ; $'\underline{x} = \Psi(\underline{x})$; $'x^\lambda = 'x^\lambda(x^\tau)$, then

$$A_{Diff}(\Psi)(x^\lambda;e^A_\mu;\ldots\ldots) = ('x^\lambda;'e^A_\mu;\ldots\ldots) \tag{5.5}$$

where

$$'\phi^{\{A\}} = \phi^{\{A\}} ; \quad 'e^A_\lambda = (\partial x^\tau/\partial 'x^\lambda)e^A_\tau ; \quad '\Gamma^{AB}_\lambda = (\partial x^\tau/\partial 'x^\lambda)\Gamma^{AB}_\tau \text{ etc.} \tag{5.6}$$

The Lie algebra of locLor consists of space time depending skew-symmetric matrices $[\delta L^{AB}(\cdot)]$, the Lie algebra of Diff M consists of all smooth vector fields $\underline{Z} = \delta x^\mu(\cdot)\, \partial/\partial x^\mu$ on M.

Every element $[\delta L^{AB}] \in$ alglocLor defines the vector field on $\mathcal{P}$

$$X_{Lor}([\delta L^{AB}]) = \delta e^A_\lambda\, \partial/\partial e^A_\lambda + \delta\Gamma^{AB}_\lambda\, \partial/\partial\Gamma^{AB}_\lambda + \delta\phi^{\{A\}}\, \partial/\partial\phi^{\{A\}} +$$

$$+ \quad \delta e^{A}_{\lambda',\tau} \; \partial/\partial e^{A}_{\lambda',\tau} \; + \; \delta \Gamma^{AB}_{\lambda\; ',\tau} \; \partial/\partial \Gamma^{AB}_{\lambda\; ',\tau} \; + \; \delta \phi^{\{A\}}{}_{,\tau} \; \partial/\partial \phi^{\{A\}}{}_{,\tau} \tag{5.7}$$

where

$$\delta e^{A}_{\lambda} = -\delta L^{A}{}_{B} e^{B}_{\lambda} \quad ; \quad \delta \Gamma^{AB}_{\lambda} = D_{\lambda} \delta L^{AB},$$

$$\delta \phi^{A} = - \delta L^{A}{}_{B} \phi^{B} \quad ; \quad \delta \phi_{A} = \delta L^{B}{}_{A} \phi_{B} \qquad \text{etc} \tag{5.8}$$

Every element $\underline{Z} \in$ algDiff M , $\underline{Z} = \delta x^{\lambda} \partial/\partial x^{\lambda}$ defines vector field on $\mathcal{P}$

$$X_{Diff}(\underline{Z}) = \delta x^{\lambda} \partial/\partial x^{\lambda} + \delta e^{A}_{\mu} \; \partial/\partial e^{A}_{\mu} + \ldots\ldots\ldots \tag{5.9}$$

where

$$\delta e^{A}_{\lambda} = - e^{A}_{\tau} \partial_{\lambda} \delta x^{\tau} \;\; ; \;\; \delta \Gamma^{AB}_{\lambda} = - \Gamma^{AB}_{\tau} \partial_{\lambda} \delta x^{\tau} \;\; ; \;\; \delta \phi^{\{A\}} = 0 \text{ etc} \tag{5.10}$$

The invariance properties of lagrangians give rise to the invariance properties of Θ_{gr} and Θ_{mat} . This means that the Lie derivatives of these 4-forms with respect to the fields X_{Lor}, X_{Diff} vanish:

$$\mathcal{L}_{X_{Lor}} \Theta_{gr} = \mathcal{L}_{X_{Diff}} \Theta_{gr} = 0$$
$$\mathcal{L}_{X_{Diff}} \Theta_{mat} = \mathcal{L}_{X_{Lor}} \Theta_{mat} = 0 \tag{5.11}$$

By means of the known formula

$$\mathcal{L}_{X} \Theta = d(X \lrcorner \Theta) + X \lrcorner d\Theta \tag{5.12}$$

one can derive from (5.11) all the identities given in Propositions 1,2 and 3.

Remark: It is natural to define the local Poincaré group as the semidirect product of the local Lorentz group and Diff M. The commutation relations for the corresponding Lie algebra are given in [7,9] . Therefore we speak about a gauge theory of the local Poincaré group.

6.NOETHER FORMS AND CONSERVATION LAWS

We prove that invariance properties of lagrangians with respect to the actions of the local Lorentz group and of the diffeomorphism group of space time give rise to the existence of two conserved quantities J^{λ} and E^{λ} , respectively. There is a relation between the conservation laws for these quantities and the field equations. Let be given a field of tetrads, a connection and a tensor field on M

$$\underline{x} \to e^{A}_{\mu}(\underline{x}) \quad ; \quad \underline{x} \to \Gamma^{AB}_{\mu}(\underline{x}) \quad ; \quad \underline{x} \to \phi^{\{A\}}(\underline{x}) \tag{6.1}$$

Therefore we have a section f of the bundle $\mathcal{P}$ satisfying

$$e^{A}_{\mu',\tau} = \partial_{\tau} e^{A}_{\mu} \quad ; \quad \Gamma^{AB}_{\mu\; ',\tau} = \partial_{\tau} \Gamma^{AB}_{\mu} \quad ; \quad \phi^{\{A\}}{}_{,\tau} = \partial_{\tau} \phi^{\{A\}} \tag{6.2}$$

Let

$$\Theta_{H-C} = \Theta_{gr} + \Theta_{mat} \tag{6.3}$$

be the Hamilton- Cartan 4-form of the system.
We define the energy-momentum 3-form on $\mathcal{P}$

$$\nu_{\underline{Z}} = - X_{Diff}(\underline{Z}) \lrcorner \Theta_{H-C} \tag{6.4}$$

where $X_{Diff}(\underline{Z})$ is defined by (5.9-10).
The pull-back of $\nu_{\underline{Z}}$ on M (i.e. we substitute (6.1) into 6.4)) defines a 3-form on M

$$\underline{E} = f^{*}(\nu_{\underline{Z}}) = \sum_{\tau} (-1)^{\tau} eE^{\tau} dx^{o} \wedge \ldots_{\tau} \ldots \wedge dx^{3} \qquad (6.5)$$

where E^{τ} is a vector field on M.

Similarly the spin 3 form on $\mathscr{P}$ is given by (see (5.7-8))

$$\mu_{[\delta L^{AB}]} = - X_{Lor}([\delta L^{AB}]) \lrcorner \Theta_{H-C} \qquad (6.6)$$

and the corresponding pull-back defines a 3 form on M

$$\underline{J} = f^{*}(\mu_{[\delta L^{AB}]}) = \sum_{\tau} (-1)^{\tau} eJ^{\tau} dx^{o} \wedge \ldots_{\tau} \ldots \wedge dx^{3} \qquad (6.7)$$

where J^{τ} is a vector field on M.

Proposition 4

Let f be a section of $\mathscr{P}$ defined by (6.1-2) and $\underline{Z} = \delta x^{\lambda} \partial/\partial x^{\lambda}$, then

$$E^{\tau} = - (Eq.I)^{\tau}_{\sigma} \delta x^{\sigma} - (Eq.II)^{\tau}_{AB} \Gamma^{AB}_{\sigma} \delta x^{\sigma} + - \nabla_{\lambda}(P^{\lambda\tau}{}_{AB} \Gamma^{AB}_{\sigma} \delta x^{\sigma}) - \nabla_{\lambda}(U^{\lambda\tau}_{\sigma} \delta x^{\sigma}) \qquad (6.8)$$

$$J^{\tau} = (Eq.II)^{\tau}_{AB} \delta L^{AB} + \nabla_{\lambda}(P^{\lambda\tau}{}_{AB} \delta L^{AB}) \qquad (6.9)$$

Proposition 5

If the field equations (3.4)-(3.6) are satisfied, then

$$\nabla_{\lambda} E^{\lambda} = 0 \quad ; \quad \nabla_{\lambda} J^{\lambda} = 0 \qquad (6.10)$$

Remark: $\nabla_{\mu}\nabla_{\nu} b^{\mu\nu} = 0$ for any skew-symmetric tensor field $b^{\mu\nu}$.

Also the converse result is true

Theorem 3

If f is a section of $\mathscr{P}$ given by (6.1),(6.2) and

(i) the matter field equations (3.6) are satisfied

(ii) for every $\underline{Z} \in$ algDiff M, $[\delta L^{AB}] \in$ alglocLor the corresponding quantities E^{τ}, J^{τ} satisfy the conservation laws (6.10), then the field equations (3.4),(3.5) are also satisfied.

At the end we would like to comment on the results of this Section.

(i) The 3-forms ν, μ were defined according to the general definition of Noether forms (cf. [6], [11]). The quantity E^{τ} generalizes that given by Komar, [13].

(ii) For a theory of gravity based on the Einstein gravitational lagrangian (with symmetric, but not metric compatible connection) Kijowski, [12] defines a E^{τ} which coincides with ours (cf.also [21]) In his approach the equation $\nabla_{\tau} E^{\tau} = 0$ is equivalent to the gravitational field equations (provided the matter field equations are satisfied). If one generalizes that result to the metric-affine theory with torsion [21], the conservation of E^{τ} does not imply the field equation. (There is no possibility to define the quantity J^{τ} in such an approach.

(iii) It would be interesting to investigate the conservation laws in the metric-affine gauge theory of gravity (with the local affine group as the gauge group). A result corresponding to that of Theorem should be valid there.
(iv) In the next paper we will show that the quantities $\underline{E}$, $\underline{J}$ are the generators of the dynamics (time evolution) of the system under consideration (cf. [21]).

REFERENCES

1. R.Adler, M.Bazin and M.Schiffer, Introduction to general relativity, McGraw Hill, New York (1965).
2. R.Arnowitt, S.Deser and C.W.Misner, The dynamics of general relativity, in: Gravitation-an introduction to current research, L.Witten. ed., John Wiley, New York (1962).
3. E.Fairchild,Jr., Phys.Rev 14D:384 (1976) and Phys.Rev. 16D:2438 (1977).
4. A.Fischer and J.Marsden, GRG 7:915 (1976).
5. A.Fischer and J.Marsden, The initial value problem and the dynamical formulation of general relativity, in: General relativity. An Einstein centenary survey, S.Hawking and W.Israel eds. Cambridge University Press (1979).
6. H.Goldschmidt and S.Sternberg, Ann.Inst.Fourier (Grenoble) 23:203 (1973).
7. F.Hehl, P.von der Heyde, G.Kerlick and J.Nester, Rev.Mod.Phys. 48 :393 (1976).
8. F.Hehl, Y.Ne´eman, J.Nitsch and P.von der Heyde, Phys.Lett. 78B: :102 (1978).
9. F.Hehl, J.Nitsch and P.von der Heyde, Gravitation and Poincaré gauge field theory with quadratic lagrangian, in: Einstein Volume Plenum Press (to appear 1979).
10. P.von der Heyde, Phys.Lett. 58A:141 (1976).
11. J.Kijowski, Comm.Math.Phys. 30:99 (1973).
12. J.Kijowski, GRG 9:857 (1978).
13. A.Komar, Phys. Rev. 113:934 (1959).
14. F.Mansouri and L.N.Chang, Phys.Rev. 13D:3192 (1976).
15. C.Misner, K.Thorne and J.Wheeler, Gravitation, W.H.Freeman, San Francisco (1973).
16. Y.Ne´eman and T.Regge, Rivista del Nuovo Cimento 1,n5 (1978).
17. J.A.Schouten, Ricci Calculus, Springer, Berlin-Heidelberg-N.Y (19
18. W.Szczyrba, Comm.Math.Phys. 51:163 (1976).
19. W.Szczyrba, Lett.Math.Phys. 2:265 (1978).
20. W.Szczyrba, Comm.Math.Phys. 60:215 (1978).
21. W.Szczyrba, A hamiltonian structure of the interacting gravitational and matter fields (preprint 1978).
22. A.Trautman, Symposia Math. 12:139 (1973).
23. A.Trautman, Elementary introduction to fibre bundles and gauge fields, Warsaw University - preprint (1978).
24. C.N.Yang, Phys.Rev.Lett. 33: 445 (1974).

THE GAUGE SYMMETRIES OF GRAVITATION

Martin A. Schweizer

Institute for Theoretical Physics
University of Zurich
Schönberggasse 9 , 8001 Zürich , Switzerland

1 Local Lorentzinvariance and the Cauchy problem

In a usual gauge theory with internal symmetry group G matter fields are represented by tensor p-forms Ψ of type (U,E) , U denoting the representation of G (in a linear space E) according to which the section Ψ is transformed under a local gauge transformation $g(x)\varepsilon G$:

$$\Psi'(x) = U(g(x))\Psi(x) \tag{1. 1}$$

If θ^ν is a local frame of 1-forms , we can expand Ψ in this frame

$$\Psi = 1/p! \, \Psi_{\mu_1 \ldots\ldots \mu_p} \theta^{\mu_1} \wedge \ldots\ldots \wedge \theta^{\mu_p} \tag{1. 2}$$

Furthermore let $X_1,\ldots,X_k$ be a basis of E , then

$$\Psi_{\mu_1 \ldots\ldots \mu_p} = \Sigma_a \Psi^a_{\mu_1 \ldots\ldots \mu_p} X_a \tag{1. 3}$$

Thus we see that Ψ carries two types of indices :

(i) External indices (μ_1 ,..., μ_p) transforming like usual covariant tensor-indices.

(ii) Internal indices (a) transforming according to the group representation $U^a_b(g(x))$.

(iii) Internal and external transformations can be performed independently one of each other .

(iv) The indices μ , ν etc. of θ^μ , $\theta^\mu \wedge \theta^\nu$ etc. and of its duals η^μ , $\eta^{\mu\nu}$ etc. do not transform under G !

(v) Internal and external symmetry are properly separated .

In such a gauge theory the dynamical elements are the gauge potential $A = A^i T_i$ (T_i a basis of Lie(G)) and the matter fields Ψ . Thus any Lagrangian for the coupled equations will be a functional of (the components of) F , Ψ , $D\Psi$, i.e. it will be a 4-form

$$L = L(F ,\Psi , D\Psi) \tag{1. 4}$$

whereby $F = dA + 1/2\, A^i \wedge A^j\, [T_i , T_j]$ is the YM-field strength . We say that (1. 4) is locally gauge invariant if

$$L(F,\Psi,D\Psi) = L(Ad_G(g(x))F, U(g(x))\Psi, U(g(x))D\Psi) \tag{1. 5}$$

for all local gauge transformations $g(x) \varepsilon G$.

Now we turn to the case where the gauge group is the orthochronous Lorentzgroup $\Lambda^{\uparrow}_{+}$ respectively its universal covering group $SL(2,\mathbb{C})$.

Let $\Lambda(x) = (\Lambda^{\mu}{}_{\nu}(x)) \varepsilon \Lambda^{\uparrow}_{+}$ always denote a local Lorentztransformation . In analogy with (1. 1) we now have a transformation law for the spin-indices of the matter fields :

$$\Psi'(x) = S(\Lambda(x))\Psi(x) \tag{1. 6}$$

From now on the frames θ^{ν} appearing explicitly in the formulas are by definition supposed to be orthonormal (the same shall be true for the duals η^{ν} , $\eta^{\mu\nu}$ etc.) . Indices are lowered and raised by the flat metric $\hat{\eta}_{\alpha\beta}$.

In contrast to usual gauge theory the index μ of θ^{μ} now does transform under $\Lambda^{\uparrow}_{+}$

$$\theta'^{\mu}(x) = \Lambda^{\mu}{}_{\nu}(x)\, \theta^{\nu}(x) \tag{1. 7}$$

The same is true for the duals η^{μ} , $\eta^{\mu\nu}$ etc. Furthermore let $\omega^{\alpha}{}_{\beta}$ denote the components of a linear connection , which is metric with respect to

$$g = \hat{\eta}_{\mu\nu}\, \theta^{\mu} \otimes \theta^{\nu} \tag{1. 8}$$

The transformation law is

$$\omega'^{\alpha}{}_{\beta}(x) = \Lambda^{\alpha}{}_{\lambda}(x)\, \omega^{\lambda}{}_{\sigma}(x)\, \overset{-1}{\Lambda}{}^{\sigma}{}_{\beta}(x) - d\Lambda^{\alpha}{}_{\lambda}(x)\, \overset{-1}{\Lambda}{}^{\lambda}{}_{\beta}(x) \tag{1. 9}$$

The corresponding curvature forms

$$\Omega^{\alpha}{}_{\beta} = d\omega^{\alpha}{}_{\beta} + \omega^{\alpha}{}_{\lambda} \wedge \omega^{\lambda}{}_{\beta} = 1/2\, R^{\alpha}{}_{\beta\mu\nu}\, \theta^{\mu} \wedge \theta^{\nu} \tag{1.10}$$

transform like

$$\Omega'^{\alpha}{}_{\beta} = \Lambda^{\alpha}{}_{\lambda}\, \overset{-1}{\Lambda}{}^{\sigma}{}_{\beta}\, \Omega^{\lambda}{}_{\sigma} \tag{1.11}$$

The torsion 2-forms

$$\Theta^{\mu} = d\theta^{\mu} + \omega^{\mu}{}_{\lambda} \wedge \theta^{\lambda} \tag{1.12}$$

transform exactly like the frames , i.e.

$$\Theta'^{\mu} = \Lambda^{\mu}{}_{\nu}\, \Theta^{\nu} \tag{1.13}$$

Finally we mention that the exterior covariant derivatives of the matter fields transform exactly like the fields themselves

$$(D\Psi)'(x) = S(\Lambda(x))\, D\Psi(x) \tag{1.14}$$

The dynamical elements in a $\Lambda^{\uparrow}_{+}$ - gauge theory are $\omega^{\alpha}{}_{\beta}$, Ψ and in addition the frames θ^{ν} , and thus any Lagrangian for the coupled equations will be a functional of (the components of) $\Omega^{\alpha}{}_{\beta}$, θ^{ν} , $D\theta^{\nu}$, Ψ , $D\Psi$, i.e. it will be a 4-form

$$L = L(\Omega^{\alpha}{}_{\beta} , \theta^{\mu} , D\theta^{\mu} , \Psi , D\Psi) \tag{1.15}$$

A comparison of (1. 4) and (1.15) shows that (1.15) has more dynamical elements . We first give a definition which is the strict analogon of (1. 5) :

a) L satisfies the *strong local Lorentzinvariance* , if for fixed θ^μ , $\theta^\mu \wedge \theta^\nu$ etc. L is invariant under the simultaneous transformations (1. 6 ; 9 ; 11 ; 14). (1.16)

An example is a massive scalar field coupled with the "YM-gravitation "

$$L = -1/2\ \Omega^\alpha{}_\beta \wedge *\Omega^\alpha{}_\beta + \kappa/2\ (d\phi \wedge *d\phi + m^2 \cdot \phi^2 \cdot \eta) \tag{1.17}$$

In (1.17) the frames and their duals just do not appear explicitly and the volume η is unchangend under Lorentzrotations of the frames. Furthermore it is easy to see that the Dirac Lagrangian $L = i\overline{\Psi}\gamma^\mu D\Psi \wedge \eta_\mu - m\ \overline{\Psi}\Psi$ does not have property (1.16) . Consequentely we must give a weaker definition :

b) L satisfies the *weak local Lorentzinvariance* , if L is invariant under the simultaneous transformations (1.6 ; 7 ; 9 ; 11 ; 14) (1.18)

The Dirac Lagrangian is invariant in this weaker sense. A transformation of the spin-indices of Ψ induces a transformation of the vector-spin index μ of γ and this latter index is contracted with η_μ. Thus we have a coupling of internal and external symmetry .
The following example

$$L = 1/2\ \eta_{\mu\nu} \wedge \Omega^{\mu\nu} + \kappa/2\ (d\phi \wedge *d\phi + m^2 \cdot \phi^2 \cdot \eta) \tag{1.19}$$

represents a massive scalar particle in GRT . This Lagrangian obviously does not have property a) but only b) , since the indices μ , ν of $\eta_{\mu\nu} = *(\theta_\mu \wedge \theta_\nu)$ cannot be kept fixed when (1.11) is applied to (1.19)

The following considerations on the level of *infinitesimal Lorentz-transformations* may help to make even more clear the difference between a) and b) . For this purpose let L fulfil property b). With the equations for the matter fields we find (modulo exact differentials)

$$\delta L = \delta\theta^\mu \wedge E_\mu + 1/2\ \delta\omega^\alpha{}_\beta \wedge S_\alpha{}^\beta = 0 \tag{1.20}$$

$E_\mu = \varepsilon_\mu + t_\mu \qquad S_{\alpha\beta} = \sigma_{\alpha\beta} + s_{\alpha\beta}$

ε_μ : energy-stress of geometry
t_μ : energy-stress of matter
$\sigma_{\alpha\beta}$: spin-density of geometry
$s_{\alpha\beta}$: spin-density of matter
$E_\mu = T_\mu{}^\nu\ \eta_\nu$: $T_\mu{}^\nu$ is the total canonical energy-stress tensor
If L satisfies condition a) , we get *no contribution* from the local (infinitesimal) Lorentzrotations of the frames , i.e.

$$\delta\theta^\mu \wedge E_\mu = 0 \tag{1.21}$$

and consequentely

$$\delta\omega^{\alpha\beta} \wedge S_{\alpha\beta} = 0 \tag{1.22}$$

Conversely if (1.21) and (1.22) hold , then L has property a) . Integration of (1.20) over compact regions yields a first identity which connects energy-stress with spin :

$$DS^{\alpha\beta} = \theta^\alpha \wedge E^\beta - \theta^\beta \wedge E^\alpha = (B^{\alpha\beta})^\mu{}_\nu \cdot T^\nu{}_\mu \cdot \eta \tag{1.23}$$

where $B^{\alpha\beta} = -B^{\beta\alpha}$ is the canonical basis in so(1,3) .If L satisfies a) , each side of (1.23) is separately equal to zero , i.e.

$$DS^{\alpha\beta} = 0 \quad \text{and} \quad T^{\mu\nu} = T^{\nu\mu}$$

Now we come to the crucial point of this discussion : If L has property a) , we know from (1.22) that

$$\delta_\omega L = 0 \tag{1.24}$$

for all variations $\delta\omega^\alpha{}_\beta = - Dh^\alpha{}_\beta$, $h_{\alpha\beta}(x) = - h_{\beta\alpha}(x)$.
There is a unique decomposition of $\omega^\alpha{}_\beta$ into the Levi-Civita part $\tilde{\Gamma}^\alpha{}_\beta$ plus the contorsion tensor 1-form $\tau^\alpha{}_\beta$, i.e.

$$\omega^\alpha{}_\beta = \tilde{\Gamma}^\alpha{}_\beta + \tau^\alpha{}_\beta$$

Now we are invited to interpret (1.24) in the following way :
L is invariant under substitutions

$$\tilde{\Gamma}^\alpha{}_\beta + \tau^\alpha{}_\beta \longrightarrow \tilde{\Gamma}^\alpha{}_\beta + (\tau^\alpha{}_\beta - Dh^\alpha{}_\beta) \tag{1.25}$$

But $Dh^\alpha{}_\beta$ is a tensor 1-form of the same type as $\tau^\alpha{}_\beta$ and so (1.25) amounts to the statement that the contorsion $\tau^\alpha{}_\beta$ can be replaced by $\tilde{\tau}^\alpha{}_\beta = \tau^\alpha{}_\beta - Dh^\alpha{}_\beta$ without affecting L ! But if this is true , the system of differential equations resulting from variations with respect to θ^ν and $\omega^\alpha{}_\beta$ cannot be complete!

Indeed the field equations resulting from (1.17) show a pathological behaviour , since the Cauchy-problem does in general not have a unique solution . Any Weitzenböck-space ($\Omega^\alpha{}_\beta = 0$) is a vacuum solution of those equations and defining in parallel frames $h^\alpha{}_\beta$ by $h^0{}_1 = h^1{}_0 = f$ (all other components are equal to zero , and f is an arbitrary function), one gets in these unchanged parallel frames a new solution $\tilde{\omega}^\alpha{}_\beta = -Dh^\alpha{}_\beta$ of the vacuum equations . For a suitable f this new solution fulfils the identical same initial conditions ! The curvature does not change but the torsion does !
Conclusion : As soon as torsion is dynamical,a Lagrange gauge theory of gravitation must be of the weak type b) .

2 The Poincaré-group as a gauge-group

In the geometry of principal fibre-bundles[1] it is possible to interpret the frames θ^ν as connection forms ; more precisely ($\omega^\alpha{}_\beta$, θ^ν) is interpreted as the connection forms of an affine Poincaré-connection . Thus the $\omega^\alpha{}_\beta$ are so(1,3)-valued rotational gauge potentials and the θ^α are $\mathbb{R}^4$-valued translational gauge potentials .
In the fibre-bundle concept the action of the translation-group is trivial and so from a physical point of view completely uninteresting . In spite of this there have been attempts[2] to gauge $\mathbb{R}^4$ in a more attractive sense. Recently Hehl et al.[3] published a paper including a general discussion of so-called Poincaré gauge field theory. The content of this paper was also presented by the author in Erice (6 - 18 May 1979). After the talk of Hehl there was a discussion in the audience about the concept of a Poincaré gauge theory . So far nobody has managed to gauge the translation group in a satisfactory way and therefore the proposal of Hehl et al. deserves some inspection .

For the component expressions we adopt the notation given in reference [3] .In addition we make again use of the elegant Cartan calculus introduced by A. Trautman[4] .

We expand orthonormal Lorentzframes in a coordinate frame

$$\theta^\nu = e^\nu{}_i dx^i \tag{2. 1}$$

The corresponding duals are ξ_ν respectively $\partial/\partial x^i$.
Furthermore let at this point

$$\Gamma^\alpha{}_\beta = \Gamma^\alpha_{\lambda\beta}\,\theta^\lambda = \Gamma^\alpha_{i\beta}\,dx^i \tag{2. 2}$$

denote a metric connection with

$$\text{curvature} \quad 1/2\, F^\alpha{}_{\beta\mu\nu}\,\theta^\mu \wedge \theta^\nu = = 1/2\, F^\alpha{}_{\beta ij}\, dx^i \wedge dx^j \tag{2. 3}$$

and with

$$\text{torsion} \quad 1/2\, F^\alpha{}_{\mu\nu}\,\theta^\mu \wedge \theta^\nu = = 1/2\, F^\alpha{}_{ij}\, dx^i \wedge dx^j \tag{2. 4}$$

Matter fields $\Psi = \Psi^\alpha E_\alpha$ are supposed to be tensor 0-forms transforming as indicated in (1. 6) . If S_* denotes the representation of so(1,3) induced by S we put

$$f^\alpha{}_\beta = 1/2\, S_*(B^\alpha{}_\beta)$$

Now let $\varepsilon(x) = \varepsilon^\nu(x)\,\xi_\nu(x)$ be an arbitrary vectorfield and $\omega(x) = 1/2\,\omega^\alpha{}_\beta(x)\,B^\beta{}_\alpha$ a so(1.3)-valued matrix (at this point the ω's do not denote the connection-forms) .

Hehl et al. define the locally Poincaré transformed field $\P\Psi(x)$ as

$$\P\Psi(x) = \Psi(x) + \delta\Psi(x) \tag{2. 5}$$

whereby

$$\delta\Psi(x) = -\,\varepsilon^\gamma(x)\,D_\gamma\Psi(x) + \omega^{\alpha\beta}(x) f_{\beta\alpha}\Psi(x) \tag{2. 6}$$

Obviously $\varepsilon^\gamma D_\gamma\Psi$ can be recognized as the infinitesimal parallel transport of Ψ along the flow lines of ε . The term $\omega^{\alpha\beta}.f_{\beta\alpha}.\Psi$ is interpreted as an infinitesimal Lorentzrotation of Ψ .
For the particular choice

$$\omega^\alpha{}_\beta(x) = \varepsilon^\gamma(x)\,\Gamma^\alpha_{\gamma\beta}(x) \tag{2. 7}$$

we can prove that (2. 6) is just a Lie-derivative , i.e.

$$\delta\Psi(x) = -\,L_\varepsilon\Psi(x) = -\,(L_\varepsilon\Psi^\alpha)(x)\,E_\alpha \tag{2. 8}$$

Applying the Cartan formula $L_\varepsilon = i_\varepsilon d + d\, i_\varepsilon$ and noticing that $i_\varepsilon\Psi^\alpha = 0$, one finds indeed

$$-\,L_\varepsilon\Psi = -\,i_\varepsilon\, d\Psi = -\,i_\varepsilon(D\Psi - \Gamma^\alpha{}_\beta f^\beta{}_\alpha\Psi) = -\,\varepsilon^\gamma D_\gamma\Psi + \varepsilon^\gamma\Gamma^\alpha_{\gamma\beta}\, f^\beta{}_\alpha\,\Psi =$$
$$= -\,\varepsilon^\gamma D_\gamma\Psi + \omega^\alpha{}_\beta f^\beta{}_\alpha\,\Psi = \delta\Psi$$

The same is true for the other infinitesimal Poincaré transformations given by Hehl et al.

$$\delta e^\alpha{}_i = -\,D_i\varepsilon^\alpha + \omega^\alpha{}_\gamma\, e^\gamma{}_i - \varepsilon^\gamma\, F^\alpha{}_{\gamma i} \tag{2. 9}$$

$$\delta\Gamma^\alpha_{i\beta} = -\,D_i\omega^\alpha{}_\beta - \varepsilon^\gamma F^\alpha{}_{\beta\gamma i} \tag{2.10}$$

Equation (2. 9) follows from the following straight-forward calculation :

$$-\,L_\varepsilon\theta^\alpha = -\,D\varepsilon^\alpha - \varepsilon^\gamma\, F^\alpha_{\gamma i}\, dx^i + \Gamma^\alpha_{\gamma\beta}\,\varepsilon^\gamma\,\theta^\beta =$$

$$= - (D_i\varepsilon^\alpha + \varepsilon^\gamma F^\alpha_{\gamma i} - \omega^\alpha_{\ \gamma} e^\gamma_{\ i}) dx^i = \delta e^\alpha_{\ i} dx^i = - L_\varepsilon \theta^\alpha \quad (2.11)$$

Similarly we find

$$- L_\varepsilon \Gamma^\alpha_{\ \beta} = \delta\Gamma^\alpha_{i\beta} dx^i \quad (2.12)$$

A total Lagrangian for gravitation and matter fields

$$L_{tot} = L(\Psi,\partial\Psi,e,\Gamma) + \upsilon(e,\partial e,\partial\Gamma) \quad (2.13)$$

is locally Poincaré invariant , if $\delta L_{tot} = 0$ for simultaneous variations (2.6;9;10) .

Actually we have proved above that for $\omega^\alpha_{\ \beta}$ of type (2. 7) local Poincaré invariance follows from the invariance under the diffeomorphism group of the space-time manifold (general covariance) .

For a general so(3.1) - matrix $\omega(x)$ we define $\hat{\omega}(x)$ by

$$\omega^\alpha_{\ \beta} =: \varepsilon^\gamma \Gamma^\alpha_{\gamma\beta} + \hat{\omega}^\alpha_{\ \beta} \quad (2.14)$$

Since $\delta L_{tot} = 0$ for $\hat{\omega}^\alpha_{\ \beta} = 0$, we find for an arbitrary $\hat{\omega}^\alpha_{\ \beta} \neq 0$

$$\hat{\omega}^\alpha_{\ \gamma} e^\gamma_{\ i} \Sigma^i_\alpha + 1/2 (D_i \hat{\omega}^\alpha_{\ \beta}) \tau^{\beta i}_\alpha = 0$$

or

$$\hat{\omega}^\alpha_{\ \beta}(e^\beta_{\ i} \Sigma^i_{\ \alpha} - 1/2 D_i\tau^\beta_{\ \alpha}{}^i) = 0 \quad (2.15)$$

since all equations are understood modulo divergences . In this notation Σ denotes the canonical energy-stress tensor and τ is the spin-density . From (2.15) we obtain the well known identity

$$D_i\tau^{\alpha\beta i} = (B^{\alpha\beta})^\mu_{\ \nu} \Sigma^\nu_{\ \mu} \quad (2.16)$$

which we have derived in section 1 as a consequence of the local Lorentzinvariance of L_{tot} (see (1.23)) .

Summarizing the above considerations we can say that local Poincaré invariance just combines general covariance (invariance under the diffeomorphism group) and local Lorentzinvariance . So there is nothing misterious about it and the question , how to gauge the translation group more attractively , is still open .

3 The Diffeomorphism Group Diff(M)

The notation in this section is the same as in part 1 .

When describing space-time by Lorentz manifolds , two models (M,g) and (M' ,g') are taken to be equivalent , if they are isometric , i.e. if there is a diffeomorphism

$$\Phi : M \to M'$$

which carries the metric g into g' , i.e. if

$$\Phi^* g' = g$$

In this context the group Diff(M) of all diffeomorphisms of a suitable Lorentz manifold (M,g) establishes a symmetry group : Starting from (M,g) let θ^ν denote local Lorentz frames , $\omega^\alpha_{\ \beta}$ the components of a metric connection and Ψ , $D\Psi$ local matter fields. For any $\Phi \in$ Diff(M) the transformed fields

$$\Phi^*\theta^\nu , \Phi^*\omega^\alpha_{\ \beta} , \Phi^*\Psi , \Phi^* D\Psi$$

describe the same physical situation on (M,Φ^*g) . This is guaranted, if the Lagrangian $L(\theta^\nu,\omega^\alpha{}_\beta,\Psi,D\Psi)$ has the following invariance property :

$$\Phi^*\{ L(\theta^\nu,\omega^\alpha{}_\beta,\Psi,D\Psi)\} = L(\Phi^*\theta^\nu,\Phi^*\omega^\alpha{}_\beta,\Phi^*\Psi,\Phi^*D\Psi) \qquad (3.\ 1)$$

for any $\Phi \in \mathrm{Diff}(M)$.
The following identity is a consequence of (3. 1)

$$Dt_\mu = Q^\nu{}_\mu \wedge t_\nu + 1/2\ R^{\alpha\beta}{}_\mu \wedge s_{\alpha\beta} \qquad (3.\ 2)$$

Hereby $Q^\nu{}_\mu = Q^\nu{}_{\mu\lambda}\ \theta^\lambda = i_\mu \Theta^\nu$ and $R^{\alpha\beta}{}_\mu = (R^{\alpha\beta})_{\mu\nu}\ \theta^\nu = i_\mu \Omega^{\alpha\beta}$
(Θ^ν is the torsion)
The idea of the proof of (3. 2) goes like this : Let D be a sufficiently small compact region with regular boundary and X a vectorfield of the following type

(i) X vanishes on M-D
(ii) If $\Phi(x,t)$ denotes the flow of X , then there is an $\varepsilon > 0$ such that $D \times [-\varepsilon,\varepsilon]$ is contained in the domain of Φ .
(iii) For any $t \in [-\varepsilon,\varepsilon]$ the flow has the property $\Phi_t . D = D$.

Obviously Φ_t may now be regarded as an element of Diff(M) , which leaves the points outside of D fixed . From (3. 1) we find

$$\begin{aligned}&\frac{d}{dt}\Big|_0 \int_D L(\Phi_t^*\theta^\nu,\Phi_t^*\omega^\alpha{}_\beta,\Phi_t^*\Psi,\Phi_t^*D\Psi) = \\ &= \frac{d}{dt}\Big|_0 \int_D \Phi_t^*\{ L(\theta^\nu,\omega^\alpha{}_\beta,\Psi,D\Psi)\} = \int_{\partial D} i_X L = 0\end{aligned} \qquad (3.\ 3)$$

In addition we notice that with the equations for the matter fields

$$\frac{d}{dt}\Big|_0 L(\Phi_t^*\theta^\nu,\Phi_t^*\omega^\alpha{}_\beta,\Phi_t^*\Psi,\Phi_t^*D\Psi) = L_X\theta^\nu \wedge t_\nu + 1/2\ L_X\omega_{\alpha\beta} \wedge s^{\alpha\beta} \qquad (3.\ 4)$$

Equation (3. 4) is understood modulo exact differentials . Now it is a mere routine to get (3. 2) from (3. 3) and (3. 4).

4. General Field Equations and the Differential Identities

Let $L = L_F + L_M$ denote a general Lagrangian for gravitation and the matter fields

$$L_F = L_F(\Omega^\alpha{}_\beta,\theta^\nu,D\theta^\nu)$$
$$L_M = L_M(\Psi,D\Psi,\theta^\nu)$$

From independent variations of $\omega^\alpha{}_\beta$ and θ^ν we find

$$\delta L_F = \delta\theta^\mu \wedge \varepsilon_\mu + 1/2\ \delta\omega_{\alpha\beta} \wedge \sigma^{\alpha\beta} \qquad (4.\ 1)$$

$$L_M = \delta\theta^\mu \wedge t_\mu + 1/2\ \delta\omega_{\alpha\beta} \wedge s^{\alpha\beta} \qquad (4.\ 2)$$

With (1.23) and (3. 2) we have the full set of differential

identities , as we shall see below :

Local Lorentz invariance

$$D\sigma^{\alpha\beta} = \theta^{\alpha} \wedge \varepsilon^{\beta} - \theta^{\beta} \wedge \varepsilon^{\alpha} \quad (4.\ 3)$$

$$Ds^{\alpha\beta} = \theta^{\alpha} \wedge t^{\beta} - \theta^{\beta} \wedge t^{\alpha} \quad (4.\ 4)$$

Diff(M) - invariance

$$D\varepsilon_{\mu} = Q^{\nu}{}_{\mu} \wedge \varepsilon_{\nu} + 1/2\ R^{\alpha}{}_{\beta\mu} \wedge \sigma_{\alpha}{}^{\beta} \quad (4.\ 5)$$

$$Dt_{\mu} = Q^{\nu}{}_{\mu} \wedge t_{\nu} + 1/2\ R^{\alpha}{}_{\beta\mu} \wedge s_{\alpha}{}^{\beta} \quad (4.\ 6)$$

General field equations

$$\varepsilon_{\mu} = -\ t_{\mu} \quad (4.\ 7)$$

$$\sigma_{\alpha\beta} = -\ s_{\alpha\beta} \quad (4.\ 8)$$

In the special context of Einstein-Cartan the following theorem has been formulated by A. Trautman [4]

> The equations resulting by covariant exterior derivation of (4. 7) and (4. 8) are algebraic consequences of the above listed differential identities and of the field equations themselves . (4. 9)

Theorem (4.9) shows that in order to get a consistent set of field equations one has to take (4. 7) and (4. 8) simultaneously .

Acknowledgements : The concept of section 1 , 3 and 4 has been elaborated with the aid of Prof. N. Straumann . The geometrical analysis of the Poincaré symmetry in section 2 was motivated by a vivid discussion I had with J. Nitsch and F.W. Hehl in Erice .

References

1 Kobayashi & Nomizu , Foundations of Differential Geometry , Vol. I , Interscience , New York (1963)
2 Y.M. Cho , Einstein Lagrangian as the translational Yang-Mills Lagrangian ,Phys Rev D 14 No 10 (1976)
3 F.W. Hehl , J. Nitsch , P. von der Heyde , Gravitation and Poincaré Gauge Field Theory with Quadratic Lagrangian , preprint , University of Cologne , West Germany (1978)
4 A. Trautman , On the structure of the Einstein-Cartan equations , Symposia Mathematica 12 , 139 NY (1973)

THE MOTION OF TEST-PARTICLES IN NON-RIEMANNIAN SPACE-TIMES

Dierck-Ekkehard Liebscher

Akademie der Wissenschaften der DDR
Zentralinstitut für Astrophysik
DDR 1502 Potsdam-Babelsberg

INTRODUCTION

The general theory of relativity was the first theory with a genuine connection between the field equations and the motion of its sources. The simplest way to see this consists in the integrability condition

(1) $$T^{ik}{}_{;k} = 0$$

of Einstein's equations

(2) $$R^{ik} - \frac{1}{2} g^{ik} R = \kappa T^{ik}.$$

To fulfil condition (1) we need the knowledge of the metric as solution of eq. (2). This intertwining tells us that we cannot choose T^{ik} freely: i.e. the motion of the sources represented by T^{ik} is restricted by the field equation.

The condition (1) alone yields equations of motion by integrating it into quasi-conservation laws over world-tubes with $T^{ik}=0$ outside. The connection between condition (1) and eq.(2) is due to differential identities implied by the general covariance of the Lagrangian. We get similiar conditions in the gauge-invariant Maxwell electrodynamics, however the identity Div j =0 leads only to the conservation of charge. Because of the fact that the energy-momentum tensor is the differential equivalent for inertia, we get equations of motion only if this tensor fulfils a differential conservation law also.

In General Relativity the connection between (1) and (2) allows two different ways to derive equations of motion. The first one uses an integration procedure outside the matter distribution and defines the structure of the objects by their gravitational field (Einstein und Grommer 1927, Einstein, Infeld and Hoffmann 1938, Infeld and Plebanski 1960, Taub 1964, Newman and Posadas 1969, Newman and Young 1970). The second one uses an integration over the matter distribution and defines the structure of the particles by moments of this distribution (Papapetrou 1951, Fock 1959, Tulczyjew 1959, Madore 1969). The main problem of the research in this direction was to derive appropriate definitions of the multipole moments, because only in the case of a pole particle the equations determine the motion unambiguously (Mathisson 1940, B. and W.Tulczyjew 1962, Taub 1964, Dixon 1964 and 1974, Ehlers and Rudolph 1976). The motion of pole particles may also be used the other way round to define or narrow the structure of space-time (Kondo 1964, Petrov 1969, Ehlers, Pirani and Schild 1972, Ross 1977). However, this structure describes particle motion and may be ampl fied before writing down field equations.

In this lesson we are interested not so much in the question of defining appropriate moments but in the question of obtaining ordinary differential equations corresponding to equations of motion from which, in the next step, these moments may be extracted. The main problem is to get an appropriate generalization of the condition (1). We do not want to consider special field equations for the gravitational field, i.e., the space-time structure. Therefore, we are bound to proceed like Papapetrou. Essentially, this method uses the equality between the ordinary divergence of the energy-momentum-tensor (and of the current) and a linear combination of these tensors themselves. It is not essential that these relations get the form of covariant divergences being zero. Therefore, in the case of non-Riemannian space-times, we look for a set of tensors q^{kA} (k denote the necessary tensor index of each tensor of the set, A denotes the other tensor indices, which number may differ in the set), which obey divergence relations of the kind

$$q^{kA}{}_{;k} + \alpha^{A}{}_{Bk}\, q^{kB} = 0. \tag{3}$$

If this set q^{kA} contains a tensor generalizing the energy-momentum tensor of Riemannian space-time then we get equations of motions by integrating these divergence relations with the method of Papapetrou like B. and W. Tulczyjew 1962.

As we will see, the coupling of non-gravitational fields to the non-Riemannian space-time structure differs from the coupling of the energy-momentum tensor. Therefore, we cannot choose freely the divergence relations in question. We have to derive these relations by the field equations for the non-gravitational fields. Because we are interested in phenomena including the spin, and because of the fact that the main part of the energy-momentum tensor of a macroscopic particle is given by Dirac matter, we derive here the divergence relations from the Dirac equation (Liebscher 1973a, Hayashi and Shirafuji 1977).

THE DIRAC EQUATION IN NON-RIEMANNIAN SPACE-TIME

We choose the Dirac equation as a general linear equation of a spinor, the coefficients representing and defining a non-Riemannian space-time structure. The general Dirac equation reads therefore

$$\gamma^k \psi_{;k} = (m + M)\psi. \tag{4}$$

Out of the coefficients γ^k we construct a metric with its geodesic (Levi-Civita) connection

$$\gamma^k \gamma^l + \gamma^l \gamma^k = -2g^{kl}, \tag{5}$$

$$(\alpha\gamma^k)^+ = \alpha\gamma^k, \quad \alpha\alpha^+ = -\alpha^2 = 1, \tag{6}$$

$$\left\{ {k \atop mn} \right\} = \frac{1}{2} g^{kl} \left(- g_{mn,l} + g_{ml,n} + g_{nl,m} \right). \tag{7}$$

The geodesic spinor connection Γ_{0m} is defined by

$$\gamma^k{}_{;m} = \gamma^k{}_{,m} + \left\{ {k \atop lm} \right\} \gamma^l + [\gamma^k, \Gamma_{0m}] = 0 \tag{8}$$

and

$$\alpha_{;m} = \alpha \Gamma_m + \Gamma_m^+ \alpha = 0, \quad \alpha_{,m} = 0. \tag{9}$$

The covariant derivative with respect to this connection will be denoted by the semicolon. We introduce a Dirac-selfadjoint base of the Clifford algebra $\mathcal{C}^A$:

$$\begin{aligned} &\mathcal{C}^0 = i, \quad \mathcal{C}^k = \gamma^k, \quad \mathcal{C}^{ik} = \sigma^{ik} = \tfrac{1}{2}[\gamma^i, \gamma^k], \\ &\mathcal{C}^{\underline{k}} = i\gamma^* \gamma^k, \quad \mathcal{C}^{\underline{0}} = i\gamma^* = \frac{i}{24} \varepsilon^{klmn} \gamma_k \gamma_l \gamma_m \gamma_n \end{aligned} \tag{10}$$

and extract the effective non-Riemannian structure of space-time by the (not unique) representation

$$M = \gamma^k (\Gamma_k - \Gamma_{0k}). \tag{11}$$

Γ_k is a spinor affinity not connected with g_{ik}. In this way, we obtain a metric space-time with an affine structure by writing down a linear spinor equation.

In Special Relativity, the energy-momentum tensor of the Dirac field has the form

$$\theta_k{}^l = \frac{i}{2} \left(\overline{\Psi}_{;k} \gamma^l \psi - \overline{\Psi} \gamma^l \psi_{;k} \right). \tag{12}$$

Of course, we cannot wait for a differential conservation law in full generality of eq.(3). But we get a set of divergence relations for the tensors constructed in analogy to the special-relativistic energy-momentum tensor and the current density

(13) $$\theta_k{}^A = \frac{i}{2}(\overline{\psi}_{;k}\tau^A\psi - \overline{\psi}\tau^A\psi_{;k}),$$

(14) $$J^A = \overline{\psi}\tau^A\psi,$$

(15) $$E^{kA} = \frac{i}{2}\overline{\psi}\{\gamma^k, \tau^A\}\psi = -D^{kA}{}_B J^B,$$

(16) $$H^{kA} = \frac{1}{2}\overline{\psi}[\gamma^k, \tau^A]\psi = C^{kA}{}_B J^B.$$

$C^{AB}{}_C$ and $D^{AB}{}_C$ are the structure constants of the algebra:

(17) $$[\tau^A, \tau^B] = 2C^{AB}{}_C\tau^C,$$
$$\{\tau^A, \tau^B\} = 2iD^{AB}{}_C\tau^C.$$

We get the relations

(18) $$\hat{\theta}^{kA}{}_{;k} + 2\varrho^{kA}{}_B\hat{\theta}_k{}^B = \Sigma^A{}_B J^B,$$

(19) $$E^{kA}{}_{;k} - 2\eta^A{}_B J^B = -2C^{kA}{}_B\hat{\theta}_k{}^B,$$

(20) $$H^{kA}{}_{;k} - 2\chi^A{}_B J^B = -2D^{kA}{}_B\hat{\theta}_k{}^B - 2mJ^A,$$

where the coefficients are given by

(21) $$\hat{\theta}^{kA} = \theta^{kA} - T^{kA}{}_B J^B,$$
$$M = \underset{\cdot}{M} + i\underset{\cdot\cdot}{M};\ \underset{\cdot}{M}, \underset{\cdot\cdot}{M} \text{ self-adjoint},$$

$$(22) \quad [[\tau^A, \gamma^k], \underset{..}{M}] - i\{[\tau^A, \gamma^k], \underset{.}{M}\} = 4\tau^{kA}{}_B \tau^B,$$

$$(23) \quad \{\{\tau^A, \gamma^k\}, \underset{..}{M}\} - i[\{\tau^A, \gamma^k\}, \underset{.}{M}] = -4\sigma^{kA}{}_B \tau^B,$$

$$(24) \quad \{\{\tau^A, \gamma^k\}, \underset{.}{M}\} + i[\{\tau^A, \gamma^k\}, \underset{..}{M}] = -4\varrho^{kA}{}_B \tau^B,$$

$$(25) \quad \Sigma^A{}_B = -\sigma^{kA}{}_{B;k} - 2\varrho^{kA}{}_C \tau_k{}^C{}_B,$$

$$(26) \quad \eta^A{}_B = \underset{0}{\eta}{}^A{}_B - C^{kA}{}_C \tau_k{}^C{}_B,$$

$$\frac{i}{2}(\overline{M}\tau^A + \tau^A M) = \underset{0}{\eta}{}^A{}_B \tau^B,$$

$$(27) \quad \chi^A{}_B = \underset{0}{\chi}{}^A{}_B - D^{kA}{}_C \tau_k{}^C{}_B,$$

$$\frac{1}{2}(\overline{M}\tau^A - \tau^A M) = \underset{0}{\chi}{}^A{}_B \tau^B.$$

The coefficients $\Sigma^A{}_B$ are curvature-like, the others correspond to non-geodesic transport coefficients. J^A contains currents and spin densities, H and E are combination of them, $\hat{\Theta}^{kA}$ are all conserved quantities in SRT if the Maxwell field is zero. $\hat{\Theta}^{kl}$ is the generalization of the energy-momentum tensor. It may be modified by

$$(28) \quad \overset{*}{\Theta}{}^{kA} = \hat{\Theta}^{kA} + \lambda^A{}_B E^{kB} + \lambda C^{kA}{}_B E^{lB}{}_{;l}.$$

In the special case of a Dirac equation connected with a given non-Riemannian space-time via Weyl's lemma of covariantly constant spin matrices (Weyl 1929) we get the expressions

(29) $$M = \mu_k \gamma^k - \nu_k \gamma \overset{*}{\gamma}^k, \; \nu_k \text{ real}, \; \mu_k = \mu_{\cdot k} + i\mu_{\cdot\cdot k},$$

and we can separate the set

(30) $$\hat{\theta}^{kl}{}_{;k} = 2\mu_{\cdot k}\hat{\theta}^{lk} + F^l{}_m j^m + \Sigma^l{}_{\underline{m}} j^{*m},$$

(31) $$j^k{}_{;k} = 2\mu_{\cdot k} j^k,$$

(32) $$-\frac{1}{2}E^{ijk}{}_{;k} = \frac{1}{2}\varepsilon^{ijlk} \overset{*}{j}_{l;k} = \hat{\theta}^{ij} - \hat{\theta}^{ji} + \nu^i j^j - \nu^j j^{*i} - \mu_{\cdot k}\varepsilon^{ijk} \overset{*}{j}_r + \mu^i j^j - \mu^j j^i.$$

If we look for a variational principle in this case, we get $\mu_{\cdot k} = 0$. In general a variational principle leads to anti-selfadjoint matrices M.

THE INTEGRATION OF THE DIVERGENCE RELATIONS

In order to integrate the set

(33) $$q^{kA}{}_{;k} + \alpha^A{}_{Bk} q^{kB} = 0$$

into equations of motion, we use the moment distributions of q^{kA} with respect to a world-line inside a world-tube, which a posteriori may be determined to be that, around which moments higher than a certain degree vanish (B. and W.Tulczyjew 1962, Madore 1969, Dixon 1964 and 1974, Liebscher 1973b). This line is chosen to fix the orientation of a system of Fermi-coordinates of the second kind:

(34) $$X^i(\tau) = (\tau, 0, 0, 0), \; g_{ik} = \eta_{ik}$$

(35) $$\left\{ {k \atop lm} \right\} = b^k u_l u_m - u^k b_l u_m - u^k u_l b_m$$

on the reference world-line. The moments may be defined by

(36) $$\int \sqrt{-g}\, F(x)\, q^{kA}\, d^3x = F(X)\, K^{kA} + \sum_{n=1}^{\infty} F_{,i_1 \ldots i_n}(X)\, K^{i_1 \ldots i_n kA},$$

where $K^{i_1 \ldots i_n kA}$ is symmetric in the indices $i_1, \ldots, i_n$ and orthogonal to u_i, if summation runs over one of these indices. Defining the symbol β^A_{Bk} by

(37) $$\beta^A{}_{Bk}\, q^{kB} = \sum_i \left\{ {m_i \atop rk} \right\} q^{km_1 \ldots r \ldots}$$

and $\alpha^A{}_{Bk} + \beta^A{}_{Bk} = \gamma^A{}_{Bk}$ we integrate the equations

(38) $$\frac{1}{\sqrt{-g}} \left(\sqrt{-g}\, q^{kA} \right)_{,k} + \gamma^A{}_{Bk}\, q^{kB} = 0,$$

(39) $$\frac{1}{\sqrt{-g}} \left(\sqrt{-g}\, q^{kA} \xi^i \right)_{,k} + \gamma^A{}_{Bk}\, q^{kB} \xi^i - \left(\delta^i_k - u^i u_k \right) q^{kA} = 0,$$

(40) $$\frac{1}{\sqrt{-g}} \left(\sqrt{-g}\, q^{kA} \xi^i \xi^j \right)_{,k} + \gamma^A{}_{Bk}\, q^{kB} \xi^i \xi^j - \left(\delta^i_k - u^i u_k \right) q^{kA} \xi^j - \left(\delta^j_k - u^j u_k \right) q^{kA} \xi^i = 0,$$

which correspond to (33) and use the three space-like Fermi coordinates $\xi^i = (0, x^1, x^2, x^3)$, and get

(41) $$\frac{d}{d\tau} u_k K^{kA} + \gamma^A{}_{Bk} K^{kB} + \gamma^A{}_{Bk,i} K^{ikB} = 0,$$

(42) $$\frac{d}{d\tau} u_k K^{ikA} + \gamma^A{}_{Bk} K^{ikB} - K^{iA} + u^i u_k K^{kA} = 0,$$

(43) $$K^{jiA} - u^i u_k K^{jkA} + K^{ijA} - u^j u_k K^{ikA} = 0.$$

We solve eq.(43) by

$$(44) \qquad K^{ijA} = M^{ijA} + u^j u_k K^{ikA},$$

where M^{ijA} is antisymmetric in i and j and orthogonal to the velocity u_i of the particle. Defining now

$$(45) \qquad P^A = u_k K^{kA}, \quad \hat{P}^A = P^A + \gamma^A{}_{Bk} L^{kB},$$

$$(46) \qquad L^{iA} = u_k K^{ikA},$$

$$(47) \qquad N^{ijA} = M^{ijA} - u^j L^{iA} + u^i L^{jA},$$

we get the equations of motion

$$(48) \qquad \frac{D}{d\tau}\hat{P}^A + \alpha^A{}_{Bk} u^k \hat{P}^B = -\frac{1}{2} R^A{}_{Bik} N^{ikB}$$

$$-\frac{1}{2}\left(\alpha^A{}_{Bk;i} - \alpha^A{}_{Bi;k} + \alpha^A{}_{Ci}\alpha^C{}_{Bk} - \alpha^A{}_{Ck}\alpha^C{}_{Bi}\right) N^{ikB},$$

which corresponds to Papapetrou's equation for the transport of the momentum vector, and

$$(49) \qquad \frac{D}{d\tau} L^{iA} + \gamma^A{}_{Bk} M^{ikB} + u^i b_k L^{kA}$$

$$= K^{iA} - u^i P^A,$$

which corresponds to Mathisson's equation for the transport of the spin tensor of the particle. Higher than first moments are neglected in all q^{kA}. If the calculation shall include also higher moments, we have only to enlarge equs.(41), (42) and (43) by introduction of additional equations with more factors ξ^i . The last one of these equations always gives a structural relation between the components of the highest moments involved, the other equations give equations of motion.

Evidently, eq.(48) contains the transport law of the momentum of the particle. We got a geodesic transport corrected by the terms represented by $\alpha^A{}_{Bk}$, which may be interpreted as a general affine transport, and a coupling of the dipole moments of the particle to the curvature of this general connection. Reducing the set q^{kA} to the symmetric energy-momentum tensor of General Relativity we get exactly the equations of motion of a pole-dipole particle as derived eg. by Papapetrou.

Equations (48) and (49) have to be analysed into the different quantities for every concrete case. The general pole particle, defined now by $K^{ijA}=0$, reduces eq.(49) to

$$K^{iA} - u^i u_k K^{kA} = 0 \tag{50}$$

and obeys therefore the transport law

$$\frac{DP^A}{d\tau} + \alpha^A{}_{Bk} u^k P^B = 0 \tag{51}$$

which corresponds again to the general affine connection mentioned above.

THE DIRAC-MATTER TEST-PARTICLE

We return to the divergence relations eqs. (18), (19) and (20). Because of θ^{ko} being connected with the orbital part of the current, θ^{kij} with the conserved spin, $\theta^{k\underline{l}}$ with the Bargmann-Wigner vector and $\theta^{k\underline{o}}$ with the chargeless pseudocurrent, only θ^{kl} and $\hat{\theta}^{kl}$ may represent the energy-momentum tensor. Picking out its divergence relation we may write

$$\hat{\theta}^{kl}{}_{;k} = 2\mu_{\cdot k}\hat{\theta}^{lk} \tag{52}$$

+ internal structure and curvature terms,

using the formula (23) in the case A:=1

(53) $$\rho^{kl}{}_{B}\,c^{B} = 2 g^{kl} \dot{M} .$$

Other definitions of the energy-momentum tensor (28) may change this law into

(54) $$\overset{*}{\Theta}{}^{kl}{}_{;k} = 2\,\mathrm{Re}\,\mu_k \overset{*}{\Theta}{}^{lk} + O(M)\,\overset{*}{\Theta}{}^{kl}_{v} + \dots$$

The main point used now is eq.(19) in the case A:=ij. It has the form

(55) $$E^{kij}{}_{;k} = \lambda\,\overset{*}{\Theta}{}^{kl}_{v} + O(M)\,E^{kA} .$$

Because of the antisymmetry of E^{kij}, it allows only the trivial solution for the pole condition eq.(50) and we get

(56) $$u^{l} p^{k} = u^{k} p^{l} .$$

Pole particles move on geodesics. Their rest mass may vary - if $\mathrm{Re}\,\mu_k \neq 0$ - but not in the case of a variational principle for the Dirac equation. In this case M is anti-selfadjoint and $\dot{M}$ vanishes. It should be noted, that pole particles allow for spin, $\hat{\Theta}^{kij}$ leads to a non-trivial solution of its pole condition, but this spin is decoupled from the outer motion of the particle. Consequently, also in theories which take into account internal spin and non-Riemannian connections (Trautman 1972, Hehl 1973), in order to feel the non-Riemannian component of the connection, a particle has to be endowed with dipole structure of its mass and spin density. The coupling of the spin to the gravitational field looks like that in GR by its structure. Only in GR the spin consists entirely of the dipole moment of the mass distribution.

Besides of this result we see by the total antisymmetry of E^{kij} that the phenomenological description by $E^{kij} = u^k S^{ij}$ is incomplete. A term $u^k S^{ij}$ alone does not describe internal spin in a worldtube but exclusively external spin. The spin flux $u^k S^{ij}$ is due to the motion of Riemannian spinning test-particles. Using instead the totally antisymmetric combination

(57) $$E^{kij} = u^k S^{ij} + u^i S^{jk} + u^j S^{ki}$$

we get

(58) $$E^{kij} = \frac{1}{2} \varepsilon^{kijl} \varepsilon_{lmnr} u^m S^{nr} = \frac{1}{2} \varepsilon^{kijl} S_l ,$$

which may be used in Friedmann models with time-like S_l or homogeneous distributions of S_l orthogonal to the velocity (Kuchowicz 1976).

References

Dixon, W.G., 1964, A covariant multipole formalism for extended test bodies in general relativity, Nuovo Cim., 34:317.
Dixon, W.G., 1974, Dynamics of extended bodies in general relativity, Phil.Trans.Roy.Soc.London, A 277:59.
Ehlers, J., Pirani, F.A.E., and Schild, A., 1972, The geometry of free fall and light propagation, in: Festschrift in honour of J.L.Synge, Oxford.
Ehlers, J., and Rudolph, E., 1977, Dynamics of extended bodies in general relativity. Center-of-mass description and quasi-rigidity, GRG Journal, 8:197.
Einstein, A., und Grommer, J., 1927, Allgemeine Relativitatstheorie und Bewegungsgesetz, Sitzungsber. Preuss.Akad.Wiss.Berlin, 1927:2.
Einstein, A., Infeld, L., and Hoffmann, B., 1938, The gravitational equations and the problem of motion, Ann.Math., 39:65.
Fock, W.A., 1959, Teorija prostranstva-vremeni i gravitacii, Leningrad.
Havas, P., 1964, The connection between conservation laws and laws of motion in affine spaces, J.math.Phys., 5:373.
Hayashi, K., and Shirafuji, T., 1977, Spacetime structure explored by elementary particles: Microscopic origin for the Riemann-Cartan geometry, Progr.Theor.Phys., 57:302.

Hehl, F.W., and van der Heyde, P., 1973, Spin and the structure of space-time, Ann.Inst.H.Poincare, 19:179.
Hehl, F.W., 1974, On the energy tensor of spinning massive matter in classical field theory and general relativity, Princeton university preprint.
Infeld, L., and Plebanski, J., 1960, Motion and relativity, New York.
Kondo, K., 1964, A statistical approach to the principle of general relativity, Tensor, 15.168.
Kuchowicz, B., 1978, Friedmann-like cosmological models without singularities, GRG Journal, 9:511.
Liebscher, D.-E., 1973a, The equivalence principle and non-Riemannian space-times, Ann.Physik (Leipzig), 30:309.
Liebscher, D.-E., 1973b, The integration of generalized systems of dynamical equations, Ann.Physik (Leipzig), 30:321.
Madore, J., 1969, The equations of motion of an extended body in general relativity, Ann.Inst.H.Poincare, 11:221.
Mathisson, M., 1940, The variational equation of relativistic dynamics, Proc.Camb.Phil.Soc., 36:331.
Newman, E.T., and Posadas, R., 1969, Equations of motion and the structure of singularities, Phys.Rev., 187:1784.
Newman, E.T., and Young, R., 1970, New approach to the motion of a pole-dipole particle, J.math.Phys., 11:3154.
Papapetrou, A., 1951, Spinning test-particles in general relativity, Proc.Roy.Soc.London, A 209:248.
Petrov, A., 1969, O modelirovanii putej probnych tel v pole gravitacii, Doklandy AN SSSR, 186:1302.
Ross, D.K., 1977, Geodesics in gauge supersymmetry, Phys.Rev. D, 16:1717.
Taub, A.H., 1964, Motion of test bodies in general relativity, J.math.Phys., 5:112.
Trautman, A., 1972, On Einstein-Cartan theories III., Bull.acad.polon.sci.,Ser.sci.math.astr.phys.,20:895.
Tulczyjew, W., 1959, Motion of multipole particles in GRT, Acta phys.polon., 18:393.
Tulczyjew, B., and Tulczyjew, W., 1962, On multipole formalism in general relativity, in: Recent developments in General Relativity, London.
Weyl, H., 1929, Elektron und Gravitation, Z.Physik, 56:330.

TORSION AND STRONG GRAVITY IN THE REALM OF ELEMENTARY PARTICLES AND COSMOLOGICAL PHYSICS

Venzo de Sabbata and Maurizio Gasperini
Istituto di Fisica dell'Università, Bologna and
Istituto di Fisica dell'Università, Ferrara, Italy

INTRODUCTION

Starting with some considerations on the minimal coupling principle between matter and torsion we illustrate a different kind of coupling on the ground of the motion equations. It is then possible to write down the Lagrangian of the torsionic contact interaction between two Dirac particles in the V - A standard form if at least one of the two fermions is massless and not necessarily both particles as in the case of minimal coupling. A complete identification of the torsionic and weak interaction could be possible however only by redefining the spin-torsion coupling constant.

We introduce then the strong gravity contributions to the nucleonic axial-vector form factor, which is evaluated in the non relativistic limit and in the approximation of the exact SU_3 symmetry. Besides finding the correct result according to the experimental value of the axial form factor in the case of neutron beta decay, we have also some agreements with the behaviour of the axial coupling constant in very light nuclei. If in this context also torsion is introduced, the spin corrections through the Einstein-Cartan equations in the case of strong gravity cannot be neglected and we are led quite naturally to some beautiful considerations regarding the Cabibbo angle.

At last the strong gravity corrections to the electromagnetic field of a particle are considered and we deduce an interesting relation between masses and radii of the elementary particles which, if extended to the Universe, is in agreement with Dirac's large numbers hypothesis. If moreover also spin and torsion are introduced in the strong gravity equations, then we are led to an enlarged Dirac law which includes a possible total angular momentum of the Universe.

CHAPTER 1

THE SPIN-TORSION COUPLING CONSTANT AND THE MINIMAL COUPLING PRINCIPLE

If the Einstein-Cartan theory[1] or, more generally, the spaces with a nonvanishing torsion are introduced into the theory of the elementary particles, some results are easily obtained which suggest a similarity between the weak and torsionic interactions[2,3,4]. If, in addition, the minimal coupling[1] between matter and torsion is substituted by a more general coupling deduced from the motion equations, then a closest analogy between the two interactions can be established so that one can think about their possible identification.

According to the 'minimal coupling principle', in order to couple a matter field ψ with torsion, one starts with the special relativistic Lagrangian density $\mathcal{L}(\psi, \partial\psi, \eta)$ ($\eta_{\mu\nu}$

$= diag(1, -1, -1, -1)$ is the flat space metric), and makes the replacements

$$\eta_{\mu\nu} \rightarrow g_{\mu\nu} \qquad \partial_\mu \rightarrow \nabla_\mu(\Gamma) \quad (1.1)$$

where $g_{\mu\nu}$ and $\nabla_\mu(\Gamma)$ are metric and covariant derivative of a Riemann-Cartan space, vhose connection Γ satisfies[1,2]

$$\begin{aligned} \Gamma^{\mu}_{\alpha\beta} &= \left\{ {\mu \atop \alpha\,\beta} \right\} - K^{\mu}_{\alpha\beta} \\ K^{\mu}_{\alpha\beta} &= -Q_{\alpha\beta}{}^{\mu} - Q^{\mu}{}_{\alpha\beta} + Q_{\beta}{}^{\mu}{}_{\alpha} \end{aligned} \quad (1.2)$$

$Q^{\mu}_{\alpha\beta} = \Gamma^{\mu}_{[\alpha\beta]}$ is the torsion tensor and $\left\{ {\mu \atop \alpha\,\beta} \right\}$ is, as usual, the Christoffel symbol (square brackets denote antisymmetrization). Following this procedure in a flat space ($g = \eta$) with torsion ($\Gamma^{\mu}_{\alpha\beta} = -K^{\mu}_{\alpha\beta}$), the 'minimal' interaction lagrangian we get for a Dirac field (1/2 spin particles) is (except for a numerical factor of order of unity)[1,2,5]

$$\mathcal{L}_I = \chi\, \varepsilon_{\mu\alpha\beta\gamma} \bar{\psi} \gamma^5 \gamma^\mu \psi\, Q^{\alpha\beta\gamma} \quad (1.3)$$

where $\varepsilon_{\mu\alpha\beta\gamma}$ is the four dimensional Levi-Civita tensor, and $\chi = 8\pi k_g$, where k_g is the Newton constant ($\hbar = c = 1$). The universality and axial character of this term, together with the weakness of the coupling constant, are analogous to the properties of the weak interaction as already pointed out[2,3,18]. Now we remember that the principle of minimal coupling is an empirical prescription taken from General Relativity, and, if used in spaces with torsion, it is not supported by any known experimental result; on the contrary, we know that it does not work when applied to photons, gravitons and any other gauge field[1,6]. Therefore it is reasonable to look for a different kind of coupling, provided that the new nonminimal terms be introduced on the ground of some physical motivation. In order to do this, we consider the motion equations of a Dirac particle in a Riemann-Cartan space[49]. From the equation of conservation, neglecting curvature ($g_{\mu\nu} \simeq \eta_{\mu\nu}$) and using Papapetrou's method[7], we get that the force F^{μ} upon the particle is, in a first approximation[8]

$$F^{\mu} = \frac{dp^{\mu}}{ds} = \left(1 - \frac{v^2}{c^2}\right)^{-1/2} \int d^3x\, B^{\mu} \qquad (1.4)$$

where

$$B^{\mu} = R^{\mu}{}_{\alpha\beta\gamma}\Sigma^{\alpha\beta\gamma} - \partial_{\alpha}T^{[\mu\alpha]} - 2Q_{\beta\alpha}{}^{\alpha}T^{\mu\beta} + (K_{\alpha\beta}{}^{\mu} - Q_{\beta}{}^{\mu}{}_{\alpha})T^{\alpha\beta} \qquad (1.5)$$

$T^{\mu\nu}$ is (in general non symmetrical)the stress tensor and $\Sigma^{\mu\alpha\beta}$ is the spin density tensor of a Dirac particle; $R^{\mu}{}_{\alpha\beta\gamma}$ is the Riemann tensor of the nonsymmetrical connection (1.2). Therefore, if H is the energy density of the interaction between particle and torsion (i.e. the Hamiltonian), we have from (1.4)

$$\frac{d}{dt}\int d^3x\, H = \int d^3x\, B^{o} \qquad (1.6)$$

Since the explicit computation of B, retaining only the contribution of the first order in the coupling constant χ, gives

$$H = -i\, Q_{\mu\alpha}{}^{\alpha}\bar{\psi}\gamma^{\mu}\psi - \Sigma^{\mu\alpha\beta}Q_{\mu\alpha\beta} \qquad (1.7)$$

we obtain this 'nonminimal' interaction lagrangian for the Dirac field

$$\mathcal{L}_I = -H = i\, V_{\mu}J^{\mu} - A_{\mu}J^{5\mu} \qquad (1.8)$$

where

$$J^{\mu} = \bar{\psi}\gamma^{\mu}\psi\,,\quad J^{5\mu} = \bar{\psi}\gamma^{5}\gamma^{\mu}\psi\,,\quad V_{\mu} = Q_{\mu\alpha}{}^{\alpha},\quad A_{\mu} = \frac{1}{4}\varepsilon_{\mu\alpha\beta\gamma}Q^{\alpha\beta\gamma}$$

(the same interaction was already given in a different context[9], but here it is directly deduced from the motion equations). It is just on the ground of this coupling that we find a formal analogy with the weak interaction lagrangian.

To this purpose let us recall briefly some particularity of the Einstein-Cartan theory. According to the field equations of this theory, there is no torsion outside the matter distribution, since torsion is rigidly bound to the spin density $\Sigma_{\alpha\beta}^{\ \ \mu}$ by an algebraic relation[1]:

$$Q_{\alpha\beta}^{\ \ \mu} = \chi\left(\Sigma_{\alpha\beta}^{\ \ \mu} - \frac{1}{2}\delta_\alpha^\mu \Sigma_{\gamma\beta}^{\ \ \gamma} - \frac{1}{2}\delta_\beta^\mu \Sigma_{\alpha\gamma}^{\ \ \gamma}\right) \qquad (1.9)$$

In empty space, where $\Sigma = 0$, there is no torsion, and the field equations are reduced to the well known general relativity equations. It is possible, however, to introduce torsion even outside spinning bodies by adding a kinetic term to the field lagrangian of the torsion[6,8] and allowing a nonminimal coupling between torsion and matter[10].

As the Einstein-Cartan theory is, like the general relativity, a classical theory with a macroscopic domain of validity, we can say that at quantum level torsion is a propagating field with a non zero vacuum expectation value, at least in the close proximity of matter. In particular, we want to picture every spinning particle surrounded by a kind of "torsionic cloud", falling down in intensity very fast with distance, like the pionic virtual cloud surrounding a nuclear charge.

If we consider only processes with very low momentum transfer, then the torsion derivatives are negligible and we can neglect the kinetic term in the torsionic field lagrangian[6]. The variation of the total action, assuming the nonminimal coupling (1.8) as interaction lagrangian and taking the usual Einstein-Cartan static term R/χ for the field lagrangian, gives then[6,8]

$$V_\mu = i\frac{3}{16}\chi J_\mu, \qquad A_\mu = \frac{3}{16}\chi J_\mu^5 \qquad (1.10)$$

In this case the torsionic contact interaction between two Dirac particles (ψ and ψ') becomes

$$\mathcal{L} = i V_\mu(\psi) J^\mu(\psi') - A_\mu(\psi) J^{5\mu}(\psi') =$$
$$= -\frac{3}{16}\chi\left[J_\mu(\psi) J^\mu(\psi') + J^5_\mu(\psi) J^{5\mu}(\psi')\right] \quad (1.11)$$

and it may be written in the V – A standard form if at least one of the two fermions is massless and it is described then by a two component spinor. In fact, if for instance $\gamma^\mu \partial_\mu \psi' = 0$ and $(1 - \gamma^5)\psi' = 2\psi'$, then from (1.11), we have

$$\mathcal{L} = -\frac{3}{32}\chi\bar{\psi}'\gamma_\mu(1 - \gamma^5)\psi'(J^\mu + J^{5\mu}) =$$
$$= -\frac{3}{32}\chi\bar{\psi}'\gamma_\mu(1 - \gamma^5)\psi'\bar{\psi}\gamma^\mu(1 - \gamma^5)\psi \quad (1.12)$$

(In the 'minimal case', only the axial term, that is the second term $A_\mu J^{5\mu}$ appeared in the lagrangian (1.8), and the V – A structure of the interaction was achieved only if both particles had a negligible mass[3]).

So we have obtained a torsionic interaction lagrangian, describing leptonic and also semileptonic processes, which is formally identical to the weak interaction lagrangian in a flat space-time, except for the value of the coupling constant.

An even more detailed analogy can be established by considering, for instance, the particular lepton-nucleon interaction[8,11]. Therefore one is led to think that the torsion may provide a geometrical model for weak interactions, just like curvature does for gravitational forces. Notice however that the coupling constant, in the lagrangian (1.12), is

$$\frac{3\chi}{32} = \frac{3\pi k_g \hbar^2}{4c^2} \sim 10^{-81}\ \mathrm{erg\cdot cm^3}$$

while the Fermi weak coupling constant is

$$\frac{G_F}{(2)^{1/2}} \sim 10^{-50}\ \mathrm{erg\cdot cm^3}$$

In order to allow a complete identification of the torsionic and weak interactions, we must postulate then a spin-torsion coupling constant χ' which differs from the mass-curvature one by the factor

$$\frac{\chi'}{\chi} = (8/9)^{1/2} \frac{G_F c^2}{\pi k_g \hbar^2} \sim 10^{31} \quad (1.13)$$

(as regards the possibility of giving an arbitrary value to the torsionic constant, independent from the value of Newton's gravitational constant, one can find some investigation[4,12]).

At last we notice that, with a spin-torsion constant χ' given by (1.13), the minimal radius of the Universe, R_o, evaluated by cosmological theories which avoid the gravitational collapse by introducing torsion and then repulsive spin forces into the cosmological equations as in the case of Trautman's Universe[13], should become $R_o \sim 10^{10}$ cm, instead of $R_o \sim 1$ cm; a number which is much more close to the result given by the strong gravity theory that is $R_o \sim 10^{13}$ cm[14].

CHAPTER 2

STRONG GRAVITY AND WEAK INTERACTION

1. The weak interaction and the form factors.

In the previous chapter we have seen that one can try to understand weak interactions as being Cartan forces mediated by torsion and acting upon the intrinsic spin of the elementary particles.

In this chapter we want to consider the influence of the strong gravity (or "f-gravity")[15,16] and of the torsion upon the weak interactions, unrispectively of their own nature and origin. With reference to this purpose, let us recall some general, but not yet well understood, properties of the weak interaction.

We know, from baryon β decays, for instance, that the axial form factor g_A in the matrix element of the hadronic weak current $J(h)$, is different from unity also at zero momentum transfer ($q^2 = 0$). Consider for example the neutron decay amplitude: it is proportional to[17]

$$\langle p|\, J(h)\,|n\rangle \equiv \langle p|\, V(h) - A(h)\,|n\rangle \propto \bar{u}_p \gamma (g_V - g_A \gamma_5) u_n \quad (2.1)$$

and, from the experiments[17], we have

$$g_V \simeq 1 \quad , \qquad g_A/g_V = 1.25 \pm 0.01 \tag{2.2}$$

We have defined

$$g_{A,V} = g_{A,V}(0) = \lim_{q^2 \to 0} g_{A,V}(q^2) \tag{2.3}$$

and we shall always restrict our consideration to the $q^2 = 0$ limit. The numerical value of g_V can be theoretically justified by assuming that, in the approximation in which SU_3 is an exact symmetry for the hadronic world, the vector part of the hadronic current is conserved (C.V.C. hypothesis[17,19]): in fact if $\partial_k V^k(h) = 0$, then $g_V = 1$. (The same situation occurs in electrodynamics, where the conservation of the electromagnetic current implies that the electric charge is not renormalized by the strong interactions).

The anomalous value of g_A can be understood as a strong interaction effect, due in particular to the pion cloud coupled to baryons, which prevent the axial current from being a conserved vector (P.C.A.C. hypothesis[20]):

$$\partial_k A^k(h) \propto m_\pi \varphi_\pi \tag{2.4}$$

where m_π is the pion mass and φ_π the pionic field. As the leptons are not coupled to the meson field, for them $g_A = 1$ too and the leptonic weak current can be written in the simple form

$$J(l) = \bar{\psi} \gamma (1 - \gamma_5) \psi \tag{2.5}$$

Even if on these grounds a good qualitative understanding of the weak interactions structure is achieved, the quantitative predictions on the value of g_A, however, are not in a perfect agreement with the experimental results.

This is the starting point of our considerations. Since the axial weak current is affected also by gravity, in addition to strong interactions (as other authors have already pointed out[21]) and the gravitational corrections to g_A are not neglegible, a priori, in the case of the hadronic strong gravity, then we want to suggest that the experimental value of g_A can fully agree with a theoretical computation only if also the strong gravity contribution to the form factor is considered.

2. Semileptonic interactions on a strong metric background.

According to the "tensor-meson dominance" hypothesis[22], leptons interact gravitationally among themselves by exchanging spin 2, massless particles (gravitons) which are coupled to matter by Newton's constant (k_g); the gravitational interaction among hadrons, on the contrary, takes place with the exchange of a sort of heavy graviton (the f meson) with a coupling constant (k_f) of the same order as the strong one

$$k_f \, m_p^2 \sim \hbar c \qquad k_f/k_g \sim 10^{39} \tag{2.6}$$

(m_p is the nucleonic mass).

This is the situation from a quantum point of view. From a classical (or geometrical) point of view, we must consider that while leptons interact in a nearly flat metric background (since the leptonic gravitational interaction is always negligible, in a first approximation, if compared with the others) the hadronic interactions, on the contrary, take place in a curved space-time whose metric tensor represents the coupling to the f meson field, which is never negligible.

Now we consider a semileptonic weak interaction, which, according to the standard current-current theory[23] is described by the flat-space lagrangian

$$\mathcal{L} = \frac{G_F}{(2)^{1/2}} \eta^{ik} J_i(h) \, J_k^+(l) + \text{h.c.} \tag{2.7}$$

(for the moment we are concerned only with strangeness-conserving processes, then we can neglect the Cabibbo angle ϑ_c, working in the limit $\vartheta_c = 0$).

In a first approximation ("naked" and pointlike particles) the currents can be exactly written in the V - A form

$$J(h) = J(l) = \bar{\psi} \gamma (1 - \gamma_5) \psi \tag{2.8}$$

We must remember, at this point, that also when the hadron we are considering is a boson, its weak interactions can be described in terms of its constituent quarks. In any case, therefore, the fields appearing in the current (2.8) are spinors satisfying the Dirac equation

$$(i\gamma^{\kappa}\partial_{\kappa} - \frac{mc}{\hbar})\psi = 0 \tag{2.9}$$

Multiplying (2.9) by $(-i\gamma^{\ell}\partial_{\ell} - mc/\hbar)$ and imposing on the new equation the condition that it reduce to the Klein-Gordon equation (in order that the energy condition $p^2 = m^2c^2$ holds), we get the well known property of the Dirac matrices in a flat space-time

$$\gamma^{(i}\gamma^{\kappa)} = \eta^{i\kappa} \tag{2.10}$$

(round brackets denote symmetrization). In a curved space-time this condition must be obviously generalized by the introductio of four new matrices γ^{μ} such that

$$\gamma^{(\mu}\gamma^{\nu)} = g^{\mu\nu} \tag{2.11}$$

This can be accomplished in a standard way, as is well known, if one defines, at each space-time point, a system of four vectors $\{e_{\kappa}{}^{\mu}\}$ (the tetrads), which span the local Minkowski space tangent to the Riemann space[50,51]. Here , and in what follows, Latin letters are Lorentz indices and denote the vectors, while Greek letters are covariant indices and denote the vector components in the Riemann space. We define, as usual, the new matrices

$$\gamma^{\mu} = \eta^{i\kappa} e_i{}^{\mu} \gamma_{\kappa} \tag{2.12}$$

Then, by tetrads properties

$$\eta^{i\kappa} e_i{}^{\mu} e_{\kappa}{}^{\nu} = g^{\mu\nu} \tag{2.13}$$

also the condition (2.11) is satisfied.

In our case, hadrons interact with strong gravity, represented by the metric tensor $f_{\mu\nu}$, while leptons interact with ordinary gravity $g_{\mu\nu}$; we have two metric tensors, and so we are compelled to introduce two systems of tetrads $\{a_\kappa{}^\mu\}$ and $\{e_\kappa{}^\mu\}$ such that

$$\eta^{i\kappa} a_i{}^\mu a_\kappa{}^\nu = f^{\mu\nu}, \qquad \eta^{i\kappa} e_i{}^\mu e_\kappa{}^\nu = g^{\mu\nu}$$
$$f_{\mu\nu} a_i{}^\mu a_\kappa{}^\nu = \eta_{i\kappa}, \qquad g_{\mu\nu} e_i{}^\mu e_\kappa{}^\nu = \eta_{i\kappa} \tag{2.14}$$

and their inverses $\{a_{\kappa\mu}\}$, $\{e_{\kappa\mu}\}$ defined by

$$\eta^{i\kappa} a_i{}^\mu a_{\kappa\nu} = \delta^\mu_\nu \quad , \quad \eta^{i\kappa} e_i{}^\mu e_{\kappa\nu} = \delta^\mu_\nu \tag{2.15}$$

The generalization of the Dirac matrices to their general covariant expression (2.12) involving the tetrads, induces then a differentiation of the two weak currents:

$$J_\kappa(1) \longrightarrow J_\mu(1) = \eta_{i\kappa} e^i{}_\mu J^\kappa(1)$$
$$J_\kappa(h) \longrightarrow J_\mu(h) = \eta_{i\kappa} a^i_\mu J^\kappa(h) \tag{2.16}$$

(the γ_5 matrix is not modified in a curved space-time, because of its definition[24]). In order to introduce the gravitational interaction into the semileptonic process (2.7), according to the minimal coupling principle, we must simply make general covariant the flat-space lagrangian.

In this case this is an ambigous prescription, however[22], owing to the existence of two metric fields: in fact, from (2.16) we have

$$J_\mu(h)\, J^\mu(1) \neq J^\mu(h)\, J_\mu(1) \tag{2.17}$$

since, in general

$$a_{\kappa\mu} e_i{}^\mu \neq a_\kappa{}^\mu e_{i\mu} \neq \eta_{i\kappa} \tag{2.18}$$

Therefore the 'minimal' semileptonic lagrangian, after introducing gravity, may be written, in general[22]

$$\mathcal{L} = G_F \left(\frac{e\,a}{2}\right)^{1/2} W_1 J_\mu(h) J^{+\mu}(l) + W_2 J^\mu(h) J^+_\mu(l) + h.c. \tag{2.19}$$

where $W_1 + W_2 = 1$ and $e \equiv \det e^\kappa{}_\mu$, $a \equiv \det a^\kappa{}_\mu$ or, written in full,

$$\mathcal{L} = G_F\left(\frac{e\,a}{2}\right)^{1/2} (W_1 g^{\mu\nu} + W_2 f^{\mu\nu}) \eta^{i\kappa} \eta^{jb} a_{i\mu} e_{j\nu} J_\kappa(h) J^+_b(l) + h.c. \tag{2.20}$$

where the two metric fields, $f^{\mu\nu}$ and $g^{\mu\nu}$, are solutions of the coupled field equations of the two-tensor theory, which in vacuum take the form[22]

$$G_{\mu\nu}(f) + \frac{1}{(-f)^{1/2}} \frac{\partial L(f,g)}{\partial f^{\mu\nu}} = 0$$
$$G_{\mu\nu}(g) + \frac{1}{(-g)^{1/2}} \frac{\partial L(f,g)}{\partial g^{\mu\nu}} = 0 \tag{2.21}$$

Here $G_{\mu\nu}$ is the usual Einstein tensor, and $L(f,g)$ is a suitable (and non linear) interaction term mixing the two tensor fields, which vanishes, in the lowest order, if both fields are massless[22].

Since the strong gravity range is much longer than the range of the weak interaction (which behave, in the low energy limit, like a contact interaction) we can regard strong gravity, in this application, as a long range force, and neglect, in a first approximation, the mass of the f meson (as pointed out also elsewhere[29]). In this limit the equations (2.21) become uncoupled

$$R_{\mu\nu}(f) = 0 \quad , \quad R_{\mu\nu}(g) = 0 \tag{2.22}$$

($R_{\mu\nu}$ is the Ricci tensor). If we neglect g-gravity, that is if we impose

$$e_{\kappa}^{\ \mu} \simeq \delta_{\kappa}^{\ \mu} \qquad (2.23)$$

the semileptonic lagrangian (2.20) becomes

$$G_F \left(\frac{a}{2}\right)^{1/2} \left[W_1 a^{\kappa}_{\ \mu} J_{\kappa}(h) L^{+\mu} + W_2 a_{\kappa}^{\ \mu} J^{\kappa}(h) L^{+}_{\mu} \right] + \text{h.c.} \qquad (2.24)$$

where $L^{\mu} = \delta^{\mu}_{\kappa} J^{\kappa}(1)$.

At last we must specify the strong metric: we choose, for simplicity, a static and spherically symmetrical solution of (2.22), so that the "strong line-element", in isotropic rectangular coordinates, becomes

$$ds^2 = \left(\frac{1 - \varphi/2}{1 + \varphi/2}\right)^2 c^2\, dt^2 - (1 - \varphi/2)^4 \left|d\vec{r}\right|^2 \qquad (2.25)$$

$$\varphi = U_o/r = U_o\,(x_1^{\,2} + x_2^{\,2} + x_3^{\,2})^{-1/2}$$

where U_o is a constant which, in general relativity, is fixed by imposing that in the weak field approximation and in the nonrelativistic limit, the gravitational force be reduced to its Newtonian expression. Such a condition cannot be imposed in our case, evidently, since the strong gravity field falls rapidly to zero outside a hadron, and there is no classical, macroscopical limit. Therefore we do not give U_o explicitly for the present, and putting

$$f_{\mu\nu} = (f^{\mu\nu})^{-1} = \left(\begin{array}{c|c} \alpha & 0 \\ \hline 0 & I\beta \end{array}\right) \qquad (2.26)$$

where

$$\alpha = \left(\frac{1 - \varphi/2}{1 + \varphi/2}\right)^2, \quad \beta = -(1 - \varphi/2)^4, \quad I = \begin{pmatrix} 1 & & 0 \\ & 1 & \\ 0 & & 1 \end{pmatrix} \qquad (2.27)$$

we choose the following system of tetrads $\{a^{\kappa}{}_{\mu}\}$ with their inverse $\{a_{\kappa}{}^{\mu}\}$:

$$a^{\kappa}{}_{\mu} = \left(\begin{array}{c|c} \alpha^{1/2} & 0 \\ \hline 0 & I(-\beta)^{1/2} \end{array}\right), \quad a^{\kappa\mu} = \left(\begin{array}{c|c} 1/\alpha^{1/2} & 0 \\ \hline 0 & -I/(-\beta)^{1/2} \end{array}\right) \tag{2.28}$$

$$a_{\kappa\mu} = \left(\begin{array}{c|c} \alpha^{1/2} & 0 \\ \hline 0 & -I(-\beta)^{1/2} \end{array}\right) \quad a_{\kappa}{}^{\mu} = \left(\begin{array}{c|c} 1/\alpha^{1/2} & 0 \\ \hline 0 & I/(-\beta)^{1/2} \end{array}\right)$$

3. Neutron beta decay.

Now we consider the special case of the neutron β decay

$$n \longrightarrow p + e^{-} + \bar{\nu}_e \tag{2.29}$$

This process is described by the semileptonic lagrangian (2.24) if we put

$$J_{\kappa}(h) = \bar{\Psi}_p \gamma_{\kappa} (1-\gamma_5)\Psi_n , \quad J_{\kappa}(l) = \bar{\Psi}_e \gamma_{\kappa} (1-\gamma_5)\Psi_{\nu_e} \tag{2.30}$$

where Ψ_p, Ψ_n, Ψ_e, Ψ_{ν_e} are the Dirac fields of proton, neutron, electron and electronic neutrino. Recalling that the hadron undergoing a β decay can be assumed to be nonrelativistic[25], we can use the well known properties of the Dirac spinor in this limit

$$\begin{aligned} \bar{\Psi}\gamma_{\kappa}\Psi &\simeq \Psi^{+}\Psi & \kappa &= 0 \\ &\simeq 0 & \kappa &= 1, 2, 3 \\ \bar{\Psi}\gamma_5\gamma_{\kappa}\Psi &\simeq 0 & \kappa &= 0 \\ &\simeq \Psi^{+}\vec{\sigma}\Psi & \kappa &= 1, 2, 3 \end{aligned} \tag{2.31}$$

and then, introducing into (2.24) the strong metric (2.26), we are led to the following lagrangian describing the neutron β decay in the strong gravity nucleonic field:

$$\mathcal{L} = \frac{G_F}{(2)^{1/2}} \left\{ \left[W_1 \alpha^{3/4} (-\beta)^{3/4} + W_2 \alpha^{-1/4} (-\beta)^{3/4} \right] \psi_p^+ L^o \psi_n + \right. \tag{2.32}$$

$$\left. + \left[W_1 \alpha^{1/4} (-\beta)^{5/4} + W_2 \alpha^{1/4} (-\beta)^{1/4} \right] \psi_p^+ \vec{\sigma} \cdot \vec{L} \psi_n \right\}$$

Notice that, if we introduce an "effective" hadronic current $\tilde{J}$

$$\tilde{J}_\kappa = \bar{\psi}_p \gamma_\kappa (G_V - G_A \gamma_5) \psi_n \tag{2.33}$$

then the flat-space lagrangian

$$\frac{G_F}{(2)^{1/2}} \tilde{J}_\kappa L^{+\kappa} \tag{2.34}$$

in the nonrelativistic limit becomes

$$\frac{G_F}{(2)^{1/2}} \left\{ G_V \psi_p^+ L^o \psi_n + G_A \psi_p^+ \vec{\sigma} \cdot \vec{L} \psi_n \right\} \tag{2.35}$$

Comparing this expression with (2.32), we see that the strong gravity effects can be reproduced, in a flat space-time, defining an effective hadronic current like (2.33), provided that we put

$$G_V = W_1 (-\alpha\beta)^{3/4} + W_2 \alpha^{-1/4} (-\beta)^{3/4}$$
$$\tag{2.36}$$
$$G_A = W_1 \alpha^{1/4} (-\beta)^{5/4} + W_2 (-\alpha\beta)^{1/4}$$

Since we are in the nonrelativistic approximation, these relations can be greatly simplified as, from virial theorem, $(v/c)^2 \sim \varphi$, and then we can also use the weak field approximation for the gravitational potential φ.

In this limit

$$\alpha \longrightarrow 1 - 2\varphi = \alpha'$$
$$\beta \longrightarrow -(1 + 2\varphi) \simeq -1/\alpha' \qquad (2.37)$$

and the relation between metric and effective form factors become

$$G_V = W_1 + W_2(\alpha')^{-1} \quad , \quad G_A = W_1(\alpha')^{-1} + W_2 \qquad (2.38)$$

In the approximation in which SU_3 can be regarded as an exact symmetry for the hadronic world, the vector part of the weak current must be conserved. Therefore, in this approximation, we can fix the value of the parameters W_1 and W_2 by imposing $G_V = 1$. The set of equations

$$W_1 + W_2(\alpha')^{-1} = 1 \quad , \quad W_1 + W_2 = 1 \qquad (2.39)$$

gives then $W_1 = 1$, $W_2 = 0$ and so the strong gravity contribution to the weak form factors is, from (2.38)

$$G_V = 1 \quad , \quad G_A = 1 + 2\varphi \qquad (2.40)$$

It is a simple matter, at this point, to get an estimate of G_A in order to check the validity of our approximations.

We consider the axial vector matrix element between two fully "dressed" nucleonic states (that is in the presence of strong interactions, in addition to strong gravity): if g_A is the axial form factor, including both pion and f meson contributions, we have

$$\langle p | A_\kappa(h) | n \rangle \propto g_A \bar{u}_p \gamma_\kappa \gamma_5 u_n \qquad (2.41)$$

In the limit $\varphi \longrightarrow 0$ (no gravitational interaction at all, $G_A = 1$) it must be

$$\left[g_A\right]_{G_A=1} = g_A(\pi) \qquad (2.42)$$

if $g_A(\pi)$ denotes only the pionic contribution. On the contrary, in the limit $g_A(\pi) = 1$ (no pion cloud, only strong gravity), the effective current (2.33) leads to

$$\langle p | A_k(h) | n \rangle = \langle \bar{\psi}_p G_A \gamma_k \gamma_5 \psi_n \rangle \propto G_A \bar{u}_p \gamma_k \gamma_5 u_n \tag{2.43}$$

namely

$$[g_A]_{g_A(\pi)=1} = G_A \tag{2.44}$$

Then we can put, for values of g_A close enough to unity

$$g_A \simeq G_A g_A(\pi) = g_A(\pi)\,(1 + 2\varphi) \tag{2.45}$$

The theoretical estimate of the pionic form factor $g_A(\pi)$, performed by means of the Adler-Weisberger sum rule, is $g_A(\pi) = 1.15$. Inserting this number into (2.45), together with the experimental result (2.2) we get, for a nucleon

$$\varphi = \frac{1}{2}\left(\frac{g_A}{g_A(\pi)} - 1\right) \simeq 4.5 \cdot 10^{-2} \tag{2.46}$$

and so our use of the weak field approximation is justified.

4. The axial form factor in very light nuclei.

An analysis of the presently available experimental results on the muon nuclear capture rates in very light nuclei, seems to suggest that the value of the axial form factor vary with the increasing nuclear complexity[26]. A similar behaviour can be easily deduced from our previous relation (2.45).

Remember, in fact, that in the absence of the pionic cloud ($g_A(\pi) = 1$) the axial form factor must approach unity in the limit $m_p \longrightarrow 0$[20]. Then we must impose on the gravitational potential instead of the newtonian limit, the following condition for fixed value of r

$$[\varphi]_{g_A(\pi)=1} \xrightarrow{m_p \to 0} 0 \tag{2.47}$$

The simplest choice is to assume, in analogy with the general relativistic expression for the potential, a linear dependence on the mass, namely $[U_0]_{g_A(\pi)=1} \simeq K\, m_p$ (2.48) where K is a constant. From (2.45) then we are led to

$$g_A - 1 \simeq 2K \frac{m_p}{r} \tag{2.49}$$

Since strong gravity has no practical effect at a distance greater than the hadron spatial extension, replacing r with the radius of the hadron itself, for a nucleus of atomic number A (mass = $m_p A$, radius = $r_0 A^{1/3}$) we have

$$g_A \simeq 1 + b A^{2/3} \tag{2.50}$$

where $b = 2K m_p / r_0 =$ const. That is: "The nuclear axial coupling constant increases together with the nuclear surface" (for small values of the mass number A only).

From muon capture experiments in hydrogen, deuterium and ^{3}He nuclei, one can obtain[26]

$$\begin{aligned} g_A &= 1.17 \pm 0.03 \qquad A = 1 \\ g_A &= 1.39 \pm 0.1 \qquad A = 2 \\ g_A &= 1.27 \pm 0.06 \qquad A = 3 \end{aligned} \tag{2.51}$$

If we take b from muon capture in hydrogen, the theoretical prediction (2.50) are not in disagreement with the experimental results (2.51), as we can see in Fig.1.

It must be noticed, however, that the incomplete experimental knowledge of the other form factors, which is needed in order to extract g_A from the experiments, has been overcome by means of theoretical assumptions which may have altered the results (2.51) as it is shown in a complete analysis of data[26].

Finally we want to check the validity of our statement that the strong gravitational contribution to g_A must be evaluated at a distance of the order of a particle's radius.

If we define $K = K_f'/c^2$, where K_f' has the same dimensions as the gravitational constant, putting in (2.49) $r = r_0 = 1.2 \cdot 10^{-13}$cm (nucleon radius) by means of the estimate (2.46) we find

$$K_f' = 4.5 \cdot 10^{-2} \frac{r_0 c^2}{m_p} = 0.29 \cdot 10^{31} \text{c.g.s.} \tag{2.52}$$

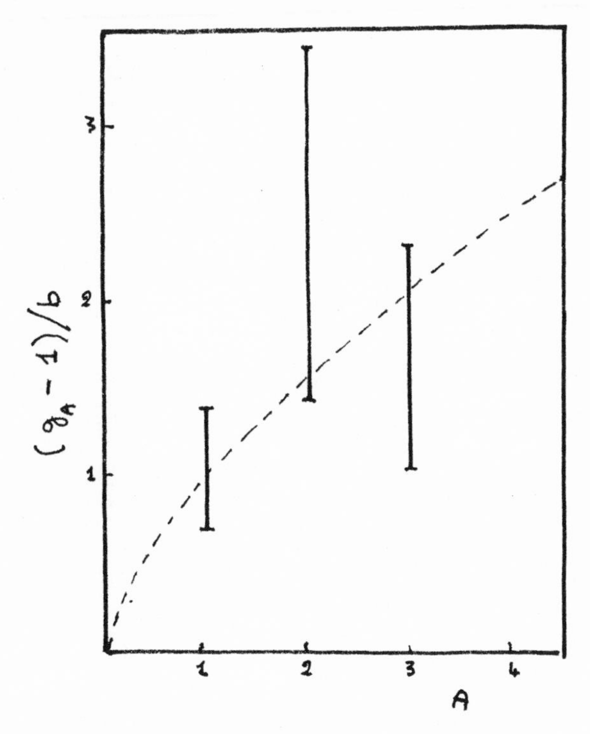

Fig.1 The dashed curve represents the theoretical prediction $A^{2/3}$: the full lines are the numerical values experimentally allowed

The standard evaluation of the strong gravitational coupling constant, on the other hand, are performed by requiring that the "strong Schwarzschild radius" of the nucleon be equal to its Compton wave lenght[16,27]. This gives

$$K_f = \frac{\hbar c}{2 m_p^2} = 0.57 \cdot 10^{31} \text{c.g.s.} \quad (2.53)$$

and this number is close enough to(2.52) to support our speculation about the influences of the strong gravity on the weak form factors.

5. Strong Gravity with Torsion.

The deviations from the general relativity equations produced by spin corrections according to the Einstein-Cartan theory, are negligible at normal matter densities[28] since, in this theory, the spin-torsion coupling constant is the same as the Newton constant K_g.

If however we introduce spin and torsion into the classical field equations of the strong gravity theory, according to the Einstein-Cartan formalism, the contribution of the hadronic spin to the strong metric cannot be neglected provided that the equality between the hadronic spin-torsion coupling constant and the strong gravity constant K_f ($\sim 10^{39} K_g$) is assumed.

Consider in fact the strong gravity field equations which, in presence of matter must be written, in general, as

$$G_{\mu\nu}(f) = \chi_f T^{(h)}_{\mu\nu} + \frac{1}{(-f)^{1/2}} \frac{\partial L}{\partial f^{\mu\nu}}$$
$$G_{\mu\nu}(g) = \chi_g T^{(1)}_{\mu\nu} + \frac{1}{(-g)^{1/2}} \frac{\partial L}{\partial g^{\mu\nu}} \qquad (2.54)$$

Now we generalize the affine connection Γ of the two Riemann spaces introducing an asymmetric part, related to the material spin as in the Einstein-Cartan theory:

$$\Gamma_{\alpha\beta}^{\mu}(f) = \left\{ {\mu \atop \alpha\beta} \right\}^{(f)} - K^{(h)}{}_{\alpha\beta}{}^{\mu}$$
$$\Gamma_{\alpha\beta}^{\mu}(g) = \left\{ {\mu \atop \alpha\beta} \right\}^{(g)} - K^{(\ell)}{}_{\alpha\beta}{}^{\mu} \qquad (2.55)$$

The field equations become then, written in terms of the system of tetrads (2.14) and (2.15)

$$G_{\mu\nu}(f) = \chi_f T^{(h)}_{\mu\nu} + \frac{1}{(-f)^{1/2}} \frac{\partial L}{\partial a^{\kappa\mu}} a^{\kappa}{}_{\nu}$$
$$Q^{(h)}_{\alpha\beta\mu} = \chi_f J^{(h)}_{\alpha\beta\mu} \qquad (2.56')$$

$$G_{\mu\nu}(g) = \chi_g T^{(l)}_{\mu\nu} + \frac{1}{(-g)^{1/2}} \frac{\partial L}{\partial e^{\kappa\mu}} e^{\kappa}{}_{\nu}$$

$$Q^{(l)}_{\alpha\beta\mu} = \chi_g J^{(l)}_{\alpha\beta\mu} \qquad (2.56\ )$$

where $G_{\mu\nu}$ is no more a symmetrical tensor, and

$$T^{(h)}_{\mu\nu} = \frac{1}{(-f)^{1/2}} \frac{\delta\Lambda}{\delta a^{\kappa\mu}} a^{\kappa}{}_{\nu}, \quad T^{(l)}_{\mu\nu} = \frac{1}{(-g)^{1/2}} \frac{\delta\Lambda}{\delta e^{\kappa\mu}} e^{\kappa}{}_{\nu} \qquad (2.57)$$

if Λ is the matter lagrangian density, and finally

$$J_{\alpha\beta}{}^{\mu} = \Sigma_{\alpha\beta}{}^{\mu} - \frac{1}{2} \delta^{\mu}_{\alpha} \Sigma_{\nu\beta}{}^{\nu} - \frac{1}{2} \delta^{\mu}_{\beta} \Sigma_{\alpha\nu}{}^{\nu} \qquad (2.58)$$

where $\Sigma_{\mu\alpha\beta}$ is the canonical spin density tensor. In our approximation the interaction term in (2.54) may be neglected and then $L = 0$; moreover $g_{\mu\nu} \simeq \eta_{\mu\nu}$, because we can disregard ordinary gravity.

The strong gravity field produced by a hadronic matter density is then determined by the usual Einstein-Cartan equations:

$$G_{\mu\nu}(f) = \chi_f T^{(h)}_{\mu\nu}$$

$$Q^{(h)}_{\alpha\beta\mu} = \chi_f J^{(h)}_{\alpha\beta\mu} \qquad (2.59)$$

and so, although these equations reduce in vacuum to Einstein's familiar equations for empty space[30,31,32] , the metric field is in general different from the solution of general relativity, because it is affected also by the spin contents of the gravitating body[33].

6. Spin correction to the strong metric.

We want to introduce this spin correction into the static and spherically symmetrical strong field of the nucleon, given by (2.25).

Using the Weyssenhoff fluid[34,52] as a semiclassical model of spinning matter, and assuming for the nucleon a radius $r \sim \hbar/m_p c$, we can see that we must replace the constant U_o in (2.25) by the new constant U[35]

$$U = U_o - U_1 = K_f \frac{m_p}{c^2} - \frac{3}{8} K_f^2 \frac{m_p^3}{\hbar c^3} \qquad (2.60)$$

Since

$$U_1/U_o = 3 K_f m_p^2/8 \hbar c \sim 1 \qquad (2.61)$$

the torsionic corrections to the strong gravity equations cannot be neglected, as previously stated, while, for ordinary gravity, replacing K_f with K_g we have

$$U_1(g)/U_o(g) = 3 K_g m_p^2/8\hbar c \sim 10^{-39} \qquad (2.62)$$

The spin effects on the metric are then reducible, with suitable approximation, to a correction of the linear relation between the hadronic mass and the potential φ. Notice however that the condition (2.47) , which must be satisfied in order that a connection between strong gravity and the axial form factors be established, still holds.

Therefore it is possible to obtain an estimate of the nucleonic strong potential, including spin contributions, from (2.46):

$$(U/r)_{r = \hbar/m_p c} \simeq 4.5 \cdot 10^{-2} \qquad (2.63)$$

This relation can be used, together with (2.60), to evaluate K_f. Since it is a quadratic equation, it provides then two solutions, $K^{(+)}$ and $K^{(-)}$:

$$K^{(\pm)} = \frac{4}{3} \frac{\hbar c}{m_p^2} (1 \pm 0.96) \qquad (2.64)$$

which are rather different $(K^{(-)}/K^{(+)} \sim 10^{-2})$ and only $K^{(+)}$

is of the same order as the usually accepted value for the strong gravity coupling constant ($K_f \sim \hbar c/m_p^2$).

7. The Cabibbo angle.

It is worth stressing, however, that from (2.64) we are naturally led to the introduction of an angle θ . In fact, defining

$$K_f = 8\hbar c/ 3m_p^2 \tag{2.65}$$

we can put

$$K^{(+)} = K_f \cos^2\theta \quad , \qquad K^{(-)} = K_f \operatorname{sen}^2\theta \tag{2.66}$$

where $\cos^2\theta = 0.98$ and $\operatorname{sen}^2\theta = 0.02$. It is amusing to observe that, with these numbers, on has

$$\theta \simeq \frac{1}{2}\theta_c \tag{2.67}$$

where θ_c is the Cabibbo angle.

In order to suggest that this is not only a numerical coincidence, let us consider once again the relation between the effective hadronic form factor and the strong gravity potential (2.36).

In the approximation of exact SU_3 symmetry, if we develop the form factors in power series of φ (weak field approximation) we get, to the first order, only a contribution to the axial charge

$$\begin{aligned} G_F\, G_V &\simeq G_F \\ G_F\, G_A &\simeq G_F\,(1 + 2\varphi) \end{aligned} \tag{2.68}$$

(see (2.40)) while, to the second order

$$\begin{aligned} G_F\, G_V &\simeq G_F\,(1 - \varphi^2/4)^{1/2} \\ G_F\, G_A &\simeq G_F\,(1 - \varphi^2/4)^{1/2}(1 + 2\varphi) \end{aligned} \tag{2.69}$$

we obtain also a renormalization of the Fermi constant:

$$G_F \longrightarrow G_F' = G_F(1 - \varphi^2/4)^{1/2} \tag{2.70}$$

Whether we choose $K^{(+)}$ or $K^{(-)}$ as being the strong gravity constant, from (2.60), (2.65) and (2.66), U can be written, in both cases

$$U = (2\hbar/3\, m_p c)\ \mathrm{sen}^2 2\theta \tag{2.71}$$

Therefore, evaluating the contribution of the potential at a distance of the order of $\hbar/m_p c$, as before, in the limit $\theta \sim 0$ we have from (2.70) and (2.71)

$$G_F' = G_F \cos 2\theta \tag{2.72}$$

On the other hand the effective coupling constant for neutron's decay according to the Cabibbo theory, is given, as is well known, by

$$G_F \cos \theta_c \tag{2.73}$$

Again we are led to the identification $\theta = \theta_c/2$ in agreement with the previous numerical suggestion, and this seems to indicate that it may be possible to interpretate the Cabibbo angle as a torsion-strong-gravity correction to the weak interaction lagrangian.

CHAPTER 3

STRONG GRAVITY CONTRIBUTIONS TO THE ELECTROMAGNETIC FIELD OF A PARTICLE : A STARTING POINT FOR COSMOLOGICAL DEDUCTIONS

According to the so called "vector meson dominance" hypothesis[15], the electromagnetic interactions among hadrons are mediated by the exchange of the ρ meson, while among leptons they are mediated, as usual, by photons γ.

Since ρ and γ are coupled by a mixing term, if an electromagnetic field is given, a probability of finding a locally nonzero density of ρ mesons is also given, and then also a strong gravity field, because ρ, being a hadron, is gravitationally coupled to the f meson[15]. Therefore it makes sense to speak about a strong gravity influence upon the electromagnetic field of a particle, whether the particle be a hadron or a lepton (obviously the influences of g-gravity are fully negligible, even at a distance of the order of nuclear radius[36]).

We shall represent this influence adopting a classical approach as before, i.e. by writing the electromagnetic equations in a curved space-time, and taking the ρ mesons produced by the electromagnetic field as the source of the strong metric, according to Salam's theory.

We assume as the basic lagrangian for the $\rho-\gamma$ coupling to the first order in e/g, the following expression[15,37]

$$\mathcal{L} = \mathcal{L}_1 + \mathcal{L}_2 \tag{3.1}$$

$$\mathcal{L}_1 = -\frac{1}{4} A_{ik} A^{ik} - e A_i \left[I^i(h) + I^i(l) \right] \tag{3.2}$$

$$\mathcal{L}_2 = -\frac{1}{4} \rho_{ik} \rho^{ik} + \frac{1}{2} m_\rho^2 \rho_i \rho^i - g \rho_i I^i(h) + \frac{e^2}{g} \rho_i I^i(l) \tag{3.3}$$

$$A_{ik} = 2\partial_{[i} A_{k]}, \qquad \rho_{ik} = 2\partial_{[i} \rho_{k]} \tag{3.4}$$

where $\hbar = c = 1$, m_ρ is the ρ mass, A_i and ρ_i are the photon and ρ meson fields, $g^2/\hbar c \simeq 14$ and $e^2/\hbar c \simeq 1/137$ are the dimensionless coupling constant for the strong and electromagnetic interactions respectively, and finally $I(h)$ and $I(l)$ are the hadronic and leptonic current densities.

The electromagnetic field produced by a hadron in flat space-time is described then by the equations

$$\partial_l A^{kl} = - e I^k(h) \tag{3.5}$$

obtained by putting $I(l) = 0$ in $\mathcal{L}_1$.

If we introduce now strong gravity, we must rewrite (3.5) in a curved space-time endowed with a metric tensor f_{ik}. So we obtain

$$\partial_\ell \left[(-f)^{1/2} f^{k\alpha} f^{\ell b} A_{\alpha b} \right] = - e (-f)^{1/2} I^k(h) \tag{3.6}$$

Consider, for simplicity, the electrostatic field of a pointlike hadron at rest. Then (throughout this chapter Greek indices run from 1 to 3) we have $I^\alpha = A_\alpha = 0$, $(-f)^{1/2} I^0 = \delta(\vec{x})$ and $\partial A_i / \partial t = 0$ and the equation (3.6) is reduced to[36]

$$\partial_\alpha \varepsilon^{\alpha\beta} \partial_\beta A_0 = - e \delta(\vec{x}) \tag{3.7}$$

$$\varepsilon^{\alpha\beta} = - (-f)^{1/2} f^{\alpha\beta} / f_{00} \tag{3.8}$$

where f_{ik} is a solution of the strong gravity equations, which in the limit $m_f \simeq 0$ and in the weak field approximation $(f_{ik} = \eta_{ik} + \lambda_{ik})$ become

$$\Box \lambda_{ik} = 2\chi_f \left(\tau_{ik} - \frac{1}{2} \eta_{ik} \tau \right) \tag{3.9}$$

where, if we take the ρ mesons as the strong gravity source, we have

$$\tau_{ik} = - \rho_{i\ell} \rho^{k\ell} + \frac{1}{4} \eta_{ik} \rho_{\ell m} \rho^{\ell m} + m_\rho^2 \left(\rho_i \rho_k - \frac{1}{2} \eta_{ik} \rho_\ell \rho^\ell \right) \tag{3.10}$$

where the ρ meson field must be determined by the following equations

$$\partial_\ell \rho^{k\ell} - m_\rho^2 \rho^k = - g I^k(h) \tag{3.11}$$

(obtained from the lagrangian $\mathcal{L}_2$ with $I(1) = 0$). Assuming also $m_\rho \simeq 0$, (since $m_\rho < m_f$ and we are in the limit $m_f = 0$), in the static case of a pointlike particle (3.11) gives $\rho_\alpha = 0$, $\rho_0 = g/r$. Then (3.8) becomes

$$\varepsilon_{\alpha\beta} = \delta_{\alpha\beta}(1 - \lambda_{oo}) + \lambda_{\alpha\beta} \tag{3.12}$$

and the field equations (3.9) are reduced to

$$\nabla^2\lambda_{oo} = -\chi_f(\rho_{\alpha o})^2$$
$$\nabla^2\lambda_{\alpha\beta} = 2\chi_f[\rho_{\alpha o}\rho_{\beta o} - \frac{1}{2}\delta_{\alpha\beta}(\rho_{\alpha o})^2] \tag{3.13}$$

By inserting now $\rho_o = g/2$ into (3.13) and the expression (3.12) into the equation (3.7), we obtain the following expression for the Coulomb field, plus the strong gravity contributions induced by ρ mesons, of a pointlike hadron[37]:

$$\vec{E} = \frac{e\vec{x}}{r^3}(1 - \frac{2K_f g^2}{c^4 r^2}) \tag{3.14}$$

It is worth stressing that as a limit case of this solution a value of r exist for which the electric potential

$$V = \frac{e}{r}(1 - \frac{2K_f g^2}{3c^4 r^2}) \tag{3.15}$$

has an equilibrium point, and the electric force upon a test charge become zero. In fact

$$\lim_{r \to g(2K_f)^{1/2}/c^2} \vec{\nabla} V = 0 \tag{3.16}$$

Taking, as usual, $K_f \sim \hbar c/m_p^2$, this hadronic characteristic radius r(h) is

$$r(h) = g\,\frac{(2\,K_f)^{1/2}}{c^2} \simeq \frac{\hbar}{m_p c}\left(\frac{g^2}{\hbar c}\right)^{1/2} \sim \frac{\hbar}{m_p c} \qquad (3.17)$$

and then: the strong gravity contributions to the electromagnetic field of a hadron become dominant at a distance of the order of the proton Compton wave-length; such a distance, incidentally, is also the equivalent of the protonic Schwarzschild radius in the strong gravity theories[38,40].

In the case in which the electric charge be carried by a lepton, developing the same computations as before, starting from the lagrangian (3.1) with $I(h) = 0$, we obtain

$$\vec{E} = \frac{e\,\vec{x}}{r^3}\left(1 - \frac{2\,K_f\,e^2}{c^4\,r^2}\right) \qquad (3.18)$$

and the corresponding leptonic characteristic radius $r(l)$ turns out to be

$$r(l) = \frac{e}{c^2}\,(2\,K_f)^{1/2} \sim \frac{\hbar}{m_p\,c}\left(\frac{e^2}{\hbar\,c}\right)^{1/2} \qquad (3.19)$$

Noticing that $r(l)/r(h) = e/g$ and that[39] $e^2/g^2 \simeq m_e/m_p$, where m_e is the electron mass, we are led to the interesting relation between masses and equilibrium radii:

$$m_e/r^2(l) = m_p/r^2(h) = \text{const.} \qquad (3.20)$$

By supposing that (3.20) be universally valid, and may be extended also to the Universe as a whole, i.e. by writing

$$m_e/r^2(l) = m_p/r^2(h) = M/R^2 \qquad (3.21)$$

where R is the Hubble radius and $M = N\, m_p$,we get that the number N of baryons in the Universe is given by

$$N = M/m_p = \left[R/r(h) \right]^2 \sim (10^{40})^2 \qquad (3.22)$$

in agreement with Dirac's large numbers hypothesis[40,41] .

On the basis of the results obtained by introducing spin and torsion into the strong gravity equations, now we shall try to extend the Dirac hypothesis[44] also to the total angular momentum of the Universe, besides its mass and radius.

If we regard the Universe and a hadron as two similar physical systems, differing only for a scale factor which carries the Newton gravity field into the strong gravity field[38,42,43], then we are led to assume that the gravitational potential of the Universe, $\varphi(U) \sim K_g M/Rc^2$, is of the same order of magnitude as the hadronic strong gravity potential $\varphi(h) \sim K_f m_p/rc^2$, (where $r \sim \hbar/m_p c$) , and we can deduce then the following relation[40,41,42,43] :

$$K_f / K_g \sim (M / m_p)(r / R) \sim 10^{40} \qquad (3.23)$$

When torsion is introduced into the strong gravity field equations, we must take into account also spin contributions to the metric (as shown in the previous chapter) and we must introduce an "effective" potential $\varphi = \varphi_0 + \varphi_1$ which, with suitable approximations, may be written as (see (2.60))

$$\varphi_0 \sim K_f\, m_p/ r\, c^2 \quad , \quad \varphi_1 \sim K_f^2 \sigma^2 r^2/ c^6 \qquad (3.24)$$

where $\sigma \sim \hbar /r^3$ is the hadronic spin density. By assuming covariance of physical laws under scale dilatations [42,43] also in the case of this spin-distorced potential, since $\varphi_0(U) \sim \varphi_0(h)$, we can deduce $\varphi_1(U) \sim \varphi_1(h)$, that is

$$K_f\, \sigma\, r \sim K_g \Sigma R \qquad (3.25)$$

where $\Sigma \sim S/R^3$ is the total spin density of the Universe.

So we have a relation connecting the spin density of a particle and the total angular momentum of the Universe as a whole.

By using the analogous relations (3.21) and (3.23) between masses, radii and coupling constants we find, from (3.25), that $\sigma \sim \Sigma$ and then

$$S \sim (R / r)^3 \hbar \sim 10^{120} \hbar \qquad (3.26)$$

that is the spin density of an elementary particle and of the Universe are the same.

This conclusion is in agreement with the semi-empirical speculation of Cavallo[45], and with analogous results[46,47] obtained on the grounds of different assumptions, with no reference to the concept of torsion.

Let us evaluate some numerical implication of the result (3.26).

Suppose that the angular momentum of the Universe be due to classical rigid rotation: then $\Sigma \sim M\,\omega/R$ and the angular velocity of rotation ω, from (3.25), turns out to be

$$\omega \sim (m_p / M)(R / r^2)c \sim c / R = 1 / T \qquad (3.27)$$

where T is a characteristic time for the Universe deduced from the Hubble constant $T \sim 10^{18}$ sec. It is a remarkable fact that this angular velocity is identical to the angular velocity of the cosmological model of Gödel[48]

$$\omega_G = 2\,(\pi G \rho)^{1/2} \sim (K_g M/R^3)^{1/2} \sim c/R = 1/T \qquad (3.28)$$

(we have used[40,41] $K_g M \sim Rc^2$).

At last we remember that the minimal radius of the Universe R_o according to Trautman's cosmological model, depends on the total spin of the Universe, and, disregarding a numerical factor of the order of unity, it is given by[4,13]

$$R_o = (K_g S^2/M\, c^4)^{1/3} \qquad (3.29)$$

If we accept the value of the total spin of the Universe given by (3.26) from (3.25), (3.21) and (3.23) we obtain

$$R_0 \sim R(K_f m_p^2/\hbar c)^{1/3} \sim R \qquad (3.30)$$

that is, paradoxically, the minimal radius of the Universe has the same numerical value of the Hubble radius.

Therefore if the spin density of an elementary particle and of the Universe are the same, one could think, at a first sight, that the dimensions of our cosmos in the past cannot have been considerably smaller than its present dimensions.

This is not the case, however, if we accept the large numbers hypothesis of Dirac[44] : in fact according to the law of variation in time, if a number is of the order of $(10^{40})^n$, then it must change with the time t proportionally to t^n.

From (3.26) it follows then $S \propto t^3$ and, from (3.29), we have $R_0 \propto t$ (as $K_g \propto t^{-1}$), i.e. also the minimal radius increases in time (since $M \propto t^2$).

In conclusion we are led to think that the angular momentum and the radius of the Universe change in time in such a way that the Universe, at any given instant of time, assumes the minimal radius allowed by the angular momentum characteristic of that age.

There is no absolute minimal radius for the Universe, but only a minimal radius relative to a given value of spin.

According to Dirac's law, spin changes in time so that the minimal radius increases and the Universe is expanding.

We think that the question of the gravitational collapse and of the initial singularity should be re-examined on these grounds.

REFERENCES

1. F.W.Hehl, P.von der Heyde, G.D.Kerlick and J.N.Nester, Rev. Mod.Phys. 48, 393 (1976); F.W.Hehl, Lectures at the "International School of Cosmology and Gravitation", Erice 1979 (this volume).
2. F.W.Hehl and B.Datta, J.Math.Phys.12, 1334 (1971).
3. C.S.Sivaram and K.Sinha, Lett.Nuovo Cimento 13, 357 (1975).
4. F.Kaempffer, Gen.Rel.Grav.7, 327 (1976).
5. V.Rodichev, Sov.Phys.JEPT 13, 1029 (1961).
6. K.Hayashi and R.Sasaki, preprint MPI-PAE/PTR 23/77 (1977).
7. A.Papapetrou, Proc.Roy.Soc. 209A, 248 (1951).

8. V.De Sabbata and M.Gasperini, Lett.Nuovo Cimento 21, 328 (1978).
9. K.Hayashi, preprint, Max Planck Institute (München 1971).
10. W.Baker, W.Atkins and W.Davis, Nuovo Cimento 44B, 1 (1978).
11. V.De Sabbata and M.Gasperini, Lett.Nuovo Cimento 23, 657 (1978).
12. Y.M.Cho, J.Phys. 11A, 2385 (1978).
13. A.Trautman, Nature 242, 7 (1973).
14. C.Isham, A.Salam and J.Strathdee, Nature 244, 82 (1973).
15. C.Isham, A.Salam and J Strathdee, Phys Rev. D3, 867 (1971).
16. A.Salam, "Nonpolynomial Lagrangians, renormalization and gravity" in Lectures from the Coral Gables Conferences on Fundamental Interactions p.3 (Gordon and Breach, New York, 1971).
17. C.Jarlskog, "Phenomenology of weak interactions" § 6, Proceedings of the 1974 Cern School of Physics, CERN 74-72 (Geneva, 1974).
18. B.Kuchowicz, Acta Cosmologica 3, 109 (1975).
19. J.Bernstein, Chaps.9 and 10 in "Elementary Particles and their Currents" (Freeman and Company, San Francisco and London, 1968).
20. J.Bernstein, Chaps.11 and 12 of ref.19.
21. R.Delbourgo and A.Salam, Phys.Lett. 40B, 381 (1972).
22. C.Isham, "Strong Gravity" in Lectures from the Coral Gables Conferences on Fundamental Interactions p.95 (Gordon and Breach, New York, 1971).
23. C.Jarlskog, § 10 of ref. 17.
24. J.B.Griffiths and R.Newing, J.Phys. 3A, 136 (1970).
25. J.Sakurai, Chap.3 in "Advanced Quantum Mechanics" p.167 (Addison Wesley, Reading, MA 1967).
26. A.Vitale, A.Bertin and G. Carboni Phys.Rev. D11, 2441 (1975).
27. C.Sivaram and K.Sinha, Lett.Nuovo Cimento 9, 704 (1974).
28. F.W.Hehl, P.von der Heyde and G.Kerlick, Phys.Rev. D10, 1066 (1974).
29. K.Tennakone, Phys.Rev. D10, 1722 (1974).
30. A.Trautman, Symp.Matem. 12, 139 (1973).
31. T.W.Kibble, J.Math.Phys. 2, 212 (1961).
32. A.Prasanna, Phys.Rev. D11, 2076 (1975).
33. F.W.Hehl, P.von der Heyde, G.D.Kerlick and J.N.Nester, Rev. Mod.Phys. 48, 393 sect.V § A.1 (1976).
34. J.Weyssenhoff and A.Raabe, Acta Phys.Pol. 9, 7 (1947).
35. V.De Sabbata and M.Gasperini, Gen.Rel.Grav. 10,731, 825 (1979).

36. V.De Sabbata and M.Gasperini, Lett.Nuovo Cimento 24, 520 (1979).
37. V.De Sabbata and M.Gasperini, Lett.Nuovo Cimento 24, 215 (1979).
38. E.Recami and P.Castorina, Lett.Nuovo Cimento 15, 347 (1976).
39. E.A.Lord, C.Sivaram and K.Sinha, Lett.Nuovo Cimento 11, 142 (1974).
40. V.De Sabbata and P.Rizzati, Lett.Nuovo Cimento 20, 525 (1977).
41. C.Sivaram and K.Sinha, Phys.Lett. 60B, 181 (1976).
42. P.Caldirola, M.Pavšič and E.Recami, Nuovo Cimento 48B, 205 (1978).
43. P.Caldirola, M.Pavšič and E.Recami, Phys.Lett A66, 9 (1978) and Lett.Nuovo Cimento 24, 531 (1979).
44. P.A.M.Dirac, Proc.Roy.Soc. 338A, 439 (1974).
45. G.Cavallo, Nature 245, 313 (1973).
46. R.M.Muradyan, preprint E2-9804 (Dubna 1976).
47. P.Caldirola, M.Pavšič and E.Recami, §6 of ref.42.
48. K.Gödel, Rev.Mod.Phys. 21, 447 (1948).
49. D.E.Liebscher, Lectures at the "International School of Cosmology and gravitation", Erice 1979 (this volume).
50. F.W.Hehl, Lectures at the "International School of Cosmology and Gravitation" Erice 1979 (this volume).
51. A.Trautman, Lectures at the "International School of Cosmology and Gravitation" Erice 1979 (this volume).
52. B.Średniawa, Lectures at the "International School of Cosmology and Gravitation", Erice 1979 (this volume).

THE FADING WORLD POINT*

Peter G. Bergmann

Department of Physics
Syracuse University
Syracuse, New York 13210

The subject of this talk is concerned with the nature of space-time, and of its elements, the world points. Classical field theory takes it for granted that space-time is a manifold, with appropriate conditions of differentiability. The same assumption holds for Poincaré-invariant quantum field theories. No matter what further structures are imposed on the space-time manifold, it has been usually taken for granted that each world point has an invariant identity, which is preserved under all symmetry transformations of a given theory.

From the point of view of Poincaré invariance, this is the only sensible attitude to take. Starting from a three-dimensional Cauchy surface (or hyperplane) one can identify the points of the four-dimensional space-time unambiguously by geometric specifications, for instance by telling through which point of the Cauchy hypersurface will a straight line pass normal to the Cauchy plane which goes through the world point in question, and how great is the distance of the world point from the Cauchy plane along this line.

In general-relativistic theories the situation is far less clear-cut. Until a specific metric field has been introduced onto the manifold, the construction that looks so straightforward in conventional field theories lacks meaning. And as the metric field is not to differ in principle from other physical fields, it looks

*Work supported by the National Science Foundation (U.S.A.) under Grant No. MPS74-15246.

as if the identity of a world point is inextricably bound up with the physics, - the totality of physical fields - , present.

Nevertheless, in the conventional manner in which we look at general relativity, and in most unitary field theories, the identity of world points is preserved by the assumed symmetry group of transformations, transformations that map one world point on another world point without regard to the metric and other fields supposed to exist. This symmetry group is usually referred to as the group of diffeomorphisms (or of curvilinear coordinate transformations).

At this Course we are being exposed to a number of speculative theories in which the role of the group of diffeomorphisms is modified. One is Penrose's twistor approach, the other the set of theories known as supersymmetry and supergravity. In these theories it is attempted to construct symmetry groups that are not only not isomorphic to pure space-time symmetry groups, but not even homomorphic. There is no need to dwell on these proposals at length; we are hearing from their initiators directly.

I consider attempts along these lines to be of great interest, for two reasons. One is the same argument that speaks in favor of unitary theories in general: If I believe that gravity is no different in principle from other physical fields, then it is reasonable to look for a fusing of the distinct symmetries characteristic of all these fields, space-time symmetries for gravitation, gauge symmetries of various kinds for the other physical fields.

The other grounds for looking with favor on these attempts is that below the Planck length there appears to exist no conceivable procedure for distinguishing experimentally among different world points; it may well be that the lower bound for separating points lies even higher, perhaps somewhere beneath the Fermi length. One is entitled to look with suspicion at formal structures that cannot conceivably be related to anything detectable. Of course, these arguments are speculative. They may not be borne out by future developments. Right now they serve to justify one's curiosity.

Aside from theories that today are still to be considered speculative, I should like to show you that in standard general relativity there is an important symmetry group that destroys the world point as an invariant object. That is the symmetry group of mappings of Cauchy data on Cauchy data.

During the early Fifties those of us interested in a Hamiltonian formulation of general relativity were frustrated by a recognition that no possible canonical transformations of the field variables could mirror four-dimensional coordinate transformations and their commutators, not even at the infinitesimal level. That is because (infinitesimal or finite) canonical transformations deal with the dynamical variables on a three-dimensional hypersurface, a Cauchy surface, and the commutator of two such infinitesimal transformations must be an infinitesimal transformation of the same kind. However, the commutator of two infinitesimal diffeomorphisms involves not only the data on a three-dimensional hypersurface but their "time"-derivatives as well. And if these data be added to those drawn on initially, then, in order to obtain first-order "time" derivatives of the commutator, one requires second-order "time" derivatives of the two commutating diffeomorphisms, and so forth. The Lie algebra simply will not close.

Dirac finally constructed a commutator in three dimensions that did not require off-surface derivatives, and thereby succeeded in completing the Hamiltonian formulation of general relativity[1]. His device, formally speaking, consisted of replacing partial derivatives of the metric along the x^0-axis ("time" derivatives) by derivatives in the direction perpendicular to the chosen Cauchy hypersurface. The generators of these infinitesimal mappings, no longer "pure" diffeomorphisms (i.e. mappings irrespective of any metric) were the so-called Hamiltonian constraints, which among themselves form a system of involution, a closed Lie algebra.

At the time of the Dirac papers the nature of the Poisson bracket commutators that he had constructed was not entirely clear. Had Dirac merely discovered a new Lie algebra, or was his Lie algebra the germ of a group? If so, what was the nature of the group?

In 1971 A. Komar and I were able to answer that question: There was a new group[2]. However, this group differed essentially from the group of diffeomorphisms. It resembled it in that the form of the theory, the general theory of relativity, was preserved in complete detail by this group of mappings; that is to say, the Hamiltonian constraints, which incorporate the dynamics of the theory, are form-invariant under the mappings generated by these same constraints. But Dirac's mappings do not map a point on another point.

Given a Cauchy surface with two different sets of permissible Cauchy data imposed on its, and one of Dirac's mappings; then a particular point of that Cauchy surface, identified by its coordinate triplet, will be mapped on two different new points. In other words, the map of a given point depends on the assumed metric and its derivatives. There is a subset of Dirac mappings for which this result can be avoided. But unless the mappings are restricted to those that map the Cauchy surface on itself (i.e. to three-dimensional mappings), the restricted subset does not form a group, and once the group is closed, we are back to the general case.

The point seems to be this. Any particular solution of Einstein's field equations (a Ricci-flat space-time) can be described uniquely in terms of Cauchy data on a space-like three-dimensional hypersurface, but the converse is, of course, not true: Any one Ricci-flat space-time can be described in terms of an infinity of distinct Cauchy data, which differ from each other in the choice of Cauchy surface and its coordinatization. Dirac's group consists of the mappings of Cauchy data on equivalent data. His mappings leave unchanged the identity of any particular Ricci-flat space-time. But if we consider that, as a transformation group, these mappings map the totality of permissible Cauchy data on itself, then each individual mapping sends a given point into a whole cloud of points, depending on the particular Cauchy data involved.

Do Dirac's mappings have anything in common with the four-dimensional diffeomorphisms that are usually considered "the" symmetry group of general relativity? The answer is that the orbits of both kinds of mappings are the range of equivalent representations of one-and-the-same Ricci-flat space-time; one in terms of four-dimensional coordinatizations, the other in terms of Cauchy data. But the two groups are not isomorphic to each other, nor is there a homomorphism. In contrast to Poincaré-invariant theories, the symmetry mappings of four-dimensional fields and of their Cauchy data in general relativity differ profoundly.

What this whole analysis may teach us is that the world point by itself possesses no physical reality. It acquires reality only to the extent that it becomes the bearer of specified properties of the physical fields imposed on the space-time manifold. Perhaps this recognition will turn out to be of value in the years to come.

References

1. P. A. M. Dirac, Roy. Soc. (London) Proc. A246, 333 (1958); Phys. Rev. 114, 924 (1959).
2. P. G. Bergmann and A. Komar, Intl. J. of Theoret. Physics 5, 15 (1972).

SUPERALGEBRAS, SUPERGROUPS AND GEOMETRIC GAUGING

Yuval Ne'eman

Tel-Aviv University, Tel-Aviv, Israel*
and
Center for Particle Theory, The University of Texas
Austin, Texas 78712†

INTRODUCTION

This lecture series falls into two parts:

I. Superalgebras, Supergroups and Supermanifolds. This is mathematical in its content, but the presentation is for physicists. The applications of the "supers" have been in the forefront in recent years, and most mathematical texts do not contain the necessary material - simply because the mathematical content is also extremely recent.

II. Forms on a (Rigid) Group Manifold, a Principal Bundle and a Soft (Dali) Group Manifold. We develop the elements of the exterior calculus, on a Lie Group Manifold, thus reproducing results going back to Cartan etc. We then develop the geometric theory of gauging. For a local internal symmetry, this is done on a Principal Bundle. For a non-internal group such as the Poincaré or Super Poincaré groups, we reproduce Gravity and Supergravity by using a "soft" Group Manifold, i.e. a manifold whose tangent is the original rigid group. We construct the relevant theory, which we have recently developed in collaboration with T. Regge and J. Thierry-Mieg.

*Supported in part by the US-Israel Binational Science Foundation.

†Supported in part by EY-76-S-05-3992.

PART I: SUPERALGEBRAS, SUPERGROUPS AND SUPERMANIFOLDS

1. "Supers" Enter Physics and Mathematics

We have reviewed the emergence of Graded Lie Algebras in Mathematics and in Physics in a paper[1] which described the situation as of 1974. The new algebraic concept had been in use in Physics for a long time in the manner in which Mr. Jourdan had been using prose. (Example: the fermion quantization algebra a_i, a_i^+, N_i, corresponding to GL(1/1).) Since 1971, particle physicists were using a "Supergauge," which led to Supersymmetry in 1973, as described in our review.[1] One important omission in that document is the paper of Golfand and Likhtman[2] in the USSR, inventing supersymmetry as early as 1971. That work had not been referred to in 1972-74 related Soviet papers, and we had missed it altogether, just as it had been missed by Wess and Zumino and Salam and Strathdee in 1973-74.

Except for ref. 2, the work on supergauges and on supersymmetry (superconformal SU(2,2/1) as in Wess-Zumino's first paper, or super-Poincaré as in ref. 2 and later work in the West) was done without realizing what the exact formal structure was. Supergroups are groups, and Berezin had broken that ground on the Mathematical side. The new symmetries were thus introduced at the group level, but the axioms defining their generator superalgebras were not known to the Physicists, even to the relatively limited extent to which they had been formulated by the Mathematicians. We supplied the mathematical formalism (including a proof of the Poincaré-Birkhoff-Witt theorem ensuring the existence of a Universal Covering Algebra, essential to their representation theory and to the classification proof). What was still missing was a classification and throughout 1974-76 Freund and Kaplansky, Rittenberg with Pais and then, with Nahm and Scheunert, and V. G. Kac worked on that problem. It was completed by Kac[3] who published a classification of all simple Lie superalgebras. Djokovic and Hochschild made important contributions (see bibliography in ref. 4).

There are several more recent reviews of Lie Superalgebras[4] and of Supersymmetry.[5] We shall give here only the main results relevant to gauging, but our material will be self-contained.

2. Superalgebras

These consist of a linear vector space (dimension n) with a grading. The latter is a classification, most commonly by $\mathbb{Z}(2)$. However, we may also have a Z grading, or some product such as $\mathbb{Z}(2) \times \mathbb{Z}(2)$ etc. The grading is preserved by the bracket operation,

$$\left.\begin{aligned} & L = \sum_{(i)}^{\oplus} L_{(i)} \\ & \forall \ell_{(i)} \in L_{(i)}\,, \quad [\ell_{(i)},\ell_{(j)}\} \subset L_{(i+j)} \end{aligned}\right\} \tag{2.1}$$

The bracket is a $\mathbb{Z}(2)$-graded bracket,

$$[\ell_{(i)},\ell_{(j)}\} = -(-1)^{ij}[\ell_{(j)},\ell_{(i)}\} \tag{2.2}$$

i.e. it is a commutator except for the case in which both i and j are odd, when it becomes an anticommutator. Note that it is the bracket which caused the return to Super algebras, Super groups, Super manifolds rather than Graded Lie algebras etc., as we had used in 1974. On the one hand, mathematicians had been using a grading on a Lie algebra without applying the $\mathbb{Z}(2)$ graded bracket (2 • 2). The conformal algebra su(2,2), for example, has a grading defined by the eigenvalues of the commutators with the dilation generator (i.e. the dimension) though it is an ordinary Lie algebra. On the other hand, one may introduce other types of brackets. The outcome was a decision to stick to the superlatives, even though they carry an impression of a Superiority complex...

The last axiom is a graded Jacobi identity,

$$\begin{aligned} [\ell_{(i)},[\ell_{(j)},\ell_{(k)}\}\} = {} & [[\ell_{(i)},\ell_{(j)}\},\ell_{(k)}\} \\ & + (-1)^{ij}[\ell_{(j)},[\ell_{(i)},\ell_{(k)}\}\} \end{aligned} \tag{2.3}$$

The sign corresponds to $\ell_{(i)}$ "passing through" $\ell_{(j)}$ in the last term. Clearly, the sum of even graded $L_{(i)}$ forms an ordinary Lie algebra L_0. Under L_0, the sum of odd gradings L_1 forms a module and behaves as a representation of L_0. We thus always have a $\mathbb{Z}(2)$ grading

$$L = L_0 + L_1 \quad . \tag{2.4}$$

3. Superconformal and Super-Poincaré Algebras

We enlarge the conformal algebras (P_μ, $J_{\mu\nu}$, D, K_μ) of translations P_μ, Lorentz transformations $J_{\mu\nu} = -J_{\nu\mu}$, dilations D and special conformal transformations K_μ (all anti-hermitian here) by adjoining two sets of (Lorentz) Majorana spinors S_α, R_α ($\alpha = 1 \cdot\cdot 4$) and one pseudoscalar generator E:

$$
\begin{aligned}
&[P_\mu, P_\nu] = 0 \\
&[J_{\mu\nu}, J_{\rho\sigma}] = \eta_{\nu\rho} J_{\mu\sigma} - \eta_{\nu\sigma} J_{\mu\rho} + \eta_{\mu\sigma} J_{\nu\rho} - \eta_{\mu\rho} J_{\nu\sigma} \\
&[J_{\mu\nu}, P_\rho] = \eta_{\nu\rho} P_\mu - \eta_{\mu\rho} P_\nu \\
&[D, J_{\mu\nu}] = 0 \\
&[D, P_\mu] = P_\mu \\
&[K_\mu, K_\nu] = 0 \\
&[J_{\mu\nu}, K_\rho] = \eta_{\nu\rho} K_\mu - \eta_{\mu\rho} K_\nu \\
&[D, K_\rho] = -K_\rho \\
&[P_\mu, K_\nu] = 2J_{\mu\nu} + 2\eta_{\mu\nu} D
\end{aligned}
\tag{3.1}
$$

$$
\begin{aligned}
&\{S_\alpha, S_\beta\} = -(\gamma^\mu C)_{\alpha\beta} P_\mu \\
&[S_\alpha, P_\mu] = 0 \\
&[S_\alpha\ J_{\mu\nu}] = \tfrac{1}{2}(\sigma_{\mu\nu})_\alpha^{\ \beta} S_\beta \ ; \\
&[D, S_\alpha] = \tfrac{1}{2} S_\alpha \\
&[K_\mu, S_\alpha] = (\gamma_\mu)_\alpha^{\ \beta} R_\beta \\
&\{R_\alpha, R_\beta\} = (\gamma^\mu C)_{\alpha\beta} K_\mu \\
&[R_\alpha, K_\mu] = 0 \\
&[R_\alpha, J_{\mu\nu}] = \tfrac{1}{2}(\sigma_{\mu\nu})_\alpha^{\ \beta} R_\beta \\
&[D, R_\alpha] = -\tfrac{1}{2} R_\alpha \\
&[P_\mu, R_\alpha] = -(\gamma_\mu)_\alpha^{\ \beta} S_\beta \\
&\{S_\alpha, R_\beta\} = \tfrac{1}{2}(\sigma^{\mu\nu})_{\alpha\beta} J_{\mu\nu} - C_{\alpha\beta} D + (i\gamma_5 C)_{\alpha\beta} E \\
&[E, P_\mu] = 0 \\
&[E, J_{\mu\nu}] = 0 \\
&[E, K_\mu] = 0 \\
&[E, D] = 0
\end{aligned}
\tag{3.2}
$$

$$
\begin{aligned}
[E,S_\alpha] &= \tfrac{3}{4} i\, (\gamma_5)_\alpha^{\ \beta} S_\beta \\
[E,R_\alpha] &= -\tfrac{3}{4} i (\gamma_5)_\alpha^{\ \beta} R_\beta
\end{aligned} \tag{3.2}
$$

We observe that the complete superalgebra has a grading[1] proportional to the eigenvalue of D (the dimension):

$$
\begin{array}{ccccccccc}
\cdots & L_{(-3)} & L_{(-2)} & L_{(-1)} & L_{(0)} & L_{(1)} & L_{(2)} & L_{(3)} & \cdots \\
 & 0 & K_\mu & R_\alpha & J_{\mu\nu} & S_\alpha & P_\mu & 0 & \\
 & & & & D & & & & \\
 & & & & E & & & &
\end{array} \tag{3.3}
$$

This is a $\mathbb{Z}$ grading, with only $L_{(i)}$, $-2 \leq i \leq 2$ containing non-trivial generators. The Lie algebra (3.1) is known to correspond to so(4,2) or to its covering group su(2,2). As we shall see, the new superalgebra is su(2,2/1). The system can be further extended by adjoining the generators F_{ij} of an internal symmetry algebra u(n) (except for $n = 4$ where we can have either su(4) or u(4)). The S_α and R_α then behave as n under that algebra and become $S_{\alpha i}$, $R_{\alpha i}$ ($i = 1 \cdot\cdot n$). The new brackets are

$$
\begin{aligned}
\{S_{\alpha i}, S_{\beta j}\} &= -\delta_{ij} (\gamma^\mu C)_{\alpha\beta} P_\mu \\
\{R_{\alpha i}, R_{\beta j}\} &= \delta_{ij} (\gamma^\mu C)_{\alpha\beta} K_\mu \\
\{S_{\alpha i}, R_{\beta j}\} &= \tfrac{1}{2}\, \delta_{ij} (\sigma^{\mu\nu})_{\alpha\beta} J_{\mu\nu} - \delta_{ij} C_{\alpha\beta} D + (i\gamma_5 C)_{\alpha\beta} F_{ij} \\
[F_{ij}, F_{k\ell}] &= \delta_{ij} F_{k\ell} + \delta_{i\ell} F_{jk} - \delta_{ik} F_{j\ell} - \delta_{j\ell} F_{ik} \\
[F_{ij}, S_{\alpha k}] &= \delta_{jk} S_{\alpha i} - \delta_{ik} S_{\alpha j} \\
[F_{ij}, R_{\alpha k}] &= \delta_{jk} R_{\alpha i} - \delta_{ik} R_{\alpha j} \\
[F_{ij}, P_\mu] &= [F_{ij}, J_{\mu\nu}] = [F_{ij}, D] = [F_{ij}, K_\mu] = 0
\end{aligned} \tag{3.4}
$$

The new superalgebra is su(2,2/n), except for $n = 4$ when it may also be msu(2,2/n) with the F_{ij} traceless and generating su(4) rather than u(4).

su(2,2/n) thus provides a way of merging internal symmetry with the Lorentz group non-trivially, something otherwise forbidden as a symmetry of the S-matrix.[6] Considering that its Lie subalgebra is the conformal algebra, only massless systems may have su(2,2/n) as an S-matrix symmetry.

Taking the Poincaré subalgebra of (3.1), i.e. $(P_\mu, J_{\mu\nu})$ and adjoining just the S_α of (3.2), we have the Super-Poincaré (GP) algebra, known as Supersymmetry. Using $S_{\alpha i}$ instead of S_α and adjoining an orthogonal algebra o(n) with generators $G_{ij} = -G_{ji}$ we get Extended Supersymmetry (EGP). This is the maximal allowed symmetry of the S-matrix,[6] except for the possibility of adding Abelian charges, which will appear on the right hand side of the $\{S_{\alpha i}, S_{\beta j}\}$ anticommutator, but either commute with the entire superalgebra ("Central Charges" C_i) or just with everything except for the G_{ij} ("Internal Translations" T_i), under which they behave as vectors.[7] Excluding C_i and T_i generators, the superalgebra EGP is thus given by

$$
\begin{aligned}
&[P_\mu, P_\nu],\ [J_{\mu\nu}, J_{\rho\sigma}],\ [J_{\mu\nu}, P_\rho] \quad \text{as in} \quad (3.1) \\
&[P_\mu, S_{\alpha i}] = 0 \\
&[J_{\mu\nu}, S_{\alpha i}] = -\frac{1}{2}\,(\sigma_{\mu\nu})_\alpha^{\ \beta} S_{\beta i} \\
&\{S_{\alpha i}, S_{\beta j}\} = -\delta_{ij}(\gamma^\mu C)_{\alpha\beta} P_\mu
\end{aligned}
\tag{3.5}
$$

We shall return to the Abelian generators after discussing the simple superalgebras osp(4/n).

A representation of GP (the massive case) is given by the action of the S_α on the "vacuum." Taking some spin j representation of the Poincaré group (j, j_z, χ, M) where χ denotes Parity, we act with two available S_α. This is because we can rewrite the four S_α as two chiral pairs, with the momenta appearing on the right hand side only when a Left and Right spinor are made to anticommute. Otherwise, the $\{S_L, S_L'\} = \{S_R, S_R'\} = 0$. We can thus regard either set as "raising" operators (and the other as "lowering"), when acting on a state at rest. The right hand side is then $\{S_L^A, S_R^B\} = \delta^{AB} M$ so that we have generated a Clifford Algebra $C_2 \otimes C_2$. We get four states,

$$
\begin{aligned}
&(\vec{j}, j_z, \chi, M),\ (\vec{j} + \tfrac{1}{2}, j_z \pm \tfrac{1}{2}, \chi\eta, M) \\
&(j, j_z, (-\chi), M)
\end{aligned}
$$

the last one being generated by the commutator of two S_L etc. ... η is the spinor parity of S.

In the massless case the two sets S_L and S_R decouple. We use a null plane instead of a rest-frame and get just C_2 with two states, j_z and $j_z + 1/2$ (or j_z and $j_z - 1/2$).

4. The $\mathbb{Z}(2)$ and $\mathbb{Z}$ Graded Grassmann Ring; Supermanifolds

Let $\mathbb{\Lambda}$ be a $\mathbb{Z}(2)$-graded ring of parameters of countable dimension over the complex field $\mathbb{C}$. $\mathbb{\Lambda}$ is the direct sum of its even (Bose-statistics) component $\mathbb{\Lambda}_0$ and odd (Fermi-statistics) component $\mathbb{\Lambda}_1$. The multiplication respects the gradation,

$$\begin{aligned} &\mathbb{\Lambda} = \mathbb{\Lambda}_0 + \mathbb{\Lambda}_1 \\ &\mathbb{\Lambda}_I \mathbb{\Lambda}_J \subset \mathbb{\Lambda}_{I+J \bmod(2)} \end{aligned} \tag{4.1}$$

$\mathbb{\Lambda}$ is at the same time also $\mathbb{Z}$ graded,

$$\mathbb{\Lambda} = \sum_{i=0}^{\infty}{}^{\oplus} \Lambda^{(i)} \quad , \quad \Lambda^{(0)} \in \mathbb{C} \tag{4.2a}$$

$$\mathbb{\Lambda}_0 = \sum_{r=0}^{\infty}{}^{\oplus} \Lambda^{(2r)} \quad , \quad \mathbb{\Lambda}_1 = \sum_{r=0}^{\infty}{}^{\oplus} \Lambda^{(2r+1)} \tag{4.2b}$$

$$\Lambda^{(r)} \Lambda^{(s)} \subset \Lambda^{(r+s)} \tag{4.2c}$$

Multiplication is associative and graded-abelian

$$\begin{aligned} &\forall \lambda^{(i)} \in \Lambda^{(i)} \;, \; \lambda^{(j)} \in \Lambda^{(j)} \;, \\ &\lambda^{(i)} \lambda^{(j)} = (-1)^{ij} \lambda^{(j)} \lambda^{(i)} \end{aligned} \tag{4.3}$$

It is further assumed that $\mathbb{\Lambda}$ is generated by the identity 1 and the lowest Fermi subspace $\Lambda^{(1)}$: given a countable basis θ^a in $\Lambda^{(1)}$, the $\Lambda^{(r)}$ are spanned by the set of $\binom{n}{r}$ (the binomial coefficient) antisymmetrized products (for finite n, the dimension of $\Lambda^{(1)}$)

$$\theta^{a_1} \theta^{a_2} \dots \theta^{a_r}$$

Hence $\mathbb{\Lambda}_1$ is nilpotent, using (4.2c) and

$$\Lambda^{(r)} = 0 \qquad \text{for} \qquad r > n \tag{4.4}$$

An element of $\mathbb{\Lambda}$ thus admits a unique decomposition over the θ^a and their products, such that only a finite number of coefficients do not vanish

$$\forall \vec{x} \in \Lambda \quad \vec{x} = x^0 + x^i\theta^i + x^{ij}\theta^i\theta^j + \cdots$$

$$x^0, x^i, x^{ij}, \cdots \in \mathbb{C}$$

$$\theta^i \in \Lambda^{(1)} \tag{4.5}$$

If x^0 does not vanish, $(\vec{x})^{-1}$ exists and is defined by its (finite) power expansion.

Given an involution $\bar{\theta}$ in $\Lambda^{(1)}$,

$$\bar{\bar{\theta}} = \pm\theta \tag{4.6}$$

we induce an involution in Λ. We choose here an action which reverses the order of the θ generators,

$$\bar{\vec{x}} = x_1^{0*} + x^{i*}\bar{\theta}^i + x^{ij*}\,\bar{\theta}^j\bar{\theta}^i + \cdots \tag{4.7a}$$

$$\overline{(\vec{x}\vec{y})} = \overline{\vec{y}\vec{x}} \tag{4.7b}$$

The coefficients x^0, x^i, x^{ij} are thus complex-conjugated so that $\bar{\bar{x}}^0 = x^0$, $\bar{\bar{x}}^i = x^i$, $\bar{\bar{x}}^{ij} = x^{ij}$ etc. As to the bases θ^i, $\theta^i\theta^j$ etc. they follow either of the (4.6) choices. Only the plus choice would make $\vec{x}$ an eigenstate of the double involution. The minus sign will be utilizable when working with matrices, where we can compensate for the sign by an appropriate definition of transposition.

A vector space $E(m/n;\Lambda;v)$ of dimension (m/n) and valency v is a vector space whose first n components belong to Λ_v, the n others to Λ_{v+1} and are therefore of opposite statistics

$$V \in E(m/n;\Lambda;v) \begin{cases} V^i \in \Lambda_v & i = 1 \cdots m \\ V^j \in \Lambda_{v+1} & j = m+1, \cdots m+n \end{cases} \tag{4.8}$$

Linear mappings of a Λ vector space V onto itself (homomorphisms which respect the valency) may be represented by matrices of type $(m\|n)$ over Λ. These are matrices of dimension $(m+n)^2$ such that the m^2 and n^2 box-diagonal elements ("A" and "D") are bosonic and in Λ_0, whereas the elements in the $m \times n$ ("γ") and $n \times m$ ("β") rectangles off the diagonal boxes are fermionic and in Λ_1. Those square supermatrices

$$M(m\|n;\Lambda) \tag{4.9a}$$

are therefore of zero valency and belong to $E(m^2+n^2/2mn;\Lambda;0)$. We shall thus denote the various statistics - components of a supermatrix $M(m\|n;\Lambda)$:

$$\left.\begin{array}{l} M = \begin{array}{c} \quad m \quad n \\ \left(\begin{array}{c|c} A & \beta \\ \hline \gamma & D \end{array}\right) \begin{array}{l} m \\ n \end{array} \end{array} \\ A,D \subset \Lambda_0 \\ \beta,\gamma \subset \Lambda_1 \end{array}\right\} \tag{4.9b}$$

If A and D are regular, M is regular and one may write

$$M = \left(\begin{array}{c|c} A & \\ \hline & D \end{array}\right) \left(\begin{array}{c|c} 1 & \xi \\ \hline \psi & 1 \end{array}\right)$$

$$\xi = A^{-1}\beta, \quad \psi = D^{-1}\gamma$$

$$M^{-1} = \begin{pmatrix} F & -F\beta D^{-1} \\ -D^{-1}\gamma F & D^{-1}(\gamma F\beta + D)D^{-1} \end{pmatrix} \tag{4.10a}$$

$$F = (A - \beta D^{-1}\gamma)^{-1} \tag{4.10b}$$

i.e. the supermatrix has an inverse.

We now introduce the "carrier space," i.e. the column supermatrix V upon which the M act to produce homomorphisms. This is a vector space $V(m/n;\Lambda;v)$.

Underlining a supermatrix will imply retaining the $\Lambda^{(0)}$ component only. We can now define the supertrace of M, which involves the valency v of V,

$$\mathrm{str}\, M = (-1)^v\ (\mathrm{Tr}\,\underline{A} - \mathrm{Tr}\,\underline{D}) \tag{4.11}$$

We can also define a superdeterminant

$$\mathrm{s\,det}\, M = \exp \mathrm{str} \log M = \det \underline{A} \,.\, \det \underline{D(M^{-1})} \tag{4.12a}$$

where, from (4.10)

$$D(M^{-1}) = D^{-1}((\gamma(A - \beta D^{-1}\gamma)^{-1}\beta + D))D^{-1} \tag{4.12b}$$

$$\mathrm{s\,det}\,(MN) = \mathrm{s\,det}\, M \,.\, \mathrm{s\,det}\, N \tag{4.13}$$

The above material was covered by refs. 4, except for the role of valency, which becomes important as a result of ref. 8.

The topology of supermanifolds was studied by various authors, who have gradually managed to add structure.[9] Take first, for example, a smooth real manifold X, and the $C^\infty(X)$ algebra of smooth functions on it. One can think of the geometry as being in X itself (this is the traditional view), but one can also extract it all from the $C^\infty(X)$ algebra; even X itself can be recovered from it. The study of $\Lambda\!\!\Lambda$ manifolds has followed both paths. Berezin himself originally, and DeWitt, use the first approach, studying $\Lambda\!\!\Lambda$ ("supermanifolds"). In the second approach, one extends the ring of functions over X to include anticommuting elements ("graded manifolds"). This was used by Berezin and Leites, Kostant and M. Batchelor who connected the two approaches. A. Rogers enlarged the ring and took a manifold of functions with a norm over a supermanifold, which comes closest to the Physics requirements.

5. The Classification of Simple Lie Superalgebras

See references 3,4 and the papers listed in these articles.

As usual, a simple superalgebra L is one which contains no non-trivial ideal (L itself and zero are the trivial ones). An ideal $I \subset L$ is defined by

$$\forall i \in I, \quad \ell \in L \qquad [i,\ell\} \in I \tag{5.1}$$

The Killing form metric is defined by (Ad stands for the adjoint representation and ℓ_a are a basis of L)

$$g_{ab} = \mathrm{str}\left(\ell_a^{Ad}\,\ell_b^{Ad}\right) = c^e{}_{fa}\,(-1)^F\,c^f{}_{eb} \tag{5.2}$$

where we have replaced the matrix elements of ℓ^{Ad} by structure constants. F is the grading of f. There is also a similar supermetric for any representation R

$$g^R_{ab} = \mathrm{str}\left(\ell_a^R\,\ell_b^R\right) = (-1)^{AB}\,g^R_{ba} \tag{5.3}$$

(A,B are the gradings of a,b).

If $\det g^R_{ab} \neq 0$, the Casimir operators

$$K_n = \mathrm{str}\left(\ell^R_{a_1}\cdots\ell^R_{a_n}\right)\ell^{a_n}\cdots\ell^{a_1} \tag{5.4a}$$

where we raise indices with the inverse of g_{ab}

$$[K_n, \ell_a] = 0 \quad . \tag{5.4b}$$

Schur's lemma is modified, since a matrix which commutes with the entire superalgebra in a given irreducible representation is either a multiple of the identity or, if dim V_0 = dim V_1 it may be a non-singular matrix permuting V_0 and V_1.

6. Classical Superalgebras - the Superlinear (or Superunitary) Sequence

In these, the odd generators $\ell_i \in L_1$ form a completely reducible representation of L_0. L_0 is then reductive (i.e. the sum of a semi-simple Lie algebra and Abelian algebras).

Class Ia. Linear supertraceless superalgebras $s\ell(m/n)$ or $su(m/n)$.

(for $m \neq n$, $m > n \geq 1$)

This is the supermatrix algebra $\underset{\sim}{M}(m/n;C)$, i.e. we represent (2.1)-(2.2) by complex matrices. This corresponds first to projecting out of $M(m\|n;\Lambda)$ the lowest even and odd powers of θ^i, so that the new supermatrix can be written

$$\underline{\underline{M}}(m\|n;\Lambda) = \Lambda^{(0)} + \Lambda^{(2r+1)} = \underline{A} + \underline{D} + (B + C)\theta^{2r+1} \tag{6.1a}$$

where (2r+1) denotes the smallest odd grading, and

$$\underline{\underline{\beta}} = B\theta^{2r+1} \quad , \quad \underline{\underline{\gamma}} = C\theta^{2r+1} \tag{6.1b}$$

We denote by $\underset{\sim}{M}$ the ordinary complex matrix of the coefficients

$$M : \begin{array}{c} \;\; m \quad n \\ \left(\begin{array}{c|c} A & B \\ \hline C & D \end{array} \right) \end{array} \begin{array}{c} \\ m \\ n \end{array} \tag{6.2}$$

For $s\ell(m/n)$,

$$\text{str}\, \underset{\sim}{M} = 0 \tag{6.3}$$

The dimensions are $m^2 + n^2 - 1$ even and 2mn odd generators. The even subalgebra is

$$L_0 \; : \; s\ell(m) \oplus s\ell(n) \oplus g\ell(1) \tag{6.4}$$

The $s\ell(m/n)$ also generate Unitary supergroups and can be denoted therefore as $su(m/n)$. We have already encountered the superconformals $su(2,2/1)$ and $su(2,2/n)$. Recently, we have introduced $su(2/1)$ as the gauge-ghost-theory of weak and electromagnetic interactions. Its representations were calculated in ref. 10.

Class Ib. Median linear supertraceless superalgebras $ms\ell(m;\mathbb{C})$ or $msu(m)$.

$$ms\ell(m) = s\ell(m/m)/e$$

e is the center, here a multiple of the identity. For $s\ell(m/m)$, the $\operatorname{str} L = 0$ condition still leaves the identity as a non-trivial ideal, since

$$\operatorname{str} e = 0 \ , \qquad \forall m = n$$

so that we have to extract it explicitly. The dimensions are thus $2m^2 - 2$ even and $2m^2$ odd generators

$$L_0 \ : \ s\ell(m) \oplus s\ell(m)$$

Note that $g_{ab} \equiv 0$ here, but

$$\det g^R_{ab} \neq 0 \qquad .$$

One example is $msu(4) = su(2.2/4)/e$, the extended superconformal algebra for $n = 4$.

7. Classical Superalgebras - the Orthosymplectic Sequences

Class IIa. Orthosymplectic superalgebras $osp(2r/n)$, $n > 1$. Assume the matrix algebra is acting on $V = E(2r/n;\Lambda;1)$ of (4.8). The metric H will be of the form (this was explored by Freund and Kaplansky)

$$\left.\begin{aligned} H &= G_{(2r)} \oplus I_{(n)} \\ G_{(2r)} &= i\sigma_2 \otimes I_r \ , \quad G^2 = -1 \ , \quad G^T = -G \end{aligned}\right\} \tag{7.1}$$

We define the inner product of two vectors

$$x,\ y \in V \ ; \qquad (x,y) = x^T H y$$

As a result, the matrices of the algebra,

$$L = \left(\begin{array}{c|c} A & B \\ \hline C & D \end{array} \right) , \quad A^T = GAG , \quad D^T = -D , \quad C = -B^T G \tag{7.2}$$

Thus the A are $2r \times 2r$ symplectic matrices, the D are $n \times n$ antisymmetric (orthogonal) matrices, and the B are arbitrary $2r \times n$ matrices which fix the C completely.

There are thus $r(2r+1) + 1/2\, n(n-1)$ even, and $2rn$ odd generators

$$L_0 : \quad sp(2r) \oplus O(n)$$

Note: osp(2/2) is the covering of $s\ell(2/1)$. det $g_{ab} \neq 0$ except for

Class IIa_2: osp(2r/2r + 2)

$$\det g_{ab} = 0$$

The orthosymplectic algebras have been much in use in the treatment of supergravity. We list the main points:

(1) It has been long known that the Poincaré group can be considered as a contraction of 5-dimensional rotations (the de Sitter group O(4,1) or O(3,2)). We rewrite the commutator

$$[J_{\mu 5}, J_{\nu 5}] = -J_{\mu\nu} \qquad \mu,\nu = 0,\cdots 3 \tag{7.3}$$

with a rescaling by a radius R, as required dimensionally

$$P_\mu = \frac{1}{R} J_{\mu 5} = \varepsilon J_{\mu 5} \tag{7.4a}$$

$$[P_\mu, P_\nu] = -\varepsilon^2 J_{\mu\nu} \xrightarrow[\varepsilon \to 0]{} 0 \tag{7.4b}$$

To get GP, we start with OSp(4/1). The even subgroup is $Sp(4) = \overline{O(4,1)}$, the covering group of the de Sitter group. The superalgebra osp(4/1) is given by

$$J_{\mu\nu} = -J_{\nu\mu} , \quad J_{\mu 5} = -J_{5\mu}$$

$$\begin{aligned} [J_{\mu\nu}, J_{\rho\sigma}] &= \eta_{\nu\rho} J_{\mu\sigma} - \eta_{\nu\sigma} J_{\mu\rho} + \eta_{\mu\sigma} J_{\nu\rho} - \eta_{\mu\rho} J_{\nu\sigma} \\ [J_{\mu 5}, J_{\rho\sigma}] &= \eta_{\mu\rho} J_{\sigma 5} - \eta_{\mu\sigma} J_{\rho 5} \\ [J_{\mu 5}, J_{\nu 5}] &= -J_{\mu\nu} \end{aligned} \tag{7.5}$$

$$\begin{aligned}
\{\hat{S}_{\alpha>},\hat{S}_{\dot{\beta}}\} &= 2i\sigma^{\mu}{}_{\alpha>\dot{\beta}>}J_{\mu 5}\\
\{\hat{S}_{\alpha>},\hat{S}_{\beta>}\} &= i\sqrt{2}\,(\sigma^{\mu}\sigma^{\nu})_{\alpha>\beta>}J_{\mu\nu}\\
\{\hat{S}_{\dot{\alpha}>},\hat{S}_{\dot{\beta}>}\} &= -i\sqrt{2}\,(\sigma^{\mu}\sigma^{\nu})_{\dot{\alpha}>\dot{\beta}>}J_{\mu\nu}\\
[J_{\mu 5},\hat{S}_{\alpha>}] &= 1/\sqrt{2}\,(\sigma_{\mu\alpha>}{}^{<\dot{\beta}})\hat{S}_{\dot{\beta}>}\\
[J_{\mu 5},\hat{S}_{\dot{\beta}>}] &= -1/\sqrt{2}\,(\sigma_{\mu\dot{\beta}>}{}^{<\alpha})\hat{S}_{\alpha>}\\
[J_{\mu\nu},\hat{S}_{\alpha>}] &= 1/2\,\hat{S}_{<\beta}(\sigma_{\mu}{}^{\beta><\dot{\gamma}}\sigma_{\nu\dot{\gamma}>\alpha>} - \sigma_{\nu}{}^{\beta><\dot{\gamma}}\sigma_{\mu\dot{\gamma}>\alpha>})\\
[J_{\mu\nu},\hat{S}_{\dot{\alpha}>}] &= 1/2\,\hat{S}_{<\dot{\beta}}(\sigma_{\mu}{}^{\dot{\beta}><\gamma}\sigma_{\nu\gamma>\dot{\alpha}>} - \sigma_{\nu}{}^{\dot{\beta}><\gamma}\sigma_{\mu\gamma>\dot{\alpha}>})
\end{aligned} \tag{7.5}$$

where we have used 2-spinor and bracket notation (see Appendix A of ref. 11), where $\hat{S}_{\dot{\alpha}>}$ is left handed, $\hat{S}_{\alpha>}$ right handed and

$$(S_{\alpha>})^{+} = -S_{<\dot{\alpha}} \tag{7.6}$$

and the bracket convention is

$$\left.\begin{aligned}
\phi^{\alpha>} &= \phi^{<\alpha}\\
\phi_{\alpha>} &= -\phi_{<\alpha}\\
\phi_{<\alpha}\psi^{\alpha>} &= \phi^{<\alpha}\psi_{\alpha>} = \psi_{<\alpha}\phi^{\alpha>}
\end{aligned}\right\} \tag{7.7}$$

(the same holds for $\dot{\alpha}$).

If we now rescale by replacing

$$S_{\alpha>} = \varepsilon^{1/2}\,\hat{S}_{\alpha>}\,, \qquad S_{\dot{\alpha}>} = \delta^{1/2}\,\hat{S}_{\dot{\alpha}>} \tag{7.8}$$

we find we can have several contractions:

(A) Using only $\varepsilon^{1/2}$, we rescale

$$P_{\mu} = \varepsilon^{1/2}\,J_{\mu 5}$$

and take $\varepsilon^{1/2} \to 0$. This yields

$$\{S_{\alpha>},S_{\beta>}\} = 0$$

$$[P_{\mu},P_{\nu}] = 0$$

but leaves a simple subalgebra $osp(2,C/1)_{Left}$, composed of the

$J_{\mu\nu}$ and $S_{\dot{\alpha}>}$. The same process could have been done using $\delta^{1/2}$ and contracting the left-handed supersymmetry generators (leaving a simple subalgebra osp(2,C1/)$_{Right}$). These are affine superalgebras, with a weakly reducible structure[11]

$$\left.\begin{aligned} &A = F \oplus H \\ &[F,F] \subset F \\ &[F,H] \subset H \end{aligned}\right\} \tag{7.9}$$

with H the contracted part and F the simple subgroup. This is also a symmetric decomposition[11]

$$[H,H] \subset F \tag{7.10}$$

Note that the original osp(4/1) was already weakly reducible and symmetric (WRS) with respect to the same choices of H and F.

(B) We can rescale both $S_{\alpha>}$ and $S_{\dot{\alpha}>}$ using both $\varepsilon^{1/2}$ and $\delta^{1/2}$ and rescale the momenta accordingly

$$P_{\mu} = \varepsilon^{1/2}\, \delta^{1/2}\, J_{\mu 5} \quad .$$

This yields GP itself. Note that GP is not WRS, it is only weakly reducible, with the $J_{\mu\nu}$ as F.

Contraction of osp(4/1) has been used to reproduce Supergravity.[11-13]

(2) Extended Supersymmetry can be derived similarly by contraction of osp(4/n), yielding an o(n) interior symmetry. All allowed extensions can be derived by starting from a semidirect product,[7] o(n) $\circledS$ osp(4/n).

(3) In the Arnowitt-Nath[14] model, the supermanifold is Riemannian, with 3 + 1 space-time and 2r internal (fermionic and spinorial) dimensions. The tangent algebra is thus the (super) affine extension of osp(3,1/2r). This does not contain GP, which can only appear as a subalgebra of the diffeomorphisms.[15]

8. Classical Hyperexceptional and Exceptional Superalgebras

Class III. The P(m), m > 3 superalgebras. In $\underset{\sim}{M}$ of (5.6), take m = n,

$$\left.\begin{aligned} A^T &= -D \\ B^T &= B \\ C^T &= -C \\ \mathrm{tr}A &= 0 \end{aligned}\right\} \qquad (8.1)$$

There are $2m^2 - 1$ generators, $m^2 - 1$ even and m^2 odd

$$L_0 : s\ell(m)$$

$$g^R_{ab} \equiv 0$$

Class IV. The Q(m), m > 3 superalgebras. We invented them as an example in ref. 1. Take in (5.6)

$$\left.\begin{aligned} m &= n \\ A &= D \\ B &= C \\ \mathrm{tr}B &= 0 \end{aligned}\right\} \qquad (8.2)$$

In fact, this implies for the defining representation a structure in which the $m \times m$ matrices λ_a of su(m) appear everywhere,

$$L_0 : \begin{pmatrix} \lambda_a & \\ & \lambda_a \end{pmatrix} \qquad L_1 : \begin{pmatrix} & \lambda_i \\ \lambda_i & \end{pmatrix} \qquad (8.3)$$

We again had to extract the identity from L_0, which is why we use traceless λ_a. Alternatively we have to divide by e. The brackets are (we take a, b, c for L_0 and i, j, k for L_1)

$$\begin{aligned} [\ell_a,\ell_b] &= if_{abc}\ell_c \\ [\ell_a,\ell_i] &= if_{aij}\ell_j \\ \{\ell_i,\ell_j\} &= d_{ija}\ell_a \end{aligned} \qquad (8.4)$$

where the f are su(m) structure constants and the d_{ija} are the symmetric coefficients produced by $\{\lambda_i,\lambda_j\} = 2d_{ijk}\lambda_k + 4/3\ \delta_{ij}$. This is thus an example in which the symmetric Lie bracket is not just an anticommutator. Of course, in the adjoint representation, the bracket relations will be again represented by commutators and anticommutators.

There are m^2 even and m^2 odd generators.

We have recently shown[16] that gauging Q(3) as a ghost-symmetry reproduces Chiral dynamics and phenomenological SU(6) symmetry.

Class V. Exceptional classical superalgebras. These are F(4) with 40 generators, G(3) with 31 and the infinite set osp(2/4;α), α a complex number, with 17 generators. The Lie subalgebras are respectively su(2) $\oplus$ O(7), su(2) $\oplus$ g(2) and su(2) $\oplus$ su(2) $\oplus$ su(2). The odd generators are thus always su(2) 2-spinors: respectively an $\underset{\sim}{8}$ spinorial representation of O(7), a $\underset{\sim}{7}$ of g(2) and a (1/2,1/2) 4-vector of su(2) $\oplus$ su(2), the covering algebra of O(4).

9. Non-Classical Superalgebras

Class VI. Cartan-type simple superalgebras. These are non-classical. They are diffeomorphisms of Λ and can be constructed by taking vector-fields in Λ. Using the Grassmann generators θ^{a_i} of section 4,

$$\overline{\theta^{a_1}\theta^{a_2}\cdots\theta^{a_r}}\ \frac{\partial}{\partial\theta^{a_j i}}$$

where the θ^a have to be antisymmetrized. There is a $\mathbb{Z}$-grading provided by the total power of θ, counting $\partial/\partial\theta$ as -1. The operators $\overline{\theta^1\theta^2}\,\partial/\partial\theta^r$, taking $n=4$ for example, are of grade 1 and fermionic. The anticommutator of two such operators are indeed of grade 2 and bosonic. For a dimension n, w(n) is the maximal such algebra, ending with the antisymmetrized product $\overline{\theta^1\theta^2\cdots\theta^n}$ $\partial/\partial\theta^i$, i.e. grading $n-1$ is the maximal non-trivial one. The even generators do not make a semi-simple algebra: for an even grade 2r, $L_{(2r)}$ itself will be Abelian if $4r > n$, for instance. However, $[L_{(2r)}, L_{(2r-2)}] \subset L_{(4r-2)} \neq L_{(2r)}$ if $r > 1$ and $n \geq 5$. This exemplifies the non-reductivity of L_0.

The w(n) thus have $n2^n$ generators. The sequence starts at $n=3$, since w(2) ~ sℓ(2/1). The grade zero Lie subalgebra is gℓ(n). The other classes are the s(n), $n \geq 3$, the $\tilde{s}$(n), $n \geq 4$ and even, and the h(n), $n > 4$. The corresponding $L_{(0)}$ are $\overline{s\ell}(n)$, sℓ(n) and so(n) respectively, the total dimensions $(n-1)2^n+1$, $(n-1)2^n+1$ and 2^n-2.

10. Supergroups

To construct a supergroup, we return to the supermatrices over Λ in section 4. The subset of $M(m\|n;\Lambda)$ with an inverse (M^{-1} as in (4.10)) forms the group $GL(m/n)$, with ordinary matrix multiplication as the product.

The power series is well-defined for the M, as it is finite for the non-zero grades. This allows the use of the exponential map and the logarithm,[17] and one may generate any group element by exponentiating the superalgebra. The ring of parameters has to be Λ, so that the even generators are multiplied by even parameters and the odd generators by odd powers of θ. Note that a finite supergroup is just a group, and can be generated by exponentiation with complex parameters provided we construct the appropriate ordinary Lie algebra. This is found by replacing the odd generators by their direct product with θ^a, taking the dimension $\Lambda_{(1)}$ to be much larger than the number of odd generators,

$$\left[\theta^{a_r} \otimes \ell_A, \theta^{a_s} \otimes \ell_B\right] = \theta^{a_r}\theta^{a_s} \otimes [\ell_A, \ell_B\}$$

which may require several generators $\theta \otimes \ell_A$ for the same ℓ_A if the right hand side should not vanish.

To define the various groups we need to define the operations of transposition and Hermitian conjugation on M.

Taking

$$M = \left(\begin{array}{c|c} A & \beta \\ \hline \gamma & D \end{array} \right) \tag{10.1}$$

the supertranspose is given by (T is the ordinary transposition)

$$M^{ST} = \left(\begin{array}{c|c} A^T & -\gamma^T \\ \hline \beta^T & D^T \end{array} \right) \tag{10.2}$$

which preserves

$$(MN)^{ST} = N^{ST} M^{ST} \tag{10.3}$$

while

$$(MN)^T \neq N^T M^T \quad .$$

Complex conjugation can be of 4 types, depending upon its action on the Grassmann elements $\theta^{a_i}\theta^{a_j} \cdots \theta^{a_r}$. It can be defined as preserving their order or inverting it, and its action on a single θ is the involution of (4.6) with either sign for $\bar{\bar{\theta}} = \pm\theta$. To get a Hermitian conjugation which will fulfill

$$(MN)^{h.c.} = N^{h.c.} M^{h.c.} \tag{10.4a}$$

$$((M)^{h.c.})^{h.c.} = M \tag{10.4b}$$

we have to restrict our choice to the following two possibilities only:

$$M^{+} = M^{T*} \tag{10.5a}$$

where the star denotes the bar of (4.6) with the plus sign

$$\theta^{*} = \bar{\theta} \qquad \theta^{**} = \theta\ , \qquad (\theta\theta')^{*} = \theta'^{*}\theta^{*} \tag{10.5b}$$

and super-hermitian conjugation,

$$M^{s+} = M^{ST\times} \tag{10.6a}$$

where the $^{\times}$ denotes the bar of (4.6) with the minus sign and with preservation of the order of the Grassmann elements

$$\theta^{\times} = \bar{\theta}\ , \qquad \theta^{\times\times} = -\theta\ , \qquad (\theta\theta')^{\times} = \theta^{\times}\theta'^{\times} \tag{10.6b}$$

We can thus have the following classical supergroups, using (4.12) for super-unimodularity, and either (10.5) or (10.6) for Unitarity:

(1) SL(n/m).

$$M \in GL(n/m), \qquad s\det M = 1 \tag{10.7}$$

(2) OSp(2r/n).

$$\left.\begin{aligned} &M \in GL(2r/n) \\ &M^{ST} H M = H \end{aligned}\right\} \tag{10.8}$$

where H is the metric of (7.1).

(3) P(n). This requires a special type of transposition for median matrices

$$(M(m\|m))^{P} = \begin{pmatrix} D^{T} & -\beta^{T} \\ \gamma^{T} & A^{T} \end{pmatrix} \tag{10.9a}$$

The elements of P(n) obey

$$M \in GL(m/m) \ , \quad MM^P = 1 \ , \quad s\det M = 1 \tag{10.9b}$$

(4) Q(n). This requires a special type of superdeterminant since $A = D$ and $\beta = \gamma$ (see (8.3)) yield $s\det M = 1$ automatically. Kac's ω-determinant $s\det_\omega$ is defined by

$$s\det_\omega M = 1 + \frac{\omega}{2} \operatorname{tr} \ell n[(A-\beta)^{-1}(A+\beta)] \tag{10.10a}$$

$$= \exp \operatorname{str}_\omega \ell n M$$

$$s\det_\omega (MN) = s\det_\omega M \,.\, s\det_\omega N \tag{10.10b}$$

$$\operatorname{str}_\omega M = \omega \operatorname{tr} \beta \tag{10.10c}$$

ω is a constant anticommuting element.

(5) Unitary Supergroups U(m/n).

$$M \in GL(m/n) \ , \quad MM^+ = 1 \tag{10.11}$$

(6) Superunitary Supergroups sU(m/n).

$$M \in GL(m/n) \ , \quad MM^{s+} = 1 \tag{10.12}$$

(7) Unitary Superunimodular Supergroups SU(m/n).

$$M \in GL(m/n) \ , \quad s\det M = 1 \ , \quad MM^+ = 1 \tag{10.13}$$

(8) Superunitary Superunimodular Supergroups SsU(m/n).

$$M \in GL(m/n) \ , \quad s\det M = 1 \ , \quad MM^{s+} = 1 \tag{10.14}$$

(9) Superunitary Orthosymplectic Groups sOSp(m/n).

$$\left.\begin{array}{l} M \in GL(m/n) \\ M^{ST} HM = H \\ MM^{s+} = 1 \end{array}\right\} \tag{10.15}$$

We do not list the exceptionals and non-classical.

PART II: FORMS ON A (RIGID) GROUP MANIFOLD, A PRINCIPAL BUNDLE AND A SOFT (DALI) GROUP MANIFOLD

11. Program

Chapter I was needed in order to double our "supply" of available groups and algebras. It was especially important since it does appear plausible that some type of Supergravity might provide a unifying principle, bringing together Gravity and other interactions. Indeed, through Supergravity, we have witnessed considerable advances in the treatment of Quantum Gravity itself. More recently, gauging a Supergroup has also yielded an interesting embedding of the Weinberg-Salam model of Asthenodynamics (Weak + E.M. Interactions) in a highly constraining aesthetic theory.[8]

Supergravity is a gauge theory of GP, and it is natural that we should thereby be led to another look at Gravity itself, as the gauge theory of P, the Poincaré group.

The material we present is the result of a collaboration, first with Tullio Regge[13,18] and then with Jean Thierry-Mieg.[11] We present a sequence of three manifolds[19]: the (rigid) Group Manifold (a structure which has been known to mathematicians and physicists for half a century), the Principal Bundle (known for a quarter of a century, but knowingly used in Physics mostly since 1974 for the study of internal gauge theories) and the Soft (Dali) Group Manifold. We conceived of the latter in developing a method for the gauging of non-internal groups, such as P or GP. We show how under certain conditions it may factorize, or undergo a Spontaneous Fibration, i.e. become equivalent on mass shell to a Principal Bundle.

12. Differential Geometry and Lie Groups

This is a vast subject of which we can only reproduce a few key definitions and results.[20] We also do not attempt to generalize systematically the theory to superalgebras[9,21] but shall provide a few examples of interest.

Let M be a differentiable manifold with covering U_α and let x_α be coordinates on U_α. A p-form on U_α is an expression of the kind,

$$\eta_\alpha = \sum_{\{A\}} dx_\alpha^{A_1} \wedge dx_\alpha^{A_2} \wedge \cdots \wedge dx_\alpha^{A_p} E^{(\alpha)}_{A_1 \cdots A_p}(x_\alpha) \tag{12.1}$$

$$\{A\} = \{A_1, \cdots, A_p\}$$

where (a,b represent the gradings of A,B when the variables are elements of a Grassmann algebra)

$$dx_\alpha^A \wedge dx_\alpha^B = -(-1)^{ab}\ dx_\alpha^B \wedge dx_\alpha^A$$

and the $E^{(\alpha)}_{\{A\}}$ are differentiable functions of the x_α. On $U_\alpha \cap U_\beta$ we identify the forms η_α and η_β provided

$$\eta_\alpha = \sum_{\{A\}}\sum_{\{B\}} dx_\alpha^{A_1} \frac{\partial x_\beta^{B_1}}{\partial x_\alpha^{A_1}} \wedge dx^{A_2} \frac{\partial x_\beta^{B_2}}{\partial x_\alpha^{A_2}} \wedge \cdots \wedge dx_\alpha^{A_p} \frac{\partial x_\beta^{B_p}}{\partial x_\alpha^{A_p}} \times$$
$$\times E^{(\beta)}_{B_1\cdots B_p}(x_\beta(x_\alpha)) \tag{12.2}$$

In this case, the set of all forms η_α defines a unique form η on M. In the sequel, unless specified, we drop the α dependence.

The ^ product of differentials can be extended by linearity in an obvious way to generic forms, and we have

$$\eta^{pa} \wedge \xi^{qb} = (-1)^{pq+ab} \qquad \xi^{qb} \wedge \eta^{pa} \tag{12.3}$$

where η^{pa}, ξ^{qb} are p,q forms respectively, and a,b are their GLA gradings.

The exterior differential $d\eta^p$ of a p-form is a (p+1)-form defined recursively by the properties:

$$\left.\begin{aligned} &d(dx^A) = 0 \qquad\qquad dQ(x) = \sum_A dx^A \frac{\partial Q}{\partial x^A} \\ &d(\eta^{pa} \wedge \xi^{qb}) = d\eta^{pa} \wedge \xi^{qb} + (-1)^p\ \eta^{pa} \wedge d\xi^{qb} \end{aligned}\right\} \tag{12.4}$$

For any form, we always have $d(d\eta) \equiv 0$. Note that the operator d behaves as a p = 1, a = 0 element in this product Grassmann manifold of differential forms.

Given manifolds M and N, a map λ, $M \xrightarrow{\lambda} N$, and a form ν^p on N, the "pull back" $\lambda^*\nu^p$ is a p-form on M, defined by the following procedure.

If ν^p is given by

$$\nu = \sum_{\{A\}} dy^{A_1} \wedge \cdots \wedge dy^{A_p}\ N_{A_1\cdots A_p}(y)\ ; \qquad y \in N \tag{12.5}$$

and the map λ is realized by $y = \lambda(x)$, then we have,

$$\overset{*}{\lambda} \nu = \sum_{\{B\}} dx^{B_1} \frac{\partial y^{A_1}}{\partial x^{B_1}} \wedge \cdots \wedge dx^{B_p} \frac{\partial y^{A_p}}{\partial x^{B_p}} N_{\{A\}}(\lambda(x)) \tag{12.6}$$

It can be checked that

$$\left.\begin{aligned} d\overset{*}{\lambda}\nu &= \overset{*}{\lambda} d\nu \\ \overset{*}{\lambda}(\nu \wedge \sigma) &= \overset{*}{\lambda}\nu \wedge \overset{*}{\lambda}\sigma \end{aligned}\right\} \tag{12.7}$$

The d,^ operations are thus coordinate independent and are preserved through maps. This property makes them ideal in treating covariant theories. We now apply the algorithms of forms to Lie groups.

Let G be a Lie group (or a supergroup) parametrized by $x^1 \cdots x^c$. We use in the sequel elements $x, y, z \cdots \in G$ whose parameters are $x^1 \cdots x^c$, $y^1 \cdots y^c$, etc. The element $x^A = 0$ denoted by e is the identity in G. The product on G is a map $G \cdot G \overset{\Lambda}{\to} G$, and we write

$$z = xy \quad \text{instead of} \quad z = \Lambda(x,y) \quad ,$$

satisfying the associativity condition

$$t(xy) = (tx)y \qquad \forall t,x,y$$

In coordinates we have for the product element z^A,

$$z^A = \Lambda^A(x^1 \cdots x^c, y^1 \cdots y^c) \tag{12.8}$$

yielding the differential forms,

$$dz^A = dz^M \; V(x,y)^A_M + dy^N \; W(x,y)^A_N \tag{12.9}$$

In the sequel we use the $c \times 1$ row matrices

$$dz = \{dz^1 \cdots dz^c\} \; , \qquad dx = \{dx^1 \cdots dx^c\} \quad \text{etc.}$$

and consider V^A_M, W^A_N as $c \times c$ matrices. Briefly, then

$$dz = dxV + dyW$$

Associativity implies $t(xy) = (tx)y$. This equation can be differentiated in all sets of variables, and the results compared. One gets:

$$\left.\begin{aligned} &V(t,x)V(tx,y) = V(t,xy) && \text{(a)}\\ &W(t,x)V(tx,y) = V(x,y)W(t,xy) && \text{(b)}\\ &W(tx,y) = W(x,y)W(t,xy) && \text{(c)} \end{aligned}\right\} \tag{12.10}$$

As a consequence of (12.10) we have $V(x,e) = W(e,x) = 1$. The product on G gives two kinds of natural maps on G, the left and right translations:

$$G \xrightarrow{l(a)} G \qquad \text{where} \qquad l(a)(x) = ax$$

$$G \xrightarrow{r(a)} G \qquad \text{where} \qquad r(a)(x) = xa$$

Consider now the ($c \times 1$-matrix) 1-forms

$$\left.\begin{aligned} \omega &= dx\, W(x^{-1},x)\\ \pi &= dx\, V(x,x^{-1}) \end{aligned}\right\} \tag{12.11}$$

Using (12.10) it is easily checked that

$$l^*(a)\omega = \omega\,, \qquad r^*(a)\pi = \pi \tag{12.12}$$

ω and π are therefore called Cartan's left and right invariant forms, respectively, or more briefly the L.I. and R.I. forms. By the properties of d and $\wedge$, any form obtained from $\omega(\pi)$ through these operations is also L.I. (R.I.). Therefore, if we expand the L.I. two-form $d\omega^A$ over the basis provided by the $\omega \wedge \omega$,

$$d\omega^A = \frac{1}{2} C^A_{\cdot BE}\, \omega^B \wedge \omega^E \tag{12.13}$$

the $C^A_{\cdot BE} = -C^A_{\cdot EB}$ must be L.I. o-forms (that is functions) and are therefore constants, the structure constants of the Lie algebra of G. Similarly one has

$$d\pi^A = -\frac{1}{2} C^A_{\cdot BE}\, \pi^B \wedge \pi^E \tag{12.14}$$

Equations (12.13), (12.14) are referred to as the Cartan-Maurer equations. Under right translations we have:

$$\left.\begin{aligned} r^*(a)\omega &= dx\, V(x,a)W(a^{-1}x^{-1},xa)\\ &= dx\, W(x^{-1},x)W(a^{-1},e)V(a^{-1},a) = \omega\,\mathrm{ad}(a^{-1}) \end{aligned}\right\} \tag{12.15}$$

where

$$\mathrm{ad}(a^{-1}) = W(a^{-1},e)V(a^{-1},a) = W(a^{-1},e)V^{-1}(e,a^{-1}) \tag{12.16}$$

(In deriving (12.15) and (12.16), use has been made of (12.10).)

Similarly,

$$1^*(a)\pi = \pi V(e,a^{-1})W(a,a^{-1}) = \pi\,\mathrm{ad}(a) \tag{12.17}$$

By using (12.10) one further finds that:

$$\mathrm{ad}(a)\mathrm{ad}(x) = \mathrm{ad}(ax) \tag{12.18}$$

so that ad(a) is a $c \times c$ representation of G, the adjoint representation. The matrix ad(a) is tied to the structure constants by useful identities.

Consider the map $G \times G \xrightarrow{\Lambda} G$. From the previous comments it is clear that $\Lambda^*\omega$, $\Lambda^*\pi$ also satisfy the Cartan-Maurer equations. If we use x,y respectively as coordinates on the first (second) copy of G, we can write $xy = \Lambda(x,y)$ and,

$$\Lambda^*\omega = \omega_y + \omega_x \mathrm{ad}(y^{-1}) \tag{12.19}$$

In (12.19), ω_y, ω_x are the Cartan L.I. forms written in the variables x,y respectively. Also, x translates y on the left and does not appear in ω_y; y translates x on the right and appears in $\mathrm{ad}(y^{-1})$ according to (12.15). By inserting (12.19) in (12.13) we find an identity relating 2-forms. Separate identification of the coefficients of $dx \wedge dx$, $dx \wedge dy$, $dy \wedge dy$ produces:

$$\left.\begin{aligned} C^{B}_{\cdot FD}\mathrm{ad}(y^{-1})_{B}^{\cdot A} &= C^{A}_{\cdot BE}\mathrm{ad}(y^{-1})_{F}^{\cdot B}\mathrm{ad}(y^{-1})_{D}^{\cdot E} \\ d(\mathrm{ad}(y^{-1})_{B}^{\cdot A} &= C^{A}_{\cdot DE}\,\omega_y^{D}\,\mathrm{ad}(y^{-1})_{B}^{\cdot E} \end{aligned}\right\} \tag{12.20}$$

These identities are very useful in establishing the gauge covariance of field equations.

The map $a \to \mathrm{ad}(a^{-1})^T = \mathrm{cd}(a)$ is also a representation, the "coadjoint" representation of G, which is equivalent to the adjoint in a number of interesting cases (^{T} stands for the transposed matrix). Tangent vectors transform according to the coadjoint representation.

1-forms are essentially synonymous to covariant vector fields used by physicists. Contravariant fields appear in the mathematical literature as "tangent vectors" and are written in the form:

$$U = u^M \frac{\partial}{\partial x^M} \tag{12.21a}$$

Given a 1-form $Q = dx^M q_M$, we can define the scalar product $(U,Q) = u^M q_M$ as the "value" Q(U) of Q on U. Clearly, $(\partial/\partial x^N)dx^M = \delta^M_N$. We shall denote this "contraction" by the symbol $\lrcorner$, e.g.

$$\partial_N \lrcorner\, dx^M = \delta^M_N \tag{12.21b}$$

Given two tangent vectors, T, U we have a natural bracket operation:

$$[T,U] = \left[T^M \frac{\partial U^N}{\partial x^M} - U^M \frac{\partial T^N}{\partial x^M} \right] \frac{\partial}{\partial x^N} \tag{12.22}$$

defining a Lie-algebra structure on tangent vectors. If we write

$$\omega^A = \omega_M^{\cdot A} dx^M$$

and introduce $\Omega^M_{\cdot A}$ through

$$\omega_{M\cdot}^{A}\Delta_{A\cdot}^{N} = \delta^N_M \qquad \text{or} \qquad \Delta_B^{\cdot M}\omega_{M\cdot}^{A} = \delta^A_{B\cdot} \tag{12.23}$$

We can then define tangent vectors D_A through

$$D_A = \Delta_A^{\cdot M} \frac{\partial}{\partial x^M} \tag{12.24a}$$

Clearly,

$$\omega^A(D_B) = \delta^A_{\cdot B} \tag{12.24b}$$

[Note that we use A, B, ••• for "Latin" or anholonomic indices, and M, N, ••• for "Greek" or holonomic indices.]

After some algebraic manipulations one finds by using (12.23), (12.13):

$$[D_A, D_B] = C^E_{\cdot AB} D_E \tag{12.25}$$

which shows that the D_A form a finite Lie algebra, i.e., by definition the Lie algebra $\mathcal{A}$ of G. In this role, the $C^E_{\cdot AB}$ reappear in a form more familiar to the physicist. However, (12.25) is fully equivalent in content to the Cartan-Maurer equations. From (12.23), (12.24) it is also clear that the D_A are invariant under left-translations.

One can similarly introduce a right-invariant vector field S_B such that $\pi^A(S_B) = \delta^A_{\cdot B}$, and

$$[S_A, S_B] = -C^E_{\cdot AB} S_E \tag{12.26}$$

The geometrical interpretation of D_A is clear when considering the identity $W(x^{-1},x)W(x,e) = W(e,x) = 1$. From (12.11) and the second equation in (12.23) we have,

$$D_A = W_A^{\cdot M}(x,e)\frac{\partial}{\partial x^M} \tag{12.27}$$

But now if $\psi(x)$ is a generic function on G, the expression

$$\begin{aligned} dy^A D_A\psi(x) &= dy^A W_A^{\cdot M}(x,e)\frac{\partial}{\partial x^M}\psi(x) \\ &= dy^A \frac{\partial x^M}{\partial y^A}\frac{\partial}{\partial x^M}\psi(x) \\ &= dx^M\frac{\partial\psi}{\partial x^M} = \delta\psi \end{aligned} \tag{12.28}$$

represents the infinitesimal change $\delta\psi$ of ψ under an infinitesimal translation to the right in the argument of ψ. In this sense, the D_A generate right translations. Similarly, the S_A generate left translations. Quite naturally then both the L.I. of D_A and the R.I. of S_A can be expressed as

$$[D_A,S_B] \equiv 0 \quad . \tag{12.29}$$

13. Examples: The Poincaré and Super-Poincaré Groups

The Poincaré group P is defined as the set of pairs $x = (\Xi,\vec{x})$, where $\Xi \in SO(3,1)$, $\vec{x} \in \mathbb{R}^4$. The product rule is, setting $t = (\Theta,\vec{t})$:

$$(Z,\vec{z}) = (\Xi\Theta,\Xi\vec{t}+\vec{x}) = (\Xi,\vec{x})(\Theta,\vec{t}) \tag{13.1}$$

Thus, the index A,B etc. of Section 12 runs in the range $1 \cdots 10$. Because of the peculiar structure of P we use $a = 1 \cdots 4$ and the pairs (a,b) with $a > b$, instead of A. In this way, the differential form of (13.1), corresponding to (12.9) reads,

$$\left.\begin{aligned} dZ &= d\Xi\Theta + \Xi d\Theta \\ d\vec{z} &= d\Xi\vec{t} + d\vec{x} + \Xi d\vec{t} \end{aligned}\right\} \tag{13.2}$$

The forms ω are obtained as in (12.11) by setting $d\Xi$, $d\vec{x} = 0$ in (13.2) and then carrying out the replacements $(\Theta,\vec{t}) \to (\Xi,\vec{x})$ and $(\Xi,\vec{x}) \to (\Xi,\vec{x})^{-1}$. The result is

$$\left.\begin{aligned} \Omega &= \Xi^{-1}d\Xi \\ \vec{o} &= \Xi^{-1}d\vec{x} \end{aligned}\right\} \tag{13.3}$$

Ω is thus a skew-symmetric matrix with elements $\omega^{ab} = -\omega^{ba}$. Componentwise we have,

$$\left.\begin{aligned} \omega^{ab} &= \Xi^{ca} d\Xi^{cb} \\ o^{a} &= \Xi^{ca} dx^{c} \end{aligned}\right\} \qquad (13.4)$$

The Cartan-Maurer equations (12.13) read:

$$\left.\begin{aligned} d\omega^{ab} + \omega^{ac} \wedge \omega^{cb} &= 0 \\ do^{a} + \omega^{ac} \wedge o^{c} &= 0 \end{aligned}\right\} \qquad (13.5)$$

Similarly, the L.I. forms are (putting $d\Theta$, $dt = 0$, and replacing $(\Theta,\vec{t})$ by $(\Xi,\vec{x})^{-1}$)

$$\pi = d\Xi\Xi^{-1}, \qquad \vec{p} = d\vec{x} - \pi\vec{x} \qquad (13.6)$$

Tangent vectors appear as operators, linear in $(\partial/\partial\Xi^{ab})$ and $(\partial/\partial x^{a})$, and tangent to the defining variety of SO(3,1), with the normalization,

$$\Xi^{ab}\Xi^{cb} = \delta^{ac}$$

A convenient set of L.I. vectors is:

$$\left.\begin{aligned} D_{ab} &= \Xi^{ca} \frac{\partial}{\partial\Xi^{cb}} - \Xi^{cb} \frac{\partial}{\partial\Xi^{ca}} \\ D_{a} &= \Xi^{ca} \frac{\partial}{\partial x^{c}} \end{aligned}\right\} \qquad (13.7)$$

Clearly

$$\left.\begin{aligned} \Omega(D_{a}) &= \vec{\omega}(D_{ab}) = 0 \\ \omega^{cf}(D_{ab}) &= \delta^{ca}\delta^{fb} - \delta^{cb}\delta^{fa} \\ \omega^{f}(D_{a}) &= \delta^{f}_{a} \end{aligned}\right\} \qquad (13.8)$$

For the R.I. vector fields we find,

$$\left.\begin{aligned} S_{ab} &= x^{a} \frac{\partial}{\partial x^{b}} - x^{b} \frac{\partial}{\partial x^{a}} + \Xi^{af} \frac{\partial}{\partial\Xi^{bf}} - \Xi^{bf} \frac{\partial}{\partial\Xi^{af}} \\ S_{a} &= \frac{\partial}{\partial x^{a}} \end{aligned}\right\} \qquad (13.9)$$

A general field $\psi(x)$ can be considered as an element of the Poincaré group, with its x^{M} dependence mentioned explicitly, and its dependence on Ξ implicit and fixing its "intrinsic" spin (except that we have to replace the Poincaré group P by its covering group). It is the usual convention (13.1) in defining the Poincaré group product that picks left-translations (i.e., S_A) for coordinates and fields: $(\Lambda,\vec{a})(\Xi,\vec{x}) = (\Lambda\Xi,\Lambda\vec{x}+\vec{a})$ which e.g. for $\Xi=0$, $\vec{x}=x^\mu$ and an infinitesimal P transformation $\Lambda = 1+\varepsilon^{[\mu\nu]}$, $\vec{a}=\alpha^\mu$ yields $(0,x^\mu) \rightarrow (0,x^\mu+\varepsilon^{[\mu\nu]}x^\nu+\alpha^\mu)$. Had we used right translation, we would have found $(0,x^\mu) \rightarrow (0,x^\mu)$. The entire picture is thus R.I., and it is natural that the Hilbert space considerations should lead to the (S_{ab},S_a) set of (13.9), with "total" angular momentum S_{ab} including both that linear action on field components ("spin") represented by the $\Xi^{[af}(\partial/\partial\Xi^{b]f})$ part, and a complementary "orbital" piece $x^{[a}(\partial/\partial x^{b]})$.

On the other hand, the procedure for "gauging" a symmetry implies the use of (coordinate-invariant) forms, which thus have to be the L.I. forms ω^A (and their orthonormal L.I. algebra D_A). The L.I. set $\{D_{ab},D_a\}$ is naturally factorized (up to a Lorentz-transformation on the D_a) and will thus appear in the infinitesimal treatment of gauging, with only D_{ab} giving rise to a gauge-invariance of the Lagrangian. The action of D_a merges with general-coordinate transformations on x space, and a modified $\hat{D}_a$ can indeed be identified with the latter.[22]

The formalism can be extended to supergroups. A most interesting example is provided by the Super-Poincaré Group (GP) presented in equation (3.5). Consider first a Grassmann algebra, $\Lambda = \Lambda_0 \oplus \Lambda_1$ (eq. (4.2)) where Λ_0, Λ_1 are respectively the even and odd components of Λ. We can then form a set GP of triples $x = (\Xi,\vec{x},\xi)$ with $x^a \in \Lambda_0$, $\xi^\alpha \in \Lambda_1$, $\Xi \in SO(3,1)$. We consider $\vec{x}$ as an SO(3,1) vector, ξ as a Majorana spinor. We define a product:

$$z = (Z,\vec{z},\zeta) = xt = \left(\Xi\Theta, \Xi\vec{t} + \vec{x} + \frac{\xi\vec{\gamma}u(\Xi)\theta}{2}, u(\Xi)\theta+\xi\right) \qquad (13.10)$$

with $t = (\Theta,\vec{t},\theta)$

This product is associative provided:

$$u^{-1}(\Xi)\gamma^a u(\Xi) = \Xi^{ab}\gamma^b \qquad (13.11)$$

or in brief, $u^{-1}(\Xi)\vec{\gamma}u(\Xi) = \Xi\vec{\gamma}$

Upon exterior differentiation of (13.10) we get:

$$\left.\begin{aligned} dz &= d\Xi\Theta + \Xi d\Theta \\ d\vec{z} &= d\Xi\vec{t} + d\vec{x} + \frac{1}{2}\, d\bar{\xi}\vec{\gamma}u(\Xi)\theta + \frac{1}{2}\,\bar{\xi}\vec{\gamma}du(\Xi)\theta \\ &\quad + \Xi d\vec{t} + \frac{1}{2}\,\bar{\xi}\vec{\gamma}u(\Xi)d\theta \\ d\zeta &= du(\Xi)\theta + d\xi + u(\Xi)d\theta \end{aligned}\right\} \qquad (13.12)$$

Repeating the procedure leading to ω we obtain

$$\left.\begin{aligned} \Omega &= \Xi^{-1} d\Xi \\ \vec{o} &= \Xi^{-1}(d\vec{x} - \tfrac{1}{2}\,\overline{\xi}\vec{\gamma} d\xi) \\ \omega &= u^{-1}(\Xi) d\xi \end{aligned}\right\} \tag{13.13}$$

The corresponding R.I. forms are

$$\begin{aligned} \Pi &= d\Xi\Xi^{-1} \\ \vec{p} &= d\vec{x} - d\Xi\Xi^{-1}\vec{x} - \tfrac{1}{2}\, d\overline{\xi}\vec{\gamma}\xi - \tfrac{1}{4}\,\overline{\xi}\vec{\gamma}\pi^{ab}\sigma^{ab}\xi \\ \pi &= d\xi - \tfrac{1}{2}\,\pi^{ab}\sigma^{ab}\xi \end{aligned} \tag{13.14}$$

In deriving these results we used the identities (see Appendix of ref. 13)

$$\left.\begin{aligned} \tfrac{1}{2}\,\pi^{ab}\sigma^{ab} &= du(\Xi)u^{-1}(\Xi) \\ \tfrac{1}{2}\,\omega^{ab}\sigma^{ab} &= u^{-1}(\Xi)du(\Xi) \end{aligned}\right\} \tag{13.15}$$

Supersymmetric theories are frequently constructed in "superspace" $\mathbb{R}^{4/4}$. Elements of $\mathbb{R}^{4/4}$ form equivalence classes of GP under right-multiplication by SO(3,1) elements. Therefore,

$$(\Xi,\vec{x},\xi) \simeq (\Xi\Theta,\vec{x},\xi)$$

and $\mathbb{R}^{4/4}$ is conveniently parametrized by (x,ξ) only. GP multiplication on the left by $(\Theta,\vec{t},\theta)$ therefore acts naturally on $\mathbb{R}^{4/4}$ and we have:

$$\left.\begin{aligned} x' &= \Theta\vec{x} + \vec{t} + \overline{\theta}\vec{\gamma}u(\Theta)\xi \\ \xi' &= u(\Theta)\xi + \theta \end{aligned}\right\} \tag{13.16}$$

The metric for P

$$ds^2 = dx^a \otimes dx^a = \vec{o}^a \otimes \vec{o}^a \tag{13.17}$$

is obviously left invariant on P and therefore P-invariant on $\mathbb{R}^4$. Similarly, we have on $\mathbb{R}^{4/4}$, using (13.13)

$$ds^2 = \vec{o}^a \times \vec{o}^a + k\overline{\omega} \times \omega \tag{13.18}$$

which is also GP-invariant. Notice however that on odd superforms the product[21] $\wedge(\otimes)$ is commutative (anticommutative) and on even superforms $\wedge(\otimes)$ is anticommutative (commutative) according to the usual convention (12.3). In this way we retrieve the well-known metric of the superspace "vacuum."[23] Also of great interest are the tangent vectors. The L.I. tangent vectors are given by:

$$\left.\begin{aligned} D_{ab} &= \Xi^{ca}\frac{\partial}{\partial\Xi^{cb}} - \Xi^{cb}\frac{\partial}{\partial\Xi^{ca}} \\ D_a &= \Xi^{ca}\frac{\partial}{\partial x^c} \\ D &= u^{-1}(\Xi)\left(\frac{\partial}{\partial\bar{\xi}} - \frac{1}{2}\frac{\partial}{\partial x^m}\gamma^m\xi\right) \end{aligned}\right\} \tag{13.19}$$

The R.I. tangent vectors are:

$$\left.\begin{aligned} S_{ab} &= x^a\frac{\partial}{\partial x^b} - x^b\frac{\partial}{\partial x^a} + \Xi^{af}\frac{\partial}{\partial\Xi^{bf}} - \Xi^{bf}\frac{\partial}{\partial\Xi^{af}} \\ &\qquad + \bar{\xi}\sigma^{ab}\frac{\partial}{\partial\bar{\xi}} \\ S_a &= \frac{\partial}{\partial x^a} \\ S &= \frac{\partial}{\partial\bar{\xi}} + \frac{1}{2}\frac{\partial}{\partial x^m}\gamma^m\xi \end{aligned}\right\} \tag{13.20}$$

By neglecting the $\partial/\partial\Xi$ terms in S_A (spin) we obtain the infinitesimal generators of motions on $\mathbb{R}^{4/4}$. The D components are related to the "covariant derivatives" of Salam and Strathdee.[1] The Maurer-Cartan equations appear as:

$$\left.\begin{aligned} &d\omega^{ab} + \omega^{ac}\wedge\omega^{cb} = 0 \\ &do^a + \omega^{ac}\wedge o^c + \frac{1}{2}\bar{\omega}\gamma^a\wedge\omega = 0 \\ &d\omega + \frac{1}{2}\omega^{ab}\sigma^{ab}\wedge\omega = 0 \end{aligned}\right\} \tag{13.21}$$

In deriving (13.19) to (13.21), care must be taken to keep a consistent ordering of Fermi variables in order to avoid confusing results.

14. Gauge Theory Over a Principal Fiber Bundle: Ghosts and BRS Equations

We remind the reader of the relevant definitions[24]:

A Principal Bundle ($\mathcal{P}$) is a quintuplet ($\mathcal{P},\mathcal{B},\pi,G,\cdot$), with $\mathcal{P}$ the Principal Bundle itself and $\mathcal{B}$ the base space, both C^∞ manifolds. π is a canonical projection of $\mathcal{P}$ onto $\mathcal{B}$, G a Lie Group, and the dot denotes the action of G in $\mathcal{P}$:

$$\mathcal{P} \times G \to \mathcal{P} \quad \text{or} \quad (p,g) \to p \cdot g \qquad \forall p \in \mathcal{P},\ g \in G \tag{14.1}$$

We emphasize in particular one of the axioms characterizing $\mathcal{P}$:

$$\forall p \in \mathcal{P},\ g \in G: \quad \pi(p \cdot g) = \pi(p) \tag{14.2}$$

i.e., the transformations of G are "vertical," with both the original and transformed elements of $\mathcal{P}$ "sitting above" the same value in $\mathcal{B}$ (hence the complications in gauging translations in space-time). Another important axiom is that of local-triviality, namely the isomorphism

$$\pi^{-1}(V_x) \sim V_x \otimes G\ , \qquad \forall x \in \mathcal{B} \tag{14.3}$$

with V_x a neighbourhood of x. The fiber F in $\mathcal{P}$ is globally identical to the whole of G. The dot is an operation inducing an isomorphism from the Lie algebra A of G onto the "vertical" Killing vector fields of $\mathcal{P}_*$, the tangent manifold (L is a Lie derivative), realized as in (12.22) by the Poisson Bracket,

$$\begin{aligned} &D : A \to \mathcal{P}_*\ , \qquad y_i \to D_{y_i}\ , \qquad y_i \in A \\ &[D_{y_i}, D_{y_j}]_{P.B.} = D_{[y_i,y_j]_{L.B.}} =: L_{y_i}(D_{y_j}) \end{aligned} \tag{14.4}$$

The D_y vectors are "vertical." Since the dimensionality of $\mathcal{P}_*$ is larger than that of A the dot map has no inverse. Horizontal vectors are defined as a kernel by the introduction of a vertical one-form field, the connection ω which maps vectors onto A. It thus provides the missing inverted map. Applying (12.21b),

$$\begin{aligned} &\omega : \mathcal{P}_* \to A \\ &\forall t \in \mathcal{P}_*\ , \qquad t \lrcorner\, \omega = \omega(t) = \omega^i(t) y_i \in A \end{aligned} \tag{14.5}$$

The connection is however not an independent entity: its restriction to the fiber is the pull-back of the Left-invariant one-forms over G, (eq. (12.11)) and as in (12.24b),

$$\forall y \in A\ , \qquad \omega(D_y) = D_y \lrcorner\, \omega = y$$

The Lie derivative of ω with respect to Killing vector fields is constrained as

$$L_y \omega = -[y,\omega] = [\omega,y] \tag{14.6a}$$

This can be shown to result from an equivariance condition obeyed by the horizontal part of $P_* = V \oplus H$

$$\begin{aligned} &\omega_u(h) = 0 , \qquad \forall u \in P \Longleftrightarrow h_u \in H_u \\ &H_{u\cdot g} = H_u \cdot g \end{aligned} \tag{14.6b}$$

This leads to the Cartan-Maurer structural equation, which guarantees that the curvature Ω (or field-strength)

$$\Omega : = d\omega + \frac{1}{2} [\omega,\omega] \tag{14.7}$$

(the bracket in the second term is a compact notation for the right hand side of eq. (12.13)) is purely horizontal,

$$\forall y \in A , \qquad D_y \lrcorner \Omega = 0 \tag{14.8}$$

As a result,[25] we may decompose ω with respect to a section Σ (with coordinates x lifted from the base space) into the sum,

$$\omega = \phi + \chi , \qquad D_y \lrcorner \phi = 0 , \qquad \partial_\mu \lrcorner \chi = 0 \tag{14.9}$$

$\phi = dx^\mu \phi_\mu$, with ϕ_μ the horizontal Yang-Mills gauge potential describing local classical Physics. χ is the Feynman-DeWitt-Fadeev-Popov[26] "ghost" $\chi = dy^i \chi_i$. (Note that in the quantized theory one uses the components ϕ_μ of the gauge one-form ϕ, but effectively works with χ as a whole, not with its components χ_i) so that χ anticommutes just because it is a one-form. A differential df can be expanded,

$$df = bf + sf , \qquad b : = dx^\mu \partial_\mu , \qquad s : = dy^i \partial_i \tag{14.10}$$

equations (14.7)-(14.8) can be rewritten as three equations (for (ij), (μi) and ($\mu\nu$) indices)

$$\Omega_{ij} = \frac{1}{2} (\partial_i \chi_j - \partial_j \chi_i + [\chi_i,\chi_j]) = 0$$

$$\Omega_{\mu i} = \partial_\mu \chi_i - \partial_i \chi_\mu + [\phi_\mu,\chi_i] = 0$$

$$\Omega_{\mu\nu} = \frac{1}{2} (\partial_\mu \phi_\nu - \partial_\nu \phi_\mu + [\phi_\mu,\phi_\nu])$$

which can be rewritten as ($B\chi = b\chi + [\phi,\chi]$)

$$\Omega_{\|} = s\chi + \frac{1}{2} [\chi,\chi] = 0 \tag{14.11}$$

$$\Omega_{|\_} = s\phi + B\chi = 0 \tag{14.12}$$

$$\Omega_{=} = b\phi + \frac{1}{2}[\phi,\phi] \quad (14.13)$$

We recover the BRS equations for the gauge and ghost fields. These equations guarantee unitarity of the quantized Lagrangian.[27] The applications of our geometric identification of the ghost fields and derivation of BRS, to path integrals and to the renormalization procedure are described elsewhere.[25,28,29]

Note however that if G is a supergroup, the components ω^i for the odd part of the superalgebra are commutative. In that case, the χ will be a bosonic scalar field, and we have identified it[8] with the Goldstone-Nambu (or Higgs-Kibble after spontaneous breakdown) fields. We have given elsewhere a detailed discussion of this mechanism and shown other examples.[16] Geometric gauging of an <u>internal</u> supergroup produces a gauge theory with Goldstone-Nambu fields and can also represent spontaneous breakdown of the gauge defined by the even subalgebra.[8]

Lastly, we remind the reader that the Yang-Mills Lagrangian makes use of the existence of a metric on $\mathcal{B}$ implicit in Hodge adjunction ("duality") and reads

$$L = \Omega \wedge {}^*\Omega \quad (14.14)$$

L is a horizontal 4-form and consequently saturates the 4-dimensional horizontal dimensionality. Its exterior differential, which is a 5-form, may be written

$$dL = sL = dy^i \frac{\partial L}{\partial y^i} \quad (14.15)$$

BRS invariance thus amounts to the closure of L in P,

$$dL = sL = 0 \quad . \quad (14.16)$$

15. The Soft Group Manifold

Gravity and its extensions involve groups with non-internal action, i.e. transforming space-time. We thus have to abolish the distinction between the Fiber and Base submanifolds. This involves a new geometrical object, the "Soft" Group Manifold.

We have studied in sections 12-13 the "rigid" Group Manifold. To <u>gauge</u> on the Group Manifold G, for G a non-internal group, we modify that manifold and add here the adjective "Soft," in the spirit of Salvador Dali's Soft Self Portrait or Watches. In G, the rigidity of the group G as fixed by the Cartan-Maurer equations for the left-invariant one-forms (eq. (12.13))

$$d\omega + \frac{1}{2}[\omega,\omega] = 0$$

is relaxed, allowing for a non-vanishing right-hand side, the curvature. Concurrently, the left-invariant generator algebra D_y orthogonal to the ω (eq. (12.24))

$$\omega^i(D_y j) = \delta^i_j$$

obeys the G commutation relations (12.25) with the structure constants defined by the associativity relations in G and applied in $[\omega,\omega]$. In the Soft Group Manifold $\mathcal{G}$, this will no longer be true and the $\tilde{D}$ basis orthogonal to ρ, the soft ω, does not close on the Lie algebra of G.

A soft group manifold $\mathcal{G}$ is a triplet (G,A_i,ρ). $\mathcal{G}$ itself is characterized by the group action (replacing (14.1)), a map $\mathcal{G}\cdot G \overset{A}{\rightarrow} \mathcal{G}$, so that we write $g_1 = gg'$ for $g_1 = (g,g')$ and we have associativity (g' and g'' belong to the <u>rigid</u> G)

$$g(g'g'') = (gg')g'' \qquad (15.1)$$

A is the right-invariant Lie algebra, and ρ is an A-valued one-form (the connection) over $\mathcal{G}$ (it can be regarded as a set of frames or "fundamental forms")

$$\rho : \mathcal{G}_* \rightarrow A \; ; \quad \forall t \in \mathcal{G}_* \, , \qquad t \lrcorner \rho = \rho(t) = \rho^i(t)y_i \qquad (15.2)$$

Whenever the ρ^i span the cotangent space $\mathcal{G}^*$, the connection defines an inverse mapping D from A onto the Killing vector fields of $\mathcal{G}_*$

$$\tilde{D} : y \in A \rightarrow \tilde{D}_y \, , \qquad \rho^i(\tilde{D}_y j) = \delta^i_j \qquad (15.3)$$

Since no assumption such as (14.2) is made with respect to the variation, the curvature R defined by

$$R := d\rho + \frac{1}{2}[\rho,\rho] \qquad (15.4)$$

is completely unrestricted. From the Jacobi identity

$$[\rho[\rho,\rho]] = 0$$

we get the Bianchi identity

$$DR = 0 \qquad (15.5)$$

and R remains linked to the second exterior covariant differential

$$D\phi = d\phi + [\rho,\phi] \qquad DD\phi = [R,\phi]$$

Gauge transformations correspond to the "local" right group action

$$xg = z \ , \quad x \overset{g}{\to} y \ , \quad x,y \in \mathcal{G} \quad g \in G$$

with invariance inducing in the connections the finite variations (using (12.6), (12.19) and ω the L.I. form of G satisfying (12.13))

$$\Gamma(\lambda)\rho^A_x = \lambda^* \omega^A_y + \rho^B_x \, \mathrm{ad}(\lambda^{-1}(x))_B^{\ A} \tag{15.6a}$$

and infinitesimally

$$\delta\rho = D\lambda \ , \qquad \delta R = -[\lambda,R] = D(\delta\rho) \tag{15.6b}$$

A second type of gauge transformation, the AGCT,[13,18,22] is defined by the (left) translations in $\mathcal{G}$ generated by the Lie derivatives along the Killing vector fields:

$$\tilde{\varepsilon} = \varepsilon^i \tilde{D}_{yi} \qquad \varepsilon = \varepsilon^i D_{yi}$$

$$\begin{aligned} \mathcal{L}_{\tilde{\varepsilon}}\rho &= d\varepsilon + \tilde{\varepsilon} \lrcorner d\rho = D\varepsilon + \tilde{\varepsilon} \lrcorner R \\ \mathcal{L}_{\tilde{\varepsilon}}R &= D(\tilde{\varepsilon} \lrcorner R) - [\varepsilon,R] \end{aligned} \tag{15.7}$$

Over $\mathcal{P}$, $\tilde{\varepsilon}$ is vertical whereas R is purely horizontal (14.11)-(14.13) which makes $\tilde{\varepsilon} \lrcorner R = 0$ and (15.6) and (15.7) coincide. Any Lagrangian expressed just in terms of differential forms over $\mathcal{G}$ is translationally invariant under (15.7) but not necessarily under (15.6).

We can now summarize these results by comparing a Lie group G, a Principal Bundle $\mathcal{P}$ with fiber $\mathcal{F}$ = G, and the "soft" group manifold $\mathcal{G}$. A Lie group G is a space in which finite motions are specified, up to global topological considerations, by the knowledge of the Lie algebra A. The connection forms over G are the Cartan Left-invariant forms ω. They provide a rigid triangulation and a vanishing curvature

$$d\omega + \frac{1}{2}[\omega,\omega] = 0$$

In $\mathcal{P}$, there are two sectors, namely a vertical fiber $\mathcal{F}$ which is identical to G, and a horizontal base-space $\mathcal{B}$. The connection forms are only vertical and glued together vertically only. Therefore, the curvature two-form Ω is purely horizontal, and $\mathcal{P}$ is specified by the knowledge of the fields over a section Σ. A new vertical direction in the cotangent is defined by orthogonality to Σ, and ω splits (in a non-orthogonal one-form basis) into the sum of the (conventional and horizontal) gauge potential and the (normal to Σ-oblique) ghost.

Lastly, in $\mathcal{G}$, we have a unified structure. Punctually, it is identical to G. However, the connection one-forms ρ are locally independent. In $\mathcal{G}$, any direction is a gauge direction (as in $G = \mathcal{F} \subset \mathcal{P}$), but all directions are curved (as in $\mathcal{B} \subset \mathcal{P}$). Two types of gauge transformations are defined, the group transformations (15.6) generated by the right group action and the AGCT (15.7) generated by the Killing motions. The AGCT Lie algebra is not isomorphic to the gauge algebra but both are closed:

$$\begin{aligned} [\tilde{D}_y, \tilde{D}_{y'}] &= \tilde{D}_{[y,y']} + (\tilde{D}_y \lrcorner \tilde{D}_{y'} \lrcorner R^i)\tilde{D}_{y^i} \\ [D_y, D_{y'}] &= D_{[y,y']} \end{aligned} \tag{15.8}$$

16. Weakly-Reducible Symmetric Groups

An important class of physical theories (including Gravity and Supergravity) corresponds to an intermediate structure: spontaneous fibration of a Group Manifold $\mathcal{G}$. $\mathcal{G}$ "spontaneously" becomes flat along some direction, which is hereafter the "vertical." Spontaneity is the state of affairs subsequent to the application of the equations of motion (i.e., it sets in at the classical level upon choosing a Lagrangian). This mechanism only occurs if G is a Weakly Reducible and Symmetric (WRS) group,[13,11] a structure we discussed in section 7.

A Lie group G is Weakly Reducible if its Lie algebra A is the direct sum of a subalgebra F (the Fiber algebra) and an F-module H, the horizontal module.

$$A = F + H \tag{16.1}$$

$$[F,F] \subset F \tag{16.2}$$

$$[F,H] \subset H \tag{16.3}$$

The decomposition is called symmetric if

$$[H,H] \subset F \tag{16.4}$$

A WRS Simple Lie Group or Supergroup will be denoted WRSS. The decomposition is not unique, because F is not an ideal in A. There may even exist two independent decompositions, which we denote as Left ("L") and Right ("R"), as discussed in section 7.

$$\begin{aligned} &A = F \oplus M \oplus N \oplus H \\ &F_L = F \oplus M \qquad H_L = N \oplus H \\ &F_R = F \oplus N \qquad H_R = M \oplus H \end{aligned} \tag{16.5}$$

which then require the commutation relations

$$\begin{aligned}
&[F,F] \subset F , \quad [H,H] \subset F , \quad [F,H] \subset H \\
&[F,M] \subset M , \quad [F,N] \subset N \\
&[H,M] \subset N , \quad [H,N] \subset M \\
&[M,N] \subset H , \quad [M,M] \subset F , \quad [N,N] \subset F
\end{aligned} \tag{16.6}$$

In what follows, we use the range of indices (a,b,c) to label generators in F, and (h,i,j) in H. There are thus three types of non-vanishing structure constants,

$$f^{a}_{\cdot bc} , \quad f^{i}_{\cdot ah} , \quad f^{a}_{\cdot hi} \quad \text{where} \quad [A_m, A_n] = A_{\ell} f^{\ell}_{\cdot mn}$$

The two-forms of curvature R of equation (15.4) can be rewritten as

$$R^{A} = d\rho^{A} - \frac{1}{2} f^{A}_{\cdot BC} \rho^{C} \wedge \rho^{B} \tag{16.7}$$

whereas the covariant derivative of η is defined as

$$\begin{aligned}
(D\eta)^{A} &= d\eta^{A} - f^{A}_{\cdot BC} \rho^{C} \wedge \eta^{B} \\
(D\eta)_{A} &= d\eta_{A} - \rho^{C} \wedge \eta_{B} f^{B}_{\ CA}
\end{aligned} \tag{16.8}$$

However, in the case of a WRSS group, we may introduce the covariant differential with respect to the F subgroup (denoted by $\wedge$)

$$\begin{aligned}
(\hat{D}\eta)^{A} &= d\eta^{A} - f^{A}_{\ Ba} \rho^{a} \wedge \eta^{B} = (D\eta)^{A} + f^{A}_{\ Bh} \rho^{h} \wedge \eta^{B} \\
(\hat{D}\eta)_{A} &= d\eta_{A} - \rho^{a} \wedge \eta_{B} f^{B}_{\ aA}
\end{aligned} \tag{16.9}$$

and a reduced curvature, taken over the F subgroup only

$$\hat{R}^{a} = d\rho^{a} - \frac{1}{2} f^{a}_{\ bc} \rho^{c} \wedge \rho^{b} \tag{16.10}$$

For the "torsions" (curvatures with a horizontal anholonomic index) we have,

$$R^{h} = d\rho^{h} - f^{h}_{\ ia} \rho^{a} \wedge \rho^{i} = \hat{D}\rho^{h} \tag{16.11}$$

and for the "true" curvature (i.e. with a vertical index) we get

$$R^{a} = \hat{R}^{a} - \frac{1}{2} f^{a}_{\ hi} \rho^{i} \wedge \rho^{h} \tag{16.12}$$

The Bianchi identities become

$$\hat{D}\hat{R}^{a} = 0 , \qquad \hat{D}R^{h} = \hat{D}\hat{D}\rho^{h} = -f^{h}_{\ ia}R^{a} \wedge \rho^{i} \tag{16.13}$$

Upon variation of the connections, equation (15.6) becomes

$$\delta R^{a} = \hat{D}(\delta\rho^{a}) - f^{a}_{\bullet ih}\rho^{h} \wedge \delta\rho^{i} \tag{16.14}$$

$$\delta R^{h} = \hat{D}(\delta\rho^{h}) - f^{h}_{\bullet ai}\rho^{i} \wedge \delta\rho^{a} \tag{16.15}$$

17. Geometric Quadratic Lagrangian for WRSS Groups

Let G be a Group-Manifold constructed over G, a WRSS Lie group. We construct the appropriate Lagrangian four-form from ρ only (i.e. geometrically).

To preserve gauge-invariance, the Lagrangian must depend on ρ^{A} through R^{A} only. The anholonomic indices must be contracted through the introduction of a gauge-invariant tensor. Furthermore, we avoid using any vector-field to lower the exterior degree by contraction. We are thus restricted to a Lagrangian quadratic in R,

$$L_{\text{INVARIANT}} = R^{A} \wedge \overset{\circ}{\iota}_{AB}R^{B} , \qquad D\overset{\circ}{\iota}_{AB} = 0 \tag{17.1}$$

Using (15.5) we find that such an L_{INV} is a closed form (and a topological invariant)

$$dL_{INV} = DL_{INV} = 0 \tag{17.2}$$

However, G being a WRSS group, we pick a decomposition (16.1) and select for ι_{AB} a regular F-invariant two-tensor

$$\hat{D}\iota_{AB} = 0 \tag{17.3}$$

Seen as an A-tensor, ι_{AB} is not invariant and

$$\iota_{ah} = 0 \qquad \iota_{hi} = 0 \qquad (D_{\iota})_{ab} = 0$$

but (17.4)

$$(D_{\iota})_{ah} = \iota_{ab}f^{b}_{\bullet ih}\rho^{i}$$

and we get a Lagrangian of the type used by MacDowell-Mansouri[12]

$$L = R^{a} \wedge \iota_{ab}R^{b} , \qquad \hat{D}\iota_{ab} = 0 \tag{17.5}$$

Using (16.12)-(16.13), the Euler-Lagrange variations of the connections produce after integration by parts two equations of motion, one from ρ^a and the other from ρ^h:

$$E_a := 2\iota_{ab}\hat{D}R^b = 0 \quad \text{or} \quad \iota_{ab}f^b{}_{hi}\rho^i \wedge R^h = 0 \tag{17.6}$$

$$E_h := 2f_h{}^a{}_i\rho^i \wedge \iota_{ab}R^b = 0 \tag{17.7}$$

(note the index transposition in $f^a{}_{hi}$). The (flat) Lie group G with connections ω as in (15.7) is itself a solution of both equations.[13] It represents the vacuum, and in the terminology of ref. 13 we note that

$$\zeta_a = \iota_{ab}R^b \,, \quad \zeta_h = 0 \,, \quad \hat{D}\zeta_A = 0 \bmod (f^a{}_{hi}) \tag{17.8}$$

In the limit of group-contraction over F, ζ_A becomes a <u>pseudo-curvature</u>,[13,18] i.e. its covariant differential vanishes.

$$D\zeta_A = 0 \bmod (f^a{}_{hi}) \tag{17.9}$$

L of (17.5) is not G gauge-invariant, it is only invariant under the F subgroup. Under an H-gauge transformation, $R^A\iota_{AB}R^B$ gets an apparent modification $R^h(\delta\iota)_{hb}R^b$,

$$\delta L = 2R^h f_h{}^a{}_i \lambda^i \iota_{ab} R^b \tag{17.10}$$

In a number of cases of interest, this quantity vanishes on-shell, i.e. in the terminology of ref. 30, L is pseudo-invariant. However, it is always true that the exterior differential of L vanishes on-shell,

$$dL = DL = \hat{D}L = 2R^a\iota_{ab}f^b{}_{hi}\rho^i \wedge R^h \overset{\Omega}{=} 0 \tag{17.11}$$

where $\overset{\Omega}{=}$ denotes equality modulo the equations of motion. Such Lagrangians may be called "pseudo-closed." Equation (17.11) replaces the usual BRS invariance (14.16) of the Yang-Mills case, and should be considered as a possible starting point towards a quantum theory on the gauged (soft) group manifold.

Except on singularities, the connections ρ span G^*, and R can be decomposed on this co-basis[13]

$$R^A = \frac{1}{2} R^A{}_{BC}\rho^C \wedge \rho^B \tag{17.12}$$

With respect to the WRS decomposition, R splits in three pieces, according to its anholonomic components,

$$\begin{aligned} R &= R_{\parallel} + R_{|-} + R_{=} \\ R^A_{\parallel} &= \frac{1}{2} R^A{}_{\cdot ab}\rho^b \wedge \rho^a \end{aligned} \tag{17.13}$$

$$R^A_{|-} = R^A_{\cdot ah}\rho^h \wedge \rho^a$$
$$R^A_{=} = \frac{1}{2} R^A_{\cdot hi}\rho^i \wedge \rho^h \qquad (17.13)$$

Inserting this expansion in (17.6)-(17.7) we get three sets of equations of motion (since ρ^a is not explicitly involved),

$$\varepsilon_{Ai}\rho^i R^A_{\|} = 0$$
$$\varepsilon_{Ai}\rho^i R^A_{|-} = 0 \qquad (17.14)$$
$$\varepsilon_{Ai}\rho^i R^A_{=} = 0$$

We have proved in ref. 11 that this implies that all pure fiber components of the torsion vanish

$$R^h_{\|} \overset{\circ}{=} 0 \qquad (17.15)$$

For the $R_{|-}^h$ and $R^h_{=}$ the situation is more involved.

For the WRS decompositions (see section 7) of $Sp(4) = \overline{O(4.1)}$ and $OSp(4/N)$,

$$R^h_{|-} \overset{\circ}{=} 0 \qquad (17.16)$$

and similarly

$$R^h_{=} \overset{\circ}{=} 0 \qquad (17.17)$$

thus ensuring the vanishing of the complete generalized torsions

$$R^h \overset{\circ}{=} 0$$

for gravity and chiral supergravity. The precise criteria for groups to fulfill (17.16)-(17.17) have not yet been determined. Taking now equation (17.7) and defining (since ι_{ab} is regular)

$$\tilde{R}_a = \iota_{ab} R^b$$

we can write for the purely vertical components,

$$\tilde{R}_{a\|} f^a_{\ hi} = 0 \qquad (17.18)$$

For the $R_{a|-}$ the analysis is more complicated since a term linear in ρ^i will occur in applying (17.12) to produce the horizontal index. We get the equation

$$\tilde{R}_{a,bj} f^{a}{}_{hi} - \tilde{R}_{a,bi} f^{a}{}_{hj} = 0$$

Rewriting this expression as an equality and using cyclic permutations of the expression on the left we have shown[11] that $\tilde{R}_{a|-}$ satisfies the same equation as $\tilde{R}_{a\|}$,

$$\tilde{R}_{a|-} f^{a}{}_{hi} = 0 \tag{17.19}$$

For WRS decompositions in which [H,H] spans the entire F, and remembering that ι_{ab} is regular, we get

$$R^{a}_{\|} = R^{a}_{|-} \overset{\circ}{=} 0 \tag{17.20}$$

The curvature two-form is thus purely horizontal on-shell. Since the torsions vanished on-shell altogether (equations (17.15)-(17.17)), the on-shell geometry is that of a principal bundle (14.8), (14.11) and we denote this phenomenon as Spontaneous Fibration. Note that BRS closure was also recovered on shell (17.11).

18. Gravity

We reproduce first de Sitter gravity,[12] by working in the soft group manifold G: $0(4.1)$ or $\overline{0(4.1)} \sim Sp(4)$. This is a WRSS group and we select a particular $0(3.1)$ for F. Accordingly, we label the generators in F as $J_{[ab]}$ $(a,b = 0,\cdots,3)$ and those in H as J_{a5} or P_a. The non-vanishing Lie brackets are given in section 7 (our $J_{[ab]} = -J_{[ab]}{}^{+}$ etc.). The [ab] indicates a Lorentz double-index. We have replaced the Greek indices μ,ν by Latin a,b as anholonomic indices.

F admits two metrics ι_{AB}: The Killing metric,

$$\iota_{[ab],[cd]} = \eta_{ac}\eta_{db} - \eta_{ad}\eta_{cb} \tag{18.1}$$

or the Levi-Civita tensor,

$$\iota_{[ab],[cd]} = \varepsilon_{abcd} \tag{18.2}$$

Thus, the most general geometrical quadratic Lagrangian is given by

$$L = \varepsilon_{abcd} R^{[ab]} \wedge R^{[cd]} + i\alpha R^{[ab]} \wedge \eta_{a[c}\eta_{d]b} R^{[cd]} \tag{18.3}$$

The first term is hermitian and an anholonomic pseudoscalar, the second an anholonomic scalar. For the action to preserve parity, we would have to pick $\alpha = 0$, since the holonomic part of the four-form is itself a pseudoscalar. On the other hand, $\alpha = \pm 1$ projects out either one of the two SU(2) subgroups of the Lorentz group, i.e. Left or Right handed Gravity, yielding an irregular ι_{AB}. However, for $\alpha \neq \pm 1$, the second term is irrelevant to the equations of motion. We take $\alpha = 0$.

Contraction to the Poincaré group results as explained in section 7 from a rescaling of J_{a5} and taking $\varepsilon^2 \to 0$. We have shown in refs. 13,18 how the two-form

$$\zeta_A = \iota_{AB} \left. \frac{\partial f^B{}_{\cdot CD}}{\partial \lambda} \right|_{\lambda=0} (\rho^C \wedge \rho^D) \tag{18.4}$$

is a pseudocurvature

$$\hat{D}\zeta_A = D\zeta_a \overset{\circ}{=} 0 \qquad \text{for} \quad \rho \to \omega \tag{18.5}$$

As a result of this contraction, one gets the Trautman form[31]

$$L = \varepsilon_{abcd}\rho^c \wedge \rho^d \wedge R^{[ab]} \tag{18.6}$$

and spontaneous fibration occurs, so that

$$\begin{aligned} &R^a \overset{\circ}{=} 0 \\ &R_{\|}^{[ab]} \overset{\circ}{=} R_{|-}^{[ab]} \overset{\circ}{=} 0 \end{aligned} \tag{18.7}$$

and the fiber variables factorize out.[13] Note that

$$\begin{aligned} &\zeta_{[ab]} = \varepsilon_{abcd}\rho^c \wedge \rho^d \\ &\zeta_a = 0 \end{aligned} \tag{18.8}$$

We thus recover on-shell a Lorentz group Principal Bundle over space-time. In the flat limit, $\rho^{[ab]} \equiv \omega^{[ab]}$ and is by definition vertical, reducing to the ghost $\chi^{[ab]}$ (14.9). The vierbein ρ^a is not a connection in that Lorentz bundle, but is purely horizontal and appears in the flat limit as the Cartan left invariant form of the translation group T^4. The ρ^a thus plays geometrically the role of the ghost of the translation group which in our picture does not act in any superimposed fiber but in spacetime itself.

19. Supergravity

The Super Poincaré group is a contraction of OSp(4/1) (the "Super de Sitter" group) as shown in section 7, equations (7.5)-(7.8).

If we use $F : O(3,1)$, the group is Weakly Reducible but not Symmetric. However, it has two independent symmetric decompositions, as shown in section 7, both parity-violating (chiral). We can follow both and then reconstitute a parity-preserving summed Lagrangian.

For the left-chiral decomposition, $F_L = (J_{[ab]}, S_{\dot{\alpha}>})$ and $H = (S_{\alpha>}, T)$, while the right-chiral splitting is $F_R = (J_{[ab]}, S_{\alpha>})$ and $H = (S_{\dot{\alpha}>}, T)$. (We denote the P_a subgroup by T.)

We now list the allowed ι_{AB}. For the left-chiral fiber group, we write first the most general Lorentz-invariant 2-tensor with A, B taking values in F_L.

$$\iota_{[ab][cd]} = \varepsilon_{abcd} + i\alpha\eta_{a[c}\eta_{d]b}$$
$$\iota_{\dot{\alpha}>,<\dot{\beta}} = \beta\varepsilon_{\dot{\alpha}><\dot{\beta}}\,, \qquad \iota_{[ab]<\dot{\alpha}} = \iota_{\dot{\alpha}>[ab]} = 0 \tag{19.1}$$

It has to be invariant under $S_{\dot{\alpha}>}$, and we can thus cancel the tensor-spinor's contribution to the differential

$$D\iota_{\dot{\alpha}>[cd]} = -\iota_{\dot{\alpha}><\dot{\beta}}\rho^{<\dot{\gamma}}f_{\dot{\gamma}>}{}^{\dot{\beta}>}{}_{[cd]} - \iota_{[ab][cd]}\rho^{<\dot{\gamma}}f_{\dot{\gamma}>}{}^{[ab]}{}_{\dot{\alpha}>} \tag{19.2}$$

which has the solution

$$\beta = -2\sqrt{2}\,(1+\alpha) \tag{19.3}$$

yielding the Left-chiral pre-contraction Supergravity Lagrangian,

$$L_L = \varepsilon_{abcd}R^{[ab]}\wedge R^{[cd]} + 2i\alpha R^{[ab]}\wedge R_{[ab]} - 2\sqrt{2}(1+\alpha)R^{<\dot{\alpha}}\wedge R_{\dot{\alpha}>} \tag{19.4}$$

We note that this includes the parity-violating singular solution of (18.3) for gravity itself, with $\alpha = 1$, i.e., Left-chiral gravity. Scaling the H generators $H \to \lambda H$, we get $[T,T] = [S,S] = [S,T] = 0$, $[S,\dot{S}] \to T$ as in the Super Poincaré group, but $[\dot{S},\dot{S}] \to J$, the fiber group is still simple.

We can write the explicit expressions for the curvatures

$$R^{[ab]} = \hat{R}^{[ab]} - \rho^a\wedge\rho^b - \frac{i}{4}\sqrt{2}\,\rho^{<\alpha}\sigma^{[a}{}_{\alpha><\dot{\beta}}\wedge\sigma^{b]\dot{\beta}><\gamma}\rho_{\gamma>}$$
$$R^{\dot{\alpha}>} = \hat{R}^{\dot{\alpha}>} + \frac{1}{\sqrt{2}}\,\sigma_a{}^{\dot{\alpha}>}{}_{<\alpha}\rho^{\alpha>}\wedge\rho^a \tag{19.5}$$

After contraction, we get

$$L_L(\lambda^2 \to 0) = -\varepsilon_{abcd}\rho^a \wedge \rho^b \wedge \hat{R}^{[cd]} - 2i\alpha\hat{R}^{[ab]} \wedge \rho^a \wedge \rho^b$$
$$+ \sqrt{2}\,(1+\alpha)\rho^{<\alpha} \wedge (\sigma_a\sigma_b)_{\alpha><\beta} R^{[ab]} \wedge \rho^{\beta>} \qquad (19.6)$$
$$- 2(1+\alpha)\rho^a \sigma_{a<\alpha<\dot{\alpha}} \hat{R}^{\dot{\alpha}>} \rho^{\alpha>}$$

This is in fact the result of the MacDowell-Mansouri procedure.[12] However, we have not recovered conventional Supergravity. For that purpose we first derive Right-chiral Supergravity, in analogy to (18.6).

We now perform a second contraction in each of L_L and L_R, namely we rescale $T \to \mu T$ and $\dot{S} \to \mu\dot{S}$ in L_L, for instance. As a result, the $\hat{R}^{[ab]}$ lose the $[\dot{S},\dot{S}]$ contributions and become ordinary Einstein $R_\varepsilon^{[ab]}$ curvatures. We perform a similar rescaling $T \to \mu T$ and $S \to \mu S$ in L_R, with similar results. We add up the two Lagrangians so as to get a Parity preserving Lagrangian,

$$L = \frac{1}{2} L_L(\lambda^2\mu^2) + \frac{1}{2} L_R(\lambda^2\mu^2)$$
$$= \varepsilon_{abcd}\rho^a \wedge \rho^b \wedge \cdot R_\varepsilon^{[cd]} + (1+\alpha)\sigma_{a<\alpha<\dot{\beta}}\rho^a \wedge (\rho^{\dot{\beta}>} \wedge \hat{\hat{R}}^{\alpha>} + \hat{\hat{R}}^{\dot{\beta}>} \wedge \rho^{\alpha>}) \qquad (19.7)$$

which is indeed the conventional Supergravity Lagrangian,[32] except that the scale of the spinor is as yet unconstrained. The choice $\alpha = -1$ kills supergravity. We choose to set $\alpha = -5$, which corresponds to the conventional Supergravity Lagrangian. Note that in these geometric theories we only use connections, i.e. fields belonging to the adjoint representation of the gauge group. To reproduce the "Stony Brook algebra" we would first have to express our anholonomic forms over a coordinate system.

In this theory, fibration is incomplete. Restricting F to SO(3,1) yields a Weakly-reducible but non-symmetric decomposition. Following the two different WRSS decompositions as we did here does not guarantee fibration with space-time as base space, as the two partial Lagrangians are fibrated over different subgroups. We thus have to check in detail which of the results of section 17 still hold.

The vector-torsion R^a is identical in its final doubly-contracted phase in either L_L or L_R. Since it should vanish in either chirality it vanishes in (19.7) as well (the $\hat{D}$ and $\hat{R}$ denote the Lorentz subgroup here)

$$R^a = \hat{D}\rho^a - 2i\sigma^a{}_{<\alpha<\dot{\beta}}\rho^{\dot{\beta}>} \wedge \rho^{\alpha>} = 0 \qquad (19.8)$$

This is not true of $R^{\alpha>}$ which is a torsion (and vanishes) in L_L

and a curvature (and doesn't vanish) in L_R. $R^{\dot{\alpha}>}$ plays the opposite roles and can't vanish too.

Since F includes SO(3,1) in both chiralities, the vertical single or double Lorentz components of R^{ab}, R^{α} and $R^{\dot{\alpha}}$ all vanish. However, we shall see that $R^{[ab]}{}_{\cdot\cdot c\dot{\alpha}}$, which belong sometime to $R^{ab}_{=}$ in one chirality and to $R^{ab}_{|-}$ in the other, does not vanish.

The equations of motion are,

$$E_{[ab]} : \varepsilon_{abcd}\rho^{c} \wedge R^{d} = 0 \rightarrow R^{a} \overset{\circ}{=} 0 \qquad (19.9)$$

yielding (19.8)

$$E_{a} : \varepsilon_{abcd}\rho^{b} \wedge R^{[cd]} - 2\sigma_{a<\alpha<\dot{\beta}}(\rho^{\dot{\beta}>} \wedge R^{\alpha>} + R^{\dot{\beta}>} \wedge \rho^{\alpha>}) \overset{\circ}{=} 0 \qquad (19.10)$$

$$E_{\alpha} : i\sigma_{a\alpha><\dot{\beta}}\rho^{a} \wedge R^{\dot{\beta}>} \overset{\circ}{=} 0 \qquad (19.11)$$

$$E_{\dot{\alpha}} : i\alpha_{a\alpha><\dot{\beta}}\rho^{\dot{\beta}>} \wedge R^{a} \overset{\circ}{=} 0 \qquad (19.12)$$

from which one can derive[13,18]

$$R^{\alpha>} \overset{\circ}{=} R^{\alpha>}{}_{ab} \neq 0 \qquad (19.13)$$

$$R^{[ab]} \overset{\circ}{=} R^{[ab]}{}_{cd} + R^{[ab]}{}_{\cdot\cdot c\dot{\alpha}>} + R^{[ab]}{}_{\cdot\cdot <\alpha c} \qquad (19.14)$$

We also check that (19.7) is indeed constructed from pseudo-curvatures, i.e. that the equations of motion (19.9)-(19.12) are solved by the "flat" Lie group solution G = G (the Graded-Poincaré). It is obvious in the equations themselves, which all vanish identically when $R^{A} \rightarrow 0$. However, we also note for further reference that the coefficients of R^{A} in (19.7) all fulfill eq. (17.10) (for $\zeta_{[ab]}$ this is known by construction (18.6))

$$\zeta_{\dot{\beta}>} = 4\sigma_{a<\alpha\dot{\beta}>}\rho^{a} \wedge \rho^{<\alpha}, \qquad \zeta_{\alpha>} = -4\sigma_{a\alpha>\dot{\beta}>}\rho^{a} \wedge \rho^{<\dot{\beta}} \qquad (19.15)$$

and taking just SO(3.1) for F, using

$$(D\zeta)_{A} = \hat{D}\zeta_{A} - \rho^{h} \wedge \zeta_{E}f^{E}{}_{hA} \qquad (19.16)$$

From (18.8) we find

$$D\zeta_{[ab]} = \hat{D}\zeta_{[ab]} + \frac{1}{2}\{\rho^{<\alpha}\zeta_{<\beta}\sigma_{[ab]}{}^{\beta>}{}_{\alpha>} + \rho^{<\dot{\alpha}}\zeta_{<\dot{\beta}}\sigma_{[ab]}{}^{\dot{\beta}>}{}_{\dot{\alpha}>}\} = E_{[ab]}$$

$$D\zeta_{\dot{\beta}>} = \hat{D}\zeta_{\dot{\beta}>} - 2i\rho^{<\alpha} \wedge \zeta_{a}\sigma^{a}{}_{\alpha>\dot{\beta}>} = \hat{D}\zeta_{\dot{\beta}>} \qquad (19.17)$$

$$D\zeta_{\alpha>} = \hat{D}\zeta_{\alpha>} - 2i\rho^{<\dot{\beta}} \wedge \zeta_a \sigma^a{}_{\dot{\beta}>\alpha>} = \hat{D}\zeta_{\alpha>}$$
$$D\zeta_a = \hat{D}\zeta_a = 0 \tag{19.17}$$

where we are using the structure constants of the fully contracted Super-Poincaré group and the value $\zeta_a = 0$. Now inserting (19.15) we note that we cannot use (16.11) since our choice of F does not correspond to a symmetric decomposition,

$$R^a = \hat{D}\rho^a + 2i\rho^{<\alpha} \wedge \sigma^a{}_{\alpha><\dot{\beta}}\rho^{\dot{\beta}>}$$

and we find that ζ_A is indeed a pseudo-curvature

$$D\zeta_{\dot{\beta}} = \sigma_{a\dot{\beta}><\alpha}(R^a + 2i\sigma^a{}_{<\gamma<\dot{\delta}}\rho^{\dot{\delta}>} \wedge \rho^{\gamma>}) \wedge \rho^{\alpha>} - \sigma_{a\dot{\alpha}><\alpha}R^{\alpha} \wedge \rho^a \stackrel{\Omega}{=} 0 \tag{19.18}$$

We may also verify that L is pseudoclosed

$$\hat{D}\ell = \hat{D}\zeta_{<A}R^{A>} + \zeta_{<A}\hat{D}R^{A>}$$

We have just verified the vanishing of the first term. The second one contains $DR^{\alpha>} = 2(\sigma_a\sigma_b)^{\alpha><\beta}R^{[ab]} \wedge \rho_{\beta>}$, and so

$$\rho^a\sigma_{a<\alpha<\dot{\beta}}\{\rho^{\dot{\beta}>}(\sigma^c\sigma^d)^{\alpha><\gamma}R^{[cd]} \wedge \rho_{\gamma>} + (\sigma^c\sigma^d)^{\dot{\beta}><\dot{\gamma}}R_{[cd]} \wedge \rho_{\dot{\gamma}>} \wedge \rho^{\alpha>}\}$$
$$\sim \rho^a\varepsilon_{abcd}R^{cd} \wedge \sigma^b{}_{<\alpha<\dot{\beta}}\rho^{\dot{\beta}>} \wedge \rho^{\alpha>} \stackrel{\Omega}{=} 0 \tag{19.19}$$

The auxiliary fields of the more conventional treatment are but a convenient parametrization of the superspace legged $R^{as}_{a,\alpha}$ which does not vanish even on-shell, and accordingly interfere with the AGCT gauge algebra (refs. 13,18,22): Off mass shell, all components of R^A exist, and a "tensor calculus" has been developed, as a parametrization of these components in terms of additional fields.

In conclusion, by adding the left and right WRS Lagrangians we have indeed recovered conventional supergravity, but the scale of the spinors remains a free parameter. It is fixed by GP itself if one constructs a ζ_A multiplet under the full GP.[13] The flat Lie group is a solution (up to a rescaling if $\alpha \neq -5$), and the theory is pseudo closed (BRS pseudo invariance).

In ref. 11 we have similarly analyzed extended supergravity.

The uncontracted group manifold is that of G : OSp(4/N). The generators now include I_{ij}, generating an internal symmetry group O(N). In addition, the spinorial generators, both left- and right-handed, span an N-vector representation of the O(N) for each spin component.

Our study has shown that for the parity-conserving case, the equations of motion are not satisfied by a "flat" vacuum. The physical vacuum of the theory does not coincide with the tangent manifold (the rigid Lie Group Manifold). This is a spontaneous symmetry breakdown. We also find that the Lagrangian is not pseudo-closed, i.e. it is not a closed form even after the application of the equations of motion. For $N = 1$ the spontaneous breakdown terms vanish by spinor anti-symmetry, and pseudo-closure is also ensured. However, Extended Geometric Supergravity does exist as a unified theory in the chiral case, either uncontracted or semi-contracted.

All along this work, the fields correspond to "first order" theories, and belong to the adjoint representation of the gauge group. The $N = 8$ limit does not show up. As for the transformations of "local supersymmetry" (the "Stony Brook algebra"), their exact emergence as AGCT has been presented in refs. 13,18. The lower spin fields of the usual component formalism arise as for simple supergravity as a parametrization of those components of the curvature tensor still involving superspace which do not vanish on shell. MacDowell[33] has recently attempted to impose constraints which would reinforce the equations of motion.

REFERENCES

1. L. Corwin, Y. Ne'eman, and S. Sternberg, Rev. Mod. Phys. 47:573 (1975).
2. Yu. A. Golfand and E. P. Likhtman, JETP Letters 13:452 (1971).
3. V. G. Kac, Functional Analysis and Applications 9:91 (1975).
4. V. Rittenberg, in: "Group Theoretical Methods in Physics," (Proceedings Tübingen, 1977), P. Kramer and A. Rieckers, eds., Lecture Notes in Physics 79, Springer Verlag (1978) pp. 3-21. V. G. Kac, Comm. Math. Phys. 53:31 (1977). M. Scheunert, "The Theory of Lie Superalgebras (An Introduction," Lecture Notes in Mathematics 716, Springer Verlag, Berlin-Heidelberg (1979).
5. P. Fayet and S. Ferrara, Physics Reports 32C:249 (1977).
6. R. Haag, J. T. Lopuszanski, and M. Sohnius, Nucl. Phys. B88: 257 (1975).
7. K. Zachos, Cal. Tech. thesis,(1979). The authors of ref. 6 had missed these additional possibilities, as noted by those of ref. 13. This thesis filled the gap.
8. Y. Ne'eman, Phys. Letters 81B:190 (1979).
9. F. A. Berezin and D. A. Leites, Dokl. Akad. Nauk. SSSR 224: 505 (1975). B. Kostant, in: "Differential Geometrical Methods in Mathematical Physics," (Bonn, 1975), Lecture Notes in Mathematics 570, K. Bleuler and A. Reetz, eds., Springer Verlag, Berlin-Heidelberg (1977). B. DeWitt, "Differential Supergeometry," in preparation. M. Batchelor,

in: "Group Theoretical Methods in Physics (Proceedings Austin 1978)," W. Beiglböck, A. Böhm, and E. Takasugi, eds., Lecture Notes in Physics 94, Springer Verlag, Berlin-Heidelberg-New York (1979). A. Rogers, Imperial College preprint.

10. M. Scheunert, W. Nahm, and V. Rittenberg, J. Math. Phys. 18: 155 (1977).
11. J. Thierry-Mieg and Y. Ne'eman, Ann. of Phys. (to be published).
12. S. W. MacDowell and F. Mansouri, Phys. Rev. Lett. 38:739 (1977).
13. Y. Ne'eman and T. Regge, Rivista del Nuovo Cim. Ser. III, 1 #5 (1978).
14. P. Nath and R. Arnowitt, Phys. Letters 65B:73 (1976).
15. Y. Ne'eman, Transactions N.Y. Acad. Sci. (I.I. Rabi Festschrift), 38:106 (1977).
16. Y. Ne'eman and J. Thierry-Mieg, in: "Proceedings of (Kiryat Anavim 1979) 8th Int. Conf. on Group Theoretical Methods in Physics," to be published in Ann. Israeli Phys. Soc. Same authors, Proc. Nat. Acad. Sci. USA (1980).
17. V. Rittenberg and M. Scheunert, J. Math. Phys. 19:709 (1978).
18. Y. Ne'eman and T. Regge, Phys. Letters 74B:54 (1978).
19. Y. Ne'eman, in: "Proc. XIX Int. Conf. High E. Phys. (Tokyo 78)," S. Homma et al., eds., p. 552.
20. See for example B. S. DeWitt in; "Proc. Les Houches 1963 Seminar," C. DeWitt and B. S. DeWitt, eds., New York (1964).
21. B. Zumino, in: "Proc. Conf. Gauge Theories and Mod. Field Th. (Northeastern Univ. Boston 1975)," R. Arnowitt and P. Nath, eds., M.I.T. Press, Cambridge, Mass. (1976).
22. P. Von der Heyde, Phys. Letters 58A:141 (1976).
23. G. Woo, Lett. Nuovo Cim. 13:546 (1975).
24. W. Drechsler and M. E. Mayer, "Fiber Bundle Techniques in Gauge Theories," Lecture Notes in Physics 67, Springer-Verlag, Berlin-Heidelberg-New York (1977); Y. Ne'eman, Symétries, Jauges et Variétés de Groupe, Presses de l'Université de Montréal (1978); Y. Choquet-Bruhat, C. DeWitt-Morette, and M. Dillard-Bleick, "Analysis, Manifolds and Physics," North-Holland, Amsterdam (1977).
25. J. Thierry-Mieg, "Geometrical Reinterpretation of the Faddeev-Popov Ghosts and BRS Transformations," Tel Aviv University report (1978), to be published in J. Math. Phys.
26. R. P. Feynman, Acta Phys. Polon. 26:697 (1963); B. S. DeWitt, "Dynamical Theory of Groups and Fields," Gordon and Breach Publ., New York, London, Paris (1965); L. D. Faddeev and V. N. Popov, Phys. Letters 25B:29 (1967).
27. C. Becchi, A. Rouet, and R. Stora, Ann. of Phys. (N.Y.) 98: 287 (1976).
28. J. Thierry-Mieg, These de Doctorat d'Etat (Paris, 1978).
29. J. Thierry-Mieg, to be published in Nuovo Cim.
30. J. Thierry-Mieg, Lett. Nuovo Cimento, (to be published).
31. A. Trautman, Symp. Math. 12 (Bologna), 30 (1973).

32. D. Z. Freedman, P. van Nieuwenhuizen, and S. Ferrara, Phys. Rev. D13:3214 (1976). S. Deser, and B. Zumino, Phys. Letters 62B:335 (1976).
33. S. W. MacDowell, Phys. Letters 80B:212 (1979).

FOUR LECTURES AT THE 1979 ERICE SCHOOL ON SPIN, TORSION, ROTATION AND SUPERGRAVITY *

P. van Nieuwenhuizen

Institute for Theoretical Physics
State University of New York
Stony Brook, L.I., N.Y.11794

INTRODUCTION

This set of four lectures was given for relativists. Consequently, we have not focused on a review of the latest results obtained in supergravity, but rather tried to give a coherent account of the fundamentals of supergravity. Many people who are keen to understand what supergravity is, fail because of always the same technical details. Thus we give, in these lectures, these details - especially a proof of gauge invariance - and not a sketchy description of how, in principle, to obtain the results. In this year of Einstein's centennial celebration, research in classical general relativity is less intense than it was in the first half of our century. There are, however, some new developments: twistors, Hawking radiation, and supergravity, to mention a few. Supergravity is a beautiful mathematical theory, but so far it has not yet contributed to our understanding of physical reality. That may come; perhaps Fermi-Bose symmetry and a spacetime with bosonic and fermionic coordinates are reality.

The lectures consist of two parts. In the first three lectures supergravity is derived and discussed as a field theory in ordinary spacetime (x,y,z,t) with fields $e_\mu{}^a$(tetrads), $\psi_\mu{}^a$ (gravitino fields) and $\omega_{\mu ab}$ (spin connection fields). This part is most accessible for field theorists. We first describe how one is led

---

*) Lectures given at the NATO Advanced Study Institute on COSMOLOGY AND GRAVITATION: SPIN, TORSION, ROTATION AND SUPERGRAVITY, held at Ettore Majorana International Centre for Scientific Culture.

to supergravity, starting from the ideas of a global supersymmetry, by using simple arguments, mainly a dimensional argument. Then we prove that the theory thus obtained is invariant under local supersymmetry transformations as well as spacetime transformations. Finally we show that this theory can be viewed as the gauge theory of a new kind of algebra, the so-called super Poincaré algebra. This algebra is an extension of the Poincare algebra and contains anticommutators as well as commutators.

The transformation rules of local supersymmetry involve objects ε^{α} (α= 1,4) which satisfy $\varepsilon^{\alpha}\varepsilon^{\beta} + \varepsilon^{\beta}\varepsilon^{\alpha} = 0$. We give an explicit matrix representation of such "anticommuting numbers" (also called "elements of a Grassmann algebra"), so that in principle one can substitute this representation whenever one sees an ε^{α} appearing. We also give an explicit 5x5 matrix representation of the super Poincare algebra in terms of ordinary numbers only, in the hope that seeing such an explicit representation will convince the reader that graded algebras (or, as they are rather called nowadays, superalgebras) are not mysterious objects.

In the last part of the lectures we consider superspace. This part of the lectures corresponds mostly to the idea of relativity. It is an entirely different starting point, but, as has been shown by many authors, leads in the end to the same results as the (xyzt)-only approach. The reader of these notes should realize that by trying to be paedagogical we had to suppress a discussion of much axciting new research. Some Literature one might consult is given at the end.

LECTURE 1. PHYSICAL ARGUMENTS

Supergravity is

(i) general relativity with an extra symmetry, namely a symmetry between bosons and fermions;

(ii) a field theory of elementary particles with a local gauge symmetry between bosons and fermions. As we shall see, (i) and (ii) are equivalent.

(iii) a gravitational theory with torsion. That is why supergravity is being part of this school;

(iv) a field theory which unifies as opposed to a direct product, spacetime symmetries and internal symmetries;

(v) puts several particles of different integer and half-integer spin into the same multiplet. Thus, the photon, electron, quarks, graviton, etc. are all different polarizations of the same superparticle;

(vi) a better quantum theory of gravity than Einstein relativity because the first and second order quantum corrections are finite in those supergravity models where all particles are combined into one superparticle;

(vii) the unique field theory for interacting massless real spin 3/2 fields, and must be coupled to gravity. It is consistent and there are 2 physical modes for these spin 3/2 fields which are called gravitinos.

We now give a short derivation of the main properties of global Fermi-Bose symmetry - also called global (or perhaps rather "rigid", as opposed to "local") supersymmetry. Suppose a boson field B(x) and a fermion field F(x) rotate into each other in an arbitrary space with two axis, - a boson axis and a fermion axis - over an "angle" ε. Thus (suppressing indices)

$$\delta B(x) = F(x)\varepsilon \ .$$

Covariance of this equation requires

(1) Spin. Since bosons have integer spin and fermions half-integer spin, the parameters ε have half-integer spin. The simplest choice is spin-½ and that is what supergravity uses. Thus, ε^α with $\alpha = 1,4$. We use four component spinors. For two component spinors see p.16.

(2) Statistics. Bosons commute and fermions satisfy the exclusion principle. Thus the ε^α are fermionic objects, as one might expect from the spin-statistics theorem. We will also need the Dirac bar on ε, denoted by $\bar{\varepsilon}^\alpha = (\varepsilon^\alpha)^+\gamma_4$ (or with γ_0, depending on a convention different from ours).

The question now arises whether such strange anticommuting ob- exist, or whether perhaps the whole scheme is internally inconsistent. Indeed, a representation exists. Consider the Jordan-Wigner matrices for the annihilation operators a_j:

$$\{a_j, a_k\} = 0, \ \{a_j^+, a_k^+\} = 0, \ \{a_j, a_k^+\} = \delta_{jk} .$$

Take only the annihilation matrices a_j, not the creation matrices a_j^+. If one puts

$$\epsilon_\alpha = \sum_j a_j \beta^j_\alpha$$

$$\bar{\epsilon}_\alpha = \sum_j a_j \bar{\beta}^j_\alpha \ , \quad \bar{\beta}^j_\alpha = (\beta^j_\gamma)^* (\gamma_4)_{\gamma\alpha}$$

then indeed the anticommutation relations are satisfied. Thus we define the bar operation to be unity when acting on a_j.

(3) Dimension. Bosons have dimension 1 and fermions 3/2, as follows

from the form of the Klein-Gordon and Dirac actions, and from the fact that the dimension of an action $I = \int d^4 x \mathcal{L}$ is zero. Thus the dimension of ε_α is $-\frac{1}{2}$. Consequently - and this is the basis of supergravity! - in the $\delta F \sim B\varepsilon$ relation one has a gap of one unit of dimension. Which objects can fill this gap? If we are in flat spacetime and consider massless fields, there is only the derivative ∂_μ. Thus, purely on dimensional grounds and omitting all kinds of indices

$$\delta F(x) = \partial_\mu B(x)\varepsilon \quad .$$

(4) Reality: To consider only irreducible fields, we will restrict ourselves, as is customary in supersymmetry, to real bosonic fields and the analogue for fermionic fields, namely Majorana fields. For any spinor λ^α one defines its Majorana conjugate $\tilde{\lambda}_\alpha = C_{\alpha\beta}\lambda^\beta$ as tha spinor which transforms in the same way as its Dirac conjugate $\bar{\lambda}_\alpha = \lambda^+{}_\beta(\gamma_4)^\beta{}_\alpha$. A Majorana spinor is a spinor whose Majorana conjugate equals its Dirac conjugate

$$\bar{\lambda}_\alpha = iC_{\alpha\beta}\lambda^\beta \quad .$$

The matrix C must satisfy $C\sigma_{\mu\nu}C^{-1} = -(\sigma_{\mu\nu})^T$ in order that λ_α transforms as $\bar{\lambda}_\alpha$. This one easily proves from the Lorentz transformation of λ^α

$$\lambda'^\alpha = (\exp \omega^{\mu\nu}\sigma_{\mu\nu})^\alpha{}_\beta\lambda^\beta, \; \sigma_{\mu\nu} = \frac{1}{4}[\gamma_\mu, \gamma_\nu].$$

This determines C up to two constants: $C = \alpha\,(1+\gamma_5)C_o + \beta(1-\gamma_5)C_o$ if $C_o\sigma_{\mu\nu}C_o^{-1} = -(\sigma_{\mu\nu})^T$. Usually one makes a stronger requirement, namely that the Majorana condition $\bar{\lambda} = C\lambda$ is preserved in time. If λ satisfies the Dirac equation $(\not\partial + M)\lambda = 0$, one finds from $\bar{\lambda}(\overleftarrow{\not\partial} - M) = 0$ that

$$C\gamma^\mu C^{-1} = -\gamma^{\mu,T}.$$

As shown in the appendix, C is unique up to a constant and the spin 1/2 particles described by λ are their own antiparticles.

The upshot of working with real (Majorana) fields is that one expects for massless fields with (half) integer spin J>0 two modes for each. Then one has equal amounts of bosonic and fermionic states if one has an equal number of fields. In order to keep the $\delta B = F\varepsilon$ relation real, our ε^α will also be Majorana spinors

$$\bar{\varepsilon}^T = iC\varepsilon \quad , \quad \bar{\varepsilon} = \Sigma\, a_j\bar{\beta}_j, \; \bar{\beta}_j = \beta_j^\dagger\,\gamma_4 \;, \bar{\beta}^T = iC\beta$$

Notice that the coefficients β^α are ordinary numbers. From now on we will only write ε, F and use

$$\{\varepsilon^\alpha, \varepsilon^\beta\} = \{\varepsilon^\alpha, \bar{\varepsilon}^\beta\} = \{\varepsilon^\alpha, F\} = [\varepsilon^\alpha, B] = 0$$

Let us now consider the commutator of two global supersymmetry transformations on B(x), with parameters $\varepsilon_{1,\alpha}$ and $\varepsilon_{2,\beta}$ which anticommute

$$\delta(\varepsilon_1)\,\delta(\varepsilon_2)B - \delta(\varepsilon_2)\,\delta(\varepsilon_1)\,B = \tfrac{1}{2}(\bar{\varepsilon}_2\gamma^\mu\varepsilon_1)\,\partial_\mu\,B(x)\,.$$

We have written down the exact formula with all indices as it is found in all models of global supersymmetry. This is a very interesting result. It shows that two global supersymmetry transformations lead to a translation - in other words: two rotations in internal Fermi-Bose space lead to a motion in spacetime.

We now make a giant step, and consider local parameters $\varepsilon_\alpha(x) = \Sigma a_j \beta^j_\alpha(x)$. The above equation leads us to expect that two local supersymmetry rotations with parameters $\varepsilon_1(x)$ and $\varepsilon_2(x)$ will lead to a translation ∂_μ over a distance $\bar{\varepsilon}_2(x)\gamma^\mu\,\varepsilon_1(x)$. In other words, we find shifts over distances which differ from point to point: general coordinate transformations and hence (we stress: this is a vague and heuristic argument only) gravity! This argument was the basis of the derivation of supergravity by Freedman, Ferrara and the author. One can rederive it by using consistency arguments (see Deser and Zumino).

We see thus that we can expect a local gauge theory of gravity with a local supersymmetry with parameters $\varepsilon_\alpha(x)$. *A priori* it is not clear which form of gravity theory to expect: Einstein's form with only tetrads e^a_μ, or Cartan's form with both tetrads and spin connection ω^{ab}_μ. As we shall see, the best formulation of supergravity lies in between, and is appropriately called 1.5 order formalism. If we start with Cartan's ideas, we expect gauge fields for the generators P_a and M_{ab} of the Poincaré group. For a particle physicist this is appealing since the Poincaré (and not the Lorentz group) has two Casimir operators corresponding to the two quantum numbers mass and spin. We expect a gauge field for the local supersymmetry fields $\psi^\alpha_\mu(x)$, because in general, gauge fields are obtained by affixing an index μ to the parameters. Hence, we have the Lie algebra valued vector field

$$V_\mu = e^a_\mu\,P_a + \omega_\mu^{ab}\,M_{ab} + \bar{\psi}_\mu^\alpha Q_\alpha\,.$$

(for Lorentz covariance we choose to write $\bar{\psi}_\mu{}^\alpha Q_\alpha$ rather than $\psi_\mu{}^\alpha Q_\alpha$. The generator Q_α is the supersymmetry generator)

Thus we expect that there are in our theory, in addition to spin-2 gravitons, also particles whose spin is 3/2 and which are called gravitinos. This follows from elementary addition of angular momenta, using that a four vector A_μ contains a spin 1 part A_j and a spin 0 part A_4

$$\text{spin } \psi_\mu = \frac{1}{2} \times (1+0) = \frac{3}{2} + \frac{1}{2} + \frac{1}{2} .$$

These two spin 1/2 parts correspond to the decomposition

$$\psi_\mu = \left(P^{3/2}_{\mu\nu} + P^{1/2,s}_{\mu\nu} + P^{1/2,t}_{\mu\nu} \right) \psi_\nu$$

$$P^{3/2} + P^{1/2,s} + P^{1/2,t} = I, \; P^{1/2,s}_{\mu\nu} = \frac{1}{3} \hat{\gamma}_\mu \hat{\gamma}_\nu$$

$$P^{1/2,t}_{\mu\nu} = \omega_\mu \omega_\nu \;, \; \omega_\mu = \partial_\mu \not\partial \Box^{-1}, \hat{\gamma}_\mu = \gamma_\mu - \omega_\mu .$$

This is thus a nonlocal decomposition into spin parts - or rather (since one cannot define spin for massless particles but only helicity) a decomposition into eigenfunctions of the orthonormal set of projection operators P^j. As far as the Lorentz group is concerned, ψ_μ transforms as (using van der Waerden notation)

$$\psi_\mu^\alpha = (\psi)_{A,B\dot{C}} + \psi_{\dot{A},B\dot{C}}$$

so that $\psi_\mu{}^\alpha$ splits into four inequivalent irreducible representations

$$\left(\tfrac{1}{2},1\right) = \psi_{B\underline{\dot{A}\dot{C}}} = (1+\gamma_5)\left(\psi_\mu - \tfrac{1}{4}\gamma_\mu \gamma\cdot\psi\right)$$

$$\left(\tfrac{1}{2},0\right) = \psi_{B\dot{A}\dot{C}}\,\epsilon^{\dot{A}\dot{C}} = (1+\gamma_5)(\gamma\cdot\psi)$$

and similarly, (1,1/2) and (0,1/2) with $1 - \gamma_5$.

Physical spin 3/2 fields contain only (1/2,1) and (1,1/2) and (a stronger statement) only spin-3/2 (this removes the longitudinal part $\partial_\mu\psi_\mu$ from (1/2,1) and (1/2,1)).

The two fields e^a_μ and ψ^α_μ in V_μ are identified with the vierbein fields which describe gravity in Einstein's theory of relativity, and with new spin-3/2 fields ψ_μ. Since we are heading for a theory with a local symmetry between bosons and fermions, we take the gravitino fields ψ^α_μ to be Majorana fields and massless.

Thus we expect that we will find an action such that it describes four particles: two massless bosons with helicities ± 2 and two massless fermions with helicities $\pm 3/2$. But what role does the field $\omega_{\mu ab}$ play? It cannot describe physical modes, because then the symmetry between the numbers of bose and fermi states is destroyed. This is reminiscent of the Palatini formulation of general relativity, where one first introduces two fields $g_{\mu\nu}$, $\omega_{\mu ab}$ (or, if fermions are present, e^a_μ and $\omega_{\mu ab}$) but then subsequently eliminates $\omega_{\mu ab}$ from the Hilbert action by solving the field equation $\delta I/\delta\omega_{\mu ab} = 0$,

$$I = \int -\tfrac{1}{2}\kappa^{-2} \det e\, R, \quad R = e^{a\nu} e^{b\mu} R_{\mu\nu ab}(\omega)$$

$$R_{\mu\nu ab}(\omega) = \left(\partial_\mu \omega_{\nu ab} + \omega_{\mu ac}\, \omega_\nu{}^c{}_b\right) - (\mu \leftrightarrow \nu)$$

If one inserts the solution $\omega_{\mu ab} = \omega_{\mu ab}(e)$ back into I, one regains Einstein theory of general relativity if there are no fermions present. Crucial for this result is that $\omega_{\mu ab}$ is nonpropagating. (An action of the form R^2 would have yielded a propagating spin connection. Propagating spin connection can still be the basis of a supergravity theory. An example is conformal supergravity, based on the super conformal group, and a kind of supersymmetric extension of twistor theory. However, one needs then extra fields to have equal amounts of bosons and fermions.)

Therefore, the program to be executed is the following. One must construct an action containing three fields e^a_μ, ψ^α_μ and $\omega_{\mu ab}$, in which $\omega_{\mu ab}$ is nonpropagating, with three gauge symmetries: the two local spacetime symmetries of general coordinate transformation and local Lorentz rotation, and local supersymmetry transformations. Since we will solve for $\omega_{\mu ab} = \omega_{\mu ab}(e,\psi)$, we only need the transformation laws for e^a_μ and ψ^α_μ. The transformation laws of $\omega_{\mu ab}$ follow then from the chain rule. The spacetime transformation laws are known from gravity theory, but the local supersymmetry laws have to be found. For the vierbein one expects $\delta e^a{}_\mu \sim \bar{\varepsilon}\psi_\mu$ and one is naturally lead to

$$\delta_S e^a{}_\mu = \kappa \bar{\epsilon} \gamma^a \psi_\mu$$

At first sight one might have chosen instead

$$\delta_s e^a{}_\mu = \kappa \bar{\varepsilon} \gamma_\mu \psi^a \quad \text{or} \quad \varepsilon^a{}_{\mu\nu}{}^\rho \kappa \bar{\varepsilon} \gamma^\nu \gamma_5 \psi_\rho$$

These laws are nonlinear ($\psi^a = \psi_\mu e^{a\mu}$ and $\gamma^\nu = \gamma_a e^{a\nu}$) but that should be allowed. These alternatives differ by Lorentz transformations; for example

$$\bar{\varepsilon}\gamma_\mu \psi^a - \bar{\varepsilon}\gamma^a \psi_\mu = \lambda^a{}_b e^b{}_\mu \quad \text{with} \quad \lambda^a{}_b = \bar{\varepsilon}\gamma_b \psi^a - \bar{\varepsilon}\gamma^a \psi_b$$

Fixing $\delta\psi_\mu$ to be proportional to $D_\mu \varepsilon$, the alternatives must be rejected since in the global linear limit they do not lead to the $\{Q, Q\} \sim P$ anticommutator.

We now must find how ψ_μ rotates under local supersymmetry. At first sight we again have an ambiguity. From global supersymmetry we expect $\delta\psi \sim (\partial e)\varepsilon$, while from general gauge principles we expect that the gauge field of supersymmetry transforms into the derivative of the parameter $\delta\psi_\mu \sim \partial_\mu \varepsilon$. Covariantizing with respect to gravity the latter result would lead one to (putting in the normalization we will later need)

$$\delta_S \psi_\mu = 2\kappa^{-1} D_\mu \epsilon, \quad D_\mu \epsilon = \left(\partial_\mu + \tfrac{1}{2}\omega_{\mu ab}(e,\psi)\sigma^{ab}\right)\epsilon$$

This result in fact confirms both proposals. As we shall see, after solving the ω -field equation, the part of ω linear in fields is of the form ∂e. The global linear part $\delta\psi_\mu = \kappa^{-1} (\omega_{\mu ab})_{lin}$ $\sigma^{ab}\varepsilon$ together with $\delta(\kappa^{-1} e^a{}_\mu) = \bar{\varepsilon}\gamma^a\psi_\mu$ belongs to the so-called spin (2,3/2) multiplet of global supersymmetry which was constructed by Grisaru, Pendleton and the author directly from the graded Poincaré algebra without using supergravity. With hindsight, one should have expected the rule for $\delta\psi_\mu$, since in Yang-Mills theory the rule $\delta W_\mu{}^a = (D_\mu\Lambda)^a = \partial_\mu \Lambda^a + \varepsilon^{abc} W_\mu{}^b \Lambda^c$ has the same dual character.

We have found by reasoning the laws $\delta_S e^a{}_\mu$ and $\delta_S \psi_\mu$, and expect for the spin 2 part of the action the Hilbert action in Cartan form. The spin 3/2 part of the action may contain $e^a{}_\mu, \psi_\mu{}^\alpha$ and

$\omega_{\mu ab}$, but $\omega_{\mu ab}$ should appear linearly and without derivatives, in order to be able to solve $\delta I/\delta\,\omega_{\mu ab} = 0$ algebraically. As we shall show elsewhere, the action for real massless free fields ψ_μ without ghosts, (i.e. with positive energy) is unique and given by the same action Rarita and Schwinger derived in 1941 when they were interested in the possibility that the neutrino in β - decay have spin 3/2. This action reads

$$\mathcal{L} = -\tfrac{1}{2}\epsilon^{\mu\nu\rho\sigma}\bar{\psi}_\mu\gamma_5\gamma_\nu\partial_\rho\psi_\sigma \qquad \text{(free fields)}$$

It clearly has a local invariance of the form $\delta\psi_\sigma = \partial_\sigma\,\varepsilon(x)$, and these authors make a cryptic remark about it at the end of their paper. As far as we are concerned, they found the flat-space limit of supergravity, and to return into curved space, one simply covariantizes

$$\mathcal{L} = -\tfrac{1}{2}\epsilon^{\mu\nu\rho\sigma}\bar{\psi}_\mu\gamma_5\gamma_\nu D_\rho\psi_\sigma$$

$$D_\rho = \partial_\rho + \tfrac{1}{2}\omega_{\rho ab}\sigma^{ab}$$

At this point it is worthwhile to point out that this covariantization is not unique. If one would have used the connection $\omega_{\mu ab}(e)$ defined by $(\partial_\mu e^a{}_\nu + \omega_\mu{}^{ab}(e)\, e_{b\nu}) - (\mu\leftrightarrow\nu) = 0$, then one would not have found an invariant action.

We turn, in the next lecture, to a proof that this action is invariant under local supersymmetry variations, provided one adds the Hilbert action for the gravitational field to it.

LECTURE 2: THE RULES OF THE GAME

The proof of the pudding is in the eating. Thus we start with perhaps the most essential calculation in supergravity: an explicit and detailed proof of the invariance of the gauge action of simple (also called "non-extended" or "N=1") supergravity. Only in lecture 3 will we put the results of this section in perspective.

The action is the sum of the Einstein (spin-2) curvature and the Rarita-Schwinger (spin 3/2) curvatures plus three squares of auxiliary fields

$$\mathcal{L} = -\tfrac{1}{2}\kappa^{-2}\,\det e\; e^{a\nu} e^{b\mu} R_{\mu\nu ab}(\omega) - \tfrac{1}{2}\bar{\psi}_\mu R^\mu$$
$$- \tfrac{1}{3}\kappa^{-2}\det e\,\left(S^2 + P^2 - A_a^2\right)$$

For the time being, we will omit the terms with S,P and A_a. A complete list of definitions of symbols follows:

- κ^2 is proportional to Newton's constant;
- det e is the determinant of the vierbein field $e^a{}_\mu$ which satisfies $e^a{}_\mu e_{a\nu} = g_{\mu\nu}$.
- $e^{a\nu}$ is the inverse of $e_{a\mu}$, thus $e^{a\mu}\, e_{b\mu} = \delta^a{}_b$ and $e^{a\mu} e_{a\nu} = \delta^\mu{}_\nu$. Also, $e^{a\mu}\, e_a{}^\nu = g^{\mu\nu}$.
- $R_{\mu\nu ab} = \partial_\mu \omega_{\nu ab} + \omega_{\mu ac}\omega_\nu{}^c{}_b - (\mu\leftrightarrow\nu)$ is the Riemann curvature and the spin connection field $\omega_{\mu ab}$ is at this point an independent field. We define $R_{\mu b} = R_{\mu\nu ab} e^{a\nu}$.
- ψ_μ is a vectorial spinor, thus (with spinor index a explicit) $\psi_\mu{}^a$ with $\mu = 1,2,3,4$ and a = 1,2,3,4.
- $R^\mu = \epsilon^{\mu\nu\rho\sigma}\gamma_5\gamma_\nu D_\rho\psi_\sigma$ with $\epsilon^{1234} = 1$ a density and $\gamma_5 = \gamma_1\gamma_2\gamma_3\gamma_4$ a constant matrix satisfying $\gamma_5^2 = 1$ and $\gamma_5^\dagger = \gamma_5$.
- $\gamma_\nu = e^a{}_\nu\gamma_a$ and γ_a with a = 1,2,3,4 are constant Dirac matrices satisfying $\{\gamma_a,\gamma_b\} = 2\delta_{ab}$. Also $\gamma_a = \gamma^a$ and $\gamma_a = \gamma_a^+$.

$$- \quad D_\rho \psi_\sigma = \partial_\rho \psi_\sigma + \tfrac{1}{2}\omega_{\rho ab}\sigma^{ab}\psi_\sigma \quad \text{and}$$

$\sigma^{ab} = \frac{1}{4}[\gamma^a, \gamma^b]$ No Christoffel symbol is needed, nor present, since only $D_\rho\psi_\sigma - D_\sigma\psi_\rho$ is present in R^μ and this curl is already a good world tensor.

- The gravitino field ψ_μ is a Majorana spinor $\bar{\psi}_\mu^T = iC\psi_\mu$ where C satisfies $C\gamma_\mu C^{-1} = -\gamma_\mu^T$, $\bar{\psi}_\mu = \psi_\mu^+ \gamma_4 (-)^{\delta_{\mu 4}}$

- The matrix C satisfies $C^T = -C$ and is unique up to a constant. However, it is only in particular representations that $C^{-1} = -C$.

Remarks: All our indices run from 1 to 4. Some physicists prefer to have world indices running from 0 to 3; this poses no problems since all indices are contracted.

Elimination of the spin connection using its field equation

The spin connection $\omega_{\mu ab}$ is eliminated from the action by using its field equation. We present here a simple treatment. We begin by noting that

$$\det e\left(-\tfrac{1}{2}R\right) = -\tfrac{1}{8}\epsilon^{\mu\nu\rho\sigma}\epsilon_{abcd}\, e^a_\mu e^b_\nu R_{\rho\sigma}{}^{cd}(\omega)$$

$$\delta R_{\rho\sigma}{}^{cd} = D_\rho \delta\omega_\sigma{}^{cd} - D_\sigma \delta\omega_\rho{}^{cd}$$

$$D_\rho \delta\omega_\sigma{}^{cd} = \partial_\rho \delta\omega_\sigma{}^{cd} + \omega_\rho{}^{ce}\delta\omega_{\sigma e}{}^{d} + \omega_\rho{}^{de}\delta\omega_{\sigma}{}^{c}{}_{e}$$

This result is based on the identity $\varepsilon^{\mu\nu\rho\sigma}\varepsilon_{\mu\nu\alpha\beta} = 2(\delta^\rho_\alpha\delta^\sigma_\beta - \delta^\rho_\beta\delta^\sigma_\alpha)$. Varying also the spin connection in the Rarita-Schwinger action, we have

$$-\tfrac{1}{2}\epsilon^{\mu\nu\rho\sigma}\left(\bar{\psi}_\mu \gamma_5 \gamma_\nu \sigma_{cd} \psi_\sigma\right)\left(\tfrac{1}{2}\delta\omega_\rho{}^{cd}\right)$$

We now recall an identity of Dirac gamma matrices

$$\gamma_{\nu}\sigma_{cd} = \tfrac{1}{2}\left(e_{c\nu}\gamma_d - e_{d\nu}\gamma_c + \epsilon_{bcda}\gamma_5\gamma_a e^b{}_\nu\right)$$

Only the last term contributes, since $\bar{\psi}_\mu\gamma_5\gamma_d\psi_\sigma$ is symmetric in μ and σ. This is explained in the appendix

The total field equation thus reads upon partial integration of $D_\sigma\delta\omega_\rho{}^{cd}$

$$\tfrac{1}{4}\left(D_\mu e^a{}_\sigma - D_\sigma e^a{}_\mu\right) + \tfrac{1}{8}\bar{\psi}_\mu\gamma^a\psi_\sigma = 0$$

The spin connection can be decomposed as

$$\omega_{\mu ab} = \omega_{\mu ab}(e) + \kappa_{\mu ab}$$

$$\partial_\mu e^a{}_\nu + \omega_\mu{}^{ab}(e)e_{b\nu} - \Gamma^\alpha_{\mu\nu}(g)e_{a\alpha} = 0$$

$$\partial_\mu e^a{}_\nu + \omega_\mu{}^{ab}e_{b\nu} - \Gamma^\alpha_{\mu\nu}e^a{}_\alpha = 0$$

The last two equations define $\omega(e)$ in terms of the Christoffel symbol and a non-symmetric symbol $\Gamma^\alpha_{\mu\nu}$ in terms of $\omega_{\mu ab}$, respectively. One might call them "vierbein postulates", although they are rather definitions.

Since $\Gamma^\alpha_{\mu\sigma}(g) = \Gamma^\alpha_{\sigma\mu}(g)$, it follows from the ω field equation that

$$\kappa_\mu{}^a{}_\sigma - \kappa_\sigma{}^a{}_\mu = \tfrac{1}{2}\bar{\psi}_\mu\gamma^a\psi_\sigma$$

Since torsion is defined by $S^\alpha_{\mu\nu} = \frac{1}{2}(\Gamma^\alpha_{\mu\nu} - \Gamma^\alpha_{\nu\mu})$ and clearly $\frac{1}{2}(\Gamma^\alpha_{\mu\nu} - \Gamma^\alpha_{\nu\mu}) = \frac{1}{2}(\kappa_{\mu\nu}^{\ \ a} - \kappa_{\nu\mu}^{\ \ a})$, it follows

that there is torsion

$$S^\alpha_{\mu\nu} = \tfrac{1}{4}\bar{\psi}_\mu\gamma^a\psi_\nu$$

In order to solve for $\kappa_{\mu ab}$, we use the same method as is employed to express $\Gamma^\rho_{\mu\nu}(g)$ in terms of $\partial_\alpha g_{\beta\gamma}$, that is, one uses the identity

$$(\kappa_{\mu\alpha\sigma} - \kappa_{\sigma\alpha\mu}) + (\kappa_{\alpha\sigma\mu} - \kappa_{\mu\sigma\alpha}) - (\kappa_{\sigma\mu\alpha} - \kappa_{\alpha\mu\sigma}) = 2\kappa_{\mu\alpha\sigma}$$

One finds for the contorsion tensor $\kappa_{\mu ab}$

$$\kappa_{\mu ab} = \tfrac{1}{4}\left(\bar{\psi}_\mu\gamma_a\psi_b - \bar{\psi}_\mu\gamma_b\psi_a + \bar{\psi}_a\gamma_\mu\psi_b\right)$$

1.5 order formalism. Since $\delta I/\delta\omega_{\mu ab} = 0$ where $I = \int d^4x \mathcal{L}$ is the action, one need not consider variations of $\omega(e,\psi)$ if one calculates δI. Indeed,

$$\delta I = \frac{\delta I}{\delta e}\bigg|_{\psi,\omega}\delta e + \frac{\delta I}{\delta\psi}\bigg|_{e,\omega}\delta\psi + \frac{\delta I}{\delta\omega}\bigg|_{e,\psi}\left(\frac{\delta\omega}{\delta e}\delta e + \frac{\delta\omega}{\delta\psi}\delta\psi\right)$$

and since $\delta I/\delta\omega$ at fixed e and ψ vanishes (since this is just the field equation) one may put anything one likes for $\delta\omega$, in particular $\delta\omega = 0$.

The result $\delta\omega_{\mu\, ab} = 0$ follows also from group theory. Thus an enormous simplification occurs. If one wants to obtain $\delta I(e,\psi,\omega(e,\psi))$, one need only to consider variation of the explicit e and ψ fields, those which are not in $\omega(e,\psi)$. This obvious but invaluable observation is called 1.5 order formalism, since it treats ω as if it were an independent field which has its own transformation law $\delta\omega = 0$, but $\omega_{\mu ab}$ is really not an independent field but a function of e and ψ, as in second order formalism.

In their proof of invariance of the gauge action based on second order formalism, Freedman, Ferrara and the author used $\omega = \omega(e,\psi)$ as given by $\delta I/\delta\omega = 0$, but expanded this $\omega(e,\psi)$ and subsequently needed a computer to prove cancellations. The first order formalism of Deser and Zumino led to a simplification of the action, but the law for $\delta\omega_{\mu ab}$ was complicated, and, as we now know, disagrees both with the results obtained from superspace as well as group theory. The proof we are going to give seems to be the simplest possible. Due to 1.5 order formalism, one has $\delta\omega = 0$, as also given by group theory. The cancellation proceeds in two simple steps: the Einstein tensors obtained by varying the Einstein action and the two gravitino fields cancel separately, and the variation of γ_ν plus a term $D_\mu\gamma_\nu$ due to partial integration cancel due to the torsion equation. Thus we will apply the ω-field equation twice: once by not varying ω and once in the cancellation of the $\delta\gamma_\nu$ and $D_\mu\gamma_\nu$ terms. This proof was found by Fradkin and Vassiliev and by Townsend and the author, based on an earlier idea by Chemseddine and West.

Invariance of the action. The proof of the invariance of the gauge action of simple supergravity under $\delta e^a{}_\mu = \kappa\bar{\varepsilon}\gamma^a\psi_\mu$ and $\delta\psi_\mu = \frac{2}{\kappa}D_\mu\varepsilon$ is, with 1.5 order formalism, exceedingly simple.

The variation of the Einstein action is, as usual, proportional to the Einstein tensor

$$\delta\mathcal{L}(E) = \det e\ \kappa^{-2}\ \bar{\epsilon}\gamma^{\nu}\psi^{b}\ G_{\nu b}$$

Here we used that $(\delta e^{a\nu}) = -e^{a\alpha}\delta e_{b\alpha}e^{b\nu}$ which follows from $\delta(e^{a\nu}e_{a\mu}) = 0$. We have defined $G_{\mu b} = R_{\mu\nu ab}e^{a\nu} - \frac{1}{2}e_{b\mu}(R_{\mu\nu cd}e^{c\nu}e^{d\mu})$.

Note that we only varied det e and $e^{a\nu}$ and $e^{b\mu}$ but not the $\omega_{\mu ab}$ in $R_{\mu\nu ab}$, thanks to 1.5 order formalism.

The variation of the spin 3/2 action comes from three sources. Recalling

$$\mathcal{L}(RS) = -\frac{1}{2}\epsilon^{\mu\nu\rho\sigma}\bar{\psi}_{\mu}\gamma_5\gamma_{\nu}(\partial_{\rho} + \tfrac{1}{2}\omega_{\rho ab}\sigma^{ab})\psi_{\sigma}$$

we see that we must vary $\bar{\psi}_{\mu}$, $e_{a\nu}$ in γ_{ν}, and ψ_{σ}, but not $\omega_{\rho ab}$

We first consider the variations of $\bar{\psi}_{\nu}$ and ψ_{σ} together. Integrating partially the term with $\delta\bar{\psi}_{\mu} = \frac{2}{\kappa}D_{\mu}\epsilon$ one has

$$\delta\mathcal{L}(RS, \delta\psi_{\sigma}, \delta\bar{\psi}_{\mu}) = \Big[\tfrac{1}{\kappa}(\bar{\epsilon}\gamma_5\gamma_a D_{\rho}\psi_{\sigma})(D_{\mu}e^{a}{}_{\nu})$$
$$-\tfrac{1}{\kappa}\bar{\psi}_{\mu}\gamma_5\gamma_{\nu}D_{\rho}D_{\sigma}\epsilon + \tfrac{1}{\kappa}\bar{\epsilon}\gamma_5\gamma_{\nu}D_{\mu}D_{\rho}\psi_{\sigma}\Big]\epsilon^{\mu\nu\rho\sigma}$$

Relabelling $D_{\mu}D_{\rho}\psi_{\sigma} = D_{\rho}D_{\sigma}\psi_{\mu}$, and using that the commutator of two covariant derivatives is a curvature, one finds

$$\delta\mathcal{L}(RS, \delta\psi_{\sigma}, \delta\bar{\psi}_{\mu}) = -(4\kappa)^{-1}\big[\bar{\psi}_{\mu}\gamma_5\gamma_{\nu}\sigma_{ab}\epsilon - \bar{\epsilon}\gamma_5\gamma_{\nu}\sigma_{ab}\psi_{\mu}\big]$$
$$\text{times } R_{\rho\sigma}{}^{ab}\epsilon^{\mu\nu\rho\sigma} + \tfrac{1}{\kappa}(\bar{\epsilon}\gamma_5\gamma_a D_{\rho}\psi_{\sigma})(D_{\mu}e^{a}{}_{\nu})\epsilon^{\mu\nu\rho\sigma}$$

Clearly, only terms with $\psi\gamma\epsilon$ but not with $\bar{\psi}\gamma_5\gamma\epsilon$ remain in the terms containing the curvature. Using

$$\epsilon^{\mu\nu\rho\sigma}\epsilon_{\nu abc}R_{\rho\sigma}{}^{ab} = -4\,G_{c\mu}$$

these terms cancel against $\delta\mathcal{L}(E)$.

All that is left is

$$\delta\mathcal{L}(RS, \text{rest}) = \Big[\tfrac{1}{\kappa}(\bar{\epsilon}\gamma_5\gamma_a D_{\rho}\psi_{\sigma})D_{\mu}e^{a}{}_{\nu} - \tfrac{\kappa}{2}(\bar{\psi}_{\mu}\gamma_5\gamma_a D_{\rho}\psi_{\sigma})\bar{\epsilon}\gamma^{a}\psi_{\nu}\Big]\epsilon^{\mu\nu\rho\sigma}$$

Making a Fierz rearrangement (see appendix)

$$(\bar{\psi}_{\mu}\gamma_a\gamma_5 D_{\rho}\psi_{\sigma})(\bar{\epsilon}\gamma^{a}\psi_{\nu}) = -\tfrac{1}{4}(\bar{\psi}_{\mu}\gamma_a O_j\gamma^a\psi_{\nu})(\bar{\epsilon}O_j\gamma_5 D_{\rho}\psi_{\sigma})$$

it is clear that terms with $O_j = 1$, γ_5 $i\gamma_\lambda$ γ_5 do not contribute since they yield $\bar{\psi}_\mu O_j \psi_\nu$ symmetric in μ,ν. However, also the tensor terms cancel since $\gamma_a \sigma_{\mu\nu} \gamma_a = 0$. Thus we are only left with

$$(\bar{\psi}_\mu \gamma_a \gamma_5 D_\rho \psi_\sigma)(\bar{\epsilon} \gamma^a \psi_\nu)\epsilon^{\mu\nu\rho\sigma} = \tfrac{1}{2}(\bar{\psi}_\mu \gamma_\lambda \psi_\nu)(\bar{\epsilon} \gamma^\lambda \gamma_5 D_\rho \psi_\sigma)\epsilon^{\mu\nu\rho\sigma}$$

Finally, using that the torsion is given by

$$D_\mu e^a{}_\nu - D_\nu e^a{}_\mu = \tfrac{1}{2}\bar{\psi}_\mu \gamma^a \psi_\nu$$

we see that all terms cancel.

APPENDIX: LECTURE 2

The charge conjugation matrix C. Consider the finite group with 32 elements spanned by the Clifford algebra $\{\gamma_\mu , \gamma_\nu\} = 2\delta_{\mu\nu}$ with $\mu,\nu = 1,4$. Using elementary group theory, one easily proves that there is only one irreducible faithful representation (all other irreducible representations are one-dimensional), which is 4x4 dimensional. Since $-\gamma_\mu^T$ as well as γ_μ^T satisfies the same group multiplication table, one has with Schur's lemma

$$C\gamma_\mu C^{-1} = -\gamma_\mu^T, \quad D\gamma_\mu D^{-1} = \gamma_\mu^T .$$

(For Pauli matrices, $-\vec{\sigma}^T$ satisfies the same group multiplication table as $\vec{\sigma}$, but not $+\vec{\sigma}^T$ since $\sigma_1\sigma_2 = i\sigma_3$. Thus in two dimensions one has only the relation with the - sign).

Clearly C is unique up to a constant, since from $C_1\gamma_\mu C_1^{-1} = C_2\gamma_\mu C_2^{-1}$ it follows that $C_2^{-1}C_1$ commutes with all group elements and is proportional to the unit matrix according to Schur's lemma.

This matrix C is antisymmetric. To prove this, apply the definition of C twice

$$C^{-1,T} C\gamma_\mu C^{-1} C^T = \gamma_\mu$$

from which it follows that $C^{-1}C^T$ is proportional to the unit matrix. Thus $C = kC^T$ and clearly $k = \pm 1$. It follows that the matrices C, $C\gamma_5$, $C\gamma_\mu \gamma_5$ are antisymmetric if $k = -1$ while $C\gamma_\mu$ and $C\sigma_{\mu\nu}$ are symmetric in this case. The opposite case, $k = +1$, is not possible, because in that case one would have 10 independent 4x4 antisymmetric matrices, namely $C\gamma_\mu$ and $C\sigma_{\mu\nu}$.

Two-component spinors. For people more accustomed to dotted and undotted spinors we give the translation rules. Consider a representation in which γ_5 is diagonal, for example

$$\gamma_k = \begin{pmatrix} 0 & -i\sigma_k \\ i\sigma_k & 0 \end{pmatrix} , \quad \gamma_4 = \begin{pmatrix} 0 & 1 \\ 1 & 0 \end{pmatrix}, \quad \gamma_5 = \begin{pmatrix} 1 & 0 \\ 0 & -1 \end{pmatrix}$$

One defines for λ^α undotted spinors by $\lambda^A = \frac{1}{2}(1+\gamma_5)\lambda$ and dotted spinors by $\lambda_{\dot{A}} = \frac{1}{2}(1-\gamma_5)\lambda$ with A, $\dot{A} = 1,2$. Clearly the indices of the Pauli matrices in γ_μ are as follows

$$(\sigma_\mu)^{A\dot{B}} = (\vec{\sigma}, iI)^{A\dot{B}} \quad \text{and} \quad (\sigma_\mu)_{\dot{A}B} = (\vec{\sigma}, -iI)_{\dot{A}B}$$

Symmetry properties for Majorana spinors.

Consider two Majorana spinors ψ and χ, thus $\psi = iC\bar{\psi}^T$ with $\bar{\psi} = \psi^+ \gamma_4$ and *idem* for χ. Then $\psi^T = i\bar{\psi}C^T$ and $\bar{\psi}\gamma_\mu\chi = -\chi^T\gamma_\mu^T\bar{\psi}^T$ (since ψ and χ anticommute) $= -i\bar{\chi}C^T(-C\gamma_\mu C^{-1})(-iC^{-1}\psi) = -(\bar{\chi}\gamma_\mu\psi)$ similarly it follows that

$$\bar{\psi}\chi = \bar{\chi}\psi , \quad \bar{\psi}\gamma_5\chi = \bar{\chi}\gamma_5\psi , \quad \bar{\psi}\gamma_\mu\chi = -\bar{\chi}\gamma_\mu\psi$$

$$\bar{\psi}\gamma_\mu\gamma_5\chi = \bar{\chi}\gamma_\mu\gamma_5\psi , \quad \bar{\psi}\sigma_{\mu\nu}\chi = -\bar{\chi}\sigma_{\mu\nu}\psi$$

Fierz-rearrangements.

Consider four spinors (not necessarily Majorana spinors) ψ,ψ,χ,λ Let

$$(\bar{\psi}M\varphi)(\bar{\lambda}N\chi) = -(\bar{\psi}_\alpha\chi_\delta)(\bar{\lambda}_\gamma\varphi_\beta)(M_{\alpha\beta}N_{\gamma\delta})$$

and consider for fixed γ,β, the matrix $O^{(\gamma\beta)}_{\alpha\delta} = M_{\alpha\beta}N_{\gamma\delta}$. Using the completeness of the 16 Dirac matrices O_j

$$O^{(\gamma\beta)}_{\alpha\delta} = \sum_{j=1}^{16} c_j^{\gamma\beta}(O_j)_{\alpha\delta} , \quad \mathrm{tr}(O_i O_j) = 4\delta_{ij}$$

It follows that

$$c_k^{\gamma\beta} = \frac{1}{4} O^{(\gamma\beta)}_{\alpha\delta} (O_k)_{\delta\alpha} = \frac{1}{4} (N O_k M)_{\gamma\beta}$$

Thus

$$(\bar{\psi} M \varphi)(\bar{\lambda} N \chi) = -\frac{1}{4} (\bar{\psi} O_j \chi)(\bar{\lambda} N O_j M \varphi)$$

$$O_j = (1,\ \gamma_5,\ \gamma_\mu,\ i\gamma_\mu\gamma_5,\ 2i\sigma_{\rho\sigma}) \text{ with } \rho > \sigma$$

LECTURE 3. SYMMETRIES AND GROUPS

One can define an extension of the concept of ordinary Lie algebras by admitting anticommutators as well as commutators in the bracket relations between generators. An amusing and useful notation is

$$[X_A, X_B\} = f_{BA}{}^C X_C$$

Note the order of the indices A and B. One calls such extensions super Lie algebras. The example we will consider first is the super Poincaré algebra. It has as even elements P_μ, $M_{\mu\nu}$ and as odd elements Q_α $(\alpha=1,4)$ The Poincaré algebra is

$$[M_{\mu\nu}, M_{\rho\sigma}] = \delta_{\mu\rho} M_{\nu\sigma} + \delta_{\mu\sigma} M_{\nu\rho} + \delta_{\nu\rho} M_{\mu\sigma} - \delta M_{\nu\sigma\mu\rho}$$

$$[M_{\mu\nu}, P_\rho] = \delta_{\nu\rho} P_\mu - \delta_{\mu\rho} P_\nu$$

$$[P_\mu, P_\nu] = 0$$

The extension consists of

$$[P_\mu, Q_\alpha] = 0$$

$$[M_{\mu\nu}, Q_\alpha] = -(\sigma_{\mu\nu})_{\alpha\beta} Q_\beta$$

$$\{Q_\alpha, Q_\beta\} = N^\mu_{\alpha\beta} P_\mu + R^{\mu\nu}_{\alpha\beta} M_{\mu\nu}$$

We will determine the constants $N^\mu_{\alpha\beta}$ and $R^{\mu\nu}_{\alpha\beta}$ shortly. Clearly, one can interpret Q_α as fermionic objects and $M_{\mu\nu}$, P_ν as bosonic objects as far as the (anti) commutation relations are concerned.

The Jacobi identities fix the relative signs of the MP and the MQ commutators. The Jacobi identity

$$[\{Q_\alpha, Q_\beta\}, P_\rho] + \{[P_\rho, Q_\alpha], Q_\beta\} + \{[P_\rho, Q_\beta], Q_\alpha\} = 0$$

immediately leads to $R^{\mu\nu}_{\alpha\beta} = 0$. We now proceed to determine $N^\mu_{\alpha\beta}$ from the same Jacobi identity but with P_ρ replaced by $M_{\mu\nu}$.

Using $\gamma_\mu{}^T = -C\gamma_\mu C^{-1}$ as we saw in lecture 2, one obtains from this Jacobi identity

$$N^{\mu} C \delta^{\nu\rho} - N^{\nu} C \delta^{\mu\rho} = \sigma^{\mu\nu} N^{\rho} C - N^{\rho,T} C \sigma^{\mu\nu}$$

Expanding $N^{\mu}_{\alpha\beta} = \sum d^{\mu}_{j} (O_j C)_{\alpha\beta}$ where O_j is the complete set of 16 Dirac matrices, and using the orthonormality relation $tr(O_j O_k) = 4\delta_{jk}$ one finds $d^{\mu}_{k}\delta^{\nu\rho} - d^{\nu}_{k}\delta^{\mu\rho} = \frac{1}{4}\sum_j tr\,[(\sigma^{\mu\nu}O_j + O_j a_j \sigma^{\mu\nu})\, O_k]\, d^{\rho}_{j}$

with $a_j = +1$ for $O_j = 1, \gamma_5, i\gamma_\mu\gamma_5$ and $a_j = -1$ for $O_j = \gamma_\mu, 2i\sigma_{\mu\nu}$

Clearly, $d^{\mu}_{k} = 0$ for $O_k = 1, \gamma_5, i\gamma_\mu\gamma_5, 2i\sigma_{\mu\nu}$, while for $O_k = \gamma_\mu$ one finds that d^{μ}_{k} is arbitrary. These results follow again from $tr(O_j O_k) = 4\delta_{jk}$. Thus

$$\{Q_\alpha, Q_\beta\} = -\frac{1}{2}(\gamma^\mu C)_{\alpha\beta} P_\mu$$

where we have chosen the normalization of Q_α in the super algebra such that it corresponds to $\delta e^{a}_{\mu} = \frac{1}{2}\bar{\varepsilon}\gamma^{a}\psi_\mu$.

Since at this point all Jacobi identities are satisfied, we have a closed algebra. The main result is that $\{Q, Q\} \sim P$ but not $\sim M$. One can give a representation in terms of 5x5 matrices

$$M_{\mu\nu} = \begin{pmatrix} \sigma_{\mu\nu} & \begin{matrix} 0 \\ 0 \\ 0 \\ 0 \end{matrix} \\ 0\;0\;0\;0 & 0 \end{pmatrix}, \quad P_\mu = \begin{pmatrix} -\frac{1}{2}\gamma_\mu(1+\gamma_5) & \begin{matrix} 0 \\ 0 \\ 0 \\ 0 \end{matrix} \\ 0\;0\;0\;0 & 0 \end{pmatrix}$$

$$Q_\alpha = \frac{1}{2}\begin{pmatrix} 0 & 0 & 0 & 0 & [(1-\gamma_5)C]_{1\alpha} \\ 0 & 0 & 0 & 0 & [(1-\gamma_5)C]_{2\alpha} \\ 0 & 0 & 0 & 0 & [(1-\gamma_5)C]_{3\alpha} \\ 0 & 0 & 0 & 0 & [(1-\gamma_5)C]_{4\alpha} \\ (1+\gamma_5)_{\alpha 1} & (1+\gamma_5)_{\alpha 2} & (1+\gamma_5)_{\alpha 3} & (1+\gamma_5)_{\alpha 4} & 0 \end{pmatrix}$$

$$\sigma_{\mu\nu} = \frac{1}{4}[\gamma_\mu, \gamma_\nu], \quad \{\gamma_\mu, \gamma_\nu\} = 2\delta_{\mu\nu}, \quad \gamma_5 = \gamma_1\gamma_2\gamma_3\gamma_4$$

The only nontrivial relation to be checked is the QQ anticommutator. It follows from the relation $[(1-\gamma_5)\ C]_{i\alpha}(1+\gamma_5)_{\beta j} + (\alpha \leftrightarrow \beta) = -(\gamma^\mu C)_{\alpha\beta}[\gamma_\mu(1+\gamma_5)]_{ij}$ which is true, since it holds when one multiplies it by any of the sixteen independent matrices $(O_k)_{ji}$.

We now "gauge this algebra", i.e., we produce a Lagrangian field theory which follows from this algebra according to a general method.

In order to gauge this super Lie algebra, one associates to each generator X_A a gauge field h_μ^A and a local gauge parameter ε^A as follows:

$$h_\mu^A X_A = \bar{\psi}_\mu^a Q_a + e_\mu^m P_m + \hat{\omega}_\mu^{mn} M_{mn}$$

$$\eta^A X_A = \bar{\epsilon}^a Q_a + \xi^r P_r + \lambda^{mn} M_{mn}$$

Clearly, the gauge field $\bar{\psi}_\mu^a$ is a vectorial spinor and contains, according to elementary addition of angular momenta, spin 3/2. So we see that the gauge fields of supergravity can be viewed as belonging to the graded Poincaré algebra. Gauge transformations are in general defined by

$$\delta_g h_\mu^A = (D_\mu \eta)^A \equiv \partial_\mu \eta^A + f_{BC}{}^A h_\mu^B \eta^C$$

For $h_\mu^A = e_\mu^a$ one has with $A = P_a$, $C = Q_\alpha$ and $B = Q_\beta$

$$\delta_g e_\mu^a = -\tfrac{1}{2}(\gamma^a C)_{\alpha\beta}\bar{\psi}_\mu^\beta \bar{\epsilon}^\alpha = \tfrac{1}{2}\bar{\epsilon}\gamma^a\psi_\mu$$

For $h_\mu^A = \bar{\psi}_\mu^\alpha$ and thus $A = Q_\alpha$, one has if $C = Q_\beta$, that $B = M_{ab}$ and thus

$$\delta\bar{\psi}_\mu^\alpha = \partial_\mu \bar{\epsilon}^\alpha + (\sigma_{ab})^{\beta\alpha}\hat{\omega}_\mu^{ab}\bar{\epsilon}^\beta \quad (a>b)$$

$$\delta\psi_\mu = \left(\partial_\mu + \tfrac{1}{2}\omega_{\mu ab}\sigma^{ab}\right)\epsilon = D_\mu\epsilon, \quad \underset{a \gtrless b}{\omega = -\hat{\omega}}$$

Note that in $f_{BC}{}^A h_\mu^B \eta^C$ one must sum over every generator once, so that one restricts the Lorentz generators M_{ab} to a>b.

For $\delta_g\omega_{\mu ab}$ we find $\delta_g\omega_{\mu ab} = 0$. This result was first obtained by Chemseddine and West, but they considered the action only on the mass-shell. We eliminated $\omega_{\mu ab}$ as an independent field by solving the ω field equation and found $\omega_{\mu ab} = \omega_{\mu ab}(e,\psi)$. Thus the actual value for the variation of $\omega_{\mu ab}(e,\psi)$ follows by applying the chain rule and differs from the group law $\delta_g \omega_{\mu ab}$ by an amount $\delta_e \omega_{\mu ab}$ (e for extra). For the invariance of the action we do not need $(\delta_g + \delta_e)\omega_{\mu ab}$, but for other purposes (the gauge algebra) we will need it.

Curvatures are generally defined by

$$R_{\mu\nu}^{\ A} = \partial_\nu h_\mu^{\ A} - \partial_\mu h_\nu^{\ A} + f^A{}_{BC}\, h_\nu^{\ B}\, h_\mu^{\ C}$$

For the graded Poincaré group one finds

$$R_{\mu\nu}^{\ a}(P) = D_\nu e^a{}_\mu - D_\mu e^a{}_\nu + \tfrac{1}{2}\bar{\psi}_\mu \gamma^a \psi_\nu$$

$$R_{\mu\nu}^{\ ab}(M) = \left(-\partial_\nu \omega_\mu{}^{ab} + \omega_\mu{}^{ac}\, \omega_{\nu c}{}^{b}\right) - (\mu \leftrightarrow \nu)$$

$$R_{\mu\nu}^{\ \alpha}(Q) = D_\nu \bar{\psi}_\mu^{\ \alpha} - D_\mu \bar{\psi}_\nu^{\ \alpha}$$

We notice that the ω -field equation found before is exactly given by

$$R_{\mu\nu}^{\ a}(P) = 0 \quad .$$

This is not an invariant statement; indeed $\delta R(P) \sim R(Q)$ according to the law of transformation of a curvatures. However, by solving $R^a_{\mu\nu}(P) = 0$ explicitly, such that $\omega = \omega(e,\psi)$, there are extra terms δ_e in the ω transformation such that always $\delta R_{\mu\nu}{}^a(P) = 0$. In fact, since $\delta R^A_{\mu\nu} = f^A{}_{BC} R^B_{\mu\nu} \epsilon^C$ one has

$$(\delta_g + \delta_e) R_{\mu\nu}^{\ a}(P) \equiv 0 = \tfrac{1}{2} R_{\mu\nu}(Q)\gamma^a \epsilon + \delta_e(\omega_{\nu a\mu} - \omega_{\mu a\nu})$$

since of course only $\delta_e \omega_{\mu ab}$ is non-zero but $\delta_e e^a{}_\mu = 0$, and solving by the same method as we solved for torsion, one finds

$$\delta_e \omega_{\mu ab} = (\delta_g + \delta_e)\omega_{\mu ab} = \tfrac{1}{4}\left(R_{\mu a}(Q)\gamma_b \epsilon - R_{\mu b}(Q)\gamma_a \epsilon - R_{ab}(Q)\gamma_\mu \epsilon\right)$$

The gauge action reads in terms of curvatures

$$\mathcal{L} = \left(-\tfrac{1}{2}\kappa^{-2} \det e\, e^{a\nu} e^{b\mu}\right) R_{\mu\nu}^{\ ab}(M) + \left(\tfrac{1}{4}\epsilon^{\mu\nu\rho\sigma}\bar{\psi}_\mu \gamma_5 \gamma_\nu\right) R_{\rho\sigma}(Q) .$$

This is not of Yang-Mills form. However, by considering a different graded algebra, one can remedy this. This is important work of MacDowell and Mansouri, to which we now turn.

To this purpose one considers a second example of a super Lie algebra, the graded de-Sitter algebra. Its even part is the de-Sitter group O(3,2)

$$[M_{AB}, M_{CD}] = g_{\beta C} M_{AD} + 3 \text{ more relations.}$$

The odd generators S_α $(\alpha=1,4)$ satisfy

$$[M_{AB}, S_\alpha] = -(m_{AB})_{\alpha\beta} S_\beta$$

$$\{S_\alpha, S_\beta\} = -(m_{CD}C)_{\alpha\beta} M^{CD} \qquad (C>D)$$

where $M^{A'B'} g_{A'A} g_{B'B} = M_{AB}$ with $g_{AB} = (+,+,+,-,-)$

and

$$m_{kl} = \sigma_{kl}, \; m_{k4} = i\sigma_{k4}, \; m_{k5} = \tfrac{1}{2}\gamma_k, \; m_{45} = \tfrac{i}{2}\gamma_4$$

That this is indeed a closed algebra follows from the fact that an explicit 5x5 matrix representation exists

$$M_{AB} = m_{AB}, \; (S_\alpha)_{k5} = \frac{1}{\sqrt{2}} C_{k\alpha}, \; (S_\alpha)_{5k} = \frac{1}{\sqrt{2}} \delta_{\alpha k}$$

where $k,\ell = 1,3$ and $\alpha,\mu = 1,4$. One can now again build curvatures $R^{AB}_{\mu\nu}(M)$ and $R^{\alpha}_{\mu\nu}(S)$ in terms of the fields

$$h_\mu{}^A X_A = \Omega_\mu{}^{AB} M_{AB} + \bar{\psi}_\mu{}^\alpha S_\alpha$$

As a Yang-Mills type of action, one takes

$$I = \int d^4x\, \epsilon^{\mu\nu\rho\sigma} \Big[\alpha R_{\mu\nu}{}^{ab}(M) R_{\rho\sigma}{}^{cd}(M) \epsilon_{abcd} + \beta R_{\mu\nu}(S) \gamma_5 C R_{\rho\sigma}(S) \Big]$$

where one only sums over $(A,B) = (a,b) = 1,4$. This action is clearly not invariant under the full set of gauge transformations $\delta_g h_\mu^A = (D_\mu \eta)^A$. Let us first see how it transforms under local supersymmetry transformations. $\delta R_{\mu\nu}(S) = -\bar{\epsilon}\, m_{AB} R_{\mu\nu}{}^{AB}(M)$

Clearly, the action is invariant if one puts $R^{a5}_{\mu\nu}(M)=0$ and chooses α and β appropriately. $R^{a5}_{\mu\nu}(M)=0$ is again the torsion field equation of Lecture 2. That this result should be the same is understandable if one asks: which (anti)commutators have a M^{a5} on the right hand side, and identifies M^{a5} with P_μ .

Writing the curvatures in more detail one has (up to constants)

$$R_{\mu\nu}^{ab}(M) = R_{\mu\nu}^{ab}(\text{Einstein}) + (e_\mu^a e_\nu^b - e_\nu^a e_\mu^b) + \bar{\psi}_\mu \sigma^{ab} \psi_\nu$$

$$R_{\mu\nu}^{a5}(P) = D_\nu e_\mu^a - D_\mu e_\nu^a + \bar{\psi}_\mu \gamma^a \psi_\nu$$

$$R_{\mu\nu}^{\alpha}(Q) = D_\nu \bar{\psi}_\mu - D_\mu \bar{\psi}_\nu + \bar{\psi}_\nu \gamma_\mu - \bar{\psi}_\mu \gamma_\nu$$

It is now clear what happens in the action.

(1) The $(D\psi)^2$ terms cancel against the $R^{ab}_{\mu\nu}(E)\bar{\psi}\sigma\psi$ terms upon partial integration. The term $\epsilon^{\cdots\cdots}\epsilon_{\cdots\cdots} R_{\cdot\cdot} R_{\cdot\cdot}$ is the Gauss-Bonnet topological invariant whose variation is of the form

$$\int \epsilon^{\mu\nu\rho\sigma} \epsilon_{abcd} D_\mu \delta\omega_{\nu ab} R_{\rho\sigma}{}^{cd}(M)\, d^4x$$

which vanishes due to the Bianchi identity $D_\mu R_{\rho\sigma}{}^{cd} \varepsilon^{\mu\rho\sigma\tau} = [D_\mu, [D_\rho, D_\sigma]]\varepsilon^{\mu\rho\sigma\tau} = 0$. Thus the parts quadratic in the curvatures vanish, and one finds the result that the gauge action of simple supergravity plus cosmological terms is the action obtained from the super de-Sitter algebra. Similarly, the Einstein action plus cosmological terms follows from the ordinary de-Sitter algebra.

(2) To go to the case without cosmological constant, one makes a group contraction. That is to say, one scales $e_\mu^a \to e_\mu^a g$ divides the action by g^2, and takes the limit $g \to 0$.

LECTURE 4. SUPERSPACE

There is a clear analogy between the generators Q_α and P_μ of the graded Poincaré algebra. Both correspond to the physical particles, in contradistinction to M_{mn} whose gauge field $\omega_{\mu\, mn}$ is not physical. In conformal supergravity one finds this symmetry in the constraints which read $R_{\mu\nu}^a(P)=0$ and $\gamma^\mu R_{\mu\nu}^\alpha(Q) = 0$ (plus a third:

$R^{ab}_{\mu\nu}(M) \sim R_{\mu\nu}(A)$).

Clearly the second constraint can be viewed as a square root of the first ($\gamma^\mu\gamma^\mu = 1$ and $QQ \sim P$). If one insists on a complete symmetry between bosonic and fermionic objects, then one would introduce fermionic coordinates in addition to bosonic coordinates. This is the idea of superspace, which is thus spinning space.

Superspace consists of a base manifold with coordinates $z = (x^\mu, \theta^\alpha)$ with $\mu = 1,4$ and $\alpha = 1,4$, and a tangent manifold. At each point of base space one has erected a local Lorentz frame and the orientation of axes of these tangent frames with respect to base space are given by Vielbein fields $V^A{}_\Lambda(z)$. (One can introduce several sets of θ^α_I, I = 1,2... but we will restrict ourselves here to the case I = 1 only.)

In the tangent frames, one considers as group the ordinary Lorentz group only, (and not some kind of graded Lorentz group), and one introduces as its gauge connections the fields $h_\Lambda{}^{rs}(z)$. The base manifold indices are Λ, Π, .. with $\Lambda = (\mu, \alpha)$ and μ, ν... are vector indices while α, β ... are spinor indices. The tangent manifold indices are A = (m,a) and m,n... are vector indices and a,b... are spinor indices. The situation is sketched in Fig.1

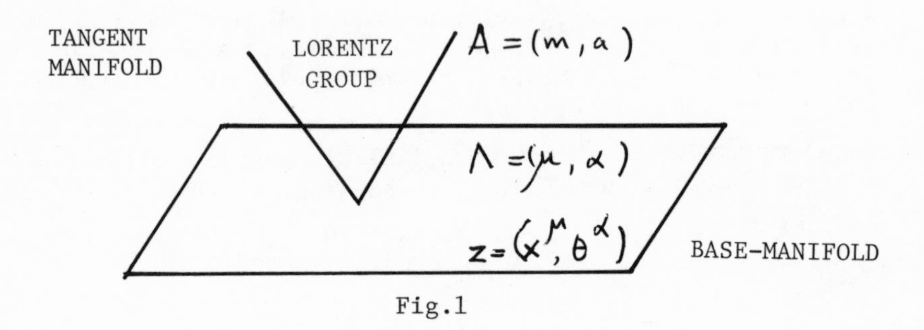

Fig.1

This structure of superspace is the same as in general relativity, except that one has an extended base-space. It is clear that an interesting "shuffling" has taken place. Whereas previously we discussed that the group in the tangent manifold consisted of P_m, Q_α, M_{mn} (or only Q_α, M_{mn}) while the base manifold is parameterized by x^μ, now the base manifold is parameterized by x^μ and θ^α. To the extent that one identifies x^μ with P_μ and Q_α with θ^α, one might say that one has brought Q_α down, from the tangent manifold to the base manifold. A more technical way of saying the same is that base space consists of the quotient $M_{mn} + P_m + Q_a$ divided by M_{mn}. This is sometimes referred to as "we live in quotient space".

Note that one has no fermionic counterpart for the Lorentz group in tangent space; in this sense there is an asymmetry between bosonic and fermionic concepts.

The fields are thus

$$V^A{}_\Lambda(z) = \begin{pmatrix} V^m{}_\mu(z) & V^m{}_\alpha(z) \\ V^a{}_\mu(z) & V^a{}_\alpha(z) \end{pmatrix}, \quad h^{rs}{}_\Lambda(z) = \left(h^{rs}{}_\mu(z), h^{rs}{}_\alpha(z) \right).$$

One now defines general coordinate transformations in the base manifold, and local Lorentz rotations in the tangent manifold. Functions $F(x,\theta)$ are per definition polynomials in θ only:

$$F(x,\theta) = F^1(x) + F^2_\alpha(x)\theta^\alpha + F^3_{\alpha\beta}\theta^\alpha\theta^\beta + \dots$$

where $\theta^\alpha\theta^\beta = -\theta^\beta\theta^\alpha$. If $F(x,\theta)$ is bosonic, $F'(x)$ is bosonic and $F^2_\alpha(x)$ is fermionic, etc. Thus $F(x,\theta)$ is either even or odd. To give an example

$$V^m{}_\mu(z) = V^m{}_\mu(x,\theta) = V^m{}_\mu(x) + \bar\psi_\mu \gamma^m \theta + \dots$$

where $V^m_\mu(x)$ is an ordinary commuting field and $\bar\psi^a{}_\mu$ is an anti-commuting field (it will be identified below with the gravitino field). Of course, one can also make a Taylor expansion in x^μ, but this series will in general not terminate after a finite number of terms. The series in θ^α terminates since the θ^α anti-commute. Note that we consider only functions at a given point $z^\Lambda = (x^\mu, \theta^\alpha)$. The parameters are $\xi^\Lambda(z) = (\xi^\mu(z), \xi^\alpha(z))$ and $\epsilon^{rs}(z)$ respectively. In complete analogy with general relativity one has for general supercoordinate transformations

$$\delta_{gc} V^A{}_\Lambda(z) = \xi^\Pi(z)\,\partial_\Pi V^A{}_\Lambda(z) + \left(\partial_\Lambda \xi^\Pi(z)\right) V^A{}_\Pi(z)$$

$$\delta_{gc} h^{rs}{}_\Lambda(z) = \xi^\Pi(z)\,\partial_\Pi h^{rs}{}_\Lambda(z) + \left(\partial_\Lambda \xi^\Pi(z)\right) h^{rs}{}_\Pi(z)$$

We will presently give an example. The local Lorentz rotations are defined by (if $A = (m,a)$)

$$\delta_{\ell L} V^A{}_\Lambda(z) = \tfrac{1}{2}\epsilon^{rs}(z) X_{rs} V^A{}_\Lambda(z) = \left(\epsilon^m{}_s V^s_\Lambda, \tfrac{1}{2}\epsilon^{rs}(\sigma_{rs})^{ab} V^b_\Lambda\right)$$

Here $[X_{rs}, X_{tn}] = \delta_{st} X_{rn}$ + 3 terms are Lorentz generators.

Still in complete analogy with general relativity one defines curved covariant derivatives and their tangent space equivalents

$$D_\Lambda \equiv \partial_\Lambda + \tfrac{1}{2} h_\Lambda{}^{rs} X_{rs} \quad , \quad D_M \equiv V_M{}^\Lambda D_\Lambda$$

The inverse vielbein satisfies

$$V_M{}^\Lambda V^B{}_\Lambda = \delta^B_M \quad \text{hence (!)} \quad V^A{}_\Pi V_A{}^\Lambda = \delta^\Lambda_\Pi$$

One can invert the vielbeins if their diagonal parts for $\theta = 0$ are non-singular.

One may define curvatures and torsion by

$$[D_A, D_B\} = -\tfrac{1}{2} R_{AB}{}^{rs} X_{rs} - 2 R_{AB}{}^C D_C$$

A representation of the algebra of infinitesimal global super-symmetry transformations on the coordinates x^μ and θ^α is given by

$$(a^m P_m)(x^\mu, \theta^\alpha) = (a^\mu, 0)$$

$$(\lambda^{mn} M_{mn})(x^\mu, \theta^\alpha) = (\lambda^\mu{}_\nu x^\nu, \tfrac{1}{2} \lambda^{ab} (\sigma_{ab})^\alpha{}_\beta \theta^\beta)$$

$$(\bar\epsilon^\alpha Q_\alpha)(x^\mu, \theta^\alpha) = (\bar\epsilon \gamma^\mu \theta, \epsilon^\alpha)$$

Indeed, $[\delta(\varepsilon_1), \delta(\varepsilon_2)]x^\mu = \frac{1}{4}\bar\varepsilon_2 \gamma^\mu \varepsilon_1 - \frac{1}{4}\bar\varepsilon_1 \gamma^\mu \varepsilon_2 = \frac{1}{2}\bar\varepsilon_2 \gamma^\mu \varepsilon_1$ and $[\delta(\varepsilon_1), \delta(\varepsilon_2)]\,\theta^\alpha = 0$. This confirms the rule

$$\{Q_\alpha, Q_\beta\} = -\tfrac{1}{2}(\gamma^\mu C)_{\alpha\beta} P_\mu$$

One may also verify that, for example, $[\delta(a), \delta(\varepsilon_1)]\, x^\mu = 0$, representing that $[P_\mu, Q_\alpha] = 0$.

Two well-known representations of the algebra of global super-symmetry in terms of the differential operators

$$D_a = \frac{\partial}{\partial \bar\theta_a} + (\not\partial \theta)^a \quad , \quad D_m = \frac{\partial}{\partial x^m} \, , \quad G_a = \frac{\partial}{\partial \bar\theta_a} - \not\partial \theta^a$$

Are given by $Q_a = +D_a$, $P_m = -D_m$; $Q_a = +G_a$, $P_m = +D_m$; $M_{mn} = x_m \partial_n - x_n \partial_m + \bar\theta \sigma_{mn} \partial/\partial\bar\theta$

The derivatives $\partial/\partial\theta^{\alpha} = \partial/\partial\theta^{b} C^{ba}$ are left derivatives. Identifying D_a and D_m with D_A, one finds the vierbeins corresponding to flat superspace (with $h_\Lambda{}^{rs} = 0$)

$$V_A{}^{\Lambda}(flat) = \begin{pmatrix} \delta^{\mu}_{m} & 0 \\ (\bar{\theta}\gamma^{\mu})_a & \delta^{\alpha}_{a} \end{pmatrix}$$

Thus the flat curvatures and torsions in $[G_A, G_B]$ with $G_A = (D_A, X^{rs})$ are the structure constants of the graded Poincaré algebra, and global supersymmetry is flat but has torsion.

We now discuss how this superspace formalism is related to simple (N=1) supergravity.

(i) Identification. One relates the expansion coefficients of lowest order in θ of superspace quantities to ordinary supergravity quantities. The choice is strongly suggested, but in principle it is arbitrary

$$V_\mu{}^m(z) = e_\mu{}^m(x) + \mathcal{O}(\theta) \qquad \xi^\mu(z) = \xi^\mu(x) + \mathcal{O}(\theta)$$
$$V_\mu{}^a(z) = \psi_\mu{}^a(x) + \mathcal{O}(\theta) \qquad \xi^\alpha(z) = \epsilon^\alpha(x) + \mathcal{O}(\theta)$$
$$h_\mu{}^{rs}(z) = \omega_\mu{}^{rs}(z) + \mathcal{O}(\theta) \qquad \epsilon^{rs}(z) = \lambda^{rs}(x) + \mathcal{O}(\theta)$$

where $\xi^\mu(x), \epsilon^\alpha(x)\ \lambda^{ab}(x)$ are the parameters of local spacetime and supersymmetry. Note that in this approach $\omega_\mu{}^{rs}(x)$ belongs to its own superfield $h_\Lambda{}^{rs}(z)$ and consequently is an independent field here. In the work of Wess and Zumino, Siegel and Gates and others, this is not so, but here we follow Brink, Gell-Mann, Ramond and Schwarz.

(ii) Compatibility. One requires that the superspace transformation rules for $e_\mu{}^m$, $\psi_\mu{}^a$, $\omega_\mu{}^{ab}$ agree with those of ordinary supergravity. For example, from superspace coordinate transformations one has

$$\delta_{gc} V_\mu{}^m(z) = \xi^\nu \partial_\nu V_\mu{}^m + \xi^\alpha \partial_\alpha V_\mu{}^m + \partial_\mu \xi^\nu V^m{}_\nu + \partial_\mu \xi^\alpha V^m{}_\alpha$$

and to order θ^0 one finds that $V^m_\mu(x) = e^m{}_\mu(x)$ is compatible provided $\xi^\alpha \partial_\alpha V^m{}_\mu = -\bar{\psi}_\mu \gamma^m \epsilon$ and $V^m{}_\alpha(x) = 0$. Hence

$$V^m{}_\mu(x,\theta) = e^m{}_\mu(x) + \bar{\psi}_\mu \gamma^m \theta + \mathcal{O}(\theta^2)$$
$$V^m{}_\alpha(x,\theta) = 0 + \mathcal{O}(\theta)$$

The condition $V^{m}_{\alpha}(x)=0$ agrees with the definition of $V^{A}{}_{\Pi}$ by $V^{A}{}_{\Pi}$ $V_{A}{}^{\Lambda} = \delta^{\Lambda}_{\Pi}$ for the case of flat superspace. But now we find that in general $V^{m}{}_{\alpha}(x) = 0$.

One repeats this process of compatibility order by order in θ, at each order identifying the expansion coefficients as functions of the quantities of ordinary supergravity. No inconsistencies are found.

(iii) Covariance of field equations. One constructs curvatures and torsions from $V^{A}{}_{\Lambda}(z)$ and $h_{\Lambda}{}^{rs}(z)$. Then one identifies at $\theta = 0$ these objects with the field equations of ordinary supergravity, and finally one requires that these relations hold for all θ as good tensor relations in superspace. For example, one finds at order θ^{0} the torsion

$$R^{r}_{ab}(x,0) \sim (\gamma^{r}C)_{ab}$$

Since the right hand side is an invariant tensor, one postulates as field equations of superspace to all orders in θ simply $R^{r}_{ab}(x,\theta) = i(\gamma^{r}C)_{ab}$ One may expand these field equations order by order in θ, and finds that they repeat the field equations of supergravity over and over. This generalization to arbitrary θ is based on the fact that the extension is unique: if, for two tensors $(T_1 - T_2)(\theta=0) = 0$, then $(T_1 - T_2)(\theta \neq 0) = 0$ as well. This follows from the fact that the terms with $\xi^{\alpha}\partial_{\alpha}$ in the super coordinate transformations connect terms of higher order in θ with terms of lower order in θ. Thus, by induction, one may prove the uniqueness of the extension.

(iv) Action. Knowing the field equations in superspace, one tries to find actions $I = \int d^4x\, d^4\theta\, \mathcal{L}(x,\theta)$ from which these equations follow by a variational principle.

We now comment on this program. There are 420 curvatures and torsions for the 112 fields $V_{\Lambda}{}^{A}$ and $h_{\Lambda}{}^{rs}$. Thus it comes as no surprise that various subsets of 112 curvatures and torsions have been found which were postulated to be the field equations and which lead to the remaining 420 field equations by means of the Bianchi identities

$$(-)^{X-Z}[[G_X, G_Y\}, G_Z\} + \text{cyclic terms} = 0$$

There are also the Bianchi identities with two G's and one fields $V^{A}{}_{\Lambda}$. The internal consistency of any particular choice of fundamental field equations is a complicated task. The linking of superspace curvatures and torsions to the ordinary supergravity field equations was done by choosing for the latter the Breitenlohner equations which were formulated in terms of many auxiliary fields such that the ordinary space gauge algebra closes. It might be simpler to consider the minimal set of auxiliary fields. There, however, $\omega_{\mu ab}$ is not an independent field, so that this might require major changes.

In the above we only defined the operations of general coordinate and local Lorentz transformations on the fields $V^{A}{}_{\Lambda}(z)$ and $h_{\Lambda}{}^{rs}(z)$. One can also define general transformation on the coordinates

$$\delta z^{\Lambda} = \xi^{\Lambda}(z)$$

The approach described above has a larger symmetry than ordinary supergravity, namely 6 local Lorentz and 8 general coordinate transformations in superspace, hence (6+8) x 16 θ-independent symmetries. This is many more than the 4+6+4 local symmetries in ordinary supergravity, but there are also many more fields in superspace: 112x16.

The possible benefits of this scheme are that it might lead to other theories than supergravity, with or without correspondences in ordinary spacetime. Also, the auxiliary fields for $N>1$ supergravity might be found from superspace. In the approach sketched above, the parameters become field-dependent, for example

$$\xi^{\alpha}(z) = \varepsilon^{\alpha}(x) + \psi_{\rho}\bar{\varepsilon}\gamma^{\rho}\theta + \ldots \; .$$

NOTE ADDED: Since last year much progress has been made. For recent literature, see Proceedings of Cargese School (July 1978, Plenum Press, M. Levy and S. Deser, eds.) and the Proceedings of the Stony Brook Workshop (Sept.1979, North-Holland, P. van Nieuwenhuizen and D. Z. Freedman, eds.).

REFERENCES

As an introduction, we refer to

1. J. Wess, Lecture notes Salamanca (1977).
2. D.Z. Freedman and P. van Nieuwenhuizen, Scientific American, Feb. 1978.
3. Proceedings 1978 summerschool at Cargese (Plenum Press).
4. P. van Nieuwenhuizen and J. Scherk, Physics Report, to be published. (This is written as a graduate course).

The gauge action of simple supergravity without auxiliary fields was found in

5. D.Z. Freedman, P. van Nieuwenhuizen and S. Ferrara Phys. Rev. D 13, 3214 (1976).
6. S. Deser and B. Zumino, Phys. Lett. 62B, 335 (1976).

The minimal set of auxiliary fields were found in

7. S. Ferrara and P. van Nieuwenhuizen, Phys. Lett. 74B, 333 (1978)
8. K. Stelle and P. C. West, Phys. Lett. 74B, 330 (1978).

An earlier but nonminimal set was found in

9. P. Breitenlohner, Nucl. Phys. B124, 500 (1977).

Some group theory we have used is discussed in

10. A. Chamseddine and P. C. West, Nucl. Phys. B129, 39 (1977).
11. S. MacDowell and F. Mansouri, Phys. Rev. Lett. 38, 739 (1977).

Conformal supergravity is discussed in

12. M. Kaku, P. K. Townsend and P. van Nieuwenhuizen, Phys. Rev. D 17, 3179 (1978) and other papers by these authors referenced therein.

Superspace is treated in

13. R. Grimm, J. Wess and B. Zumino, Nucl. Phys. B152, 255 (1979) and other papers by Wess and Zumino referenced therein.
14. L. Brink, M. Gell-Mann, P. Ramond and J. Schwarz, Phys. Lett. 74B, 336 (1978) and Phys. Lett. 76B, 417 (1978).
15. W. Siegel and J. Gates, Nucl. Phys. B147, 77 (1979).
16. J. G. Taylor, Phys. Lett. 80B, 52 (1978).

The spin (2,3/2) global multiplet was found in

17. M. T. Grisaru, H. Pendleton et al., Phys. Rev. D15, 996 (1977).

The spin 3/2 action was discussed in

18. W. Rarita and J. Schwinger, Phys. Rev. 60, 61(1941).

SELF DUAL FIELDS

Joshua N. Goldberg

Department of Physics
Syracuse University
Syracuse, New York 13210

I. Introduction

In the course of their long review paper on $\mathcal{H}$ space[1], E. T. Newman and his co-workers show that one can construct a self-dual solution of Maxwell's equations by starting from an equation which is similar to the good cut equation. Their original construction depends on the properties of $\mathscr{I}^+$ in an asymptotically flat space-time. However, the argument can be reconstructed without reference to $\mathscr{I}^+$. One merely asks for a self-dual solution of the complexified Maxwell equations on $\mathbb{C}M$, complexified Min- n-kowski space. The argument is the same as that we shall use in the following for the Yang-Mills fields.

The interest in the construction of self-dual fields for the Yang-Mills case comes from two directions. One is that these gauge fields appear to be important for the description of particle physics.[2] Therefore, their structure and eventual quantization may be necessary for our understanding of nature. The second reason, and the more important one to me, is that these fields form a model for studying the quantization of general relativity. These fields are in general nonlinear. They possess a local invariance group which gives rise to constraints in a canonical formalism. While these constraints are not geometrical constraints for a space-time, they are constraints on the geometry of a fiber bundle. The eventual goal, then, would be to use these self-dual fields, or their representatives, to constructs a quantized field. Some of these ideas should then be transferable to the quantization of general relativity.

In these lectures I shall first give a brief review of the spinor formalism as applied to the study of fields in space-time.[3] Then I shall present the construction of anti-self-dual Yang-Mills fields in terms of a two-complex parameter family of the invariance group which defines the field. Finally, these ideas will be transferred to the construction of anti-self-dual solutions of the Einstein equations. In both of these cases, the results obtained will be used to describe briefly the construction of a deformed twistor space which encodes the solutions. These constructions are equivalent to those of Richard Ward[4] and Roger Penrose[5], respectively.

II. THE SPINOR FORMALISM

It will be useful to take some time to review the relationship between spinors and geometric objects on $\mathbb{C}M$.[3] Spinors are elements of a two dimensional complex linear vector space $\tilde{\mathcal{S}}$ which is endowed with a symplectic structure. Thus, if $\tilde{\xi}^A$ and $\tilde{\eta}^A$ $(A = 0, 1)$ are the components of two independent spinors in $\tilde{\mathcal{S}}$, their inner product with respect to the symplectic structure is

$$(\tilde{\xi}, \tilde{\eta}) = \varepsilon_{AB}\, \tilde{\xi}^A \tilde{\eta}^B = \tilde{\xi}_A \tilde{\eta}^A = -\tilde{\xi}^A \tilde{\eta}_A \quad , \tag{1}$$

$$\varepsilon_{AB} = \begin{pmatrix} 0 & 1 \\ -1 & 0 \end{pmatrix} = \varepsilon^{AB}$$

It follows that

$$\tilde{\xi}_A = \tilde{\xi}^B \varepsilon_{BA} \quad ; \quad \tilde{\xi}^A = \varepsilon^{AB} \tilde{\xi}_B \quad .$$

In particular, note that $\tilde{\xi}^A \tilde{\xi}_A = 0$.

The group of transformations which leaves the inner product and the symplectic structure invariant is $\widetilde{SL(2\mathbb{C})}$. Thus for $\tilde{S}^A{}_{\bar{A}} \in \widetilde{SL(2C)}$

$$\tilde{\xi}^A = \tilde{S}^A{}_{\bar{A}}\, \tilde{\xi}^{\bar{A}}$$

and

$$\varepsilon_{\bar{A}\bar{B}} = \varepsilon_{AB}\, \tilde{S}^A{}_{\bar{A}}\, \tilde{S}^B{}_{\bar{B}} \tag{2}$$

To make the connection with $\mathbb{C}M$, we must introduce a second independent spinor space $\mathcal{S}$ with elements $\xi^{A'}$ $(A' = 0', 1')$ and symplectic structure $\varepsilon_{A'B'} = \begin{pmatrix} 0 & 1 \\ -1 & 0 \end{pmatrix} = \varepsilon^{A'B'}$.

Elements of $\mathcal{S}$ transform under $S^{A'}{}_{\bar{A}'} \in SL(2\mathbb{C})$ which is independent of the transformation of $\widetilde{SL}(2C)$ on $\tilde{\mathcal{S}}$. $SL(2\mathbb{C}) \times \widetilde{SL}(2\mathbb{C})$ depends on six complex parameters which is sufficient to give a double covering of the complex Lorentz group on $\mathbb{C}M$.

The connection between vectors in $\mathbb{C}M$ and $\mathcal{S} \times \tilde{\mathcal{S}}$ is made through the Pauli spin matrices (μ = 0, 1, 2, 3)

$$\sqrt{2}\, \sigma^{\mu}{}_{AA'} = \begin{pmatrix} 1 & 0 \\ 0 & 1 \end{pmatrix}, \begin{pmatrix} 0 & 1 \\ 1 & 0 \end{pmatrix}, \begin{pmatrix} 0 & -i \\ i & 0 \end{pmatrix}, \begin{pmatrix} 1 & 0 \\ 0 & -1 \end{pmatrix} \quad (3)$$

Under a Lorentz transformation,

$$\eta_{\mu\nu}\, \Lambda^{\mu}{}_{\bar{\mu}}\, \Lambda^{\nu}{}_{\bar{\nu}} = \eta_{\bar{\mu}\bar{\nu}} \quad (4)$$

$$\eta_{\mu\nu} = \text{diag.}\ (1, -1, -1. -1) = \eta_{\bar{\mu}\bar{\nu}}$$

and one can show that there exist matrices $S^{A'}{}_{\bar{A}'} \in SL(2\mathbb{C})$ and $\tilde{S}^{A}{}_{\bar{A}} \in \widetilde{SL}(2\mathbb{C})$ such that

$$\Lambda^{\mu}{}_{\bar{\mu}}\, \sigma^{\bar{\mu}}{}_{\bar{A}\bar{A}'} = \sigma^{\mu}{}_{AA'}\, \tilde{S}^{A}{}_{\bar{A}}\, S^{A'}{}_{\bar{A}'} \quad (5)$$

The double covering is assured because $\tilde{S} \times S = (-\tilde{S}) \times (-S)$. By lowering and raising indices with $\eta_{\mu\nu}$, ε^{AB}, and $\varepsilon^{A'B'}$ one can introduce $\sigma_{\mu}{}^{AA'}$ such that

$$\sigma^{\mu}{}_{AA'}\, \sigma_{\nu}{}^{AA'} = \delta^{\mu}{}_{\nu}$$

and

$$\sigma^{\mu}{}_{AA'}\, \sigma_{\mu}{}^{BB'} = \delta_{A}{}^{B}\, \delta_{A'}{}^{B'} \quad (6)$$

so that

$$\eta_{\mu\nu}\, \sigma^{\mu}{}_{AA'}\, \sigma^{\nu}{}_{BB'} = \varepsilon_{AB}\, \varepsilon_{A'B'} \quad (7)$$

Vectors in $\mathbb{C}M$ can be represented in terms of spinors by

$$V^{\mu}\, \sigma_{\mu}{}^{AA'} = V^{AA'} \quad , \quad (8)$$

and covectors by

$$V_\mu \, \sigma^\mu{}_{AA'} = V_{AA'} = V^{BB'} \, \varepsilon_{BA} \, \varepsilon_{B'A'} \tag{9}$$

The magnitude

$$V^\mu V_\mu = V^{AA'} \, V_{AA'} \quad . \tag{10}$$

This mapping can be extended to arbitrary tensors by

$$T^{\mu\nu\ldots}{}_{\rho\ldots} \, \sigma_\mu{}^{AA'} \, \sigma_\nu{}^{BB'} \ldots \sigma^\rho{}_{CC'} = T^{AA'BB'\ldots}{}_{CC'\ldots} \tag{11}$$

In particular, an antisymmetric tensor $T_{\mu\nu} = -T_{\nu\mu}$ has the representation

$$T_{\mu\nu} \, \sigma^\mu{}_{AA'} \, \sigma^\nu{}_{BB'} = \varepsilon_{A'B'} \, \tilde{T}_{AB} + \varepsilon_{AB} \, T_{A'B'} \tag{12}$$

$$T_{A'B'} = T_{B'A'} \quad , \quad \tilde{T}_{AB} = \tilde{T}_{BA} \quad .$$

One introduces the self-dual and anti-self dual parts through

$$T^\pm{}_{\mu\nu} = T_{\mu\nu} \pm \frac{1}{2} \, i \, \varepsilon_{\mu\nu\rho\sigma} \, T^{\rho\sigma} \tag{13}$$

with totally skew tensor $\varepsilon_{\mu\nu\rho\sigma}$ being defined by

$$\varepsilon_{0123} = -\varepsilon^{0123} = 1 \quad . \tag{14}$$

Then one finds that

$$\text{(a)} \quad T^+{}_{\mu\nu} \, \sigma^\mu{}_{AA'} \, \sigma^\nu{}_{BB'} = \varepsilon_{AB} \, T_{A'B'}$$

and

$$\text{(b)} \quad T^-{}_{\mu\nu} \, \sigma^\mu{}_{AA'} \, \sigma^\nu{}_{BB'} = \varepsilon_{A'B'} \, \tilde{T}_{AB} \quad . \tag{15}$$

With the orientation given above, $T^{+}_{\mu\nu}$ has a right-handed helicity and $T^{-}_{\mu\nu}$ a left-handed helicity.

It is also convenient to introduce the position vector

$$x^{AA'} = x^{\mu}\sigma_{\mu}^{\ AA'} \tag{16}$$

and the operator of partial differentiation

$$\partial_{AA'} = \sigma^{\mu}_{\ AA'}\,\partial_{\mu} \quad . \tag{17}$$

A null vector ℓ^{μ}, $\ell^{\rho}\ell_{\rho} = 0$, has the representation

$$\ell^{\mu} \qquad \tilde{\lambda}^{A}\,\pi^{A'} \quad . \tag{18}$$

The null vector ℓ^{μ} at the point $x^{\mu} \in \mathbb{C}M$ lies in the intersection of two totally null two-planes through the point x^{μ} (Fig. 1).

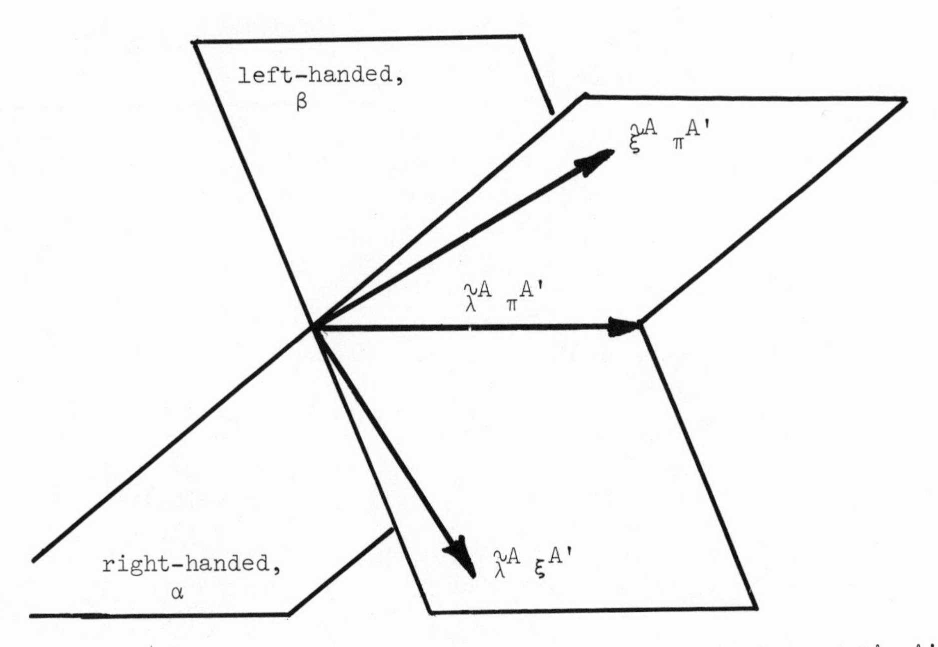

Figure 1. α and β planes containing the null vector $\ell^{\mu} \leftrightarrow \tilde{\lambda}^{A}\,\pi^{A'}$.

One of the planes is right-handed (an "α-plane") and contains vectors of the form

$$v_+^{\mu} \longleftrightarrow \tilde{\xi}^A \pi^{A'} \tag{19}$$

for all possible $\tilde{\xi}^A$, $\pi^{A'}$ fixed. The other is left-handed (a "β-plane") and contains vectors of the form

$$v_-^{\mu} \longleftrightarrow \tilde{\lambda}^A \xi^{A'} \tag{20}$$

for all possible $\xi^{A'}$, $\tilde{\lambda}^A$ fixed.

Through each point $x^{\mu} \in \mathbb{C}M$, there is a one parameter family of α-planes determined by the choice of $\pi^{A'}$ up to an overall scale; and there is a one parameter family of β-planes determined by the choice of $\tilde{\lambda}^A$ up to an overall scale.

III. YANG-MILLS FIELDS

The Yang-Mills[6,7] potential $A_{AA'}$ (spinor notation) is an element of the Lie algebra of an invariance group $\mathfrak{G}$. Under the action of an element of the group $G \in \mathfrak{G}$, the potential transforms according to

$$\bar{A}_{AA'} = G^{-1} A_{AA'} G + G^{-1} \partial_{AA'} G \quad . \tag{3.1}$$

From the potential we obtain the gauge covariant field through

$$F_{AA'BB'} = \varepsilon_{A'B'} \tilde{\phi}_{AB} + \varepsilon_{AB} \phi_{A'B'} \tag{3.2}$$

with

$$\text{(a)} \qquad \tilde{\phi}_{AB} = 2\, \partial_{(A}{}^{A'} A_{B)A'} + [A_{(A}{}^{A'}, A_{B)A'}] ,$$

$$\text{(b)} \qquad \phi_{A'B'} = 2\, \partial^{A}{}_{(A'} A_{|A|B')} + [A^{A}{}_{(A'}, A_{|A|B')}]; \tag{3.3}$$

$\phi_{A'B'}$ and $\tilde{\phi}_{AB}$ both transform according to $\bar{\phi} = G^{-1} \phi\, G$.

$\tilde{\phi}_{AB}$ and $\phi_{A'B'}$ are, respectively, the anti-self dual (left-handed) and self-dual (right-handed) parts of the field. The field equations are then

$$(a)\qquad \partial^{A}{}_{A'}\tilde{\phi}_{AB} + [A^{A}{}_{A'}\, \tilde{\phi}_{AB}] = 0$$

$$(b)\qquad \partial_{A}{}^{A'}\phi_{A'B'} + [A_{A}{}^{A'}, \phi_{A'B'}] = 0 \qquad (3.4)$$

The condition for a solution to be left-handed is $\phi_{A'B'} = 0$. Let $\pi^{A'}$ be a constant spinor and transvect Eq. (2b) with $\pi^{A'}\pi^{B'}$:

$$2\quad \pi^{A'}\partial_{AA'}A^{A}{}_{B'}\pi^{B'} + [A_{AA'}\pi,^{A'} A^{A}{}_{B'}\pi^{B'}] = 0. \qquad (3.5)$$

This looks at that part of the potential which does not lie in the α-plane defined by $\pi^{A'}$ and identifies that part as pure gauge. Thus,

$$A^{(x)}_{AA'}\pi^{A'} = \pi^{A'} G^{-1}\partial_{AA'}G \qquad (3.6)$$

because the condition (5) depends on $\pi_{A'}$. The potential itself does not depend on $\pi_{A'}$. Therefore, it follows that we may take G to be homogeneous of degree zero in $\pi^{A'}$.

From Eq. (6) we conclude that

$$A_{AA'} = G^{-1}\nabla_{AA'}G - B_{A}\pi_{A'} \qquad (3.7)$$

where B_A are position dependent matrices which are to be determined. Introduce the following parametrization of the spinors on $S^2 \times S^2$:

$$(a)\quad \pi^{A'} = \sqrt{\frac{1}{1+\zeta\tilde{\zeta}}}\begin{pmatrix}\check{\zeta}\\ 1\end{pmatrix} \qquad (c)\quad \tilde{\pi}^{A} = \sqrt{\frac{1}{1+\zeta\tilde{\zeta}}}\begin{pmatrix}\zeta\\ 1\end{pmatrix}$$

$$(b)\quad \eta^{A'} = \sqrt{\frac{1}{1+\zeta\tilde{\zeta}}}\begin{pmatrix}1\\ -\zeta\end{pmatrix} \qquad (d)\quad \tilde{\eta}^{A} = \sqrt{\frac{1}{1+\zeta\tilde{\zeta}}}\begin{pmatrix}1\\ -\tilde{\zeta}\end{pmatrix} \qquad (3.8)$$

Assign spin weight[8] $+\frac{1}{2}$ to $\pi^{A'}$ and $\tilde{\eta}^{A}$ and $-\frac{1}{2}$ to $\eta^{A'}$ and $\tilde{\pi}^{A}$ we have

$$
\begin{aligned}
&(a)\ \eth\,\pi^{A'} = \eth\,\tilde{\eta}^{A} = 0 && (d)\ \tilde{\eth}\,\tilde{\pi}^{A} = \tilde{\eth}\,\eta^{A'} = 0 \\
&(b)\ \eth\,\tilde{\pi}^{A} = \tilde{\eta}^{A} && (e)\ \tilde{\eth}\,\pi^{A'} = \eta^{A'} \\
&(c)\ \eth\,\eta^{A'} = -\,\pi^{A'} && (f)\ \tilde{\eth}\,\eta^{A} = -\,\tilde{\pi}^{A}
\end{aligned} \tag{3.9}
$$

The differential operators $\eth$ and $\tilde{\eth}$ act on functions of assigned spin weight s as follows:

$$
\begin{aligned}
&(a)\ \eth\, f(\zeta,\tilde{\zeta}) = (1 + \zeta\tilde{\zeta})^{1-s}\, \frac{\partial}{\partial\zeta}\, (1 + \zeta\tilde{\zeta})^{s}\, f(\zeta,\tilde{\zeta}) \\
&(b)\ \tilde{\eth}\, f(\zeta,\tilde{\zeta}) = (1 + \zeta\tilde{\zeta})^{1+s}\, \frac{\partial}{\partial\tilde{\zeta}}\, (1 + \zeta\tilde{\zeta})^{-s}\, f(\zeta,\tilde{\zeta})
\end{aligned} \tag{3.10}
$$

The spinor $\pi^{A'}$ depends on $\tilde{\zeta}$ so that the α-planes through x^{μ} defined by $\pi^{A'}$ are functions of $\tilde{\zeta}$. Thus the group element G in Eq. (7) is a function of $\tilde{\zeta}$. Because we want G to be regular on $S^2 \times S^2$, G will be a function of both ζ and $\tilde{\zeta}$, in general.

However, the potential $A_{AA'}$ does not depend on ζ or $\tilde{\zeta}$. It follows that

$$
\begin{aligned}
&(a)\quad \pi^{A'}\ \eth\,(G^{-1}\,\partial_{AA'}\,G) = 0 \\
&(b)\quad \eta^{A'}\ \eth\,(G^{-1}\,\partial_{AA'}\,G) = -\,\eth\, B_{A} \\
&(c)\quad \pi^{A'}\ \tilde{\eth}\,(G^{-1}\,\partial_{AA'}\,G) = +\,B_{A} \\
&(d)\quad \eta^{A'}\ \tilde{\eth}\,(G^{-1}\,\partial_{AA'}\,G) = -\,\tilde{\eth}\, B_{A}\ .
\end{aligned} \tag{3.11}
$$

The freedom in the choice of G,

$$G' = HG \quad ,$$

such that Eq. (6) is satisfied, requires

$$\pi^{A'}\, H^{-1}\, \nabla_{AA'}\, H = 0 \quad . \tag{3.12}$$

This freedom is sufficient so that one can choose either

$$\text{(a)} \qquad \tilde{\pi}^{A} \quad \eth B_A = 0$$

or

$$\text{(b)} \qquad \tilde{\eta}^{A} \quad \eth B_A = 0 \quad . \tag{3.13}$$

With the choice (10a) we find from Eq. (8) that G satisfies the Sparling equation:

$$\eth G \, G^{-1} = V\,(x^{AA'} \, \tilde{\pi}_A \, \pi_{A'} \; , \; \pi_{A'}) \quad . \tag{3.14}$$

At this point the argument can be turned around. Assume that one has G as a solution of Eq. (11). One can then show that

$$\eth(\pi^{A'} \, G^{-1} \, \partial_{AA'} \, G) = 0 \tag{3.15}$$

The quantity in parenthesis has spin weight $+ \frac{1}{2}$. With the assumption of regularity on $S^2 \times S^2$, it follows that there exist $A_{AA'}(x)$ such that

$$\pi^{A'} \, G^{-1} \, \partial_{AA'} \, G = A_{AA'}(x) \; \pi^{A'} \tag{3.16}$$

This is the starting point of our previous argument and it follows that $A_{AA'}$ defines a left-handed solution of the Yang-Mills solution.

The question arises as to the connection of this construction with a non-trivial fiber bundle over the projective twistor space $\mathbb{P}T$. Without going into details, the argument is as follows. Through each point of $\mathbb{C}M$ there is a one parameter family of right-handed twistor surfaces parametrized by $\tilde{\zeta}$(or $\pi^{A'}$), $\alpha(x, \tilde{\zeta})$. Through the origin of $\mathbb{C}M$ we consider a one parameter family of standard right-handed surfaces parametrized by ζ(or $\eta^{A'}$), $\alpha(\zeta)$. Each of the surfaces $\alpha(x, \tilde{\zeta})$ intersects a standard surface $\alpha(\zeta)$ in a single point characterized by

$$x^{AA'} = \lambda^{A} \, \eta^{A'} \; (\zeta) \quad . \tag{3.17}$$

For fixed $\zeta = \zeta_1$, $(i\lambda_1^A, \tilde{\zeta})$, are coordinates on $\mathbb{PT}$. These coordinates cover PT except for those points where $\tilde{\zeta} = -\frac{1}{\zeta_1}$. For these points the two-planes through all $x^{AA'}$ are parallel to the planes $\alpha(\zeta_1)$ through the origin. To cover these points we choose a second standard surface $\alpha(\zeta_2)$ which defines a second set of coordinates $(i\lambda_2^A, \tilde{\zeta})$. Together, the two patches $U(\zeta_1) \cup U(\zeta_2)$ cover all of $\mathbb{PT}$.

Erect over $\mathbb{PT}$ an n dimensional fiber $\mathbb{C}^n$ (n is the dimension of the representation of the group $\mathcal{G}$). The coordinates of the fiber are n independent scalings of $\pi^{A'} = \gamma \begin{pmatrix} \tilde{\zeta} \\ 1 \end{pmatrix}$. The n values of γ form an n-vector with components of $\gamma(\zeta_1)$ in $U(\zeta_1)$ and $\gamma(\zeta_2)$ in $U(\zeta_2)$. These can be related to one another through the solutions of Sparling's equation:

$$\gamma(\zeta_2) = G\,(x^{AA'}, \zeta, \tilde{\zeta})\, G^{-1}\,(x^{AA'}, \zeta_1, \tilde{\zeta})\,\gamma(\zeta_1) \qquad (3.18)$$

From the defining equation for G one can show that

$$\pi^{A'}\,\partial_{AA'}\,(G\,(\zeta_2)\,G^{-1}\,(\zeta_1)) = 0$$

which is necessary for the consistency of (18). Furthermore, this implies that

$$G(x^{AA'}, \zeta_2, \tilde{\zeta})\, G^{-1}\,(x^{AA'}, \zeta_1, \tilde{\zeta}) = F(x^{AA'}\,\pi_{A'}, \pi_{A'}) \qquad (3.19)$$

where F is homogeneous of degree zero in $\pi_{A'}$. With twistor coordinates defined through

$$Z^\alpha = (i\, x^{AA'}\,\pi_{A'}, \pi_{a'})$$

it follows that

$$F(x^{AA'}\,\pi_{A'}, \pi_{A'}) = f(Z^\alpha)$$

is a twistor function which is homogeneous of degree zero in its arguments. This is basically Ward's construction of the fiber bundle for a Yang-Mills gauge field.[4]

IV. GENERAL RELATIVITY

For the Yang-Mills field in CM, anti-self-duality is the integrability condition for the Yang-Mills connection, A_{AA}, to be trivial for displacements in right-handed null two planes, α-planes. In a general complex space-time, neither α-planes nor β-planes exist. However, in a complex space-time $\mathbb{C}\mathcal{M}$ which is described by an anti-self-dual solution of the Einstein equations, a one-parameter family of α-planes exist through each point of $\mathbb{C}\mathcal{M}$. Furthermore, an arbitrary vector is unchanged when carried by parallel transport around any closed path lying wholly in an α-plane. That is, for ℓ^μ and m^μ any two independent vectors lying in an α-plane,

$$2V^\mu{}_{;[\rho\sigma]}\,\ell^\rho m^\sigma = -R^\mu{}_{\nu\rho\sigma}\,\ell^\rho m^\sigma = 0 \tag{4.1}$$

This suggests that for $a\ell^\mu + bm^\mu$ any vector in the null blade

$\ell^{[a}m^{\sigma]}$, $\left\{{\mu \atop \rho\sigma}\right\}$ $(a\ell^\sigma + bm^\sigma)$ may be trivial. However, since

this result must hold in all α-planes, it can be true only if the space is flat. A weaker implication must hold if non-trivial left-handed space-times exist.

The Riemann tensor has the decomposition

$$R^{\mu\nu}{}_{\rho\sigma} = C^{\mu\nu}{}_{\rho\sigma} - 2\;\delta^{[\mu}_{[\rho}\,S^{\nu]}_{\sigma]} - \frac{1}{6}\,\delta^{[\mu}_{[\rho}\,\delta^{\nu]}_{\sigma]}\,R\;, \tag{4.2}$$

$$S^\nu{}_\sigma = R^\nu{}_\sigma - \frac{1}{4}\,\delta^\nu_\sigma\,R\;,$$

$$R^\nu{}_\rho = R^{\sigma\nu}{}_{\rho\sigma}\;,\qquad R = R^\rho{}_\rho\;.$$

Integrability in an arbitrary right-handed null two-blade through each point of $\mathbb{C}\mathcal{M}$ implies that

$$R_{\mu\nu} = 0 \tag{4.3}$$

and

$$C^{+\mu\nu}{}_{\rho\sigma} \equiv C^{\mu\nu}{}_{\rho\sigma} + \frac{1}{2} i \varepsilon_{\rho\sigma\iota\kappa} C^{\mu\nu\iota\kappa} = 0 \tag{4.4}$$

One can show from this that the blades can be integrated up into α-planes. Thus the space-time may be anti-self-dual with

$$C^{-\mu\nu}{}_{\rho\sigma} \equiv C^{\mu\nu}{}_{\rho\sigma} - \frac{1}{2} i \varepsilon_{\rho\sigma\iota\kappa} C^{\mu\nu\iota\kappa} \neq 0 \quad . \tag{4.5}$$

To proceed further, it is convenient to introduce spinors[3], as before. Consider an arbitrary null tetrad on the space time: (μ, a = 0, 1, 2, 3)

$$k^a = k^a{}_\mu \, dx^\mu \tag{4.6}$$

$$ds^2 = \eta_{ab} \, k^a k^b$$

$$\eta_{ab} = \left(\begin{array}{cc|cc} & 1 & & \\ 1 & & & 0 \\ \hline & 0 & & -1 \\ & & -1 & \end{array} \right)$$

The rotation coefficients $\gamma_{abc} = -\gamma_{bac}$ are defined by

$$dk^a = \omega^a{}_b \wedge k^b \tag{4.7}$$

with

$$\omega^a{}_b = \gamma^a{}_{b\,c} \, k^c \qquad :$$

and the tetrad components of the Riemann tensor are defined through

$$d\,\omega^a{}_b - \omega^a{}_c \wedge \omega^c{}_b = \frac{1}{2} R^a{}_{b\,c\,d} \, k^d \wedge k^c \tag{4.8}$$

The null basis $k_a{}^\mu$ dual to $k^a{}_\mu$ is defined by

$$k_a{}^\mu \, k^b{}_\mu = \delta_a{}^b \; ; \quad k_a{}^\mu \, k^a{}_\nu = \delta^\mu{}_\nu \tag{4.9}$$

An arbitrary tensor in the tangent space of $x^\mu \in \mathbb{C}\mathcal{M}$ is then expressed in terms of the basis vectors by

$$T^{\mu\nu\ldots}{}_{\rho\sigma\ldots} = T^{ab\ldots}{}_{rs\ldots}\, k_a{}^\mu\, k_b{}^\nu \ldots k_r{}^\rho\, k_s{}^\sigma \tag{4.10}$$

A connection between these tetrad components in the tangent space of $x^\mu \in \mathbb{C}\mathcal{M}$ and spinors is made through the Pauli matrices adapted to the null tetrad:

$$\sigma^a{}_{AA'} = \begin{pmatrix} 0 & 0 \\ 0 & 1 \end{pmatrix},\ \begin{pmatrix} 1 & 0 \\ 0 & 0 \end{pmatrix},\ \begin{pmatrix} 0 & 1 \\ 0 & 0 \end{pmatrix},\ \begin{pmatrix} 0 & 0 \\ 1 & 0 \end{pmatrix} \tag{4.11}$$

The spinor connection coefficients are then given by

$$\gamma_{abc}\, \sigma^\alpha{}_{AA'}\, \sigma^b{}_{BB'}\, \sigma^c{}_{CC'} = \varepsilon_{AB}\, \Gamma_{A'B'CC'} + \varepsilon_{A'B'}\, \Gamma_{ABCC'} \tag{4.12}$$

One can change the null tetrad by an arbitrary Lorentz transformation, $\Lambda^a{}_{\bar{a}}(x)$, at each point $x^\mu \in \mathbb{C}\mathcal{M}$.

$$k^a = \Lambda^a{}_{\bar{a}}\, k^{\bar{a}} \tag{4.13}$$

From

$$dk^a = d\,\Lambda^a{}_{\bar{a}} \wedge k^{\bar{a}} + \Lambda^a{}_{\bar{a}}\, \omega^{\bar{a}}{}_{\bar{b}} \wedge k^{\bar{b}} \tag{4.14}$$

we find

$$\gamma^a{}_{bc} = \Lambda^a{}_{\bar{a}}\, \gamma^{\bar{a}}{}_{\bar{b}\bar{c}}\, \Lambda^{\bar{b}}{}_b\, \Lambda^{\bar{c}}{}_c$$
$$- \partial_c\, \Lambda^a{}_{\bar{a}}\, \Lambda^{\bar{a}}{}_b \quad , \tag{4.15}$$

$$\Lambda^a{}_{\bar{a}}\, \Lambda^{\bar{a}}{}_b = \delta^a{}_b \quad , \quad \Lambda^{\bar{a}}{}_a\, \Lambda^a{}_{\bar{b}} = \delta^{\bar{a}}{}_{\bar{b}}$$

and

$$\partial_c = k_c{}^\mu\, \partial_\mu \quad .$$

As before, the Lorentz transformation is correlated to transformations in $SL(2\,\mathbb{C}) \times \widetilde{SL}(2\,\mathbb{C})$ by

$$\Lambda^{a}{}_{\bar{a}}\,\sigma^{\bar{a}}{}_{\bar{A}\bar{A}'} = \sigma^{a}{}_{AA'}\,\tilde{S}^{A}{}_{\bar{A}}\,S^{A'}{}_{\bar{A}'} \quad , \tag{4.16}$$

$$S^{A'}{}_{\bar{A}'} \in SL(2\,\mathbb{C}) \ ,$$

$$\tilde{S}^{A}{}_{\bar{A}} \in \widetilde{SL}(2\,\mathbb{C}) \quad .$$

It follows that the transformation law for the spinor connection coefficient is given by

$$\text{(a)} \qquad \Gamma^{A'}{}_{B'CC'} = S^{A'}{}_{\bar{A}'}\,\Gamma^{\bar{A}'}{}_{\bar{B}'\bar{C}\bar{C}'}\,S^{-1\bar{B}'}{}_{B'}\,S^{-1\bar{C}'}{}_{C'}\,\tilde{S}^{-1\bar{C}}{}_{C}$$

$$+\partial_{CC'}\,S^{A'}{}_{\bar{A}'}\,S^{-1\bar{A}'}{}_{B'}$$

and

$$\text{(b)} \qquad \Gamma^{A}{}_{BCC'} = \tilde{S}^{A}{}_{\bar{A}}\,\Gamma^{\bar{A}}{}_{\bar{B}\bar{C}\bar{C}'}\,\tilde{S}^{-1\bar{B}}{}_{B}\,\tilde{S}^{-1\bar{C}}{}_{C}\,S^{-1\bar{C}'}{}_{C'}$$

$$+\partial_{CC'}\,\tilde{S}^{A}{}_{\bar{A}}\,\tilde{S}^{-1\bar{A}}{}_{B} \tag{4.17}$$

Anti-self-duality, Eqs. (3) and (4), implies that we can choose $S^{A'}{}_{\bar{A}'}$ so that

$$\Gamma^{\bar{A}'}{}_{\bar{B}'\bar{C}\bar{C}} = 0 \quad . \tag{4.18}$$

We shall assume that the original tetrad has been chosen so that $\Gamma_{A'B'CC'} = 0$. Only global transformations, $S^{A'}{}_{\bar{A}'}$ = constant matrix, preserve this condition. Furthermore, anti-self-duality is the integrability condition for the existence of $\tilde{S}^{A}{}_{\bar{A}} \in \widetilde{SL}(2\,\mathbb{C})$ such that for a particular $\pi^{\bar{A}'} \in \mathcal{S}$,

$$\Gamma^{\bar{A}}{}_{\bar{B}\bar{C}\bar{C}'}\,\pi^{\bar{C}'} = 0 \quad . \tag{4.19}$$

Thus

$$\Gamma^{A}{}_{BCC'}\,\pi^{C'} = \pi^{C'}\,\partial_{CC'}\,\tilde{S}^{A}{}_{\bar{A}}\,\tilde{S}^{-1\bar{A}}{}_{B} \tag{4.20}$$

and

$$\Gamma^{A}{}_{BCC'} = \partial_{CC'} \tilde{S}^{A}{}_{\bar{A}} \tilde{S}^{-1\bar{A}}{}_{B} - B^{A}{}_{BC}\, \pi_{C'} \tag{4.21}$$

This is now in the same form as the equation for the connection in the case of the Yang-Mills field. As before, the transformation depends on the choice of $\pi_{C'}$, hence

$$\text{(a)} \quad \tilde{S}^{A}{}_{\bar{A}} = \tilde{S}^{A}{}_{\bar{A}}\,(x, \pi_{A'}) = \tilde{S}^{A}{}_{\bar{A}}\,(x, \zeta, \tilde{\zeta}) ,$$

$$\text{(b)} \quad B^{A}{}_{BC} = B^{A}{}_{BC}\,(x, \zeta, \tilde{\zeta}) \tag{4.22}$$

Acting with $ð$ and $\tilde{ð}$ and then transvecting in turn with $\pi^{A'}$ and $\eta^{A'}$, we find

$$\text{(a)} \quad \pi^{C'} \; ð\,(\partial_{CC'} \tilde{S}^{A}{}_{\bar{A}} \tilde{S}^{-1\bar{A}}{}_{B}) = 0$$

$$\text{(b)} \quad \eta^{C'} \; ð\,(\partial_{CC'} \tilde{S}^{A}{}_{\bar{A}} \tilde{S}^{-1\bar{A}}{}_{B}) \; - ð\, B^{A}{}_{BC}$$

$$\text{(c)} \quad \pi^{C'} \; \tilde{ð}\,(\partial_{CC'} \tilde{S}^{A}{}_{\bar{A}} \tilde{S}^{-1\bar{A}}{}_{B}) = + B^{A}{}_{BC}$$

$$\text{(d)} \quad \eta^{C'} \; \tilde{ð}\,(\partial_{CC'} \tilde{S}^{A}{}_{\bar{A}} \tilde{S}^{-1\bar{A}}{}_{B}) = - \tilde{ð}\, B^{A}{}_{BC} \tag{4.23}$$

There is enough freedom in the choice of $\tilde{S}^{A}{}_{\bar{A}}$ to set

$$\tilde{\pi}^{C} \; ð\, B^{A}{}_{BC} = 0 \quad ,$$

so that once again we find that

$$ð\, \tilde{S}^{A}{}_{\bar{A}} = \tilde{S}^{A}{}_{\bar{B}}\, V^{\bar{B}}{}_{\bar{A}} \quad , \tag{4.24}$$

$$V^{\bar{B}}{}_{\bar{A}} = V^{\bar{B}}{}_{\bar{A}}\,(u, \zeta, \tilde{\zeta}),$$

$$\pi^{A'}\, \partial_{AA'}\, u = 0 \quad , \quad \tilde{\pi}^{A}\, \eta^{A'}\, \partial_{AA'}\, u = 0 \tag{4.25}$$

This argument can also be inverted so that a solution of the above equations leads to a spinor connection of the form

$$\text{(a)} \qquad \Gamma^{A}{}_{BCC'} = \partial_{CC'} \tilde{S}^{A}{}_{\bar{A}} \tilde{S}^{-1\bar{A}}{}_{B} - B^{A}{}_{BC} \pi_{C}$$

$$\text{(b)} \qquad \Gamma^{A'}{}_{B'CC'} = 0 \quad . \tag{4.26}$$

At this point, however, one must show that the tetrad, and therefore the metric, can be determined with the above knowledge. From the defining equation for the rotation coefficients, Eq. (7) we get

$$k_{c}{}^{\mu}{}_{,\nu}\, k_{b}{}^{\nu} - k_{b}{}^{\mu}{}_{,\nu}\, k_{c}{}^{\nu} = 2\, k_{a}{}^{\mu}\, \gamma^{a}{}_{[bc]}$$

$$= -\, 2k_{a}{}^{\mu}\, \sigma^{a}{}_{AA'}\, \sigma_{[b}{}^{BA'}\, \sigma_{c]}{}^{CC'}\, \Gamma^{A}{}_{BCC'} \tag{4.27}$$

The integrability conditions for the equations are satisfied by the cyclic identity on the Riemann tensor.

In general, the null tetrad defined above will not involve surface-forming vectors. However, if one introduces the tetrad $k_{\bar{a}}{}^{\mu}$ with $\Gamma^{\bar{A}}{}_{\bar{B}\bar{C}\bar{C}'}$, $\pi^{\bar{C}'} = 0$, then one can show easily that any pair of vectors which lie in an α-plane are surface-forming. This allows one to introduce special coordinates which are adapted to these planes. This has been done by Plebanski[9] who has studied self-dual solutions directly. In terms of the Plebanski coordinates, the metric takes the form ($\Omega_p = \partial\Omega/\partial p$, etc.)

$$ds^2 = 2(\Omega_{pr}\, dr + \Omega_{ps}\, ds)\, dp + 2(\Omega_{qr}\, dr + \Omega_{qs}\, ds)\, dq \tag{4.28}$$

$$\Omega_{qr}\, \Omega_{ps} - \Omega_{ps}\, \Omega_{qs} = 1$$

The surfaces defined by p = const., q = const. and r = const., s = const. are two separate families of "parallel" α-planes.

Using the Plebanski coordinates, we can construct the deformed twistor space corresponding to the non-linear graviton of Penrose[5]. The argument will only be sketched here. Take a standard α-plane through the origin; for example, r = s = 0. For each value of $\tilde{\zeta}$, the α-planes through the points (p, q) of the standard plane will define points in $\mathbb{P}\mathcal{J}$, $(p, q, \tilde{\zeta})$, the deformed projective twistor space. For a given standard plane, not all values of $\tilde{\zeta}$ will, in fact, allow this construction. Because of the curvature of the space-time, caustics may develop between α-planes so that their intersections are not defined by a single point.

However, the coordinates (p, q, ζ) define a neighborhood $\mathcal{U}_1$ in $\mathbb{P}\mathcal{J}$ for those planes which do intersect the standard plane in a unique point. To obtain another neighborhood $\mathcal{U}_2$, choose a second standard α-plane through the origin and go through the same construction. This construction may be continued until all of the α-planes in the anti-self-dual space-time have been assigned at least to one neighborhood. One can show that in the overlap region of two neighborhoods, the coordinates are related by a canonical transformation. This preserves the structure Penrose[5] demands of $\mathbb{P}\mathcal{J}$ so that he can define the metric of the space-time from the properties of the deformed twistor space.

V. CONCLUSION

The details of these results are being prepared for publication. Further developments of this investigation would be to find solutions of the anti-self-dual fields in the manner outlined in these lectures. Then one would like to determine how information about the deformed twistor space can be used directly to study the quantization of free Yang-Mills fields and the gravitational field. Only after that has been understood would one hope to deal with interacting fields.

REFERENCES

1. M. Ko, M. Ludvigsen, E. T. Newman, and P. K. Tod, "The Theory of $\mathcal{H}$ -Space", Physics Reports, to appear.
2. R. Jackiw, Rev. Mod. Phys. 49, 681 (1977).
3. R. Penrose, Proc. Roy. Soc. Lond. A284, 159 (1965).
4. R. Ward, Phys. Lett. 61A, 81 (1977).
   R. Ward, "Curved Twistor Spaces", Thesis, University of Oxford, (1977).
5. R. Penrose, GRG 7, 31 (1976).
6. T. T. Wu and C. N. Yang, Phys. Rev. D12, 3845 (1975).
7. E. T. Newman, Phys. Rev. D18, 2901 (1978).
8. E. T. Newman and R. Penrose, J. Math. Phys. 7, 863 (1966).
9. J. Plebanski, J. Math. Phys. 16, 2395 (1975).

AN INTRODUCTION TO COMPLEX MANIFOLDS

E. T. Newman

University of Pittsburgh
Department of Physics and Astronomy
Pittsburgh, PA 15260

INTRODUCTION

We will try to give here an introduction to the theory of complex manifolds. This introduction though brief, with most proofs omitted, will hopefully contain many of the essential ideas that would be useful to physicists exploring this beautiful branch of mathematics. In the first lecture we will confine ourselves to a study of linear vector spaces--in particular to the introduction of what are called a "complex structure" and a "hermitian structure" on a real linear vector space. In the second lecture, we introduce the basic idea of a complex manifold and use the results of the first lecture to study the tangent space of a complex manifold.

Lecture I

We begin with a brief review of complex linear vector spaces. This is done to establish a certain point of view which will be useful later.

A complex linear vector space V_c consists of elements v, u . . with two operations, addition of elements e.g. v + u and multiplication by complex scalars α with the properties (1) V_c is an Abelian group with the addition as group operation; (2) $\alpha(v + u) = \alpha v + \alpha u$ and $(\alpha + \beta)v = \alpha v + \beta v$; (3) $(\alpha\beta)v = \alpha(\beta v)$ and $1 \cdot v = v$ with $\alpha, \beta \in C$. Assuming that V_c is M dimensional, i.e. that M is the maximal number of linearly independent vectors, one can choose a basis set $e_\mu \epsilon V_c$ $\mu = 1. . . M$ such that $v = \alpha^\mu e_\mu$ with $\alpha^\mu \epsilon C$ for each μ. Each $v \in V_c$ can thus be written (for v_1^μ and $v_2^\mu \in R$ for each μ)

$$v = (v_1^\mu + iv_2^\mu)\, e_\mu = v_1^\mu\, e_\mu + iv_2^\mu\, e_\mu \tag{1}$$

$$v = v_1 + iv_2$$

with v_1 and v_2 being vectors in a <u>real</u> M dimensional vector space. Since (with $\alpha = a + ib$)

$$\alpha\, v = (av_1 - bv_2) + i(av_2 + bv_1) \tag{2}$$

the complex vector space can be thought of as the direct sum of two real vector spaces $V_1 \oplus V_2$ i.e.

$$v = (v_1, v_2) \tag{3}$$

with

$$\alpha\, v = (av_1 - bv_2, av_2 + bv_1) \tag{4}$$

as the <u>definition</u> of multiplication by complex scalars. In particular multiplication by i is given by

$$iv = (-v_2, v_1) \text{ and } i^2 v = -v. \tag{5}$$

Notice that with this point of view multiplication by i is a real linear operation on $V_1 \oplus V_2$.
Remark I. We wish to emphasize that there is no canonical splitting of V_c into $V_1 \oplus V_2$.

We now begin the main portion of the lecture.

Consider a <u>real</u> linear vector space V with an even number dimension n = 2m. There are two different (but related) ways in which V can be made into a complex vector space.

1) <u>By complexification</u>. Considering v_1 and $v_2 \in V$ then

$$v_c = v_1 + iv_2$$

defines $v_c \in V_c$. (See (1) and (2)). If e_i (i = 1..n) are a basis for V then $v_c \cong (v_1^i + iv_2^i)e_i$ with $v_1 = v_1^i e_i$ and $v_2 = v_2^i e_i$. After complexification the real dim goes from n to 2n.

2) <u>By the introduction of a complex structure</u>. Consider the linear mapping of $J: V \to V$ such that $JJ = -I$ or in components relative to some basis

$$J_i{}^j\, J_j{}^k = -\,\delta_i{}^k. \tag{6}$$

Multiplication by a complex number a + ib on V can be <u>defined</u> by the real linear operation

$$(a + ib)v = av + ibJv \tag{7}$$

or $iv = Jv$. (See (5).)

Essentially what J does is to split V into two real m dimensional parts V_1 and V_2 such that $v_2 = Jv_1$ with $v_1 \in V_1$ and $v_2 \in V_2$ and with $V = V_1 \oplus V_2$. This splitting is in no sense canonical; if e_a, $a = 1. . . m$ span V_1 and $e'_a = Je_a$ span V_2 there exist other splittings $\tilde{V}_1 \oplus \tilde{V}_2$ with for example $(e'_1, e_2 \ldots, e_m)$ spanning $\tilde{V}_1$ and $(-e_1, e'_2, \ldots e'_m)$ spanning $\tilde{V}_2$. (See Remark I.)

This leads to the obvious <u>Theorem I</u>. Given a real 2m linear vector space V with complex structure J, there $\exists$ m linearly independent vectors e_a such that e_a and $e'_a = Je_a$ $(a = 1\ldots m)$ form a basis set for V.

Thus $v \in V$

$$v = v^a e_a + v'^a e'_a = (v^a e_a, v'^a e'_a)$$

$$Jv = v^a e'_a - v'^a e_a = (-v'^a e_a, v^a e'_a). \tag{8}$$

There exists a close relationship between the complexification $V \to V_c$ and the introduction of J. For $v \in V$, we define v^+ and v^- $\in V_c$ (v^+ and v^- are respectively refered to as (1, 0) and (0, 1) vectors in the literature.) by

$$v^+ = \frac{1}{2}(v - iJv), \quad v^- = \frac{1}{2}(v + iJv). \tag{9}$$

with

$$v = v^+ + v^-.$$

Note that the important fact in this decomposition of v is that

$$Jv^+ = \frac{1}{2}(Jv + iv) = i\frac{1}{2}(v - iJv) = iv^+$$
$$Jv^- = \frac{1}{2}(Jv - iv) = -i\frac{1}{2}(v + iJv) = -iv^- \tag{10}$$

i.e. v^+ and v^- are eigenvectors of J with eigenvalues i and -i.

There is the straightforward <u>theorem II</u>. $V_c = V^+ \oplus V^-$ where V^+ and V^- are respectively the spaces of eigenvectors of J with eigenvalues +i and -i.

With e_a and $e'_a = Je_a$ being a basis for V there is a natural choice of basis for V_c which explicitly exhibits the decomposition $V^+ \oplus V^-$ namely

$$\lambda_a^+ = \frac{1}{2}(e_a - iJe_a)$$
$$\lambda_a^- = \frac{1}{2}(e_a + iJ_a) \qquad (11)$$

from which it follows that $v \varepsilon V$

$$v = v^a e_a + v'^a e'_a$$

implies that

$$v^+ = (v^a + iv'^a)\lambda_a^+ . \qquad (12)$$

Essentially what we have shown is that the real 2m dimensional V with J is the same as the m complex dimensional vector space V^+. An element of V is uniquely represented by an element of V^+; multiplication of an element of V by J (see (8)) is the same as multiplying the corresponding element of V^+ by i.

As an example of the above consider V to be a two dimensional real vector space with a complex structure. Choosing any $e \neq 0 \varepsilon V$ then e and $e' = Je$ form a basis set and $v \varepsilon V$ has the form

$$v = xe + ye' \qquad (13)$$

The complexification yields

$$v_c = (x_1 + ix_2)\, e + (y_1 + iy_2)e',$$

and

$$\lambda^+ = \frac{1}{2}(e - ie'),\ \lambda^- = \frac{1}{2}(e + ie').$$

Eq. (13) can now be written

$$v = v^+ + v^- \text{ with}$$
$$v^- = \overline{v^+}$$
$$v^+ = (x + iy)\lambda^+ \qquad (14)$$

Notice that vectors (13) are equivalent to vectors (14) in the sense that each determines the other and

$$Jv = -ye + xe',\ iv^+ = (-y + ix)\lambda^+. \qquad (15)$$

In addition to the complex structure, we now wish to introduce the Hermitian structure on our real 2m dimensional vector space V. A Hermitian structure on a real vector space is a complex valued function on V x V, $h(v,w) = c$, $c \in C$ which is linear in each slot, i.e.

$$h(av + bu, w) = ah(v,w) + bh(u,w) \tag{16}$$

satisfying the further conditions

$$\overline{h(v,w)} = h(w,v) \tag{17}$$

$$h(Jv,w) = ih(v,w). \tag{18}$$

If h is written as a real plus imaginary part

$$h = g - iw$$

or defining $h_{ij} = h(e_i, e_j)$, e_i being a basis for V

$$h_{ij} = h_{ij} - iw_{ij}$$

it follows immediately from (17) that

$$g_{ij} = g_{ji} \text{ and } w_{ij} = - w_{ji} \tag{19}$$

and from (18) that

$$w_{ij} = g_{\ell j} J^{\ell}_{i} \text{ and } g_{iy} = - w_{\ell j} J^{\ell}_{i}. \tag{20}$$

From (17) and (18) it follows that

$$g_{ij} J^{i}_{\ell} J^{j}_{m} = g_{\ell m} \text{ and } w_{ij} J^{i}_{\ell} J^{j}_{m} = w_{\ell m} . \tag{21}$$

Eqs (19), (20) and (21) play an important role. Remark II. An even dimensional vector space endowed with any two of the three real structures g, w or J automatically possesses the other (assuming of course that the g and w are non-singular and satisfy (21)) by virtue of (19) and (21).

Note that if $v = v^{+} + v^{-}$ and $u = u^{+} + u^{-}$ then

$$h(v,u) = 2g(v^{+},u^{-}). \tag{22}$$

We see that the right side of (22) is a normal complex hermitian scalar product on the complex space V^{+}, the left side of (22) is the equivalent scalar product on the real vector space V. This in fact is the reason for the definition of h via (16), (17) and (18).

A final fact about g should be pointed out, namely

$$g(v^+, u^+) = 0 \tag{22}$$

for all v and u. The proof is as follows;

$$g(v^+, u^+) = g(V - iJv, u - iJu) =$$

$$g(v, u) - g(Jv, Jv) - ig(v, Ju) - ig(Jv, u) = 0$$

where we have used (19), (20) and (21).

Eq (22) leads immediately to the fact that g, using the λ^+_a and λ^-_a basis, has the simple form

$$\begin{aligned} g_{ab} &\equiv g(\lambda^+_a, \lambda^+_b) = 0 \\ g_{\overline{ab}} &\equiv g(\lambda^-_a, \lambda^-_b) = 0 \\ g_{a\overline{b}} &\equiv g(\lambda^+_a, \lambda^-_b) = \overline{g(\lambda^+_b, \lambda^-_a)} = \overline{g_{b\overline{a}}} \,. \end{aligned} \tag{23}$$

A simple but profound application of these ideas is to quantum field theory. For example, taking the vector space V (infinite dimensional) of sufficiently regular real solutions of the flat-space Klein-Gordon equation $\Box\phi = \mu^2\phi$, we can define the skew form on pairs of solutions ϕ and ψ

$$w(\phi, \psi) = \int (\phi\nabla_a\psi - \psi\nabla_a\phi)\, dS^a \tag{24}$$

the integral taken over a Cauchy surface. If real solutions are decomposed into their positive and negative frequency parts by

$$\phi = \phi^+ + \phi^-$$

Then the complex structure J on V is defined by

$$J\phi = i(\phi^+ - \phi^-) \tag{25}$$

J maps real solutions into real solutions. Using the ideas just developed we see that (24) and (25) lead to a hermitian scalar product on V. After a brief calculation we obtain

$$h(\phi,\psi) = \int (\phi^+\nabla_a\psi^- - \psi^-\nabla_a\phi^+)dS^a$$

the usual scalar product assigned to the Klein-Gordon field. The Ashtakar's have beautifully exploited this idea in their discussion of quantum field theory on curved space-times.

Lecture II

In this lecture we will show how the ideas of the previous lecture can be extended from vector spaces to manifolds.

The most important idea for us is that of a complex manifold. A complex manifold is a real $n = 2m$ dimensional manifold that can be covered with open sets U_α which are each one to one related to an open set of R^{2m} with coordinates $x^i_\alpha = \{x^a_\alpha, x'^a_\alpha\}$ (α labeling the patch) such that in the overlap region between any two patches $U_\alpha \cap U_\beta$ the coordinates are related by $z^a_\beta = f^a(z^b_\alpha)$ where $z^a_\alpha = x^a_\alpha + ix'^a_\alpha$ and $z^a_\beta = x^a_\beta + ix'^a_\beta$ with $\left|\frac{\partial f^a}{\partial z^b_\alpha}\right| \neq 0$, f^a being complex analytic or holomorphic in the overlap region.

Examples

1) A trivial example is C^m covered by one patch by R^{2m} with $z^a = x^a + ix'^a$.

2) A far more important example is $CP(n)$ complex projective n space. In particular $CP(3)$ is C^4 with points on lines through the origin identified, the origin being deleted, i.e. $z^a \equiv \lambda z^a$.

$CP(3)$ can be covered by four patches defined respectively by $z^1 \neq 0$, $z^2 \neq 0$, $z^3 \neq 0$, and $z^4 \neq 0$. The coordinates on each patch are

$$\zeta^a_{(1)} = \frac{z^a}{z^1}, \quad \zeta^a_{(2)} = \frac{z^a}{z^2}, \quad \zeta^a_{(3)} = \frac{z^a}{z^3}, \quad \zeta^a_{(4)} = \frac{z^a}{z^4}$$

with the subscripts in parentheses labeling the patches. The transformation in the overlap regions is given by

$$\zeta^a_{(b)} = \frac{\zeta^a_{(c)}}{\zeta^b_{(c)}}$$

or for example $\zeta^a_{(1)} = \frac{z^a}{z^1} = \frac{\zeta^a_{(2)}}{\zeta^1_{(2)}} = \frac{z^a/z^2}{z^1/z^2}$.

3) The third example is that of a sphere. Using complex sterographic coordinates projected from first the north pole and then the south pole we get ζ and ζ' as the coordinates of the sphere in the two patches excluding the north and then the south poles. In the overlap region $\zeta' = \zeta^{-1}$.

4) A simple and natural way to construct a large family of complex manifolds is to consider a real analytic manifold (of any dimension) i.e. a manifold with real coordinates such that

the overlap transformations are real analytic. By analytically continuing the real coordinates into the complex, a complex manifold is created - the overlap transformations are now complex analytic. The complexification of real Minkowski space or any real analytic Einstein space-time are useful examples of this process.

We next consider the tangent space at a point of a complex manifold. Remembering that the complex manifold is a real 2m dim. manifold with natural coordinates (up to holomorphic transformations) $x^i = \{x^a, x'^a\}$ such that $z^a = x^a + ix'^a$. A natural basis is then $\frac{\partial}{\partial x^i} = \left\{ \frac{\partial}{\partial x^a}, \frac{\partial}{\partial x'^a} \right\}$ and a vector v becomes

$$v = v^a \frac{\partial}{\partial x^a} + v'^a \frac{\partial}{\partial x'^a} .$$

The complexified tangent space has vectors of the form

$$v = (v_1^a + iv_2^a) \frac{\partial}{\partial x^a} + (v_1'^a + iv_2'^a) \frac{\partial}{\partial x'^a} .$$

There exists a natural tangent-space complex structure

$$J \frac{\partial}{\partial x^a} = \frac{\partial}{\partial x'^a} \tag{26}$$

$$J \frac{\partial}{\partial x'^a} = - \frac{\partial}{\partial x^a} .$$

Introducing the complex basis vectors

$$\lambda_a = \frac{1}{2} \left(\frac{\partial}{\partial x^a} - iJ\frac{\partial}{\partial x^a}\right) = \frac{\partial}{\partial z^a} \tag{27}$$

$$\bar{\lambda}_a = \frac{1}{2} \left(\frac{\partial}{\partial x^a} + iJ\frac{\partial}{\partial x^a}\right) = \frac{\partial}{\partial \bar{z}^a}$$

we have

$$J \frac{\partial}{\partial z^a} = i \frac{\partial}{\partial z^a} \tag{28}$$

$$J \frac{\partial}{\partial \bar{z}^a} = - \frac{\partial}{\partial \bar{z}^a}$$

Thus using the complex basis $\frac{\partial}{\partial z^a}$ and $\frac{\partial}{\partial \bar{z}^a}$, J becomes

$$J_i{}^j = \begin{pmatrix} i\delta_a{}^b & 0 \\ 0 & -i\delta_{\bar{a}}{}^{\bar{b}} \end{pmatrix} \tag{29}$$

This form is preserved under holomorphic coordinate transformations. Remark III. Given any even dimensional manifold, on any patch one to one with R^{2m} there exists a complex structure. The problem is to find one that can be glued together over the patches to yield a global complex structure. In general, manifolds do not possess a complex structure.

A weaker structure known as an almost complex structure is a mixed tensor field J^i_j on an even dimensional manifold such that $J_i^{\ j} J_j^{\ k} = -\delta^k_i$. There is no implication here that the manifold is a complex manifold. Not every even dimensional manifold possesses even an almost complex structure, e.g. S^4 does not possess an almost complex structure.

A natural question arises; when is an almost complex manifold (i.e. a manifold with an almost complex structure) really a complex manifold so that complex holomorphic coordinates z^a can be introduced. This is answered by the remarkable and difficult Newlander and Nirenberg Theorem; If the torsion of J vanishes, i.e. if $N_{jk}^{\ i} \equiv J_j^{\ h} J^i_{k,h} - J^h_k J^i_{j,h} - (J^i_h J^h_{k,j} - J^i_j J^h_{j,k}) = 0$
then the almost complex manifold is a complex manifold.

Returning to complex manifolds, one can inquire about possible metric structures or geometric structures naturally associated with complex manifolds. There are two in common use,

1) The first is a complex "Riemannian metric"

$$ds^2 = g_{ab}(z^c)dz^a\,dz^b \tag{30}$$

i.e. a metric tensor which is a holomorphic function of the z^a and a complex-valued line element. One can now formally do complex "Riemannian geometry", i.e. construct the holomorphic Levi-Civita connection and Riemann tensor. A great deal of work has gone into solving the holomorphic Einstein equations. If in addition to $Rab = 0$, the Weyl tensor is self or anti-self-dual one is led into the subject of H-space or the non-linear graviton.

2) The second type of metric structure naturally associated with complex manifolds is referred to as a hermitian structure. The manifold is then referred to as a hermitian manifold. A hermitian manifold is a complex manifold with a real metric g_{ij} (it is best, at least in the beginning, to think of it in terms of the real coordinates $x^i = \{x^a, x'^a\}$ satisfying

$$g_{ij}\, J^i_\ell\, J^j_m = g_{\ell m}\ . \tag{31}$$

Using the ideas of the first lecture we see that a skew-form w_{ij} can be introduced by

$$w_{ij} = g_{\ell j} J^{\ell}_{i} = - w_{ji} \tag{32}$$

or

$$w = w_{ij} \, dx^i \wedge dx^j \tag{33}$$

and hence the hermitian metric

$$h_{ij} = g_{ij} - iw_{ij} \tag{34}$$

is naturally given. It should be pointed out that strictly speaking the g, w and h are defined to act on real tangent vectors but by linearity they can be extended to act on the complex tangent space.

From Eq (23), we see that if the complex coordinates $z^a = x^a + ix'^a$ and $\overline{z^a}$ are introduced then

$$ds^2 = g_{a\bar{b}} (z^e, z^{\bar{d}}) \, dz^a \, dz^{\bar{b}} \tag{35}$$

with $\overline{g_{a\bar{b}}} = g_{b\bar{a}}$. It is not difficult to show that in the same complex basis

$$w = w_{a\bar{b}} \, dz^a \wedge dz^{\bar{b}} \tag{36}$$

$w_{a\bar{b}} = ig_{a\bar{b}}$. The hermitian structure thus becomes

$$h_{a\bar{b}} = g_{a\bar{b}} - iw_{a\bar{b}} = 2g_{a\bar{b}}$$

or

$$h_{ij} = 2 \begin{pmatrix} 0 & g_{a\bar{b}} \\ g_{a\bar{b}} & 0 \end{pmatrix}$$

Given a hermitian manifold there are two different natural connections one can introduce;

1) The Levi-Civita connection obtained from the g_{ij}.

2) A natural connection which arises from the fact that

$$\frac{\partial v^a}{\partial \bar{z}^b}\ ,\ g^{a\bar{b}}\ \frac{\partial(g_{d\bar{b}}\ v^d)}{\partial z^c}\ \ ,\ \frac{\partial v^{\bar{a}}}{\partial z^b}\ ,\ g^{b\bar{a}}\ \frac{\partial(g_{b\bar{d}}\ v^{\bar{d}})}{z^c} \qquad (37)$$

transforms as a tensor. This can be seen by remembering that the transformation matrices for v^a (and $v^{\bar{a}}$) are functions only of z^a (and $z^{\bar{a}}$) and hence are not affected by the differentiation. In fact (37) can be taken as the definition of covariant differentiation of $v^i = \{v^a, v^{\bar{a}}\}$.

A necessary and sufficient condition for the two connections to be identical is that two-form w (the Kahler form) be closed, i.e.

$$dw = 0 \qquad (38)$$

or

$$w_{[ij,\ k]} = 0.$$

A hermitian manifold with a closed Kahler form is known as a Kahler manifold. From (38) it can be shown that locally

$$g_{a\bar{b}} = \frac{\partial^2 \Omega(z,\ \bar{z})}{\partial z^a\ \partial z^b} \qquad (39)$$

where Ω is a real function.

The subject of complex manifolds and in particular Kahler manifolds is large and we have hence only touched on the fringes of the subject. Some of the major topics that appear to have relevence to physics which we have not even mentioned are the theory of deformations of a complex manifold and sheaf-cohomology theory.

In conclusion we list some references which include both a much deeper treatment of the material of these lectures, and deformation and sheaf-cohomology theory as well as other more specialized topics.

REFERENCE

1. J. Morrow and K. Kodaiva, Complex Manifolds Holt-Rinehart-Winston, NY 1979.
2. S.S. Chern, Complex Manifolds without Potential Theory, Van Nostrand, Princeton, NJ 1967.
3. K. Yano, Differential Geometry on Complex and Almost Complex Spaces (Pergamon) Mac Millan NY 1965.
4. E.J. Flaherty, Hermitian and Kahlerian Geometry in Relativity, Springer-Verlag, Berlin 1976.
5. Kobayashi and Nomizu, Vol II, Foundations of Differential Geometry, Interscience, NY 1963.

A BRIEF OUTLINE OF TWISTOR THEORY

Roger Penrose

Mathematical Institute
Oxford, OX1 3LB
England

1. Motivation

The normal space-time formalism for physics has proved remarkably successful. There are, nevertheless, several reasons for believing that, at a certain level of understanding, these space-time concepts will give way to others that are even more basic and fundamental. In the first place, the concept of space-time "point" is not very directly physical. For example, the particles that constitute matter as we know it are, even at the classical level, one-dimensional objects (world-lines) rather than zero-dimensional. Then, when quantum theory enters the picture, the uncertainty princple serves to obscure even further the relation between physical particles and space-time points. Do points really "exist" in a precisely defined sense at 10^{-13}cm? The question becomes even more pertinent at the quantum gravity level of 10^{-33}cm.

It is worth while to point out, also, the conflict of formalisms between that of relativity, which uses a four-dimensional (+---) pseudo-Riemannian real space, whose basic constitutes are "events", and that of quantum mechanics, which uses a positive-definite complex Hilbert space (of unspecified dimension) whose basic constituents are "states." What would seem desirable is a formalism which treats the space-time and quantum aspects of physics in a more unified way. Twistor theory, to a large extent achieves this. It uses the complex field $\mathbb{C}$ as its basic number system and the normal real space-time geometry may be derived. Furthermore, the space-time geometry that emerges is four-dimensional, with the correct (+---) signature, so one may say that the twistor formalism is specially tailored to the particular space-time structure needed for physics rather than being a general mathematical formalism.

In the first instance, twistor theory merely provides a reformulation of the basic concepts and, in this sense is not a "new" theory. But it suggests different ways to proceed and new hypotheses suggest themselves at a later stage. (A possible parallel would be with Cartan's reformulation of Newtonian gravitation as a space-time theory. In this form, the modifications needed to change the theory into full-blown general relativity would have seemed very simple and natural indeed.) One of the hopes, of course, is that by providing a new framework for the description of space-time and quantum physics some of the difficulties with existing theory, such as the infinities of quantum field theory and the interpretational problems of ordinary quantum theory may eventually be removed.

The "successes" of twistor theory, so far, have been mainly of a mathematical nature. The theory provides a nice representation of Minkowski-space geometry, of classical spinning massless particles and of free massless fields. It provides a programme for studying massive particles and their internal symmetry groups. Moreover, it provides a general method for solving the non-linear Einstein and Yang-Mills field equations in the (anti-) self-dual case. There is some prospect that this (anti-) self-dual restriction can be ultimately removed.

The complex numbers of quantum linear supersition have a relation to the geometry of space-time, and it is this that may be regarded as the starting point of the twistor approach. Consider, first, a non-relativistic particle of spin $\frac{1}{2}$. The complex linear combinations of the two states "spin up" and "spin down" provide the states of spin in all other spatial directions. This relates the sphere of directions in physical three-dimensional space to the Riemann sphere of ratios of pairs of complex numbers:

$$\lambda \, (\uparrow) + \mu \, (\downarrow) = (\nearrow) \qquad \text{(spin } \tfrac{1}{2}\text{)}.$$

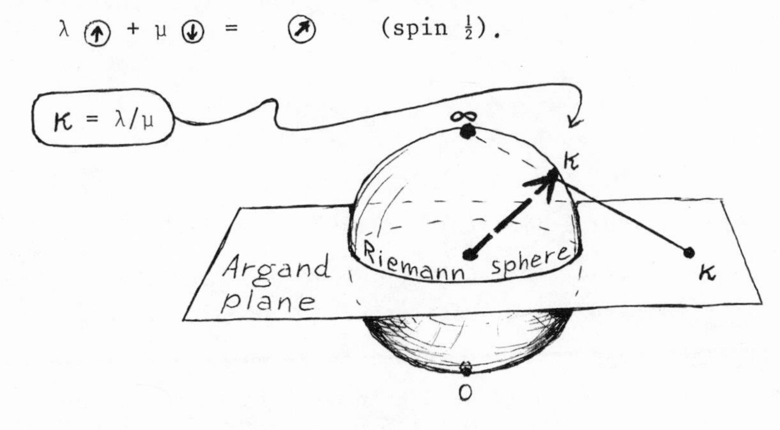

The rotations of space correspond, according to the 2-1 map $SU(2) \to SO(3)$, to the unitary fractional linear transformations. Dropping the unitary condition we get the relativistic version of this, which is the 2-1 map $SL(2,\mathbb{C}) \to O_{+}^{\uparrow}(1,3)$. The holomorphic transformations

$$\kappa \longmapsto \frac{\alpha\kappa + \beta}{\gamma\kappa + \delta}$$

of the Riemannian sphere of κ to itself thus provide the elements of the restricted Lorentz group. In terms of space-time geometry, this Riemann sphere is the celestial sphere of an observer, namely the family of all light-rays through a point of Minkowski space-time:

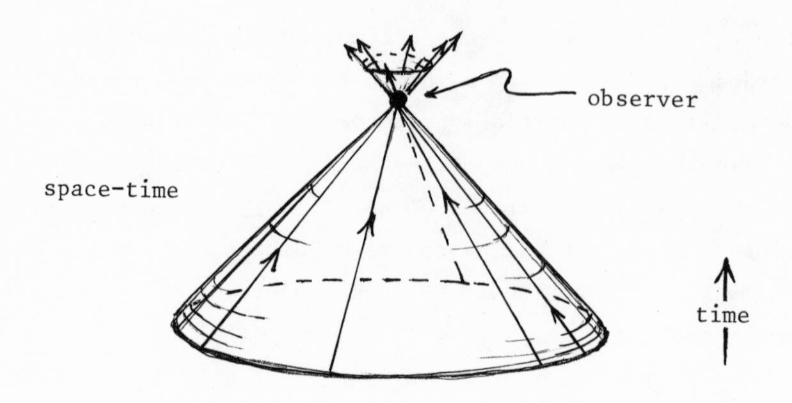

Thus, as an abstract manifold, the space of light-rays through a fixed point is, in a natural way, a complex 1-dimensional manifold (namely the Riemann sphere). Consider, now, the space of all light-rays (null straight lines) in Minkowski space, not merely the ones through a fixed point. This is a real 5-dimensional space and so cannot, as a whole, be a complex manifold. Yet, remarkably, it is not far from being one. If we consider not just light-rays but photons with spin and energy, then fixing the helicity s to take the value $s = +\hbar$ we obtain a real 6-dimensional manifold whose structure is, indeed, that of a complex 3-dimensional manifold. As we shall see shortly, this manifold, which we label $\mathbb{PT}^+$, is one (open) half of a complex projective 3-space $\mathbb{CP}^3$ (labelled $\mathbb{PT}$) the other (open) half $\mathbb{PT}^-$ representing the $s = -\hbar$ photons. The real 5-dimensional boundary $\mathbb{PN}$ between the two can be thought of as the space of light-rays in Minkowski space, $\mathbb{PN}$ containing a special projective line ($\mathbb{CP}^1$) labelled I, that represents the family of idealized light-rays that generate the light-cone $\mathscr{I}$ at infinity.

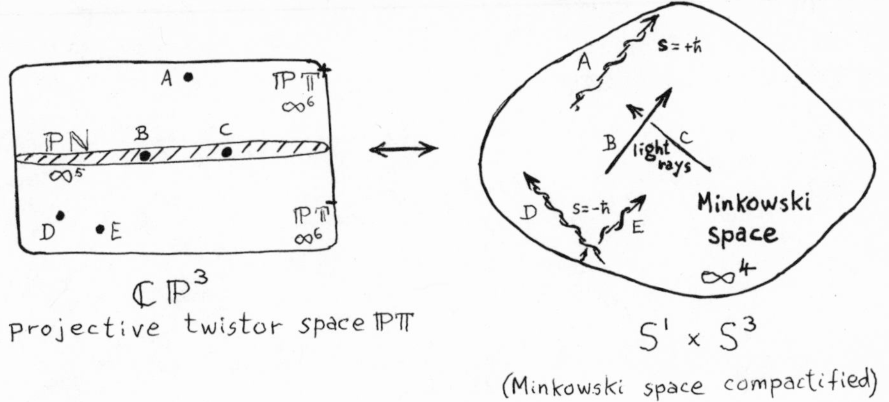

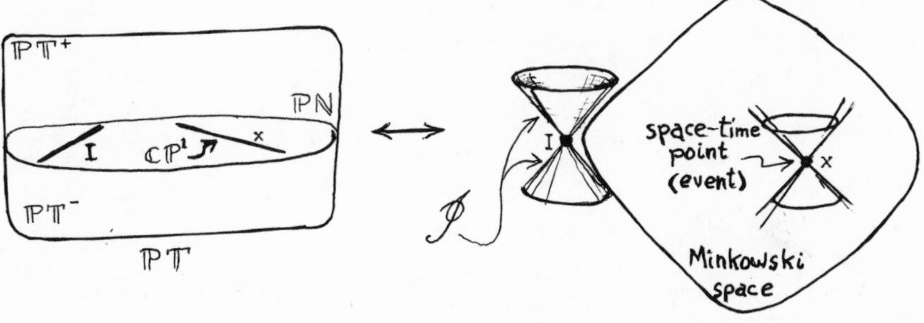

2. Explicit Description of a Twistor

A twistor Z^α is an element of a complex 4-dimensional vector space $\mathbb{T}$ (twistor space). The components $(Z^0, Z^1, Z^2, Z^3) \in \mathbb{C}^4$ of Z^α, together with the components

$$(\bar{Z}_0, \bar{Z}_1, \bar{Z}_2, \bar{Z}_3) = (\overline{Z^2}, \overline{Z^3}, \overline{Z^0}, \overline{Z^1})$$

of its complex conjugate dual twistor $\bar{Z}_\alpha$, determine the components of the momentum p^a and angular momentum M^{ab} of a massless free particle (in Minkowski space) according to:

$$P^0 = \frac{1}{\sqrt{2}} (Z^3\bar{Z}_1 + Z^2\bar{Z}_0),\ P^1 = \frac{1}{\sqrt{2}} (Z^3\bar{Z}_1 - Z^2\bar{Z}_0)$$

$$P^2 = \frac{1}{\sqrt{2}}(-Z^2\bar{Z}_1 - Z^3\bar{Z}0),P^3 = \frac{i}{\sqrt{2}} (Z^2\bar{Z}_1 - Z^3\bar{Z}_0)$$

$$M^{01} = -M^{10} = \frac{i}{2}(Z^0\bar{Z}_0 - Z^1\bar{Z}_1 - Z^2\bar{Z}_2 + Z^3\bar{Z}_3)$$
$$M^{02} = -M^{20} = \frac{i}{2}(Z^0\bar{Z}_1 + Z^1\bar{Z}_0 - Z^2\bar{Z}_3 - Z^3\bar{Z}_2)$$
$$M^{03} = -M^{30} = \frac{1}{2}(Z^0\bar{Z}_1 - Z^1\bar{Z}_0 - Z^2\bar{Z}_3 + Z^3\bar{Z}_2)$$
$$M^{12} = -M^{21} = \frac{i}{2}(Z^0\bar{Z}_1 - Z^1\bar{Z}_0 + Z^2\bar{Z}_3 - Z^3\bar{Z}_2)$$
$$M^{32} = -M^{23} = \frac{1}{2}(Z^0\bar{Z}_0 - Z^1\bar{Z}_1 + Z^2\bar{Z}_2 - Z^3\bar{Z}_3)$$
$$M^{13} = -M^{31} = \frac{1}{2}(Z^0\bar{Z}_1 + Z^1\bar{Z}_0 + Z^2\bar{Z}_3 + Z^3\bar{Z}_2)$$
$$M^{00} = M^{11} = M^{22} = M^{33} = 0.$$

In addition to the required properties $p_a p^a = 0$, $p^0 > 0$ automatically holding, the Pauli-Łubanski spin vector

$$S_a = \tfrac{1}{2}\, e_{abcd}\, p^b\, M^{cd}$$

also automatically satisfies the necessary relation

$$S_a = s\, p_a\,,$$

where the <u>helicity</u> s is given by

$$s = \tfrac{1}{2}\, Z^\alpha \bar{Z}_\alpha \quad .$$

The twistor Z^α is determined up to the phase transformation $Z^\alpha \longmapsto e^{i\theta} Z^\alpha$ (θ real) by the p^a, M^{ab} of a massless particle and, conversely, every twistor (except those for which $Z^2 = Z^3 = 0$) is related to a (finite) massless particle in this way.

<u>Projective twistor space</u> $\mathbb{PT}$ is the space of proportionality classes of non-zero twistors. The standard coordinates for $\mathbb{PT}$ are the three independent complex ratios $Z^0:Z^1:Z^2:Z^3$. $\mathbb{PT}$ is divided into two complex 3-manifolds $\mathbb{PT}^+$ (given by $Z^\alpha\bar{Z}_\alpha > 0$) and $\mathbb{PT}^-$ (given by $Z^\alpha\bar{Z}_\alpha < 0$), by their common 5-real-dimensional boundary $\mathbb{PN}$ (given by $Z^\alpha\bar{Z}_\alpha = 0$). Similarly, twistor space $\mathbb{T}$ is divided into $\mathbb{T}^+$ and $\mathbb{T}^-$ by the 7-real-dimensional cone $\mathbb{N}$.

In order to make detailed calculations within the twistor

formalism, it is convenient to adopt the 2-spinor notation according to which a vector V^a, with standard orthonormal components (V^0, V^1, V^2, V^3), corresponds to the spinor $V^{AA'}$, whose standard components are given by

$$\begin{pmatrix} V^{00'} & V^{01'} \\ V^{10'} & V^{11'} \end{pmatrix} = \frac{1}{\sqrt{2}} \begin{pmatrix} V^0+V^1 & V^2+V^3 i \\ V^2-V^3 i & V^0-V^1 \end{pmatrix}.$$

The reality of V^a corresponds to the Hermiticity of this (2 x 2) matrix. Furthermore,

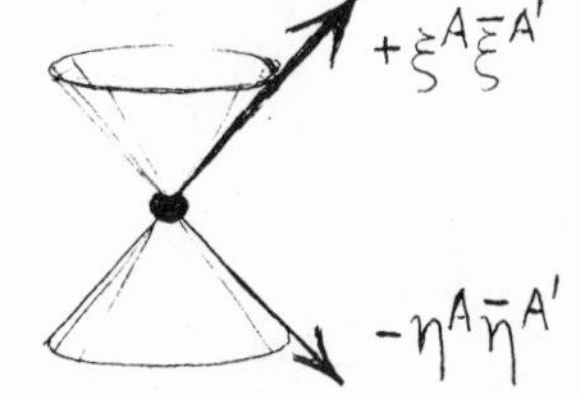

$$V^a \text{ null} \Leftrightarrow \det(V^{AA'}) = 0$$
$$\Leftrightarrow V^{AA'} = \xi^A \eta^{A'}$$
$$= \pm\xi^A \bar{\xi}^{A'} \text{ if } V^a \text{ real}.$$

The metric tensors g_{ab}, g^{ab} have components diag(1,-1,-1,-1) and take the spinor form $g_{AA'BB'} = \epsilon_{AB}\,\epsilon_{A'B'}$, $g^{AA'BB'} = \epsilon^{AB}\epsilon^{A'B'}$ where $\epsilon_{AB}, \epsilon_{A'B'}, \epsilon^{AB}, \epsilon^{A'B'}$ all have components $\begin{pmatrix} 0 & 1 \\ -1 & 0 \end{pmatrix}$. The skew ϵ's raise and lower indices according to $\xi_B = \xi^A\epsilon_{AB}$, $\xi^A = \epsilon^{AB}\xi_B$, $\eta_{B'} = \eta^{A'}\epsilon_{A'B'}$, $\eta^{A'} = \epsilon^{A'B'}\eta_{B'}$. In this notation we have, for the momentum and angular momentum;

$$P_{AA'} = \bar{\pi}_A \pi_{A'}$$

$$M^{AA'BB'} = i\,\omega^{(A}\bar{\pi}^{B)}\epsilon^{A'B'} - i\,\epsilon^{AB}\bar{\omega}^{(A'}\pi^{B')} \quad ,$$

where (with $\pi_{A'} \neq 0$)

$$Z^\alpha = (\omega^A, \pi_{A'})$$

$$\bar{Z}_\alpha = (\bar{\pi}_A, \bar{\omega}^{A'}) \quad .$$

We can also use twistors to describe massive particles (or systems of particles), but now more than one twistor $Z_1^\alpha, \ldots, Z_n^\alpha$ must be used. We have, for the total momentum p_a and angular momentum M^{ab},

$$p_{AA'} = \bar{\underset{r}{\pi}}_A \underset{r}{\pi}_{A'}$$

$$M^{AA'BB'} = i \underset{r}{\omega}^{(A} \bar{\underset{r}{\pi}}^{B)} \epsilon^{A'B'} - i \epsilon^{AB} \bar{\underset{r}{\omega}}^{(A'} \underset{r}{\pi}^{B')}$$

(summation over r = 1,...,n assumed). The linear freedom in $\underset{r}{Z}^\alpha$ and $\bar{\underset{r}{Z}}_\alpha$ is given by the n-twistor internal symmetry transformations:

$$\begin{pmatrix} \underset{r}{Z}^\alpha \\ \hline \bar{\underset{r}{Z}}_\alpha \end{pmatrix} \longmapsto \left(\begin{array}{c|c} \underset{r,s}{U}\, \delta^\alpha_\beta & \underset{r,t}{\Lambda}\, \underset{t,s}{U} \begin{pmatrix} \epsilon^{AB} & 0 \\ 0 & 0 \end{pmatrix} \\ \hline \bar{\underset{r,t}{\Lambda}}\, \underset{t,s}{U} \begin{pmatrix} 0 & 0 \\ 0 & \epsilon_{A'B'} \end{pmatrix} & \bar{\underset{r,s}{U}}\, \delta_\alpha^\beta \end{array}\right) \begin{pmatrix} \underset{s}{Z}^\beta \\ \hline \bar{\underset{s}{Z}}_\beta \end{pmatrix}$$

Here U is an arbitrary n x n unitary matrix and Λ an arbitrary n x n complex skew matrix. The group multiplication law for the n-twistor internal group is

$$(U, \Lambda) \times (\tilde{U}, \tilde{\Lambda}) = (U\tilde{U}, \Lambda + U \tilde{\Lambda} U^T) .$$

According to (naïve) twistor particle theory, the (quantized) infinitesimal generators of these groups provide an algebra of quantum numbers for elementary particles. I shall not discuss this theory here, however.

3. Change of Origin

The momentum and angular momentum of a system in special relativity have very specific transformation properties under change of origin:

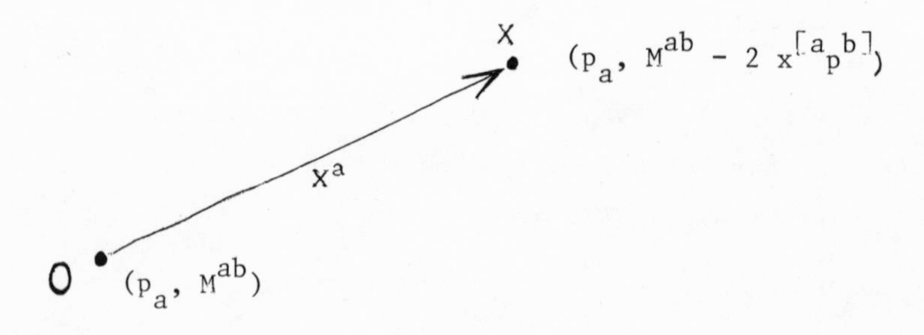

Thus, we can, if desired, think of p_a and M^{ab} as fields (i.e. functions of position) given by

$$p_a(x) = p_a ,$$
$$M^{ab}(x) = M^{ab} - 2\, x^{[a}p^{b]} ,$$

where p_a and M^{ab} denote the values at the original fixed origin O. Correspondingly, the twistor Z^α determines spinor fields:

$$\pi_{A'}(x) = \pi_{A'}$$
$$\omega^A(x) = \omega^A - i\, x^{AB'}\pi_{B'}$$

these yielding the correct $p_a(x)$ and $M^{ab}(x)$ in any one- or many-twistor system. It is easy to show that the ω-fields of this form are precisely the solutions of the twistor equation:

$$\nabla_{A'}^{(A}\,\omega^{B)}(x) = 0 .$$

A null twistor Z^α satisfies $Z^\alpha\bar{Z}_\alpha = 0$ and (provided that $\pi_{A'} \neq 0$) defines a null straight line in space-time as the locus of x^a for which $\omega^A(x) = 0$, i.e. for which

$$\omega^A = ix^{AB'}\pi_{B'}$$

and we say that the point x and the twistor Z^α are incident whenever this equation holds. Any real null straight line is obtainable in this way.

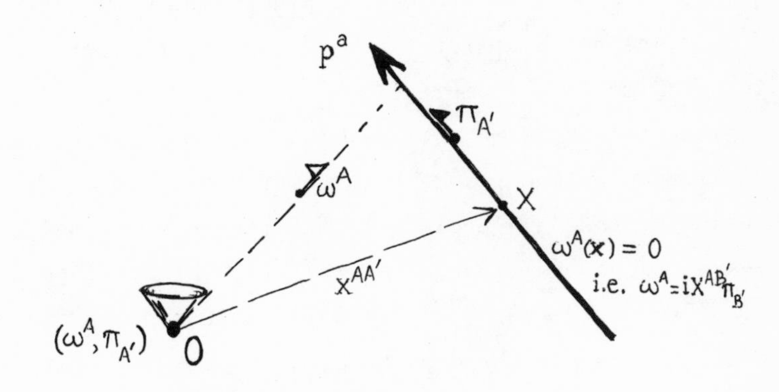

In <u>complexified</u> Minkowski space $\mathbb{CM}$, $x^{AB'}$ need not be Hermitian and the complex solutions of the above equation in x, for fixed ω and π - where Z^α need not now be null - define an <u>α-plane</u>

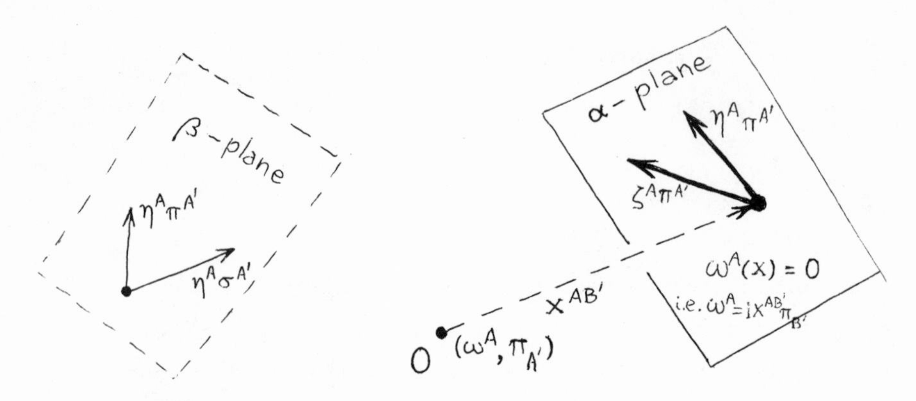

which is a 2-complex-dimensional totally null plane in $\mathbb{CM}$. All the tangent vectors to an α-plane have the (spinor) form $\eta^A\pi^{A'}$, where π is fixed and η varies. These vectors are all null and mutually orthogonal. Similarly a <u>β-plane</u> may be defined whose tangent vectors have the form $\eta^A\pi^{A'}$ where now it is the <u>unprimed</u> spinor η that is fixed and π that varies. There is an important geometrical correspondence between $\mathbb{CM}$ and $\mathbb{PT}$, namely: <u>α-planes</u> in $\mathbb{CM}$ correspond to <u>points</u> in $\mathbb{PT}$; <u>points</u> in $\mathbb{CM}$ correspond to <u>projective lines</u> in $\mathbb{PT}$; β-planes in $\mathbb{CM}$ correspond to <u>projective planes</u> in $\mathbb{PT}$; two points in $\mathbb{CM}$ are <u>null separated</u> if and only if the corresponding lines in $\mathbb{PT}$ meet; a (complex) <u>null geodesic</u> in $\mathbb{CM}$ corresponds to a <u>plane pencil</u> in $\mathbb{PT}$, i.e. the system of lines through a point and lying in one plane, this point and plane in $\mathbb{PT}$ defining a unique α-plane and β-plane in $\mathbb{CM}$ through the null geodesic. The incidence relation $\omega^A = i\,x^{AB'}\pi_{B'}$, between a point x of $\mathbb{CM}$ and a twistor $Z^\alpha = (\omega^A, \pi_{A'})$ is interpreted in $\mathbb{CM}$ as the point x lying on the α-plane Z, while in $\mathbb{PT}$ it is interpreted as the line x passing through the point Z.

The space $\mathbb{CM}$ considered here is the compactified complex Minkowski space. It contains a complex light cone at infinity denoted $\mathbb{C}\mathscr{I}$. The generators of $\mathbb{C}\mathscr{I}$ correspond to the points of a special line I in $\mathbb{PT}$, the line I itself being represented in $\mathbb{CM}$ by the vertex of $\mathbb{C}\mathscr{I}$. The points of $\mathbb{C}\mathscr{I}$ correspond to lines in $\mathbb{PT}$ meeting I. The real points of $\mathbb{C}\mathscr{I}$, i.e. the points of $\mathscr{I}$, correspond to lines in $\mathbb{PN}$ meeting I. The finite real points of Minkowski space correspond to lines in $\mathbb{PN}$ not meeting I. A complex (finite) point in $\mathbb{CM}$ corresponds to a general line in $\mathbb{PT}$ not meeting I. Complex conjugation in $\mathbb{CM}$ interchanges α-planes and β-planes. Thus, this induces a duality correspondence on $\mathbb{PT}$ whereby points and planes are interchanged. In fact this is simply the (projective) correspondence $Z^\alpha \longleftrightarrow \bar{Z}_\alpha$ which interchanges twistors with dual twistors. This correspondence sends a line in $\mathbb{PT}$ (thought of as a point locus) to another line in $\mathbb{PT}$ (thought of as the axis of a pencil of planes) - the complex-conjugate line. Only the lines in $\mathbb{PN}$ are their own complex conjugates, these corresponding to the real points of $\mathbb{CM}$. Lines in $\mathbb{PT}$, not in $\mathbb{PN}$, are of five different kinds according as they lie entirely in $\mathbb{PT}^+$, or in the closure $\overline{\mathbb{PT}}^+$ ($= \mathbb{PT}^+ \cup \mathbb{PN}$) but not in $\mathbb{PT}^+$, or meet all three of $\mathbb{PT}^+$ and $\mathbb{PN}$ and $\mathbb{PT}^-$, or lie in the closure $\overline{\mathbb{PT}}^-$ but not in $\mathbb{PT}^-$, or lie entirely in $\mathbb{PT}^-$. The corresponding points of $\mathbb{CM}$ have position vectors whose imaginary parts are past-timelike, past-null, spacelike, furture-null, or future-timelike, respectively. Thus, in particular, the forward tube (points of $\mathbb{CM}$ with past-timelike imaginary parts) corresponds to (lines in) $\mathbb{PT}^+$.

The (unimodular) linear transformations of $\mathbb{T}$ which preserve the twistor norm $Z^\alpha \bar{Z}_\alpha$, and so preserve the division of $\mathbb{T}$ into $\mathbb{T}^+$, $\mathbb{N}$ and $\mathbb{T}^-$, consitute a realization of the group SU(2,2). This induces a group of projective transformations of $\mathbb{PT}$ for which $\mathbb{PN}$ is invariant and, consequently, a group of point transformations of (compactified) Minkowski space to itself – in fact, the restricted conformal group, to which SU(2,2) is 4 - 1 related.

4. Poincaré Invariance

To express Poincaré invariance in $\mathbb{CM}$, and not just conformal invariance, we need to preserve more of the structure of $\mathbb{T}$. SU(2,2) preserves the twistor norm $Z^{\alpha}\bar{Z}_{\alpha}$ and, for unimodularity, the twistor Levi Civita symbols $\epsilon_{\alpha\beta\gamma\delta}$ and $\epsilon^{\alpha\beta\gamma\delta}$. The subgroup of SU(2,2) corresponding to the restricted Poincaré group also preserves the infinity twistors

$$I^{\alpha\beta} = \begin{pmatrix} \epsilon^{AB} & 0 \\ 0 & 0 \end{pmatrix}, \quad I_{\alpha\beta} = \begin{pmatrix} 0 & 0 \\ 0 & \epsilon_{A'B'} \end{pmatrix}.$$

These twistors serve to define the line I in $\mathbb{PT}$.

An alternative way to think of this additional Poincaré-invariant structure for $\mathbb{T}$ is as the exact sequence

$$0 \longrightarrow \mathbb{S} \longrightarrow \mathbb{T} \longrightarrow \tilde{\mathbb{S}}^* \longrightarrow 0$$

$$\omega^A \longmapsto (\omega^A, 0)$$

$$(\omega^A, \pi_{A'}) \longmapsto \pi_{A'} \;.$$

Here $\mathbb{S}$ denotes unprimed spin-space, $\tilde{\mathbb{S}}$ denotes primed spin-space, and the arterisk means dual.

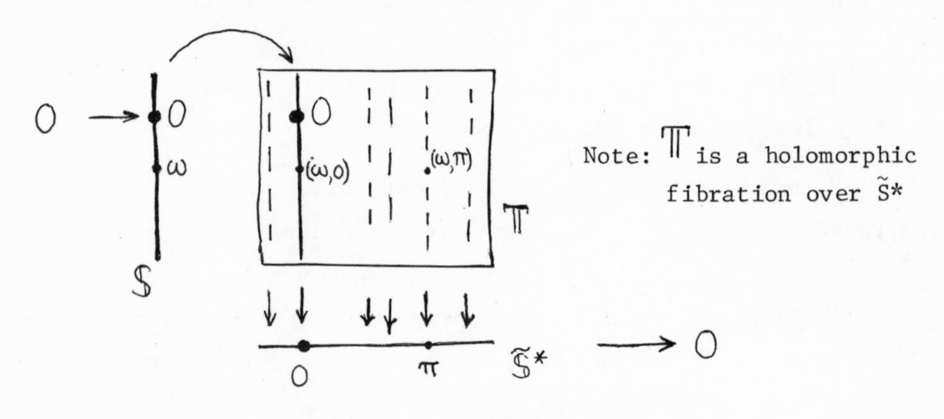

The dual of this sequence is also, when read in reverse order, its complex conjugate, and is

$$0 \longleftarrow \mathbb{S}^* \longleftarrow \mathbb{T}^* \longleftarrow \tilde{\mathbb{S}} \longleftarrow 0$$

$$(0,\tilde{\omega}^{A'}) \longleftarrow\!\shortmid \tilde{\omega}^{A'}$$

$$\tilde{\pi}_A \longleftarrow\!\shortmid (\tilde{\pi}_A,\tilde{\omega}^{A'}) .$$

We have the following invariant structure defined on $\mathbb{T}$:

holomorphic, conformally invariant

$\Upsilon = Z^\alpha \dfrac{\partial}{\partial Z^\alpha}$, Euler vector field (homogeneity operator);

$$d^4Z = \frac{1}{24}\epsilon_{\alpha\beta\gamma\delta}\, dZ^\alpha \wedge dZ^\beta \wedge dZ^\gamma \wedge dZ^\delta$$

$$= dZ^0 \wedge dZ^1 \wedge dZ^2 \wedge dZ^3$$

$= \frac{1}{4}\, d(\mathcal{D} Z)$, complex volume 4-form;

$$\mathcal{D}Z = \frac{1}{6}\epsilon_{\alpha\beta\gamma\delta}\, Z^\alpha\, dZ^\beta \wedge dZ^\gamma \wedge dZ^\delta$$

$$= Z^0 dZ^1 \wedge dZ^2 \wedge dZ^3 - Z^1 dZ^0 \wedge dZ^2 \wedge dZ^3 + Z^2 dZ^0 \wedge dZ^1 \wedge dZ^3 - Z^3 dZ^0 \wedge dZ^1 \wedge dZ^2$$

$= \Upsilon \lrcorner (d^4Z)$, projective volume 3-form;

holomorphic, Poincaré invariant

$$d^2Z = d^2\pi = \frac{1}{2} I_{\alpha\beta}\, dZ^\alpha \wedge dZ^\beta = \frac{1}{2}\epsilon^{A'B'}\, d\pi_{A'} \wedge d\pi_{B'}$$

$$= dZ^2 \wedge dZ^3 = d\pi_{0'} \wedge d\pi_{1'}$$

$= \frac{1}{2}\, d(\Delta Z)$, pull-back of "volume" (area) 2-form on $\tilde{\mathbb{S}}^*$;

$$\Delta Z = I_{\alpha\beta}\, Z^\alpha dZ^\beta = \epsilon^{A'B'}\, \pi_{A'}\, d\pi_{B'} = \pi_{0'}\, d\pi_{1'} - \pi_{1'}\, d\pi_{0'}$$

$= \Upsilon \lrcorner (d^2Z)$, pull-back of projective "volume" 1-form on $\mathbb{P}\tilde{\mathbb{S}}^*$;

also $\epsilon^{AB} \dfrac{\partial}{\partial \omega^A} \otimes \dfrac{\partial}{\partial \omega^B}$, which invariantly defines a symplectic 2-form d^4Z/d^2Z on each fibre;

non-holomorphic, conformally invariant

$\Sigma = Z^{\alpha}\bar{Z}_{\alpha} = 2s$ (twice the helicity),

$d\Sigma = i\bar{\Phi} - i\Phi$, $\Theta = d\Phi = d\bar{\Phi}$;

$\Phi = i\, Z^{\alpha} d\bar{Z}_{\alpha}$;

$\Theta = i\, dZ^{\alpha} \wedge d\bar{Z}_{\alpha}$, real symplectic structure (Kähler 2-form).

5. Twistor ("first") Quantization

To pass to a quantized theory, we may allow our twistor variables $\underset{1}{Z}^{\alpha},\ldots,\underset{n}{Z}^{\alpha}$, and their complex conjugates $\bar{Z}_{1\alpha},\ldots,\bar{Z}_{n\alpha}$ to become non-commuting quantum operators, subject to commutation rules:

$$[\underset{r}{Z}^{\alpha}, \underset{t}{Z}^{\beta}] = 0 \ , \ [\underset{r}{Z}^{\alpha}, \bar{Z}_{t\beta}] = \delta^{\alpha}_{\beta}\, \delta_{rt} \ , \ [\bar{Z}_{r\alpha}, \bar{Z}_{t\beta}] = 0 \ ,$$

taking $\hbar = 1$. This agrees with the standard x^a, p_a quantization when $s = 0$; it is compatible with the twistor symplectic structure; it gives the correct "Poincaré" commutation rules for p^a, M^{ab}; it also gives the SU(2,2) Lie algebra. No factor-ordering problems arise for the Poincaré generators p^a, M^{ab} , but for the helicity operator (for a single twistor system), we now find

$$s = \frac{1}{4}\,(Z^{\alpha}\bar{Z}_{\alpha} + \bar{Z}_{\alpha}Z^{\alpha}) \ .$$

For a twistor wave-function (for a one-twistor system), in the $\mathbb{T}$-representation we use a function $f(Z^{\alpha})$ "independent of $\bar{Z}_{\alpha}$", i.e. holomorphic in Z^{α}. Then the "operator Z^{α}" is represented simply as "multiplication by the twistor variable Z^{α}" , while the "operator $\bar{Z}_{\alpha}$" is represented by "$-\partial/\partial Z^{\alpha}$". Thus, the helicity operator becomes:

$$s = -\frac{1}{2}\, Z^{\alpha}\, \partial/\partial Z^{\alpha} - 1 = -\frac{1}{2}\,\Upsilon - 1 \ .$$

In the $\mathbb{T}^{*}$-representation (i.e. in terms of dual twistor variables) we can use $f(\bar{Z}_{\alpha})$ "independent of Z^{α}" , i.e. anti-holomorphic in Z^α. Then, in this case, the "operator Z^α" is represented by "$\partial/\partial\bar{Z}_\alpha$" and the "operator $\bar{Z}_\alpha$" by "multiplication by the variable $\bar{Z}_\alpha$". The helicity operator becomes:

$$s = \frac{1}{2}\bar{Z}_\alpha \frac{\partial}{\partial\bar{Z}_\alpha} + 1 = \frac{1}{2}\bar{T} + 1 .$$

Wave-functions for many-twistor systems are similar. For example, we can take $f(Z^\alpha_r)$ holomorphic. Then we can calculate the operators for m^2, J etc. (In fact $m^2 = 2 \sum_{i<j} I_{\alpha\beta} Z^\alpha_i Z^\beta_j I^{\rho\sigma} \partial^2/\partial Z^\rho_i \partial Z^\sigma_j$).

For a 1-twistor system the mass m necessarily vanishes and the helicity is the only scalar quantum number. (See table).

Field	Helicity	$\mathbb{T}$-representation homogeneity	$\mathbb{T}^*$-representation homogeneity
self-dual linear-gravity	+2	-6	2
self-dual Maxwell	+1	-4	0
anti-neutrino	$+\frac{1}{2}$	-3	-1
scalar wave	0	-2	-2
neutrino	$-\frac{1}{2}$	-1	-3
anti-self-dual Maxwell	-1	0	-3
anti-self-dual linear gravity	-2	2	-6

For definiteness, let us stick to the $\mathbb{T}$-representation. For a free wave-function in x-space, we have holomorphicity in the forward tube ($\mathrm{Im}(x^a)$ past-timelike), which corresponds, in $\mathbb{T}$-space, to holomorphicity in $\mathbb{T}^+$ - or in $\mathbb{PT}^+$. But (apart from polynomials, which are trivial in this context) there are <u>no</u> homogeneous functions that are globally holomorphic on $\mathbb{T}^+$. What we need is something a little more subtle. We can cover $\mathbb{T}^+$ (or $\mathbb{P}\mathbb{T}^+$) with open sets, say $\mathcal{U}_1, \mathcal{U}_2$, so that there exist such functions on the <u>intersection</u> $\mathcal{U}_1 \cap \mathcal{U}_2$. To obtain a space-time wave-function, we perform a contour integral within the complex line ($\mathbb{C}\mathbb{P}^1$) in $\mathbb{P}\mathbb{T}^+$ that represents the point in the forward tube at which the wave-function is to be evaluated.

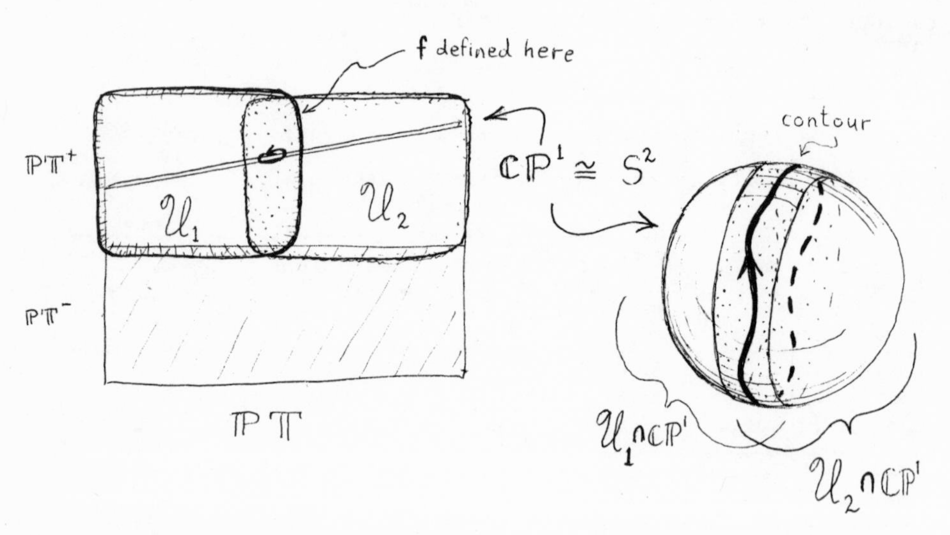

Then it turns out that the massless free-field equations are <u>automatically satisfied</u>.

<u>Case s = 0 (hom. = −2)</u>

$$\psi(x) = \frac{1}{(2\pi i)^2} \oint f(Z) d^2 Z$$

$$= \frac{1}{(2\pi i)^2} \oint f(ix^{AA'}\pi_{A'}, \pi_{A'}) d^2\pi = \frac{1}{2\pi i} \oint f(\ldots)\Delta Z$$

$$\Rightarrow \quad \Box \psi = 0 \ .$$

<u>Case $s = \frac{n}{2} > 0$ (hom. $= -n-2$)</u>

$$\psi_{A'B'\ldots L'}(x) = \frac{1}{(2\pi i)^2} \oint \pi_{A'}\pi_{B'}\ldots\pi_{L'} \, f(ix^{AA'}\pi_{A'}, \pi_{A'}) d^2\pi$$

$$\Rightarrow \psi_{A'B'\ldots L'} = \psi_{(A'B'\ldots L')} \ , \quad \nabla^{AA'}\psi_{A'B'\ldots L'} = 0 \ .$$

Case $s = -\frac{n}{2} < 0$ (hom. $= n-2$)

$$\psi_{AB\ldots L}(x) = \frac{1}{(2\pi i)^2} \oint \frac{\partial}{\partial \omega^A} \frac{\partial}{\partial \omega^B} \cdots \frac{\partial}{\partial \omega^L} f(ix^{AA'}\pi_{A'}, \pi_{A'}) d^2\pi$$

$$\Rightarrow \quad \psi_{AB\ldots L} = \psi_{(AB..L)} , \quad \nabla^{AA'} \psi_{AB\ldots L} = 0$$

The special cases $s = \pm\frac{1}{2}$ give the Dirac-Weyl neutrino equation. When $s = \pm 1$ we get the Maxwell field $F_{ab} (= -F_{ba})$:

$$F_{AA'BB'} = \underbrace{\psi_{AB}\, \epsilon_{A'B'}}_{\text{anti-self-dual (left-handed)}} + \underbrace{\epsilon_{AB}\, \tilde{\psi}_{A'B'}}_{\text{self-dual (right-handed)}}$$

the Maxwell equations $\nabla^a F_{ab} = 0$, $\nabla_{[a}F_{bc]} = 0$ becoming $\nabla^{AA'}\psi_{AB} = 0$, $\nabla^{AA'}\tilde{\psi}_{A'B'}$. When $s = \pm 2$ we get the linearized Einstein curvature field K_{abcd}:

$$K_{AA'BB'CC'DD'} = \underbrace{\Psi_{ABCD}\, \varepsilon_{A'B'} \varepsilon_{C'D'}}_{\text{anti-self-dual (left-handed)}} + \underbrace{\varepsilon_{AB}\varepsilon_{CD}\, \tilde{\Psi}_{A'B'C'D'}}_{\text{self-dual (right-handed)}} .$$

We have $K_{abcd} = K_{[cd][ab]}$, $K_{[abc]d} = 0$, $K^a{}_{bad} = 0$. The "Bianchi identity" $\nabla^a K_{abcd} = 0$, or equivalently $\nabla_{[a}K_{bc]de} = 0$, becomes $\nabla^{AA'}\Psi_{ABCD} = 0$ and $\nabla^{AA'}\Psi_{A'B'C'D'} = 0$.

All these fields enjoy a certain conformal invariance. For rescaling the metric according to $\hat{g}_{ab} = \Omega^2 g_{ab}$, with $\hat{\varepsilon}_{AB} = \Omega\varepsilon_{AB}$, $\hat{\varepsilon}_{A'B'} = \Omega\varepsilon_{A'B'}$, $\hat{\varepsilon}^{AB} = \Omega^{-1}\varepsilon^{AB}$, $\hat{\varepsilon}^{A'B'} = \Omega^{-1}\varepsilon^{A'B'}$, we get preservation of the field equations if $\hat{\psi}_{A\ldots D} = \Omega^{-1}\psi_{A\ldots D}$, $\hat{\tilde{\psi}}_{A'\ldots D'} = \Omega^{-1}\tilde{\psi}_{A'\ldots D'}$ i.e. if $\hat{F}_{ab} = F_{ab}$, $\hat{K}_{abcd} = \Omega K_{abcd}$, etc .

6. Contour integrals and sheaf cohomology

To understand the invariance properties of these contour integrals we need some basic sheaf cohomology theory. Consider first a closed contour integral of a holomorphic function of one variable z. We have:

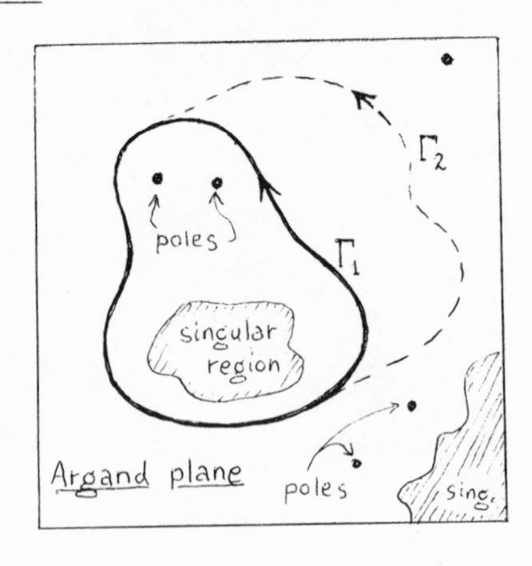

(i) $\oint_{\Gamma_1} f(z)dz = \oint_{\Gamma_2} f(z)dz$

where f is non-singular between Γ_1 and Γ_2 ,

(ii) $\oint_{\Gamma} f(z)dz = \oint_{\Gamma} g(z)dz$

where f - g is non-singular inside Γ .

In the case of the twistor integrals, the Argand plane is replace by the Riemann sphere $\mathbb{CP}^1$. By repeated applications of (i) and (ii) above, we can change the contour and function completely.

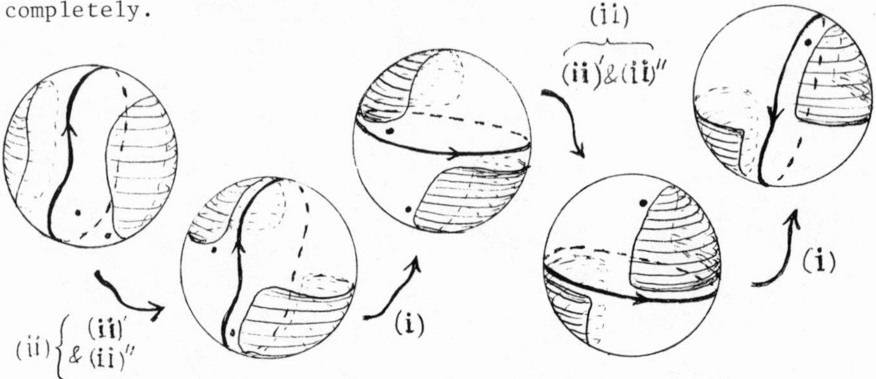

Transformation (ii), on the Rieman sphere S^2, takes the following form:

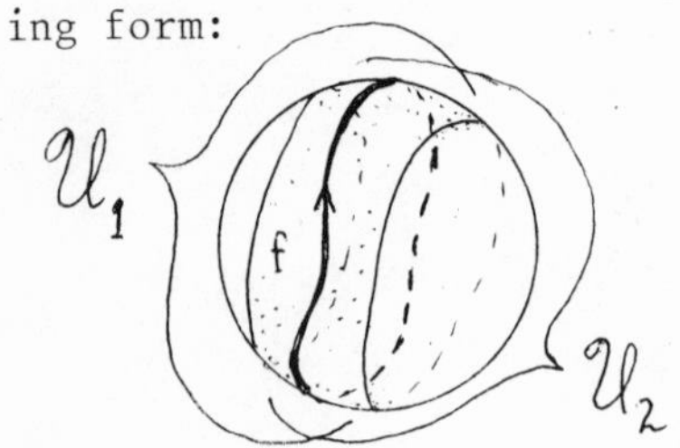

$f \equiv g$ if $f - g = h_2 - h_1$ (with (ii)′ pointing to h_2 and (ii)″ pointing to h_1)

with h_1 regular in $\mathcal{U}_1$, h_2 regular in $\mathcal{U}_2$ where

$\mathcal{U}_1$ and $\mathcal{U}_2$ are open sets which together cover S^2.

But how do we add a function f to a function g, when f and g are defined on different regions? This leads us into (Čech) sheaf cohomology.

Suppose, more generally, that we have some complex n-space Q that is Hausdorff and paracompact. Let $\{\mathcal{U}_i\}$ be an open, locally finite covering of Q. We define a <u>(p-1)-cochain</u> f to be a collection of functions

$$\underbrace{f_{ij\ldots\ell}}_{p} = f_{[ij\ldots\ell]}$$

defined on the various p-fold intersections $\mathcal{U}_i \cap \mathcal{U}_j \cap \ldots \cap \mathcal{U}_\ell = \mathcal{U}_{ij\ldots\ell}$

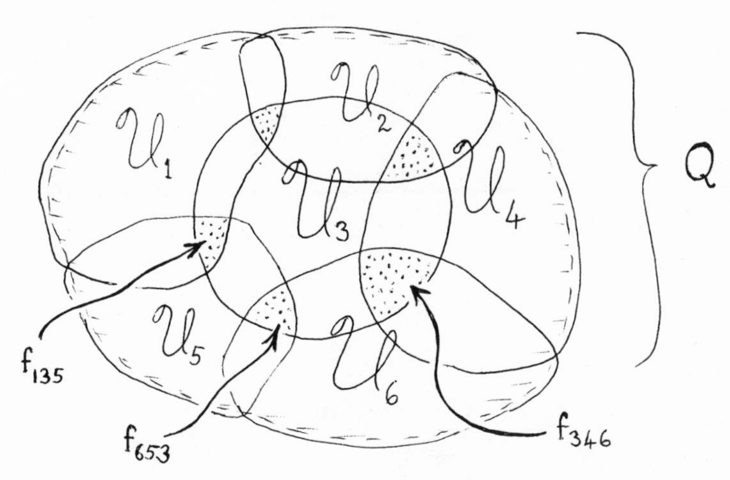

The functions belong to some suitable class (called "sections of a sheaf") defined on open sets of Q. For example, the functions could be variously specified as continuous, or C^k, or holomorphic or they could be local cross-sections of some bundle over Q (e.g. homogeneous functions, regarded as "functions" on $\mathbb{CP}^n$ - "twisted functions"). We define the <u>restriction</u> map ρ_i as follows: for $f_{j\ldots\ell}$ defined on $\mathcal{U}_{j\ldots\ell}$, we write

$\rho_i f_{j\ldots\ell}$ for $f_{j\ldots\ell}$ restricted to $\mathcal{U}_i \cap \mathcal{U}_{j\ldots\ell}$.

This satisfies $\rho_i \rho_j = \rho_j \rho_i =: \rho_{ij}$.

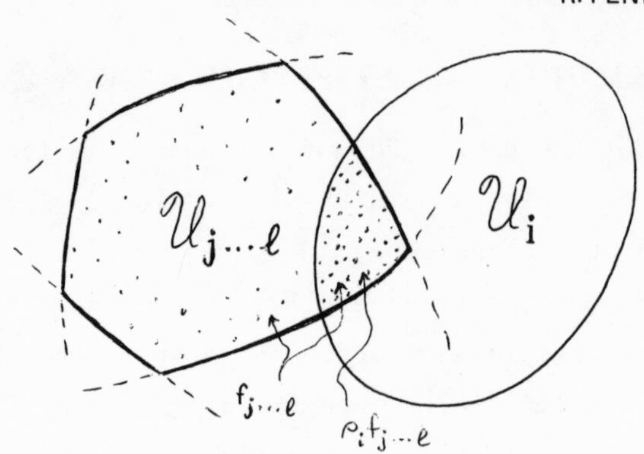

The sheaf properties that our funtions must satisfy are:

(i) if $\rho_i f = \rho_i g$ for all i, then $f = g$

(ii) if $\rho_{[i} f_{j]} = 0$ (i.e. $\rho_i f_j = \rho_j f_i$) for all i,j, then

$f_i = \rho_i g$ for some g.

Provided that these hold, our "functions" can be very general and, indeed, need not be functions in the ordinary sense - just such that an additive group is defined for each open set.

For cohomology, we need the concept of coboundary. Let $f_{j\ldots\ell} = f_{[j\ldots\ell]}$ be a (p-1)-cochain $f = \{f_{j\ldots\ell}\}$ then the coboundary operator $\overset{p}{\delta}$ maps f to a p-cochain

$$\delta f = \{\rho_{[i} f_{j\ldots\ell]}\}$$

We say that f is a cocycle if $\delta f = 0$ and that f is a coboundary if there is a cochain g such that $f = \delta g$. All coboundaries are cocycles since $\delta^2 = 0$ ($\rho_{[i}\rho_{[j} f_{k\ldots m]]} = 0$ since $\rho_{[i}\rho_{j]} = 0$). Then we define the pth cohomology group, with coefficients in the sheaf $\mathcal{S}$ (class of "functions") to be:

$$H^p_{\{\mathcal{U}_i\}}(Q, \mathcal{S}) = \{\text{p-cocycles}\} / \{\text{p-coboundaries}\} .$$

This is the cohomology group with respect to the particular covering $\{\mathcal{U}_i\}$. To remove the dependence on the covering take finer and finer coverings and form the direct limit.

This gives the required $H^p(Q,\mathcal{S})$. In practice, however, this limit usually need not be taken, provided that the covering is fine enough. In our case, $\mathcal{S}$ will be a sheaf of holomorphic functions. Then it is sufficient that each $\mathcal{U}_i$ be a Stein manifold (or domain of holomorphy).

In the special case, p = 1, where Q = $\mathbb{C}\mathbb{P}^1$ and, possibly, twisted $\mathcal{S}$= holomorphic functions, two sets $\mathcal{U}_1, \mathcal{U}_2$ covering Q are sufficient. Any holomorphic function f_{12} defined on $\mathcal{U}_1 \cap \mathcal{U}_2$ defines a cochain (with $f_{12} = -f_{12}$, $f_{11} = f_{22} = 0$). A similar function g_{12} defines the same cohomology group element if, for some h_1 holomorphic on $\mathcal{U}_1$ and h_2 holomorphic on $\mathcal{U}_2$, we have

$$f_{12} - g_{12} = \rho_{[1}h_{2]} = \tfrac{1}{2}(\rho_1 h_2 - \rho_2 h_1) \quad .$$

This is just the kind of equivalence that we encountered before, in connection with closed contour integrals on the Riemann sphere. More complicated coverings lead to

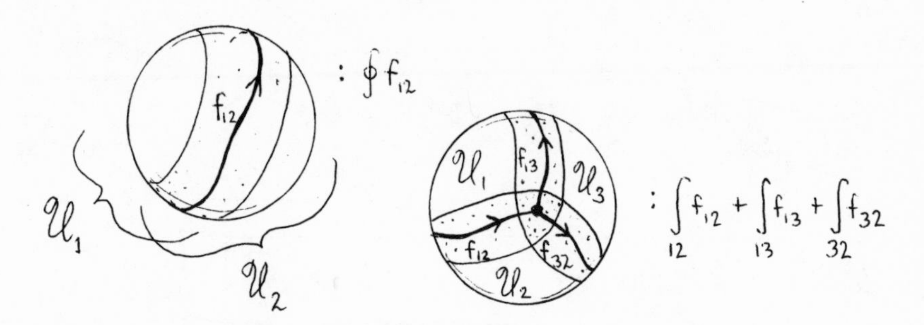

branched contour integrals.

Now, it turns out that a twistor wave-function for a massless free field of helicity s is an element of

$$H^1(\mathbb{P}\mathbb{T}^+, \mathcal{O}(-2s-2)) \text{ or } H^1(\mathbb{T}^+, \mathcal{O}\{-2s-2\}) \quad .$$

means freely holomorphic but "twisted by -2s-2"	means holomorphic & homogeneous deg. -2s-2 ($\Upsilon f = (-2s-2)f$)

A twistor function $f_{ij}(Z^\alpha)$ is a "representative cocycle" with respect to a covering $\{\mathcal{U}_i\}$ of $\mathbb{P}\mathbb{T}^+$ or $\mathbb{T}^+$; $f_{ij} = -f_{ji}$,

$\rho_{[i}f_{jk]} = 0$, and f_{ij} is taken modulo expressions of the form $\rho_{[i}h_{j]}$. The functions f_{ij} are defined on $\mathcal{U}_i \cap \mathcal{U}_j$. The condition $\rho_{[i}f_{jk]} = 0$ is a compatibility requirement on $\mathcal{U}_i \cap \mathcal{U}_j \cap \mathcal{U}_k$ and f_{ij} is taken modulo differences between functions defined on entire $\mathcal{U}_i$'s.

To add two f's defined for different coverings, we need to take a common "refinement". Thus, to form the sum of $\{f_{ij}\}$, for the covering $\{\mathcal{U}_i\}$, and $\{g_{IJ}\}$, for the covering $\{\mathcal{U}'_I\}$, take

$$\rho_{IJ}\, f_{ij} + \rho_{ij}\, g_{IJ} = F_{iJ,jJ}$$

for the covering $\{\mathcal{U}''_{iI}\}$, with $\mathcal{U}''_{iI} = \mathcal{U}_i \cap \mathcal{U}'_I$ ($\rho_{IJ} = \rho_I\rho_J = \rho_{JI}$, etc.). For a simple example of a twistor function, we can take a scalar wave "elementary state" for which

$$f_{12}(Z) = \frac{1}{A_\alpha Z^\alpha B_\beta Z^\beta}$$

where

$$\mathcal{U}_1 = \mathbb{PT}^+ - B\ , \quad \mathcal{U}_2 = \mathbb{PT}^+ - A\ ,$$

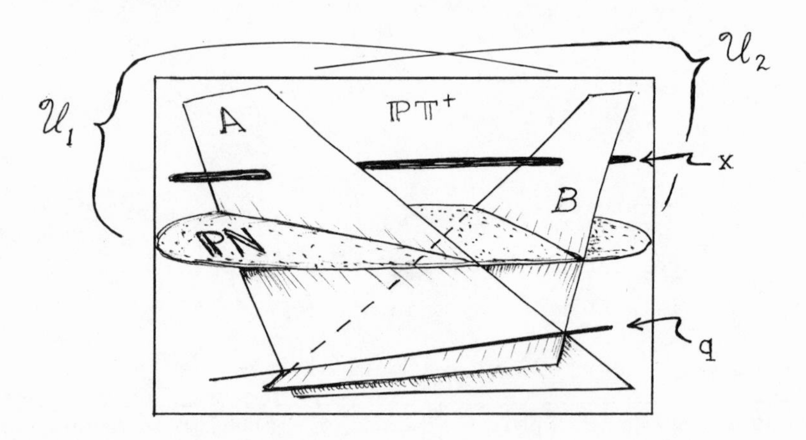

the set B having equation $B_\alpha Z^\alpha = 0$ and A having equation $A_\alpha Z^\alpha = 0$. Writing $A_\alpha = (\alpha_A, a^{A'})$, $B_\alpha = (\beta_A, b^{A'})$ we have, for the space-time field,

$$\psi(x) = \frac{1}{(2\pi i)^2} \oint f_{12}(ix^{AA'}\pi_{A'}, \pi_{A'})d^2\pi$$

$$= \frac{1}{(2\pi i)^2} \oint \frac{d^2\pi}{(i\,\alpha_A x^{AA'}\pi_{A'} + a^{A'}\pi_{A'})(i\,\beta_B x^{BB'}\pi_{B'} + b^{B'}\pi_{B'})}$$

$$= \frac{1}{(i\,\alpha_A x^A_{C'} + a_{C'})(i\,\beta_B x^{BC'} + b^{C'})} = \frac{2}{\alpha^A\beta_A(x^d - q^d)^2}$$

where

$$q^{DD'} = \frac{i\,\alpha^D b^{D'} - i\,\beta^D a^{D'}}{\alpha^A\,\beta_A}$$

This uses

$$\frac{1}{(2\pi i)^2} \oint \frac{d^2\pi}{(\lambda^{A'}\pi_{A'})(\mu^{B'}\pi_{B'})} = \frac{1}{\lambda_{A'}\mu^{A'}}$$

which follows at once from a double application of Cauchy's theorem applied to be variables $\pi_{0'}$ and $\pi_{1'}$. Note that the solution depends only on $B \cap A = q$ (and scaling), not on A and B individually. We can, for example, replace B_α by $\hat{B}_\alpha = B_\alpha + kA_\alpha$ to give

$$g_{12}(Z) = \frac{1}{A_\alpha Z^\alpha\ \hat{B}_\beta Z^\beta},$$

where the covering is now $\{\hat{\mathcal{U}}_1, \mathcal{U}_2\}$ with $\hat{\mathcal{U}}_1 = \mathbb{PT}^+ - B - \hat{B}$.

Then $f_{12} \equiv g_{12}$ since

$$f_{12} - g_{12} = \frac{\hat{B}\cdot Z - B\cdot Z}{A\cdot Z\ B\cdot Z\ \hat{B}\cdot Z} = \frac{k}{B\cdot Z\ \hat{B}\cdot Z}$$

is non-singular on $\hat{\mathcal{U}}_1$.

7. Deformations of complex manifolds

As an example, first consider some simple compact complex 1-manifolds.

Case (i) genus 0 : the sphere S^2. All deformations are trivial. There is only one complex sturcture on S^2.

Case (ii) genus 1 : the torus T^2. Now there are non-trivial deformations, parameterized by 1 complex parameter.

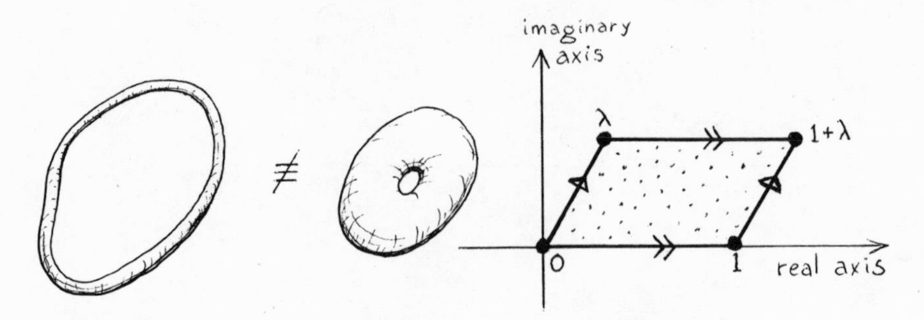

This paramter measures the "shape" of the relevant period parallelogram in the complex plane.

Now consider general complex manifolds. The connection with sheaf cohomology is through infinitesimal deformations. These are described by

$H^1(Q, \Theta)$,

provided that $H^2(Q, \Theta) = 0$ (the "normal" situation) where Θ denotes the sheaf of holomorphic tangent vector fields on Q. To see this (roughly), let V_{ij} be a tangent vector field on $\mathcal{U}_i \cap \mathcal{U}_j$ and "slide" the patches along the vector fields

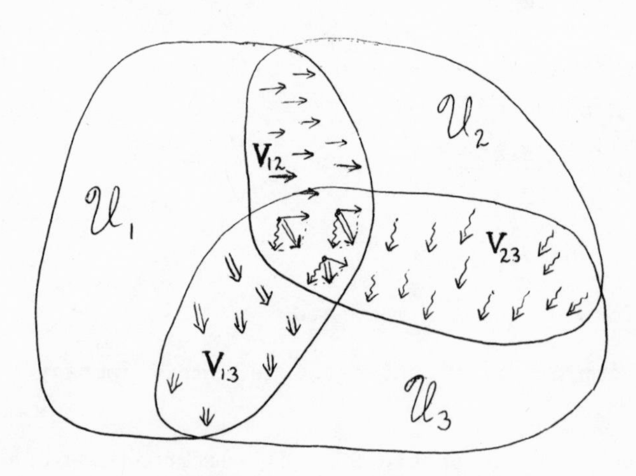

to give new patches. This is "trivial" if $V_{ij} = \rho_{[i}V_{j]}$, since this corresponds simply to sliding the patches in their entirity over Q.

8. Curved Twistor Spaces and Einstein's Equations

There are no nonltrivial deformations of the whole of $\mathbb{CP}^3$, but we can deform $\mathbb{PT}^+$ (or a neighbourhood of a line in $\mathbb{CP}^3$). Does the corresponding "space-time" still exist? Is there still a suitable 4(-complex)-parameter family of "lines?" These "lines" are to be completely characterized by the fact that they are (i) holomorphic, (ii) compact, and (iii) belong to the correct homology class. Fortunately, there are theorems (due to Kodaira) that say YES (so long as the deformation is not too large). Also, the null-cone structure is necessarily quadratic.

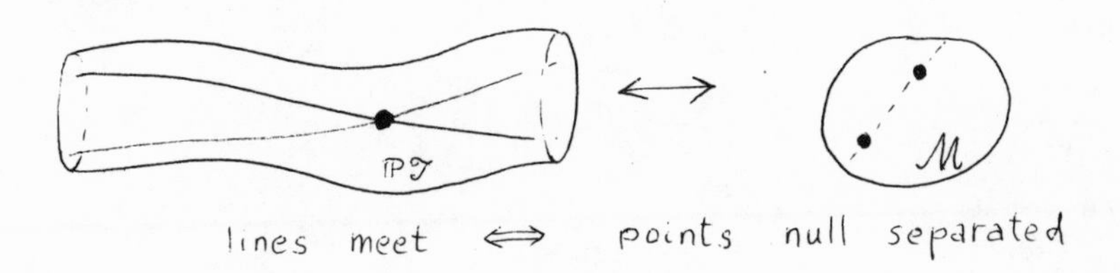

If we fix a point in the (deformed) projective twistor space, we get a 2-parameter family of lines. This is an α-surface. The existence of such a 3-parameter family of α-surfaces implies that the "space-time" is conformally (anti-) self-dual. Thus, in the spinor decomposition of the Weyl curvature tensor C_{abcd}:

$$C_{AA'BB'CC'DD'} = \Psi_{ABCD}\,\varepsilon_{A'B'}\,\varepsilon_{C'D'} + \varepsilon_{AB}\,\varepsilon_{CD}\,\tilde{\Psi}_{A'B'C'D'}$$

we have $\tilde{\Psi}_{A'B'C'D'} = 0$. The Weyl curvature is anti-self-dual in this sense. For the Ricci tensor (which needs a connection to be defined) also to vanish, we have to have a global parallelism for "primed" spinors. This is achieved by having

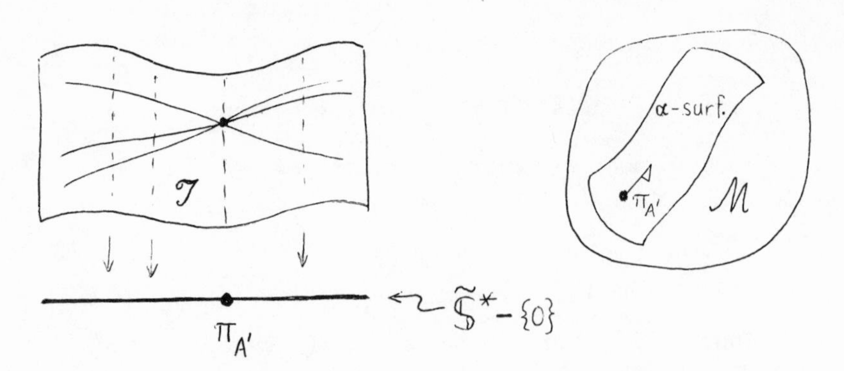

a projection from the deformed twistor space $\mathcal{T}$ to $\tilde{\mathfrak{S}}^* - \{0\}$ ($\pi_{A'}$-space). Then $\mathcal{T}$ has local holomorphic sturcture identical with that for $\mathbb{T}$, namely:

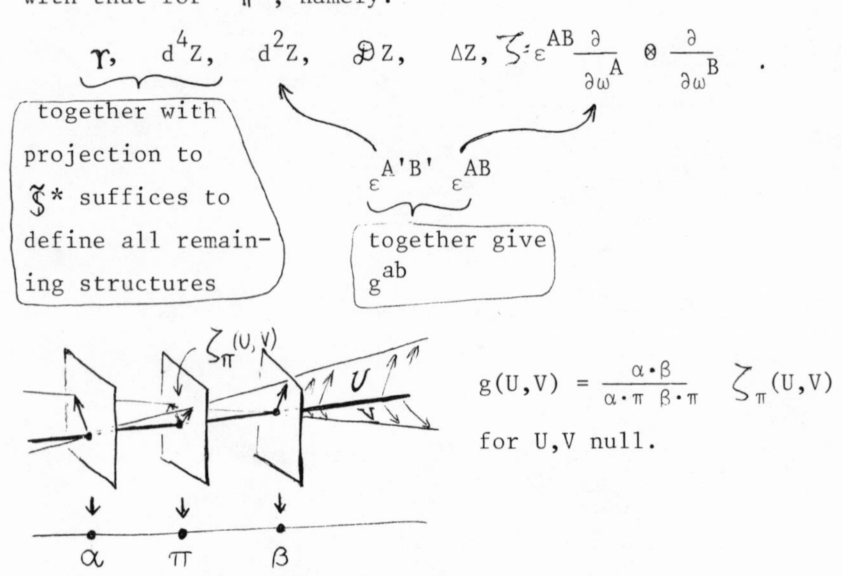

In fact, this construction gives the general anti-self-dual "vacuum space-time" $\mathcal{M}$. The metric for $\mathcal{M}$, and hence also the connection, can be calculated from the structure of $\mathcal{T}$, the quantity $g(U,V) = g_{ab}U^aV^b$ being most easily obtainable when U and V are null vectors in $\mathcal{M}$ (see fig.).

In the case of an infinitesimal deformation of $\mathbb{T}$, we can piece together the space $\mathcal{T}$ out of a number of patches with twistor coordinates Z^α_i, using a twistor function f_{ij}

(representative cocycle for an element of $H^1(\ldots, \mathcal{O}(2))$), so that f_{ij} is homogeneous of degree 2, with $\rho_{[i}f_{jk]} = 0$. Taking ε as infinitesimal ($\varepsilon^2 = 0$) we can express the "patching" as:

$$\rho_j Z_i^\alpha = \rho_i Z_j^\alpha + \varepsilon\, I^{\alpha\beta} \frac{\partial f_{ij}}{\partial Z^\beta}(Z^\alpha)$$

We can then "exponentiate" such a deformation and pass from linear gravity to non-linear Einstein theory, in the anti-self-dual case.

How about the f_{ij}'s of homogeity degree -6? Can they be correspondingly "exponentiated" to give a self-dual "vacuum"? The problem is that the definition of a point of $\mathcal{M}$ and of its conformal structures, in terms of $\mathcal{T}$, imply that $\mathcal{M}$ is anti-self-dual (because they imply that a 3-parameter family of α-planes exist in $\mathcal{M}$). We need a new definition of space-time point in terms of $\mathcal{T}$ in order to obtain a self-dual "vacuum space-time" $\tilde{\mathcal{M}}$. Now recall our exact sequence expressing Poincaré invariance in the case of flat twistor space $\mathbb{T}$:

$$0 \longrightarrow \mathbb{S} \longrightarrow \mathbb{T} \underset{x}{\overset{\longrightarrow}{\leftarrow\text{-}\text{-}\text{-}}} \tilde{\mathbb{S}}^* \longrightarrow 0\,.$$

The original definition for a point of $\mathcal{M}$ may be thought of as a cross-section of the fibration $\mathbb{T} \to \tilde{\mathbb{S}}^*$, i.e. as a map $x : \tilde{\mathbb{S}}^* \to \mathbb{T}$ whose composition with the original projection $\mathbb{T} \to \tilde{\mathbb{S}}^*$ gives the identity on $\tilde{\mathbb{S}}^*$.

$$(\omega^A \pi_{A'}) \longmapsto \pi_{A'}$$

$$(ix^{AB'}\pi_{B'}, \pi_{A'}) \underset{x}{\leftarrow\text{-}\text{-}\text{-}\dashv} \pi_{A'}$$

This holds also when $\mathbb{T}$ is deformed to $\mathcal{T}$. In order to construct $\tilde{\mathcal{M}}$, we must try the dual thing (which, equivalently, is the "complex conjugate" thing).

$$0 \longrightarrow \mathbb{S} \underset{\tilde{x}}{\overset{\longrightarrow}{\leftarrow\text{-}\text{-}\text{-}\text{-}\text{-}}} \mathbb{T} \longrightarrow \tilde{\mathbb{S}}^* \longrightarrow 0$$

We now interpret a point $\tilde{x}$ (of $\tilde{\mathcal{M}}$) as a map $\tilde{x}\colon \mathbb{T} \to \mathbb{S}$ which, when applied to the image of the injection $\mathbb{S} \to \mathbb{T}$ yields the identity on $\mathbb{S}$:

$$\omega^A \longmapsto (\omega^A, 0)$$

$$\omega^A - i\tilde{x}^{AB'}\pi_{B'} \overset{\tilde{x}}{\longleftarrow\!-\!-\!-\!-} (\omega^A, \pi_{A'}) .$$

For a "deformed" $\mathcal{T}$ and curved $\tilde{\mathcal{M}}$, this requires a suitable regularity condition at I so that the appropriate set of $\tilde{x}$ - maps can be singled out. But I can't be just "regular" in the ordinary sense or the $\tilde{x}$-maps would simply provide a flat $\tilde{\mathcal{M}}$. The most hopeful suggestion, so far, for coding the information of the curvature of $\tilde{\mathcal{M}}$ into the structure of $\mathcal{T}$ appears to be to allow $\mathcal{T}$ to have certain non-Hausdorff properties at I, so as to impose appropriate regularity conditions on the $\tilde{x}$-maps. But the matter remains unresolved, so far.

Now suppose we are given an f_{ij} cocycle on $\mathcal{T}$ (with homogeneity degree ≤ -2), then we can use the x-maps to define a massless field on $\mathcal{M}$. This may be done by just using the old contour integral definition on $x[\tilde{\mathbb{S}}^*]$. The resulting field correctly repsonds to $\mathcal{M}$'s curvature. Can we use $\tilde{x}$-maps to define massless fields on $\tilde{\mathcal{M}}$ (f_{ij} having homogeneity degree ≥ -2)? Apparently we can (assuming the $\tilde{x}$-maps are given). We need a canonical H^1 2-form Ω_{ij} of homogeneity 0 on I ($\mathcal{U}_1$ being $\omega^1 \neq 0$ and $\mathcal{U}_2$ being $\omega^0 \neq 0$):

$$\Omega_{12} = \frac{d\omega^0 \wedge d\omega^1}{\omega^0 \omega^1} .$$

This pulls back, by $\tilde{x}^{-1}$, to an H^1 2-form, denoted $\Omega_{ij}(\tilde{x})$, on $\mathcal{T}$. Also $\partial/\partial\omega^A$ is well-defined since a flat structure is induced on the constant $\pi_{A'}$ fibres by the $\tilde{x}$-map. We then evaluate

$$\frac{1}{(2\pi i)^4} \oint \frac{\partial}{\partial\omega^A} \cdots \frac{\partial}{\partial\omega^C} f_{[ij} d^2Z \wedge \Omega_{k\ell]}(\tilde{x}) .$$

Assuming that the construction of $\tilde{\mathcal{M}}$ can be carried out satisfactorily, a natural next step would be to construct vacuum space-times that are neither anti-self-dual nor self-dual (and preferably real). One may anticipate some "average" or "combined" x-map and $\tilde{x}$-map to yield a general solution of Einstein's vacuum equations. The hope would be that starting (for example) from an asymptotically simple, singularity-free real analytic vacuum space-time $\mathcal{V}$, we could construct its future null infinity $\mathbb{C}\mathscr{I}^+$, and then its asymptotic twistor space $\mathcal{T}$. Newman's $\tilde{\mathcal{H}}$-space would then be $\mathcal{M}$ and $\mathcal{H}$-space should be $\tilde{\mathcal{M}}$. The combined construction would be supposed to yield $\mathcal{V}$ back again. Some further ideas would also be needed in order to incorporate sources for the gravitational field.

References and Bibliography

Curtis, G.E., Proc. Roy. Soc. A359 (1978) 133.

Curtis, G.E., Gen. Rel. Grav. 9 (1978) 987.

Curtis, W.D., Lerner, D.E. & Miller, F.R., Gen. Rel. Grav. 10 (1979) 557.

Hansen, R.O. & Newman, E.T., Gen. Rel. Grav. 6 (1975) 361.

Hansen, R.O., Newman, E.T., Penrose, R. & Tod, K.P., Proc. Roy. Soc. A363 (1978) 445.

Hitchin, N.J., Math. Proc. Camb. Phil. Soc. 85 (1979) 465.

Hughston, L.P., Twistors and Particles (Springer Lecture Notes in Physics 97, Springer-Verlag, Berlin 1979).

Hughston, L.P. & Ward, R.S. (Eds.), Advances in Twistor Theory (Pitman Research Notes in Mathwmatics 37, Pitman Press, San Francisco 1979).

Isham, C.J., Penrose, R. & Sciama, D.W., (Eds.), Quantum Gravity: An Oxford Symposium (Clarendon Press, Oxford 1975).

Lerner, D.E. & Sommers, P.D. (Eds.), Complex Manifold Techniques in Theoretical Physics (Pitman Research Notes in Mathematics 32, Pitman Press, San Francisco 1979).

Penrose, R., J. Math. Phys. 8 (1967) 345.

Penrose, R., Int. J. Theor. Phys. 1 (1968) 61.

Penrose, R., J. Math. Phys. 10 (1969) 38.

Penrose, R., Gen. Rel. Grav. 7 (1976) 31.

Penrose, R., Repts. Math. Phys. 12 (1977) 65.

Penrose, R. & MacCallum, M.A.H., Phys. Repts. 6C (1973) 241.

Perjés, Z., Phys. Rev. D11 (1975) 2031.

Perjés, Z., Repts. Math. Phys. 12 (1977) 193.

Tod, K.P., Repts. Math. Phys. 11 (1977) 339.

Tod, K.P. & Perjés, Z., Gen. Rel. Grav. 7 (1976) 903.

Tod, K.P. & Ward, R.S., Proc. Roy. Soc. A368 (1979) 411

Ward, R.S., Phys. Lett. A61 (1977) 81.

Ward, R.S., Proc. Roy. Soc. A363 (1978) 289.

Wells, R.O. Jr., Bull. Amer. Math. Soc. (New Ser.) 1 (1979) 296

Woodhouse, N.M.J., in Group Theoretical Methods in Physics (Eds. Janner, A., Jansen, T. & Boon, M., Springer Lecture Notes in Physics 50, Springer-Verlag, Berlin 1976).

EXPERIMENTAL GRAVITATION WITH MEASUREMENTS MADE FROM WITHIN A PLANETARY SYSTEM

Robert D. Reasenberg

Department of Earth and Planetary Sciences
Massachusetts Institute of Technology
Cambridge, Massachusetts 02139

PART I

These notes are organized into three parts and roughly follow the organization of my four lectures. The first two parts deal with solar-system tests of relativity that have been or are currently being conducted at MIT by the radio physics group: 1) solar-system data and the corresponding mathematical models; 2) analysis, estimation, and results. The third part deals with three possible future missions: The Solar Probe is a proposed NASA mission to fly within a few solar radii of the Sun; VOIR is a Venus radar mapping mission that may provide high-quality Earth-Venus delay measurements; POINTS, Precision Optical INTerferometry in Space, is an instrument concept in an early stage of development at MIT and the Charles Stark Draper Laboratory. POINTS is intended to measure the second-order bending of light by solar gravity.

Introduction

The solar system is a highly inappropriate laboratory for testing relativity. Its one redeeming feature is that no better laboratory has yet been found. The solar potential at the limb of the Sun is 2×10^{-6} when expressed in the usual dimensionless units in which $G = c = 1$. At Earth's orbit, the potential is 10^{-8}. The solar system is wondrously complex and filled with myriad bodies, the vast majority of which serve only to complicate the task of those who would use this laboratory to test theories of gravitation. When we consider the "controlled environment" of a terrestrial laboratory, the situation looks less promising, in part because the gravitational coupling constant is exceedingly small. The terrestrial laboratory is hostile to sensitive mechanical experiments. Airborne and mechanically transmitted accoustic signals must be filtered out and the local gravitational field varies as a result of numerous effects ranging from the rise and fall of the ground-water table to the close passage of massive bodies such as laboratory technicians. In contrast to the electromagnetic case, gravitational shielding is not possible. Thus the inability to isolate from disturbance the apparatus of an experiment makes testing theories of gravitation difficult near the surface of the Earth.*

Another possible laboratory for gravity research is the pulsar in a binary system, PSR1913+16. This is a remote and as yet not well understood system in which gravitational interaction is strong. In principle, observations of this system should permit higher order tests of theories of gravitation. Important results have, and I expect will continue to come from the analysis of observation of this system. However, only one such system has been observed and only a single observation tool is available, the precise timing on Earth of the arrival of pulses. For these reasons a reliable description of the system is difficult to achieve and, as is the case for the solar system, many significant parameters will be hard to estimate. Thus, we are not free to abandon the solar-system laboratory. It is accessible, it has been observed for a long period of time, and its basic features are well understood. Observations are made from within the system, thus breaking many types of geometric degeneracy that would plague observations were they made from far away. Finally, numerous types of observations are possible, further breaking degeneracies. Thus for now, as in the past, the solar system is an important laboratory for testing theories of gravitation.

In the classical physics experiment, an apparatus is assembled which allows a single physical quantity to be measured. Similarly, astronomers studying the dynamics of the solar system have classically chosen a set of data which pertain particularly to one or at most a few parameters of their solar-system description. Such an approach was necessary in the era before high-speed digital computation, and is aesthetically appealing today although not always practical. The experiments that I will describe provide a marked contrast to this classical procedure, since in our analyses we may simultaneously estimate hundreds of parameters. Whereas the classical astronomer made selective use of data particularly sensitive to a quantity of interest, we gain a synergistic advantage by combining a wide variety of data in a single-solution analysis. This

*Considerable progress has been made in isolating some kinds of gravity experiments. Detectors of kHz gravity waves are accoustically isolated. Some experiments have been proposed in which isolation is achieved by rotating the instrument at a few revolutions per minute. (Ritter and Beams, 1978.)

approach requires that we work with a global model of the solar system.

Our model is based on the weak-field approximation to general relativity and includes terms to first order in the post-Newtonian correction. The system of units is based on four defined quantities: 1) the mass of the Sun, 2) the Gaussian value for the gravitational constant, 3) the AU (astronomical unit), and 4) the day of 86,400 (ephemeris) seconds. The speed of light in AU/s is an estimated quantity while its engineering value, $c = 2.99792458 \times 10^5$ km/s, is used to define the kilometer.

The ephemerides of solar-system bodies (Earth, Moon, planets and their satellites) are obtained by the numerical integration of the appropriate differential equations. A numerical technique is chosen because, for the required level of accuracy, this is less expensive to formulate and use than would be a comparably accurate analytic expansion. By contrast, the precession and nutation of Earth is adequately described by Wollard's (1953) classical analytic series. To make use of this, we numerically evaluate that series at intervals of one day and work with the tabulated numerical values in machine readable form.

Our solar-system model is parametric and contains three classes of quantities that are estimated through the analysis of data: (1) relativistic quantities including some Parametrized Post-Newtonian (PPN) parameters (see, for example, Will, 1973) and the *ad hoc* coefficients of some special effects such as the postulated secular variation in the gravitational constant and a possible violation of the principle of equivalence for massive bodies, (2) classical astronomical quantities such as planetary masses, orbital initial conditions, libration coefficients and the speed of light, (3) nuisance parameters including coefficients for models of the topography of the inner planets, locations of observing stations, engineering bias parameters, and the density of plasma around the Sun. This solar-system model is realized in a computer code (Ash, 1972) known as the Planetary Ephemeris Program (PEP) which is used in the analysis of a wide variety of astrometric data types.

Radar

The first of the modern astrometric techniques to become available was planetary radar. In 1946, the first radar reflection was received from the Moon using techniques developed during the Second World War. In 1967, the state of the art had advanced to the point where planetary radar observations taken in that year were of sufficient quality to still be important today.

The ability to accurately measure the round-trip propagation time of radar signals depends in part on the signal-to-noise ratio, SNR = P_r/P_n. The power received, P_r, is proportional to the power transmitted, P_t, the transmitting gain, G, the effective receiving antenna area, $A = \kappa A_o = \lambda^2 G/4\pi$, and inversely proportional to the path loss, $L = \sigma/(4\pi r^2)^2$, where λ is the radar wavelength, A_o is the geometric area of the receiving antenna, r is the distance from the radar to the target, and σ is the radar cross-section of the target. The noise power, P_n, depends on the receiver bandwidth and the effective temperature of the contributing components.

$$P_n = kT\Delta f \tag{1}$$

$$T = T(\text{receiver}) + T(\text{target}) + T(\text{galactic}) \tag{2}$$

where $k = 1.380 \times 10^{-23}\ J(^{o}K)^{-1}$ is Boltzman's constant and Δf is the receiver's effective bandwidth.

For a radar to measure round-trip propagation time, it must transmit a modulated signal. The modulation is often binary (e.g., on/off; phase-shifted/unshifted). The modulation is extracted from the weak echo signal and cross-correlated with a replica of the original code. For binary modulation, the shortest time interval between successive possible changes of code is called the baud length. Figure 1 shows a modulation sequence and a corresponding auto-correlation. Since the auto-correlation function is repetitive the modulation sequence must be made longer than the *a priori* uncertainty in the time delay to be measured.

Only a few radars are suitable for planetary observations. One of these is at the Arecibo Ionospheric Observatory in Puerto Rico. The characteristics of this radar are shown in Table 1. The baud length shown in column 2 sets the scale for the accuracy of the time-delay measurement. An important use of the planetary radar is the study of planets *per se*. For a review of this subject, see Pettengill (1978).

Figure 2 shows the geometry of the radar scattering from a planet. The first point contacted by a wave is called the subradar point. As the wave passes beyond the subradar point, it contacts the planet along a circle of growing diameter. As the wave advances along the planet, the angle of incidence changes as does the ratio of retroreflected power to impinging power, σ_o.

$$\sigma_o(\theta) = \frac{C\rho_o}{2}(\cos^4\theta + C\sin^2\theta)^{-3/2} \tag{3}$$

MAXIMUM LENGTH SHIFT REGISTER SEQUENCE

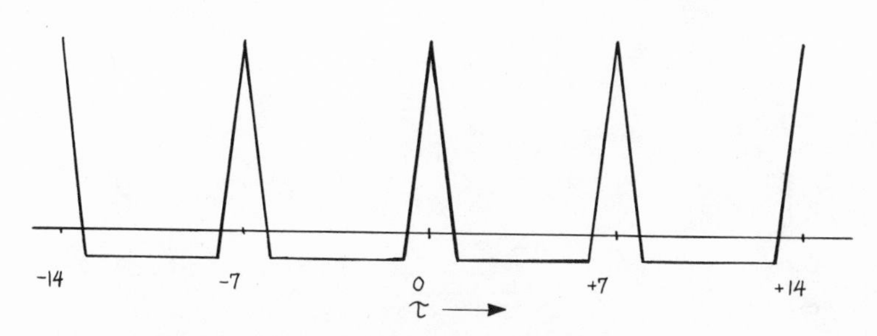

AUTOCORRELATION FUNCTION OF SEQUENCE

Fig. 1. A modulation sequence and its autocorrelation (from Evans, 1968).

Table 1. Characteristics of the Arecibo Planetary Radar

Band*	t (μs)	$D^{\dagger}$(m)	G (db)	T_R (^oK)	$\bar{P}$ (kW)$^{\Psi}$
U	6 - 10	305	61	100	150 (P)
S	1 - 2	210	72	45	400 (CW)

*UHF ∿ 0.4 GHz, S-band ∿ 2.1 GHz

†Diameter illuminated

ΨPulsed or Continuous Wave as indicated

$$\rho_o = (\sqrt{\varepsilon} - 1)^2/(\sqrt{\varepsilon} + 1)^2 \tag{4}$$

where ρ_o is the Fresnel coefficient and ε is the dielectric constant of the planet's surface. When C, the Hagfors constant, is large (say C > 1000), then $C^{-1/2} \sim \gamma \equiv$ rms slope of surface, where the slope is calculated after smoothing irregularities smaller than about 10λ and subtracting the mean slope on about a 1 km scale.

For a spinning target, the rotation produces a Doppler broadening in the returning signal. Regions of a planet producing the same

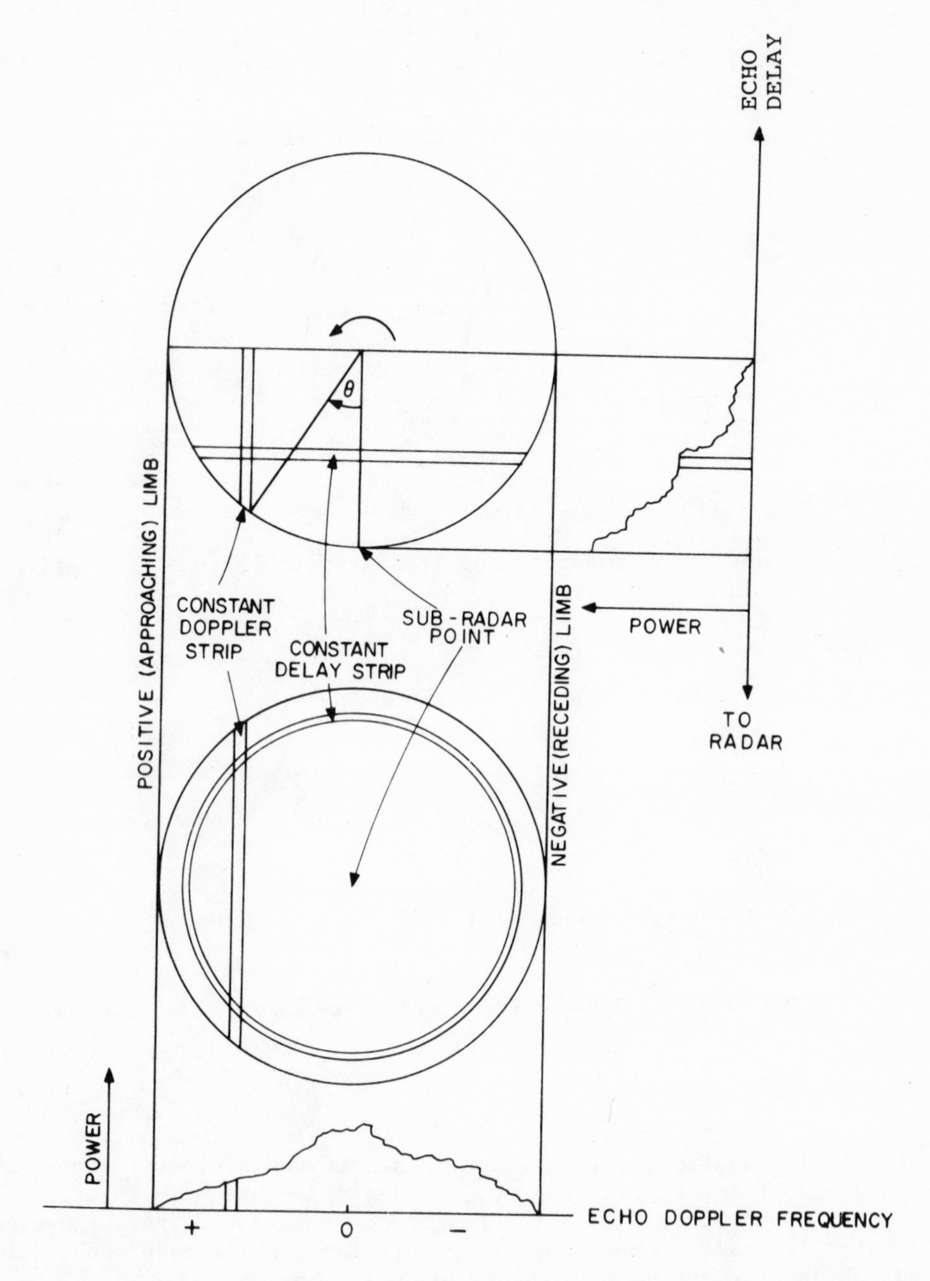

Fig. 2. Delay and Doppler resolution of a spherical target (from Ostro, 1978).

Doppler shift from a straight line that is parallel to the spin vector as viewed from the radar. The combination of delay and Doppler spreading complicates and degrades the detection of the radar signal. (See Evans and Hagfors, 1968.)

The remaining portion of this section deals with the application of radar to the study of solar-system dynamics. Figure 3a is a clear example of the principal problem that one encounters in this application, topography. To be specific I will discuss the topography of Mars, but similar problems are found in the use of radar observations of other planets. The largest topographic constructs are of continental proportions. The corresponding vertical relief is of the order of 5 km which corresponds to 35 μs of round-trip propagation time. This is large compared to submicrosecond uncertainty in the radar time delay measurement.

How can one mitigate the deleterious effects of topography on the study of solar-system dynamics? Several solutions to this problem have been tried; many of those involve a model of the topography. The first such model was a triaxial elipsoid having 9 free parameters. The second model used surface harmonics. Our current topographic model uses a two-dimensional Fourier transform and has up to 123 free parameters.

Figure 3b shows the unmodeled portion of the topography of Mars in the region corresponding to the topographic display in Figure 3a. The model used was the two-dimensional Fourier transform. From a comparison between the figures, it can be seen that the model has removed the gross topographic effects but that many significant topographic constructs have not been adequately modeled, particularly at the level of 15 m which corresponds to the 0.1 μs precision available with the best observations.

Thus even the Fourier transform model is inadequate to properly represent the surface of Mars. Because in this model each coefficient bears on the topography over the entire surface of the planet, it is undesirable (i.e., computationally burdensome) to increase the dimensionality above its current value. Therefore, to gain higher resolution, it is necessary to introduce a local model. Such a model has been incorporated in PEP. A grid of points is established on the observed, equatorial region of the planet. The model height at a given point is a linear combination of the heights at the surrounding four grid points. Our current plans call for using this model with between 500 and 1000 points on the surface of each of the inner planets.

The above deterministic models may be supplemented by a statistical treatment of the unmodeled components of the topography. For example, if two observations are made of the same target area of a planet but at widely separated times, then the topographic contribu-

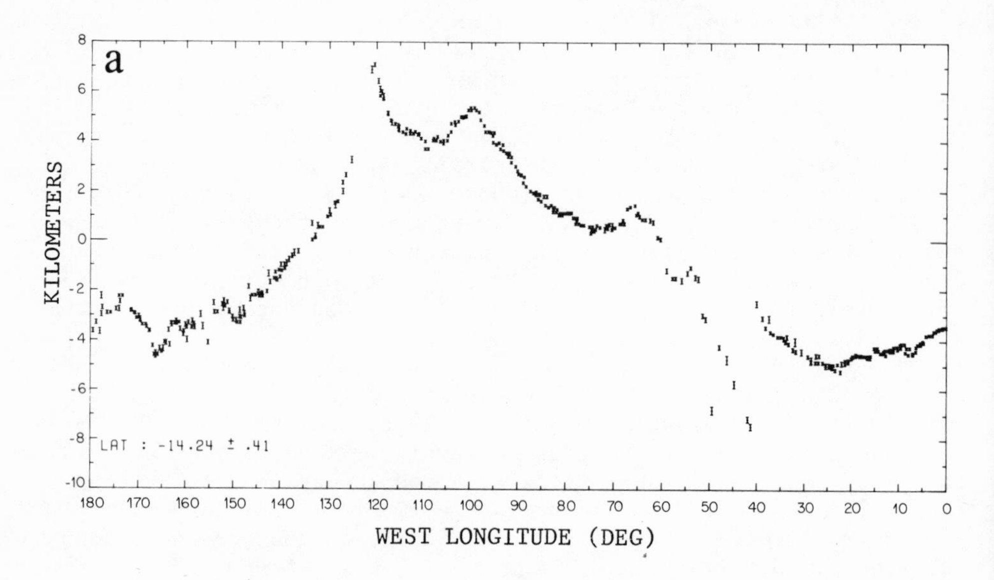

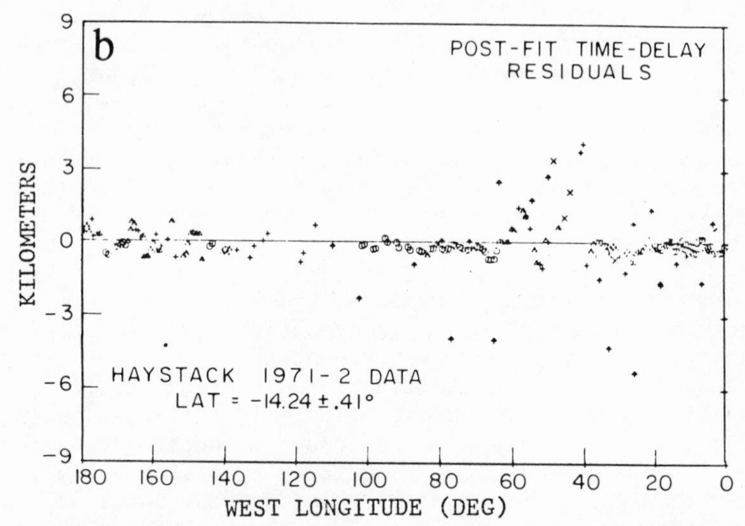

Fig. 3. Radar residuals showing the large-scale topography of Mars. (a) With no topographic model (from Pettengill *et al.*, 1973) (b) With the 123 parameter Fourier transform model.

tion to these two observations would be similar. (The topography would be identical; the topographic contribution to the radar delay could differ if the radar characteristics, such as the "foot print," were not the same for the two observations.) Such pairs of observations are known as closure points. If we treat the unmodeled component of the topography as noise, then the two observations of a closure point have correlated noise. A statistically more likely situation is for a pair of observations made at widely separated times to be of target areas that are close to each other. To estimate the corresponding statistical correlation, one must assume that the surface topography has a stationary auto-correlation and estimate that auto-correlation. Figure 4 shows such a surface correlation function empirically derived for Mars by Steve Brody (1977). It can easily be shown that the usual weighted-least-squares estimator can be generalized to include such correlations. In this generalization, the noise covariance matrix must be formed and inverted. This matrix has a linear dimension equal to the number of observations to be analyzed. For Mars, there are 25,000 observations. Thus, the matrix formed is prohibitively large and cannot be inverted at a reasonable cost with current computer technology.

To circumvent these economic problems, Brody and I developed a heuristic algorithm for finding pairs of near closure points. By including in his estimator the correlations in these pairs, Brody was able to realize a 20% reduction in the standard deviation of many of the interesting estimated parameters. The next logical step is to extend this work from pairs of points to clusters which naturally contain between 50 and 150 points. Because of limited resources and alternate, highly attractive opportunities, this approach may never be taken.

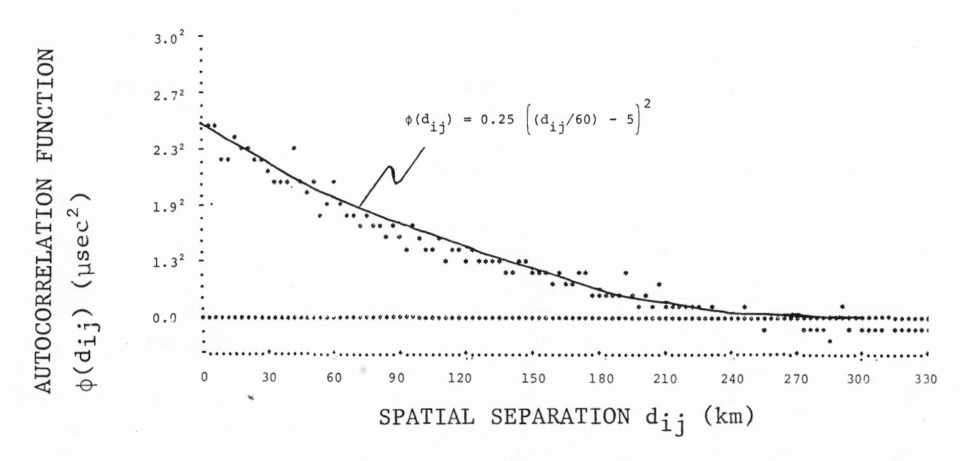

Fig. 4. Empirical surface correlation function for Mars, showing a parabolic fit to the auto-correlation histogram (from Brody, 1977).

Radio Observations of Spacecraft

Radio observations of spacecraft have many attributes and problems in common with radar observations of planets. Both techniques permit an interplanetary round-trip propagation time to be measured with an uncertainty substantially under 1 μs. The contribution of the interplanetary plasma to the round-trip propagation time confounds the interpretation of both kinds of measurements. Although there is no topographic problem per se for spacecraft measurements, the corresponding problem is the determination of the location of the spacecraft.

Interplanetary spacecraft are tracked by means of two complimentary radio observables. The radio signal Doppler shift, or phase rate observable, is used principally for determining the spacecraft orbit. The corresponding radar observable is generally not used for solar-system dynamics. The round-trip propagation time, or group delay observable, is useful for studies of solar-system dynamics.

The Doppler portion of the tracking system comprises four principal components: (1) transmitter, (2) transponder (spacecraft receiver and transmitter), (3) receiver, and (4) extractor-resolver. The design of this system is nearly optimal for tracking under ordinary conditions for which the signal-to-noise ratio is large and the phase jitter is small. The phase-coherent electronic components of this system permit an accurate Doppler-shift cycle count to be made and the fractional cycle resolution is to a few degrees of electrical phase. However, under the poor conditions encountered when, for example, the tracking signals pass near the Sun, the system performance is far below optimal. This degraded performance adversely impacts the time-delay test of relativity that is discussed in Part II.

The tracking stations use parabolic dish antennas that range from 26 to 64 m in diameter. The signals sent from these stations are at S-band (2.1 GHz) and some of the stations are equipped to receive returning signals both at S-band and at X-band (8 GHz).

Spacecraft time-delay measurements, like those made with radar, require that modulation be applied to the signal. The current spacecraft ranging system uses phase modulation by a series of single-period squarewave ranging codes. The highest frequency code, which is typically about 500 kHz, is used to make the precision measurement of time delay. That measurement has an ambiguity equal to the code's period. That ambiguity is resolved with the aid of a series of codes, each having a period of twice that of the preceding code. The ambiguity resolving codes need to be transmitted for only a brief period of time since they are each used to make a single binary decision. In contrast, the highest frequency code must often be transmitted for several minutes in order that the signal-to-noise of the accumulated returning signal be sufficiently high to permit a high-precision determination of the round-trip propagation time. Although the spacecraft ranging subsystem depends implicitly on the Doppler tracking system, the former is often able to produce useful data even when the latter is malfunctioning due to poor signal conditions.

The interplanetary plasma contributes to the time delay and Doppler observables in a way that is particularly important when the signals pass close to the Sun. This plasma contribution must be compensated in order to make full use of the precision of the observables. There are three standard methods for estimating the plasma contribution: (1) The radio systems of the more recent interplanetary spacecraft have contained a subsystem that transmits an X-band signal which is phase coherent with the S-band signal being transponded. When ranging modulation appears on the dual-frequency downlink signals, the plasma contribution to the delay can be estimated from the frequency dependence of that plasma contribution. (2) Dual-frequency downlink Doppler signals can be used to determine the rate of change of the plasma contribution to the (phase) delay. The corresponding change in the group delay is of the same magnitude but of opposite sign. (3) By comparing integrated Doppler shift with the difference between two delay observations, it is possible to determine the change in the plasma contribution during the interval between delay observations. DRVID (Differential Range _Versus_ Integrated Doppler) provides a measure of the change in plasma delay when only a single frequency band is in use.

The problem of determining spacecraft location is best considered separately for each of three classes of spacecraft. We first consider interplanetary probes. These marvelous vehicles have a mass of the order of 10^6 g and an area-to-mass ratio of the order of $0.1\ cm^2/g$ (cf, Earth at $0.2 \times 10^{-9}\ cm^2/g$). Spacecraft often carry attitude-control and propulsion systems, both of which are capable of leaking small quantities of volatile material. The resulting thrust, along with solar radiation pressure and solar wind drag, causes a significant non-gravitational acceleration of the spacecraft. Although it is possible to model such accelerations, the models are notoriously unreliable and many components of the acceleration are stochastic in nature. Thus the trajectory of an interplanetary probe is a random walk. Least-squares fitting and other common estimation techniques are not applicable to the analysis of such a trajectory. The orbit determination problem is often further complicated by the poor observation geometry which leads to degenerate estimators.

A spacecraft in orbit around a distant planet provides a more reliable target, even though the model errors and stochastic accelerations described for an interplanetary probe affect such an orbiter. The effects of model errors and stochastic accelerations can be detected on a time scale comparable to the orbital period of the body being observed. For interplanetary probes, that orbital period is of the order of a year; for a typical planetary orbiter, the period is the order of a day. Thus, for the planetary orbiter, a greatly restricted amount of time is available for the undetected buildup of error due to random motion.

There are, however, further complications associated with the tracking and analysis of the orbit of a planetary orbiter. The gravitational potential of the host planet may not be well enough known *a priori*. In the case of Mars, the non-spherical part of this potential is more important than for Earth or the Moon. Thus, even in the absence of non-gravitational forces, the trajectory of a planetary orbiter may be highly perturbed. It requires a massive data processing effort to determine accurately the potential of the host planet, assuming that a suitable data set exists, which need not be the case.

Additional radiation pressure results from light reflected or re-radiated from the host planet. For Mariner 9, which was placed in orbit around Mars, this radiation pressure had a peak value of approximately 0.1 times that of the direct solar pressure, but was highly variable. Since this component of radiation pressure depends in part on the albedo markings of the host planet, it is an extremely difficult acceleration to model. For spacecraft that pass near the surface of a planet, atmospheric drag may become an important consideration. For some preicenter passages of the Pioneer Venus Orbiter, atmospheric drag causes more than a 10 s change in the approximately one day period of the spacecraft.

Time-delay observations of an orbiter may be applied to the study of solar-system dynamics only to the extent that the planetocentric spacecraft orbit may be determined. In order to reduce the overall computational burden associated with the analysis of such observations, each orbiter time-delay observation is converted to an equivalent Earth-planet center-of-mass to center-of-mass time delay. Such pseudo-observations are called time-delay normal points. Their determination represents a massive computational effort which is complicated by all of the problems associated with spacecraft orbit determination. Orbiter normal points typically have errors that are dominated by the contribution from the spacecraft orbit determination.

A third kind of spacecraft is the planetary lander. The planetocentric orbit of such an object is relatively easy to describe. While lander "orbit determination" poses some special difficulties, the problems encountered are at a much finer scale than the corresponding problems for an interplanetary probe or planetary orbiter. Geophysical effects are believed to contribute no more than a few tens of meters to the location of the Viking landers. The more important of the effects that have been postulated make slowly varying contributions to the lander's position. These effects admit of a simple parametric representation.

The longitude and distance from the planetary spin axis of a lander are easily determined because they contribute to the observable a distinguishable signature with period equal to the apparent rotational period of the planet. However, the distance of the lander from the planet's equatorial plane produces no such distinct signature. The contribution to the time-delay observable from this "*z* coordinate" varies on the time scale with which the Earth-planet line of sight rotates with respect to the direction defined by the ascending node of the planet's equator on the ecliptic. This time scale is comparable to that of the planetary orbits and, thus, the corresponding contribution is not easily separated from orbital effects. In the case of the Viking lander, it is the orbiter normal points that provide the additional information necessary to break the degeneracy just described. The accuracy of the determination of the lander's *z* coordinate is thus dependent on the accuracy of the orbiter normal points and almost totally independent of the accuracy of the lander observations, *per se*.

Radio Interferometry

There are two principal types of radio interferometers. In the operation of each type, a radio signal from a distant source is sampled at two ends of a "baseline." In optical interferometry, the samples would be brought together to form fringes. In radio interferometry, the samples are brought together and cross-correlated in order to determine the relative phase and delay. A conventional radio interferometer uses cables to bring the signals from the antennas to the cross-correlator, thus the cross-correlation can (and usually must) be performed in real time. Very Long Baseline Interferometry (VLBI) is performed with antennas too far apart to permit cable connections. Each antenna is equipped with a high-stability oscillator that is used to convert the incoming signal to baseband. The resulting video signal is recorded on magnetic tape and the magnetic tapes are later brought together and cross-correlated. For a review of radio interferometry, see Counselman (1973, 1976).

Radio interferometry can be used for high-precision astrometry including the light bending test of general relativity. Other astrometric studies have included the location of the ALSEP packages on the Moon and the determination of the angular separation between a quasar and a spacecraft in planetary orbit. The latter study is intended to determine the orientation of the solar system and may soon provide bounds on the rate of the possible anomalous rotation of that system. Interferometry can also be used to make brightness maps of radio luminous objects such as quasars. (See, for example,

Fomalont and Wright, 1974.) These brightness maps have provided clues to the structure of many astrophysically interesting objects.

As with other radio observation techniques, there are both phase delay and group delay observables. Even for a moderately weak source, differential phase delay can be determined with an uncertainty of a few centiradian of electrical phase, modulo 2π. As Earth rotates, the differential phase changes, providing information from which the source position is ultimately extracted. Most astrometric interferometric studies require that more than one source be observed. Under good conditions in a well planned experiment, it is possible to connect successive observations of a given source: The residual differential phases from the near ends of two successive observations are extrapolated to a common epoch (between the observations) and compared; if they match, the phase is said to be connected. If the phase is connected over each gap in a series of observations of a particular source, then only a single 2π ambiguity need exist for that series. Phase connection of the observations dramatically increases the information content of the data. However, the procedure is difficult and makes stringent demands on the instrumentation, especially on the stability of the local oscillator.

The group delay observable, which depends on an extended observation bandwidth, is generally less accurate than the phase delay observable. However, it is easier to work with.

$$\tau = \frac{1}{2\pi} \frac{d\phi}{df} = B \cdot S/c \tag{5}$$

where τ is the differential delay, ϕ, the differential phase, f, the frequency, B, the baseline vector, and S, a unit vector pointing toward the source. The accuracy of the measurement depends on both the bandwidth and the signal-to-noise ratio. The Mark I VLBI system provided a recorded video bandwidth of 360 kHz. The Mark II system recorded 2 MHz. The recently developed Mark III system can simultaneously record 28 channels each of 2 MHz bandwidth.

The differential time delay observable can be used directly in astrometric studies. It can also be used as an aid in phase connection. For this purpose, heuristic algorithms have worked well in some experiments. However, the proper use of the delay data in phase connection results in a highly nonlinear estimation problem which has not been solved.

There are four principal difficulties encountered in astrometric interferometry: 1) Source structure that is resolved by the interferometer, although of astrophysical interest, decreases the astrometric usefulness of the source. 2) Instabilities in the local oscillator make phase connection difficult. The preferred local oscillator is a hydrogen maser. Currently available masers have stabilities of the order of 10^{-13} on the time scale of interest. However, these oscillators show discontinuous changes of frequency and phase at random intervals. Such discontinuities create breaks in the phase connection pattern. 3) The observable contribution of the Earth's ionosphere is highly variable. It is approximately 10 cm of phase delay for zenith observations at 8 GHz made during the day and is frequency dependent: $\tau_\phi \propto f^{-2}$. The plasma contribution from the solar corona becomes important for observations of sources having a small angular separation from the Sun. 4) The Earth's atmosphere makes a contribution to the one-station delay of approximately 2.5 m for observations made at zenith. The contribution from the dry atmosphere is proportional to the barometric pressure. However, about 10% of the delay is due to water vapor which is less easily measured. The water vapor content of the ray path can be measured with a dual-frequency radiometer operating near 23 GHz. Such instruments are able, in principle, to determine the water vapor contribution to delay with an uncertainty of a few mm. However, radiometers of this kind have not yet been used in interferometry experiments.

Lunar Laser Ranging

Through a series of fortuitous events, an optical retroreflector was carried to the Moon by Apollo 11 instead of Apollo 12. In July 1969, Astronauts Aldrin and Armstrong emplaced the retroreflector on the lunar surface. The first detection of that retroreflector occurred only a few days later, on 1 August, at the Lick Observatory which had a temporary laser ranging system. The first delay measurement was made with a 7 m uncertainty in the one-way distance to the Moon. That spectacular accuracy appears poor when compared to modern observations.

The laser ranging system comprises 5 subsystems: (1) a laser which produces a short pulse; (2) a telescope which beams the pulse toward the Moon and (2½ s later) gathers the returning photons;* (3) a retroreflector which sends the pulse back from the surface of the Moon; (4) a photomultiplier detector; and (5) timing electronics. There are now five retroreflectors on the lunar surface. Three of them were put there by the Apollo program, specifically by the Apollo 11, 14, and 15 missions. The first two of these had 100 corner-cube retroreflectors, each 2.5 cm in diameter. Designed to be able to function during the lunar day, they were recessed 1.9 cm in their support frame so as to shade them from the direct light of the Sun. The third Apollo retroreflector is a factor of three larger and is therefore used by the ranging stations more often than the other two. Finally, there are two French-built retroreflectors that were placed on the Moon by the USSR. The first one landed has been seen, at

---

*It is, of course, not necessary to use the same telescope for sending and receiving.

most, on two nights and is not currently part of the observing schedule. The other, which is not far from the Apollo 17 landing site, is part of a regular observing schedule and has functioned quite nicely for night-time observations. It has 14 unrecessed corner cubes, each 11 cm across. Because the corner cubes are large and exposed, they succumb to thermal distortion and do not function during the lunar day.

There are now at least eight observatories in five countries making observations of retroreflectors on the Moon. The McDonald Observatory has the longest history of laser observations; it has been functioning since 1969. Mcdonald transmits a 3 J laser pulse of approximately 3 ns duration; transmitted power is 10^9W. The beam of 6943 Å radiation diverges by about 2" so that it illuminates a portion of the Moon from 4 to 6 km in diameter. Although a retroreflector array receives a very small fraction of this light, the returning signal from the retroreflector is a factor of between 10 and 100 greater than the returning signal from the lunar surface. Detection is by means of a photomultiplier that is shielded by a 1 Å bandpass filter. A spatial filter of about 6" reduces stray light. The detection is marginal: approximately 1 photoelectron is detected per five laser shots. A ranging operation is considered successful if from 10 to 20 photoelectrons are collected for each observed retroreflector.

Let us look at the error budget for the current system (Silverberg, 1976). The 3 ns pulse contributes an uncertainty of 1.6 ns to the measurement. Timing electronics, detector jitter, and source geometry contribute 0.2, 0.3, and 0.2 ns, respectively. Thus a series of 10 measurements yields a precision of 0.6 ns. Timing calibration and atmospheric model error (bias) degrade the accuracy to 0.7 ns. The plans for improvements call for that error to be reduced by a factor of 3. Presumably the calibrations on the new equipment will be much better than those on the older equipment.

What must one do in order to use such high-precision observations? The answer is that one has to build a comprehensive model of the $\sim 2\frac{1}{2}$ s round-trip propagation time of the laser pulse. The stations are located on the Earth's surface; we must know where. Station coordinates must be estimated from the data. The Earth and Moon rotate unevenly; Earth wobbles and its spin rate changes. Fortunately, classical optical observations currently being made are almost good enough to provide that portion of the model. The rotation of the Moon presents a more complicated problem because until recently, we have known much less about it. The accuracy with which we must model the lunar rotation for the purpose of analyzing the laser data exceeds by three orders of magnitude the accuracy that was necessary for analyzing all of the optical data taken prior to the advent of laser ranging. In order to substantially improve our understanding of the rotation of the Moon, it is necessary to understand its (nonspherical) mass distribution since that contributes to the driving torques. Thus the model of lunar rotation involves ratios of spherical harmonics, absolute values of spherical harmonics, and the initial conditions of the rotational equations of motion. Recently, the lunar rotation equations have been augmented to include elasticity and damping, but the results of these improvements are not discussed here.

Of course the lunar orbit dominates the round-trip propagation time, and this is modeled. The motion model includes: the nonspherical distribution of Earth's mass; the motion and masses of the planets which produce perturbing forces and the motion of the Earth-Moon system around the Sun which governs the largest perturbing force.

The retroreflector laser-radar system offers one very important advantage over ordinary radar systems. There is no topography to contend with (cf, spacecraft observations which require an active system on the spacecraft). The nominal observing schedule calls for three observation periods each day, one at lunar zenith and one each at three hours before and after zenith. The observations are made every day except at new Moon, in bad weather, and when there is an unresolved conflict with other observing programs. With the current models, the postfit residuals of lunar laser ranging are down to about 25 cm. They are predominantly systematic, much larger than the nominal, 10 cm noise level, and under careful scrutiny now. There are a few types of data which are useful to use in conjunction with the laser data. Among these are the interferometry observations discussed in the previous section and spacecraft observations which give a measure of the nonspherical mass distribution of the Moon.

Having obtained some very nice results from laser ranging to the Moon, one naturally asks: Could we extend this to Mars? The answer is a resounding no. The radar equation gives an eight order of magnitude degradation in signal strength when the target moves from lunar distance to the minimum Earth-Mars distance. Because of the aberration of the signal returning from Mars, due to the relative velocity of Earth and Mars, one would have to have a retroreflector that produced a larger cone of returning light. A Mars retroreflector laser ranging system thus has a twelve order of magnitude engineering problem and there is no chance in the forseeable future of that being overcome. An active repeater on Mars, although more plausible, would be expensive and would require some advance in the state of the art.

PART II

Parameter Estimation

Estimation is central to the analysis of an experiment, but is often given minimal attention. Here we consider this critical link in the chain of information flow. The object of the process is to combine data with a parametric model so as to produce reliable estimates of the parameters, at least some of which are physically meaningful and interesting. Also required are realistic estimates of the uncertainties in the parameter estimates. Crudely speaking, the model may be divided into two classes or portions. The first describes a physical system which is to be observed. The second describes the observing process. Imbedded in each portion are the laws of nature, some of which may be the subject of the investigation.

Bear in mind that an experiment can never prove a theory. It can either contradict it or fail to contradict it. How, then, shall we go about testing general relativity, or for that matter, any other theory? An experiment can be considered useful for testing a theory only if the results of that experiment are predicted by the theory. But which of many possible experiments should be done? The first approach is to choose experiments capable of high precision. The second approach is to attempt to measure "named effects." These effects often fall into one of two classes. In the first class are the deviations of the predictions of a current theory from those of one of its predecessors. In the second class are results of violations of the assumptions under which the theory was derived. The third approach is exemplified by the parametrized post-Newtonian (PPN) formalism. Several theories are described parametrically by a single (phonomonological) supertheory. Estimates of the parameters would determine which, if any, of the theories has a chance of being correct.

In practice one may follow all three of the avenues described above when testing general relativity in the solar system. Although certain observations are particularly important to a particular "test," it will be seen below that the entire ensemble of data is advantageously used for each of the possible tests.

Bayesian Maximum *A Posteriori* Probability Estimation

Bayes Theorem is a fundamental probablistic relation that can be used to derive valid estimators. Let A, B and C each represent one or more random variables. Then Bayes Theorem reads:

$$[A,B/C] = [A/B,C]\,[B/C] \tag{6}$$

That is, the joint probability of A and B, given C is equal to the probability of A, given B and C times the probability of B, given C. In many standard texts, this Theorem is used to derive the usual weighted least-squares estimator. (See, for example, Jazwinski, 1970.) The derivation includes several assumptions, the violation of any one of which invalidates the estimator. Among these assumptions is that the data are adequately represented by the mathematical model except for the unknown values of the parameters that are to be estimated and the observation noise whose statistical description is correctly embodied in the noise covariance matrix. This assumption is substantially violated in most if not all studies of solar-system dynamics.

The solar system is vast and wondrously complex. It comprises the Sun, nine major planets, over thirty moons, countless asteroids and comets, dust, and streaming plasma. Unlike the hydrogen atom, which can be described by a beautiful mathematical theory, the numerous masses and orbital elements of the solar-system bodies must be determined emperically.

Almost all of the modern high-accuracy observations are limited to the inner solar system, that is the region of the solar system inside of the asteroid belt. It would, therefore, be convenient if we could ignore all bodies outside of the orbit of Mars. However, the perturbing effects of the giant planets cannot be ignored. We next consider the larger asteroids and find that their perturbations cannot be ignored and the remaining asteroids are collectively important. Some data are so sensitive to the motion of the Moon that the lunar motion model must include nonspherical components of the gravity fields of both the Earth and Moon. It soon becomes apparent that there is no clear cutoff to the size effects that should be included in the model. However, with each new effect, there generally comes one or more new parameters to be estimated. Although many of these parameters are interesting themselves, in the context of testing theories of gravitation, they must be considered nuisance parameters. A solar-system model that includes all the effects that are likely to be important carries with it a vast number of these nuisance parameters.

Given the available limited set of data, it is not possible to simultaneously estimate all the parameters of a solar-system model that includes all of the effects that are likely to be manifested in the data at or above the data-noise level. The resulting estimator would be degenerate. Auxiliary data can be used advantageously to estimate some of the parameters and thus relieve degeneracy. However, in the study of solar-system dynamics, one must eventually face the dilemma: If the model is kept simple, the model errors will lead to biased parameter estimates. If the model is made complete, degeneracy is to be expected. A compromise model will suffer in part from both problems.

Although there is no way to guarantee a "right answer" in such circumstances, it is possible to obtain useful bounds and a realisti estimate of the uncertainties in the estimated quantities. This can be done by performing a series of numerical experiments with the data. The object of these experiments is to uncover the deleterious effects of model errors and degeneracy. Each numerical experiment consists of solving the normal equations (i.e., the system of linear equations that results from the linearized weighted least-squares estimator) to produce a set of parameter estimates. In some experiments, the weights assigned to the data are changed and some of the data are deleted from the set. In other experiments, the parameter set is changed and the separate or combined effects of many "obscure parameters" can be determined. Although an ensemble of such experiments is useful for assessing the reliability of a parameter estimate, one must always be aware of the possibility that a class of model errors not yet considered may prove to be of considerable importance.

A time plot of postfit residuals is useful for finding blunder points among the data. Such plots often support the assertion that there are model errors by showing correlations or "structures" in th residuals. Although they may be an aid in determining the nature of the model error, as may a Fourier transform of the residuals, often a substantial portion of the model error is absorbed by a combinatio of estimated parameters. The resulting residuals bear little resemb lance to the signature of the model error.

In preparation for a large or expensive experiment, it is now common practice to perform a sensitivity study. In such studies, th observation schedule and data noise are postulated; the correspondin uncertainties for the estimated parameters can then be calculated. This approach tends to underestimate parameter uncertainties for 3 reasons. First, there is by definition no model error. Second, the idealized observing schedules are often not matched during the exper iment. Third, real data often contain significant correlated source of noise not taken into account in most sensitivity studies. To replace the usual, idealized study with one that is realistic would at least be expensive and in general would not be possible. Therefore, the interpretation of sensitivity studies requires careful analysis of the simplifying assumptions on which they are based.

Principle of Equivalence

The Principle of Equivalence is one of the cornerstones of general relativity. It has been checked in laboratory experiments to the level of 10^{-12} for the relative contributions to inertial and gravitational mass from neutrons, weak binding energy, strong bindin energy, electromagnetic binding energy, and kinetic energy. The Principle of Equivalence is found to hold to within experimental accuracy for all of these types of quantum mechanical energy (cf, experiments to test general relativity at the quantum level, discussed by Ephraim Fishbach in these proceedings).

When we consider gravitational self-energy we find the situation is quite different. For each body or system, we define the fractional gravitational self-energy, Δ, as

$$\Delta = M_\sigma/M \tag{7}$$

where $M_\sigma c^2$ is gravitational self-energy and M, the rest mass. For a spherical test body one meter in diameter, $\Delta \sim 10^{-23}$ and thus far too small for a laboratory test of the Principle of Equivalence. Since $\Delta \propto R^2\rho$, where R is the radius and ρ, the density of the test body, we are led to consider planets as test bodies. This approach was taken by Kenneth Nordtvedt, Jr. (1968a) who showed that to see an effect requires a minimum of three large bodies. He discussed experiments that use the Sun-Earth-Moon system and the Sun-Jupiter-Earth-Mars system. Nordtvedt (1968b) has alos discussed the relation between the PPN formalism and η, where η is the coefficient of the violation of the Principle of Equivalence for gravitational self-energy

$$M^g = M^i(1 + \eta\Delta) \tag{8}$$

In Equation (8), M^g and M^i are, respectively, the gratIvational and inertial masses of a body; $\eta = 0$ implies no violation.

We consider the motions of the Earth and Moon around the Sun

$$\ddot{\rho}_e = GM_s(1 + \eta\Delta_e)\frac{\rho_e}{|\rho_e|^3} + GM_m\left[1 + \eta(\Delta_e + \Delta_m)\right]\frac{r}{|r|^3} \tag{9}$$

$$\ddot{\rho}_m = GM_s(1 + \eta\Delta_m)\frac{\rho_m}{|\rho_m|^3} - GM_e\left[1 + \eta(\Delta_e + \Delta_m)\right]\frac{r}{|r|^3} \tag{10}$$

where M_s, M_e, and M_m are, respectively, the inertial masses of the Sun, Earth, and Moon; ρ_e and ρ_m are, respectively, the positions of the Earth and Moon with respect to the Sun; G is the constant of gravitation; and $r = \rho_m - \rho_e$. By rearranging terms, we obtain

$$\ddot{r} + G\frac{r}{|r|^3}M_b = -GM_s\left[\frac{\rho_m}{|\rho_m|^3} - \frac{\rho_e}{|\rho_e|^3}\right] - \eta GM_s(\Delta_e - \Delta_m)\frac{\rho_e}{|\rho_e|^3} \tag{11}$$

where $M_b = M_e + M_m$ and if the effect is to be observed, we require that $\Delta_e \neq \Delta_m$: $\Delta_e \sim 4.6 \times 10^{-10}$ and $\Delta_m \sim 0.2 \times 10^{-10}$. The coefficient of η on the right-hand side of Equation (5) leads to a 10 m perturbation in the Earth-Moon distance.

$$\delta|r| = (A\eta + B) \sin\Omega t \qquad (12)$$

where B is the coefficient of the classical perturbation having the same signature as the η term and Ω is the Sun-Earth-Moon angle: $B \sim A \times 1.1 \times 10^4$! Thus the Nordtvedt effect for the Sun-Earth-Moon system is buried under a large classical perturbation. Fortunately, the terms that contribute to B are either known or can be estimated simultaneously with η to very great accuracy.

Lunar laser ranging data have been used to check the Principle of Equivalence for gravitational self-energy in the Earth-Sun-Moon system (Shapiro et al. 1976, Williams et al. 1976). The data set comprised approximately 2000 laser normal points from observations made over a time span of 4.5 years. The parameter set included η, 6 orbital elements each for the motion of the Moon around Earth and for the Earth-Moon barycenter around the Sun, the locations of the McDonald Observatory and the lunar retroreflectors on the Moon, 6 coefficients of the spherical harmonic expansion of the gravitational field of the Moon, two instrumental biases, three coordinate system rotations, and numerous terms to represent irregularities in the rotation of Earth. The parametric model was able to fit the data with an rms error of 44 cm. (Subsequently, the rms error has been reduced to 25 cm.) The least squares solution gave $\eta = 0 \pm 1.5\%$, where the 1σ error stated makes allowance for possible systematic error. Differential very Long Baseline Interferometry (DLBI) has been used to independently check and confirm the estimated separation of the lunar retroreflectors.

The system comprising Earth, Mars, Jupiter and the Sun can also be used to check the Principle of Equivalence with time delay observations of Mars from Earth. Sensitivity studies have shown that the Viking lander data, in conjunction with previous time delay observations, should permit the Principle of Equivalence for these bodies to be checked at the 1% level or better.

The Large Numbers Hypothesis and $\dot{G}$

Dirac (1937, 1938, 1974, 1979) and others have suggested that the apparent strength of the gravitational interaction may be changing with time. This suggestion was originally motivated by the observation that dimensionless numbers in physics tend to be within 3 or 4 orders of magnitude of unity. However, the ratio of electrostatic-to-gravitational forces between an electron and a proton is $e^2/Gm_e m_p \sim 7 \times 10^{39}$ where _e_ is the charge of the electron and m_e and

m_p are the masses of the electron and proton, respectively. Similarly, if we measure time in atomic units, say $e^2/m_e c^3$, then the age of the universe is $\sim 2 \times 10^{39}$. Finally, if we express mass in units of a proton mass, then that portion of the universe receding from us at less than half the speed of light has a mass of approximately 10^{78}. The first two of these large numbers are approximately equal to each other, and the last is approximately the square of the others. That this numerological result suggests an underlying connection is the Large Numbers Hypothesis (LNH).

If the age of the universe is increasing (as one ordinarily supposes to be the case), then the LNH requires that one or more of the "constants" above must be varying with time. To check this, we assume G to be varying and write

$$G = G_o + \dot{G}(t - t_o) \tag{13}$$

where $(t - t_o)$ is the time in years since some arbitrary but recent reference epoch. For a discussion of the bounds on other possibly varying constants, see Dyson (1972). For a gravitationally controlled orbit, it follows from Equation (13) that

$$\frac{2\dot{G}}{G} = -\frac{2\dot{a}}{a} = -\frac{\dot{p}}{p} = \frac{\dot{n}}{n} \tag{14}$$

where $\underline{a}$, $\underline{p}$, and $\underline{n}$ are, respectively, the semi-major axis, period, and mean motion of the orbit. Other cosmological interpretations of the LNH give rise to equations which differ from Equation (14) by small factors. A planet's orbital longitude, L, is the time integral of its mean motion, $\underline{n}$. A linear contribution to the mean motion thus adds a quadratic contribution to the longitude. For radar time delay observations with the line of sight parallel to the orbital motion of the target planet, we find

$$\Delta\tau = \frac{4\pi a}{pc}\frac{\dot{G}}{G}(t - t_o)^2 \tag{15}$$

where $\Delta\tau$ is the time delay contribution from $\dot{G}$.

We next consider the special case of radar observations of the planet Mercury. We assume $\dot{G}/G = 10^{-11}\ y^{-1}$. If $t - t_o = 9$ years, then $\Delta\tau = 8\ \mu s$. For a thousand evenly spaced observations with a measurement uncertainty of $0.2\ \mu s$, $\sigma(\dot{G}/G) = 0.8 \times 10^{-14}\ y^{-1}$ if we assume that no other parameter needs to be estimated. As we shall see below, this estimate is naively optimistic for two separate reasons.

If in addition to estimating $\dot{G}$, we simultaneously estimate the planet's initial longitude and mean motion, then the uncertainty increases by a factor of 6. If we include the complete set of quantities that must be estimated from the data, then the masking factor rises from 6 to 80. Although it is true that the time delay uncertainty in the best radar delay measurements is approximately 0.2 μs, not all the available radar observations are of this quality. Further, the unmodeled portion of the planetary topography contributes a noise-like component to the time delay observations. Thus, a more realistic number for the measurement uncertainty is 3 μs. From our analysis of the radar data, we are able to conclude that $\dot{G}/G$ is consistent with 0 with an uncertainty of 10^{-10} y^{-1} (Reasenberg and Shapiro, 1976). If we extrapolate forward our experiences in gathering and analyzing the data, then we may estimate that by 1985 the uncertainty will be 10^{-11} y^{-1}. This extrapolation, however, contains many assumptions and the predicted uncertainty is hardly firm.

Metric Parameter β

One of the early triumphs of the general theory of relativity was the calculation of the non-classical advance of the perihelion of the orbit of Mercury. The relativistic advance of 43" per century must be separated from the effects of planetary perturbations and the precession of the Earth's spin axis which are 10^1 and 10^2 times greater respectively. Today the availability of high-speed digital computers and radar and radio observations make it *relatively* easy to separate these classical effects from the relativistic advance.

The situation is not so simple when one considers the possible large quadrupole moment of the Sun (Dicke and Goldenberg, 1967). Equation (16) gives the secular rate of advance of a planetary perihelion due to the combined effects of relativity and the solar quadrupole moment.

$$\delta\phi = \frac{3\pi r_\odot}{p}\left[\frac{2 + 2\gamma - \beta}{3} + J_2\frac{R_\odot^2}{r_\odot p}\right]\frac{\text{radians}}{\text{revolution}} \tag{16}$$

where $p = a(1 - e^2)$ is the planet's semilatus rectum; J_2 is the (dimensionless) coefficient of the quadrupole term in the Sun's mass distribution; $r_\odot \sim 3$ km is the gravitational radius of the Sun; $R_\odot \sim 7 \times 10^5$ km is its physical radius; and β and γ are Eddington or PPN parameters.

The perihelion advance makes a maximum contribution to radar measurements of echo delays which is represented approximately by $|\Delta\tau|_{max} = 2ae\delta\phi c^{-1}$. From Table 2, which gives $\Delta\tau|_{max}$ for the terrestrial planets, Mercury is found to be the best body to observe; Mars is a poor second.

Equation (16) contains β, γ, and J_2. How can we hope to separate the contributions of each? For γ, we rely on the time-delay test which is to be discussed below. The separation of the contri-

Table 2. Predicted Relativistic Contribution to the Perihelion Advance of the Inner Planets

Planet	$\delta\phi$ (arc sec y^{-1})	a (AU)	e	$\|\Delta\tau\|_{max}=2ea\delta\phi/c$ (μs y^{-1})
Mercury	.43	.4	.2	160
Venus	.09	.7	.007	2
Earth	.04	1	.017	3
Mars	.014	1.5	.09	10

butions of β and J_2 is more difficult; two possibilities exist. Either a value of J_2 can be derived from a model of the Sun, or J_2 and β can be estimated simultaneously. The solar models, and the oblateness observations on which they rest, are controversial. The solar neutrino problem further degrades the credibility of solar models. In regard to the second possibility, it can be seen from Equation (16) that the contributions of β and J_2 to the secular advance, $\delta\phi$, have different dependences on p. Observations of Mars might, therefore, seem to allow separation of β and J_2. There is a difficulty, however; both effects drop off sharply with increasing p and so neither makes much of an impact on the Mars orbit.

Another means of separation is provided by the effect of β on Kepler's third law which makes the data from all of the inner planets useful in this regard. The proposed VOIR (Venus) mission in 1985 should yield particularly useful data for this purpose.

Finally, we note that there are short-period terms in planetary orbits associated with both J_2 and β. Further analysis is needed to assess the importance of each of these differences in the effects of β and J_2 for given sets of data.

Analysis of radar data alone yielded $(2 + 2\gamma - \beta)/3 = 1.00_5 \pm 0.02$ (Shapiro et al., 1971) under the assumption that $J_2 = 0$. A later (as yet unpublished) analysis, based on an enlarged radar data set, reduced the uncertainty to 1%. However, simultaneous estimates of J_2 and β are so highly correlated, and thereby so sensitive to deficiencies in the theoretical model employed, that we have not yet obtained results that are both reliable and useful, i.e., with a true uncertainty for the estimate of J_2 of 5×10^{-6} or less.

Metric Parameter γ -- Deflection

The observation of the predicted solar deflection of starlight was one of the early successes of general relativity. Within the PPN formalism this effect depends on $(1+\gamma)/2$. Thus, for over half a century, there have been experiments from which γ could be estimated.

However, experiments that rely on visible wavelengths have tradition ally been constrained to the short intervals during a solar eclipse. Seeing from the ground is generally worse than 0".1, making it difficult to make a precise estimate of the magnitude of this 1".75 effect

As suggested by Shapiro (1967), a considerable improvement is possible when observations are made by radio interferometry. In a VLBI experiment by Counselman et al. (1974), observations were made at 8.1 GHz of the apparent position of 3C279 and 3C273B. The experi menters had to contend with the deleterious effects of the solar corona which not only produced a ray-path bending approaching 1% of the relativistic effect, but also added phase jitter to the signals, making phase connection difficult and placing a lower bound on the signal impact parameter for which phase connection was possible. Although not as severe a problem, the Earth's atmosphere also contributed both to the unmodeled part of the phase-delay and to phase jitter.

Two antennas were used at each end of an 845 km baseline: the 37 m Haystack and the 18 m Westford antennas in Massachusetts, and two 26 m antennas at NRAO in West Virginia. With four antennas it was possible to track each source nearly continuously, thus improving the chances for successful phase connection. The use of a singl hydrogen maser at each end of the baseline also facilitated the phas connection process. Fringe separation was about 0".009.

The estimate obtained for the coefficient of the light bending effect $(1+\gamma)/2$ was 0.99 with an uncertainty of 3%. This uncertainty makes allowance for possible systematic problems, biases, and model errors and is 6 times the formal uncertainty from the least squares fit to the data.

A short baseline, connected-element interferometry experiment is described by Fomalant and Sramek (1975). In this experiment, three sources -- 0111+02, 0116+08, and 0119+11 -- which nearly lie on a 10° segment of a great circle in the sky, were observed at 2.695 GHz and 8.085 GHz. The interferometer used three 26 m antenna at one end and one 14 m antenna at the other end of a 35 km baseline at NRAO. The object of the experiment was to measure the deflection of the central source which passed close to the limb of the Sun, the two other sources serving as references. During the experiment, the sources were observed sequentially with half the observations being made of the central source and the other half evenly divided between the two reference sources. Each observation lasted six minutes, fol lowed by $1\frac{1}{2}$ minutes for antenna motion. Thus a total cycle required 30 minutes.

Observations were made for $9\frac{1}{2}$ hours per day on 12 days during the 30 days surrounding 11 April 1974. The combined results from tw polarizations and three baselines led to the estimate $(1+\gamma)/2=1.015$

with an uncertainty of 1.1%. This uncertainty, which is larger than the formal σ of 0.8%, is derived in part from an analysis that used Monte Carlo and Fourier transform techniques.

Metric Parameter γ -- Delay

It was shown by Shapiro (1964) that as a consequence of general relativity, signals travelling between Earth and another planet will have an extra contribution, $\Delta\tau$, due to the direct effect of solar gravity on space-time.

$$\Delta\tau = \frac{2r_{\odot}}{c} \frac{1+\gamma}{2} \ln \frac{r_e+r_p+R}{r_e+r_p-R} \tag{17}$$

where $r_{\odot} \equiv 2GM_{\odot}/c^2 \sim 3$ km is the gravitational radius of the Sun; $M_{\odot}$, the mass of the Sun; r_e and r_p, the distances from the Sun to Earth and to the target planet, respectively; and R, the distance between Earth and the target planet. The Shapiro time delay effect has been the basis for the most accurate solar-system tests of general relativity. In such experiments there are three principal problems: 1) the determination of the locations of the end points of the transmission; 2) the measurement of the time delay; and 3) the correction of the time delay for the effects of the propagation media. Particularly troublesome is the solar corona, the contribution from which becomes largest when the signals contain the largest contribution from the Shapiro effect (i.e., for signals that pass close to the Sun). Significant fluctuations in the coronal delay exist on all scales thus far investigated; a factor of 2 increase can occur in a few hours. Thus, it is not useful to describe the corona interms of a deterministic (even parametric) model. However, the scale of the problem can be described by a simple expression

$$\tau_c = 300/f^2 d \ \mu s \tag{18}$$

where $\underline{f}$ is the signal frequency in GHz and $\underline{d}$ is the impact parameter in units of solar radii, $5 < \underline{d} < 20$. A more realistic model would include latitude dependence which, along with the mean amplitude, would vary with the 22 year solar cycle. For an S-band signal (2.2 GHz) with an impact parameter, $\underline{d} = 6$, Equation (18) yields $\tau_c = 10\ \mu s$. The corresponding relativistic delay is 170 μs. Thus, this randomly varying delay appears to place stringent limits on the accuracy with which the time delay test can be performed.

Fortunately, there are several means by which we can compensate for the effects of the solar corona on time delay. The frequency dependence of the plasma delay suggests two approaches. The first is to use a high frequency signal, thus making the delay insignificant. The second is to make measurements at two frequencies simultaneously. From the difference in the measured delays, the plasma contribution can be calculated. A potentially important limit on the accuracy with which this approach can be applied comes from the refraction of the plasma which causes signals at different frequencies to travel along slightly different paths. However, this limitation has not yet been encountered in an experiment.

The Viking Relativity Experiment has produced by far the most stringent test of the Shapiro effect. Before considering this experiment, I discuss briefly the pre-Viking time delay results. Radar observations made before 1972, both at X-band (8 GHz) and at UHF (0.4 GHz), yielded $(1+\gamma)/2 = 1.01\pm.05$. When the 1972 observations were included in the analysis, the result was $(1+\gamma)/2 = 1.00\pm.04$. Analyses of the S-band time delay observations of the Mariner 6 and 7 spacecraft yielded $(1+\gamma)/2 = 1\pm.03$. (This uncertainty was not so conservatively estimated as the others discussed in this section.)

In 1971, Mariner 9 was inserted into Mars orbit. The Navigation Team used the S-band Doppler tracking data to determine the spacecraft orbit and, in turn, to convert the time delay observations of the spacecraft into equivalent Earth-Mars center-of-mass time delays. The uncertainty associated with these normal point (pseudo-observations) ranged from 0.2 μs for observations taken far from superior conjunction to 0.7 μs at about 10 days from conjunction. The principal contribution to these uncertainties is from the orbit determination process. A combined analysis of the Mariner 9 normal points and the radar data yielded $(1+\gamma)/2 = 1\pm.02$. This uncertainty is dominated by a generous allowance for systematic error expected in part because the Mariner 9 S-band observations were severely corrupted by solar corona, as indicated for example by the systematic trends in the postfit residuals, Figure 5.

A substantially improved time delay test has been made possible by the Viking mission. The launches for the two interplanetary spacecraft were on 20 August and 9 September 1975 and the orbit insertions at Mars were on 19 June and 7 August 1976. On 20 July 1976 orbiter 1 separated from lander 1 which descended through the atmosphere to rest safely on the surface of Mars. On 3 September 1976, the second lander was also successfully deployed.

Each lander was equipped with a high-gain antenna and an S-band radio system. This S-band radio included a ranging transponder able to support time delay observations with uncertainties of 10 ns. Each orbiter was equipped with a similar radio system, including S-band transponder and high-gain antenna. In addition, each orbiter was equipped with an X-band transmitter. The X-band and S-band downlink signals are coherent and their frequencies are in a ratio of 11 to 3. Downlink signals can be used to determine the plasma columnar content along the ray path.

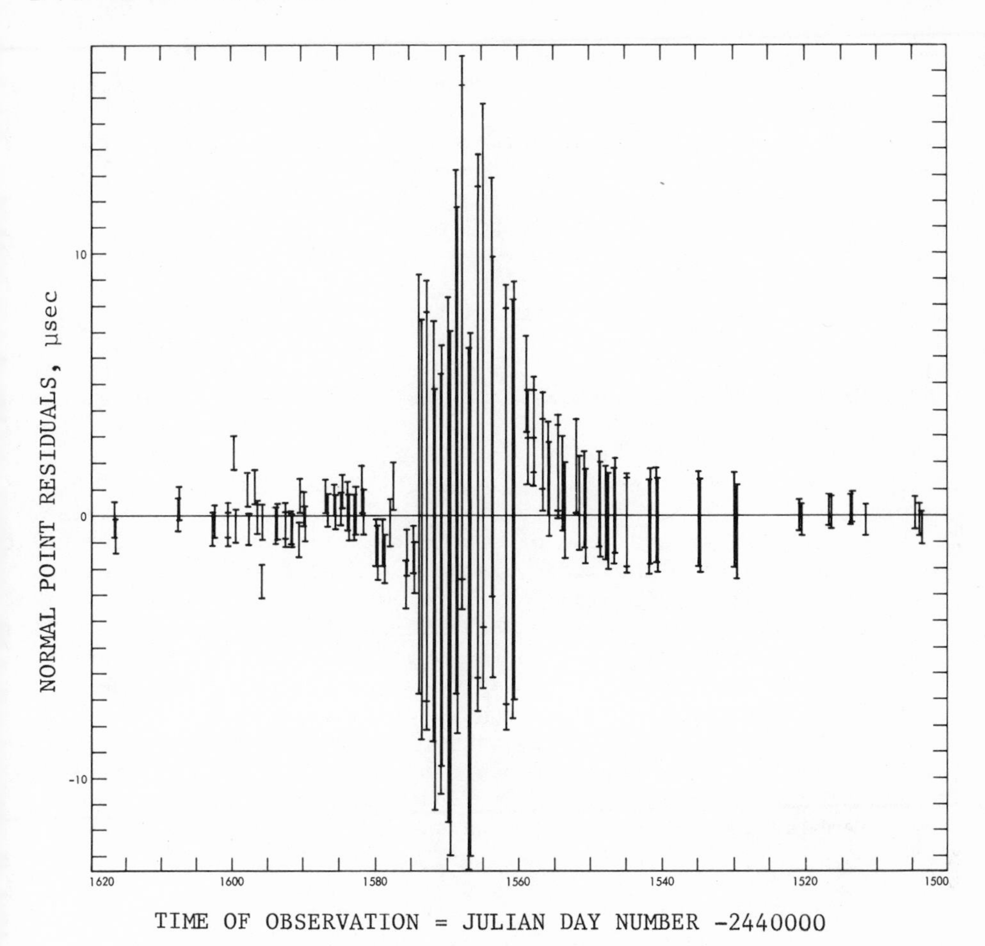

Fig. 5. Postfit residuals of the Mariner 9 time delay observations made near the epoch of superior conjunction. The error bars shown represent the assigned data uncertainties in one of the typical numerical experiments.

The landers serve as the prime points of observation. By comparison with an orbiter, the position of a lander with respect to the center of mass of its host body is relatively easy to describe. Because of the distinct (Mars diurnal) signatures with which they are associated, the lander's longitude and its distance from the Mars spin axis are both easily determined from the analysis of the tracking data. The displacement of the lander from the Mars equitorial plane is both less critical and less well determined. See the discussion of planetary landers in Part I.

Mars precesses at the rate of about $7''\!.6\ y^{-1}$. Of this, about 45 m y^{-1} is distinguishable from rotation. Both the precession and astronomical nutation (i.e., that part of the precession due to time dependent torque) are included in our model of the rotation of Mars. Two other classes of rotation terms need to be considered. The firs consists of variations in the spin rate. For Mars, seasonal condensation of the atmosphere at the poles causes a change in the moment of inertia and a corresponding change in both the spin rate and rota tional phase. The latter is estimated to be tens of meters (peak-to peak) at the equator. The terms of the second class concern the motion of the spin axis with respect to the rigid body axes. Such motion includes both wobble (i.e., Eulerian nutation) and long-term polar motion. Such effects have neither been detected nor estimated to be large enough for Mars to warrant consideration at this time.

Figure 6 shows the geometry of the plasma calibration. The plasma is measured along the orbiter downlink path and from this we must obtain estimates of the plasma contribution to the lander uplin and downlink paths. Thus there is both a spatial and temporal separation between the path along which the measurements are made and th paths along which the measurements must be applied. We use the "thi screen" approximation, according to which the entire contribution of the plasma to the propagation time occurs at the point on the ray path closest to the Sun. Using this approximation, we can correct the uplink portion of a lander round-trip propagation time using orbiter downlink plasma measurements made on Earth one Sun-Mars-Sun propagation time earlier than the lander measurement. We have not yet attempted to compensate for the spatial separation between the paths.

Figure 7 shows the relativistic delay for the 6 months around the superior conjunction of 25 November 1976. Also shown in this Figure is the S-band contribution to time delay that results from a model of the solar corona. The analysis so far has used about 330 time delay observations made during the 14 months following the first landing. The postfit residuals from these data are shown in Figure 8. The rms (one-way) postfit residual is 7 m. We have estimated the coefficient $(1+\gamma)/2$ of the Shapiro effect and find that it is equal to 1 ± 10^{-3}. This constrains the coefficient, ω, of Brans-Dicke theory to $\omega > 500$.

What hope is there of improving the accuracy of this estimate? Here I discuss briefly several possible sources of improvement: A sensitivity study has shown that the lander Doppler data, which have not yet been included in the analysis, could reduce the uncertainty in the time delay coefficient by a factor of 2. The lander observations have been extended through a second superior conjunction (January 1979). When included in the analysis, these observations may support another factor of 2 improvement both through improved

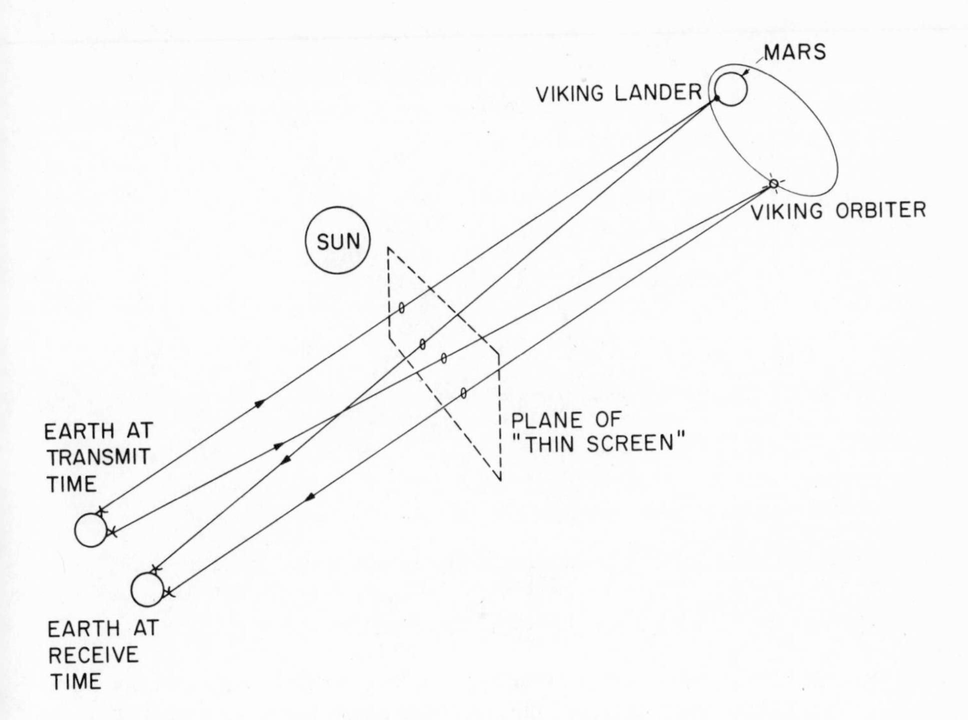

Fig. 6. Geometry of the fictitious thin screen.

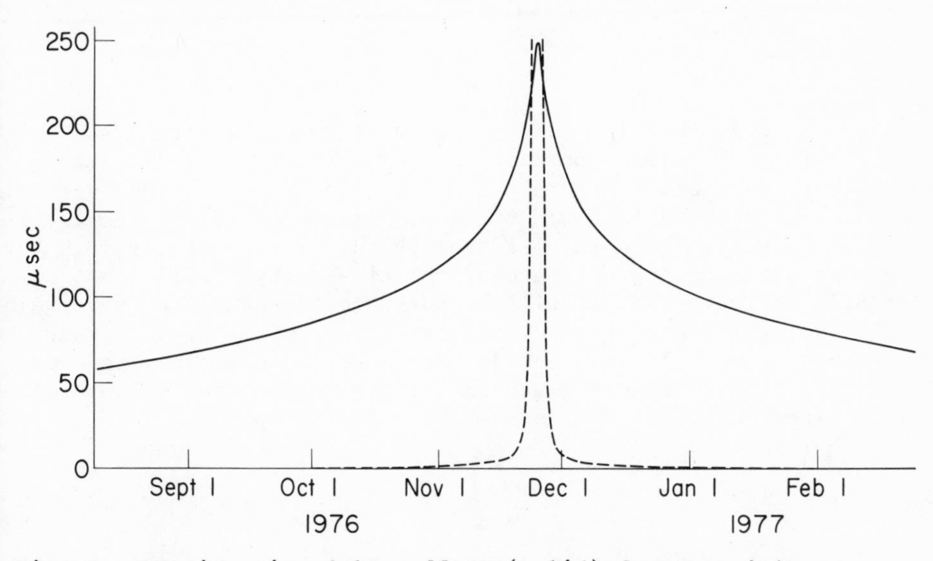

Fig. 7. Shapiro time-delay effect (solid) for the Vikings around the time of the 25 November 1976 superior conjunction. A model of the S-band plasma delay (dashed) is shown for comparison.

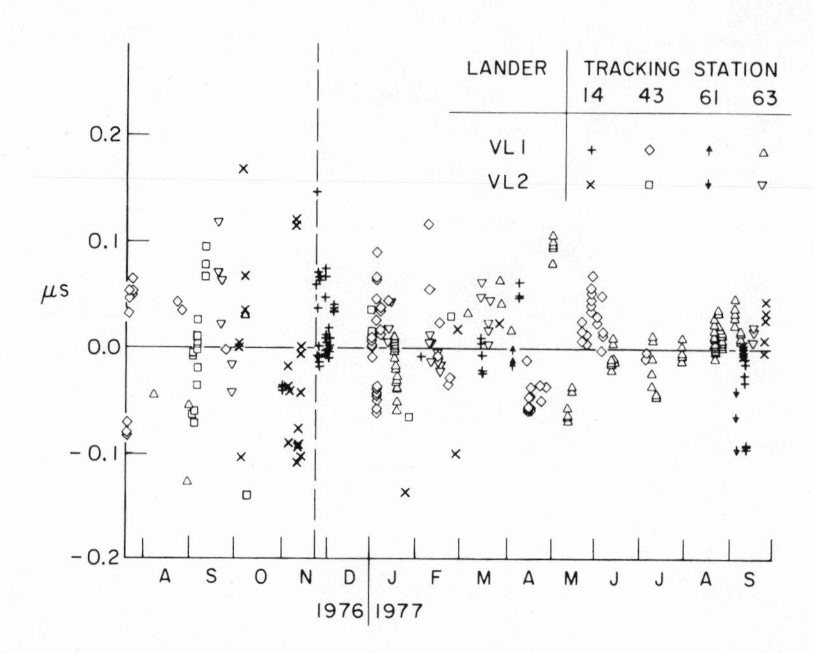

Fig. 8. Postfit residuals from the Viking lander time delay observations made during the 14 months following the first landing.

statistics, and, more importantly, by a reduction of the masking factor through clearer separation of effects. In a similar way we expect both the normal points and radar observations to be useful in reducing the masking factor. Finally, improved solar-system modeling, made possible to a large extent by the Viking data, may in turn make more accurate the estimate of the time-delay coefficient. I currently estimate that it will be possible to reduce the uncertaint in the estimate of the time-delay coefficient at least to 3×10^{-4} and perhaps lower. However, reduction of the uncertainty to below 10^{-4} does not seem likely given the limitations of the Viking relativity experiment.

PART III

This part is devoted to a discussion of three possible experiments, each of which would strengthen our experimental tests of general relativity. The first is the Solar Probe Mission which would fly a spacecraft close to the surface of the Sun. The second is VOI which would permit accurate Earth-Venus time delay measurements to be made. The third is POINTS, an astrometric instrument that could provide a second-order test of light bending by the Sun.

Solar Probe Mission

Our present experimental knowledge of the interior of the Sun is based almost exclusively on measurements of its radius, mass, luminosity, surface composition, rotation rate, and neutrino flux. Models have been used to derive the remainder of our "knowledge." Until recently, the standard models predicted a neutrino flux 4 times larger than that observed (Bahcall and Davis, 1976); changes in both theory and experimental results have reduced the disparity. Short-period solar oscillations (Hill et al., 1979), the 22 year sunspot cycle, and the Maunder minimum (Eddy, 1976) all need to be explained.

The United States National Aeronautics and Space Administration is considering a mission that would place a spacecraft close to the Sun. This Solar Probe would carry a variety of instruments and support a multitude of scientific investigations (Neugebauer and Davies, 1978). The radio tracking of this spacecraft could provide three types of information useful for checking theories of gravitation. First, because the trajectory passes close to the Sun, the spacecraft motion will be strongly influenced by both relativistic gravitation and the quadrupole moment of the solar potential. Second, the tracking signals will also pass close to the Sun and they also will be influenced by the solar potential. Third, should the spacecraft carry a stable clock, preferably a hydrogen maser, it would be possible to conduct a redshift experiment in a gravitational potential far more intense than any previously sampled (Vessot, 1975, 1978). However, it would be a formidable task to build a hydrogen maser capable of functioning unattended for five years during space flight. Therefore, this redshift experiment will not be further considered here.

What attributes of the mission will be important for the proposed studies of the spacecraft motion and signal propagation? Clearly a small perihelion distance is important for the study of $J_{2\odot}$ and β. Current mission studies are based on the perihelion distance of 4 solar radii; at this distance a gray body reaches an equilibrium temperature of $2400^{\circ}K$ (cf, melting point of steel $\approx 1600^{\circ}K$). A principal feature of the spacecraft design is a very white ceramic shield which must remain between the Sun and the rest of the spacecraft during its close encounter of the hottest kind.

For this mission, nongravitational spacecraft acceleration is quite significant. Acceleration caused by direct radiation pressure will vary as the spacecraft rotates within the attitude control cycle limits. Thermal radiation from both sides of the shield will vary throughout the encounter and produce accelerations not linearl proportional to the solar flux. Oblation of the heat shield and changes in its reflectivity would give rise to slowly changing acce erations. Finally, the highly variable solar wind would cause spac craft acceleration not in the Sun-spacecraft direction. A calculation of these components of acceleration would not be reliable.

To combat the deleterious effects of nongravitational accelera tion, it has been proposed that the spacecraft include a "drag-free system (Everitt and DeBra, 1978). In such a system, the spacecraft shields and protects a small "proof mass" which is thus free to follow a geodesic. A system of sensors determines the relative pos tion of the spacecraft with respect to the proof mass, and thruster are used to force the spacecraft to follow the motion of the proof mass. Although a drag-free system has been flown in Earth orbit, many questions need to be answered about the performance of such a system for the proposed Solar Probe Mission.

Because of the proximity of the spacecraft trajectory to the Sun, the radio tracking system will be required to operate under adverse conditions. (See the discussion of the effects of plasma on signal propagation under Viking Relativity Experiment.) The pre ferred tracking system would have dual-band uplink and downlink. I only one frequency can be used for the uplink, it should be in the higher of the two tracking bands.

Assuming that the formidable engineering problems can be solve what will we learn from this mission? Sensitivity studies have bee done (Anderson and Lau, 1978; Reasenberg and Shapiro, 1978a) to investigate this subject. (See earlier comments on the lack of reliability of formal sensitivity studies.) Table 3 shows the char acteristics of the Solar Probe trajectory used in our study; Table shows the schedule of observation; and Table 5 shows the list of model parameters estimated. Note that this study, which was done early in 1976, is of a "Sun Grazer" and thus not entirely applicabl to the proposed mission. We assumed a perfect drag-free system and X-band observations compensated for plasma corruption. The assumed observation uncertainties were 1 μs for time delay and 1 mHz for Doppler. From this study we concluded tentatively that for a singl fly-by of the Sun, $J_{2\odot}$ could be estimated with an uncertainty of less than 10^{-8}; estimates of β and γ would not be better than those already available. However, should it be possible for the spacecraft to run drag free for several solar encounters, a useful estimate of β could be obtained. In any case, a reliable estimate of $J_{2\odot}$ would contribute significantly to the determination of β from the advance of planetary perihelia.

Table 3. Characteristics of Solar-Probe Trajectory

Orbital Element	Value (1950.0 Equator and Equinox System)
Semimajor Axis	2.6 a.u.
Eccentricity	0.9982
Inclination	113.5 deg
Argument of Perihelion	0.0 deg
Longitude of Ascending Node	0.0 deg
Time of Perihelion Passage	4 February 1972
Minimum Distance from Sun	1.006 solar radii
Inclination to Ecliptic	90 deg
Angle Between Earth-Probe line and Sun-Probe line at Probe Perihelion	$\sim$45 deg

Table 4. Schedule of Probe Observations

Rate (observations/day)	Total Number		Time Interval
	Range	Doppler Shift	
0.2	148	148	763 to 23 days before encounter
8	265	265	23 days before to 10 days after encounter

VOIR

Another proposed NASA mission which may bear on $J_{2\odot}$ and β is the Venus Orbiting Imaging Radar (VOIR). This mission is intended to provide an intense study of the planet Venus; the spacecraft will carry a synthetic aperture radar which will be used to image a major portion of the planet's surface.

The mission plan calls for a nearly polar and circular orbit with a 93 minute period. This period is sufficiently short to permit the spacecraft orbit to be determined without substantial corruption from process noise, provided that the unmodeled space-

Table 5. Model Parameters

Initial Conditions of Probe Orbit	6
Initial Conditions of Earth's Orbit	6
Light-time Equivalent of the Astronomical Unit	1
Coefficient of Model of Solar Corona	1
Effect of General Relativity on Trajectory	1
Effect of General Relativity on Signal Propagation	1
Bias in Range Measurements	1
$J_{2\odot}$	1
TOTAL	18

craft acceleration is not much larger than it has been for previous spacecraft. If, as expected, a ranging transponder is provided on the spacecraft, time delay observations can be made and converted into high-accuracy time delay normal points. We expect to be able to obtain a daily normal point with a time delay uncertainty of unde 100 ns.

A set of such Venus normal points, taken over a period of at least a year, could be used to measure the violation of Kepler's third law, which is required by general relativity. The Venus semi-major axis would be most accurately determined by the normal point data. The orbital period would be determined by a combination of the normal point data and Venus radar data extending back over a decade. The remaining portion of the solar-system model, including classical perturbations of the Venus orbit and the other four elements of that orbit, are already well enough understood to have no adverse effect on the study of β with VOIR.

POINTS -- Its All Done With Mirrors

In this section, we consider a highly speculative proposal, that of Precision Optical INTerferometry in Space (Reasenberg and Shapiro 1978b). The object is to design, build, deploy, and use an optical interferometric astrometric instrument such as the one shown in Figure 9. The instrument would measure the angular separation between point sources of light nominally separated by 90°. It would employ two optical interferometers, each having a pair of ½ m telescopes separated by 10 m. Internal laser metrology would continuously determine separations and differential separations among components to the required ($\sim$0.1 Å) level. Such an instrument would advance astrometric capability by $\sim 10^5$, and open numerous and diverse areas of research.

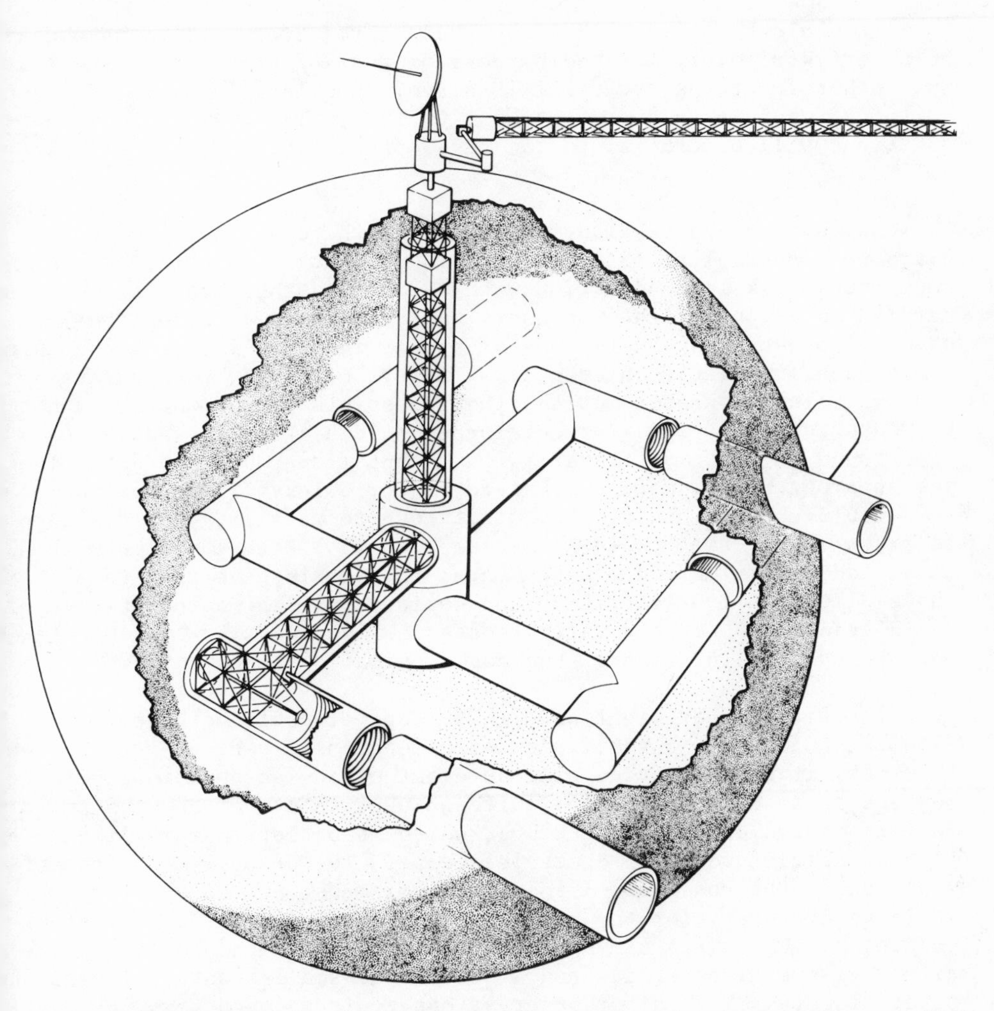

Fig. 9. An artist's rendition of the proposed astrometric satellite, POINTS.

The bending of light by the Sun reaches 1".75 for ray paths that graze the surface. This deflection is ordinarily calculated to first order in solar potential. With POINTS, it should be possible to detect and even measure the 1" x 10^{-5} second-order contribution to this bending. Such a second-order test is likely to provide a useful distinction among the currently viable relativistic theories of gravitation.

It would be naive to believe that a project so ambitious could be undertaken solely for the purpose of testing gravitation theory. What other functions could POINTS serve? How large a scientific constituency could such a project hope to find? The answers to these questions will be crucial to the development of the astrometric satellite.

An obvious application for POINTS is the development of a new ultra-precise stellar catalogue. This project would comprise the background task in conjunction with which all other projects would be carried out. The brighter quasars would be included among the observed objects, and detection of proper relative motion among them would have important consequences. If, as expected, the quasars show no relative proper motion (their absolute proper motion cannot be measured), then the star catalogue would be tied to what we presume to be an inertial frame. Within that frame we could determine the apparent proper motion of stars in our Galaxy. Some portion of this would show an annual periodicity due to paralax. The distance to stars 10 kpc away could be measured to 1%. When combined with angle and proper motion measurements, this would give rise to a three-dimensional Galactic map and would permit the determination of the distances to cepheid variable stars which in turn would provide an important calibration of the cosmological distance scale.

The first-order light bending by Jupiter could be observed. Although this would have little consequence as a test of gravitation theories, it would provide a measurement of Jovian position, with respect to an Earth-star line, with an uncertainty of about 5 km. This is equivalent to about 0".001, which is better than is likely to be possible even with Space Telescope. The proposed measurement would be of the position of the center of mass, and no correction would be necessary for the unknown and time dependent optical properties of the Jovian atmosphere. The interferometer could observe directly several asteroids and a few of the smaller satellites of the outer planets. The latter observations could be used to study the motions of these satellites as well as their parent planets.

The apparent position of a star varies periodically as a result of the motion of a dark companion. POINTS would permit us to determine whether a star of the size of the Sun, at the distance of the Galactic center, has a "Jupiter" in orbit around it. An "Earth" around a "Sun" could be detected at a distance of 5 to 10 pc. The detection of planetary systems, or the demonstration that they are rare, would have important consequences for the study of the origin of the solar system and would play a major role in any future search for extraterrestrial intelligence. The latter possibility is likely to have a significant and positive effect on the probability that POINTS is deployed.

On what schedule could we hope to see POINTS developed? A small team is now working part time on the development of the instrument concept. As yet the project is not funded. Several years of further analysis will be necessary before the actual satellite design can be undertaken. Although we have identified many difficult engineering problems, none seems to be beyond the current state of the art. Nonetheless, the earliest conceivable deployment date would be in the late 1980's and the 1990's seem a more realistic target.

Acknowledgement

I thank Mrs. G. McDonough for her careful typing of the text. This work was supported in part by the National Aeronautics and Space Administration (NASA) through the Mars Data Analysis Program Contract NSG-7556, and in part by the National Science Foundation through Grant NSFPHY78-07760.

References

Anderson, J. D., and E. L. Lau, 1978, Gravitational experiments on solar probe, in: "A Close Up of the Sun," M. Neugebauer and R. W. Davies, ed's, JPL publication 78-70, Jet Propulsion Laboratory, Pasadena.

Ash, M. E., 1972, "Determination of earth satellite orbits," Mass. Inst. of Tech., Lincoln Lab. Tech Note 1972-5.

Bahcall, J. N., and Davis, R., 1976, Science, 191:264.

Brody, S., 1977, "Some aspects of the use of planetary radar in the study of solar system dynamics," M. S. Thesis, Mass. Inst. of Tech., Cambridge.

Counselman, C. C., III, 1973, Proc. IEEE, 61:1225.

Counselman, C. C., III, S. M. Kent, C. A. Knight, I. I. Shapiro, T. A. Clark, H. F. Hinteregger, A. E. E. Rogers, and A. R. Whitney, 1974, Phys. Rev. Letters, 33:1621.

Counselman, C. C., III, 1976, Ann. Rev. of Astron. and Astrophys., 14:197.

Dicke, R. H. and H. M. Goldenberg, 1967, Phys. Rev. Letters, 18:313.

Dirac, P. A. M., 1937, Nature, 139:323.

Dirac, P. A. M., 1938, Proc. R. Soc. Lond., A165:199.

Dirac, P. A. M., 1974, Proc. R. Soc. Lond., A338:439.

Dirac, P. A. M., 1979, Proc. R. Soc. Lond., A365:19.

Dyson, F. F., 1972, The fundamental constants and their time variation, in: "Aspects of Quantum Theory," A. Salam and E. P. Wigner, ed's, Cambridge University Press.

Eddy, J. A., 1976, Science, 192:1189.

Evans, J. V., and T. Hagfors, 1968, "Radar Astronomy," McGraw-Hill, New York.

Evans, J. V., 1968, Modulation, demodulation, and data processing applied to radar astronomy, in: "Radar Astronomy," J. V. Evans

and T. Hagfors, ed's, McGraw-Hill, New York.
Everitt, C. W. F. and D. B. DeBra, 1978, Comments on the Drag-free control of a solar probe relativity mission, in: "A Close Up of the Sun," M. Neugebauer and R. W. Davies, ed's, JPL publication 78-70, Jet Propulsion Laboratory, Pasadena.
Fomalant, E. B. and M. C. H. Wright, 1974, Interferometry and Aperture synthesis, in: "Galactic and Extra-Galactic Radio Astronomy," G. L. Verschuur and K. I. Kellermann, ed's, Springer-Verlay, New York.
Fomalant, E. B. and R. A. Sramek, 1975, Astrophys. J., 199:749.
Hill, A. H. and T. P. Caudell, 1979, Mon. Not. R. Astr. Soc., 186: 327.
Jazwinski, A. H., 1970, "Stochastic Processes and Filtering Theory," Academic Press, New York.
Neugebauer, M., and R. W. Davies, ed's., 1978, "A Close Up of the Sun," JPL publication 78-70, Jet Propulsion Laboratory, Pasadena.
Nordtvedt, K., 1968a, Phys. Rev., 169:1014.
Nordtvedt, K., 1968b, Phys. Rev., 169:1017.
Ostro, S. J., 1978, "The Structure of Saturn's Rings and the Surfaces of the Galilean Satellites as Inferred from Radar Observations," Ph. D. Thesis. Mass. Inst. of Tech., Cambridge.
Pettengill, G. H., I. I. Shapiro and A. E. E. Rogers, 1973, Icarus 5. 18:22.
Reasenberg, R. D., and I. I. Shapiro, 1976, Bound on the secular variation of the gravitational interaction, in: "Atomic Masses and Fundamental Constants 5," J. H. Sanders and A. H. Wapstra, ed's, Plenum Press, New York.
Reasenberg, R. D. and I. I. Shapiro, 1978a, Possible measurements of $J_{2\odot}$ with a solar probe, in: "A Close Up of the Sun," M. Neugebauer and R. W. Davies, ed's, JPL publication 78-70, Jet Propulsion Laboratory, Pasadena.
Reasenberg, R. D. and I. I. Shapiro, 1978b, 5th International Symposium on Space Relativity at IAF Congress XXIX, Dubrovnik (submitted to proceedings to be published in Acta Astronautica ca. 1980).
Ritter, R. C. and J. W. Beams, 1978, A Laboratory measurement of the constancy of G, in: "On the Measurement of Cosmological Variations of the Gravitational Constant," L. Halpern, ed., University Presses of Florida, Gainesville.
Shapiro, I. I., 1964, Phys. Rev. Letters, 13:789.
Shapiro, I. I., 1967, Science, 157:806.
Shapiro, I. I., M. E. Ash, R. P. Ingalls, W. B. Smith, D. B. Campbell, R. B. Dyce, R. F. Jurgens, and G. H. Pettengill, 1971, Phys. Rev. Letters, 26:1132.
Shapiro, I. I., C. C. Counselman III, and R. W. King, 1976, Phys. Rev. Letters, 36:555.
Silverberg, E. C., 1976, SPIE (High Speed Optical Techniques) 94:83.
Vessot, R. F. C., 1975, Applications of frequency standards, in: "Atomic Masses and Fundamental Constants 5," J. H. Sanders and A. H. Wapstra, ed's, Plenum Press, New York.

Vessot, R. F. C. and M. W. Levine, A time-correlated four-link Doppler tracking system, in: "A Close Up of the Sun," M. Neugebauer and R. W. Davies, ed's, JPL Publication 78-70, Jet Propulsion Laboratory, Pasadena.

Will, C. M., 1973, The theoretical tools of experimental gravitation, in: "Experimental Gravitation," B. Bertolli, ed., Academic Press, New York.

Williams, J. G., R. H. Dicke, P. L. Bender, C. O. Alley, W. E. Carter, D. G. Currie, D. H. Eckhardt, J. E. Faller, W. M. Kaula, J. D. Mulholland, H. H. Plotkin, S. K. Poultney, P. J. Shelus, E. C. Silverberg, W. S. Sinclair, M. A. Slade, and D. T. Wilkinson, 1976, Phys. Rev. Letters, 36:551.

Woolard, E. W., 1953, The rotation of the earth around its center of mass, in: "Astronomical Papers of the American Ephemeris and Nautical Almanac," U. S. Naval Observatory Nautical Almanac Office, Washington, 15:1.

TESTS OF GENERAL RELATIVITY AT THE QUANTUM LEVEL

Ephraim Fischbach

Institute for Theoretical Physics, State University of New York at Stony Brook
Stony Brook, New York 11794
*Physics Dept., Perdue Univ., W. Lafayette, IN 47907

ABSTRACT

We consider several tests of general relativity at the quantum level. One of these is based on the observation that an effect of a gravitational field on a hydrogen atom is to admix states of opposite parity such as $2S_{1/2}$ and $2P_{1/2}$. This leads to a circular polarization P_γ of emitted radiation which could be sufficiently large to be detected in white dwarfs or in certain binary systems. Since the magnitude of P_γ varies from theory to theory, it is possible that a study of this effect could provide a feasible means of testing general relativity at the quantum level. We also discuss the phenomenon of gravity-induced CP-violation in fermion-antifermion systems and possible applications to K_L decays.

I. INTRODUCTION

As we celebrate this year the centennial of Einstein's birth, the experimental support for general relativity as a *macroscopic* theory is quite compelling[1]. By contrast there is very little, if any, direct evidence which bears on the question of the validity of general relativity at the *quantum* level. This is, of course, not surprising considering the smallness of the force that would characterize, for example, the interaction between a macroscopic object (such as the Earth) and a quantum system (such as a hydrogen atom). Nonetheless it is extremely important to know *experimentally*

---

*Permanent address as of September 1, 1979.

what the couplings of the gravitational field are to quantum systems (for example to a single fermion) since these couplings constitute the starting point of any attempt to develop a consistent quantum treatment of gravity. We must also bear in mind the possibility that new gravitational effects could manifest themselves at the quantum level whose existence would not have been anticipated by an extrapolation from the macroscopic realm.

At present our knowledge of the effects of gravity on quantum systems is limited to the results of the COW experiment[2] which measures the quantum mechanical phase difference of two neutron beams induced by an external gravitational field. If two coherent beams are separated by a vertical distance r and are allowed to propagate for a horizontal distance s, then the accumulated phase difference $\Delta\phi$ between the two beams will be given by [2]

$$\Delta\phi = \frac{\lambda m_n^2 g(rs)}{2\pi\hbar^2} \tag{1.1}$$

where λ is the (initial) de Broglie wavelength of the neutrons, m_n their mass, and g is the local acceleration of gravity. $\Delta\phi$ can be measured experimentally by detecting the fringe shift resulting when an initial neutron beam is split and then recombined a distance s later. Clearly (1.1) describes a quantum gravitational effect as is apparent from the fact that the expression for $\Delta\phi$ depends on g and $\hbar$ in a non-trivial way. This makes sense physically since the COW effect results from the fact that neutrons which are particles behave quantum-mechanically as waves, and that their de Broglie wavelengths can be affected by a gravitational field.

For purposes of testing general relativity at the quantum level however, we must go beyond the COW experiment. The reason for this is that the expression for $\Delta\phi$ in (1.1) depends only on Newtonian gravity to which all quantum theories of gravity reduce in the non-relativistic limit. What we must look for are <u>relativistic</u> quantum gravity (RQG) effects, which emerge from relativistic wave equations several examples of which are discussed below. We will focus primarily on one of these, gravity-induced parity violation in hydrogen, because it may represent the most practicable means of testing general relativity at the present time, and also as an illustration of the ideas that underlie tests of general relativity at the quantum level. We follow this analysis with a comparison of various terrestrial experiments, and then with a discussion of gravity-induced CP-violation in fermion-antifermion systems.

II. HYDROGEN ATOM IN A GRAVITATIONAL FIELD

We proceed to show that a hydrogen atom in an external gravitational field experiences a mixing of opposite parity states which may be viewed as the gravitational analog of the Stark effect. There is, in addition, an analog of the Zeeman effect which is, however, of lesser importance. We begin with the Dirac equation for a single fermion of mass m (say an electron) in a gravitational field which is given by

$$\{\gamma^{\mu}(x)[\frac{\partial}{\partial x^{\mu}} + \frac{i}{4}\ \sigma^{ab}e_{a}^{\nu}(x)e_{b\nu;\mu}(x)] + \kappa\}\psi(x) = 0, \quad (2.1)$$

where $\kappa = mc/\hbar$, and σ^{ab} are a set of constant Dirac matrices,

$$\sigma^{ab} = \frac{1}{2i}(\gamma^{a}\gamma^{b} - \gamma^{b}\gamma^{a}) \quad , \quad a = 1,2,3,0. \quad (2.2)$$

The tetrad fields $e_{a}^{\nu}(x)$, and their covariant derivatives $e_{b\nu;\mu}(x)$, are specified as follows: If dx^{μ} and dx^{a} denote the coordinate differentials at a given point in a general world coordinate system and in a local Minkowski system respectively, then

$$dx^{a} = e_{\mu}^{a}(x)dx^{\mu}. \quad (2.3)$$

Using Eq. (2.3) it is straightforward to show that the metric tensor $g_{\mu\nu}(x)$ and the Minkowski metric $\eta_{ab} = \mathrm{diag}(1,1,1,-1)$ are related by

$$g_{\mu\nu}(x) = \eta_{ab}e_{\mu}^{a}(x)e_{\nu}^{b}(x) = e_{b\mu}(x)e_{\nu}^{b}(x) . \quad (2.4)$$

The coordinate-dependent matrices $\gamma^{\mu}(x)$ and the covariant derivative $e_{b\nu;\mu}(x)$ are then given by

$$\gamma^{\mu}(x) = e_{a}^{\mu}(x)\gamma^{a}, \quad (2.5)$$

and

$$e_{b\nu;\mu}(x) = \frac{\partial}{\partial x^{\mu}} e_{b\nu}(x) - \Gamma_{\nu\mu}^{\lambda}e_{b\lambda}(x) , \quad (2.6)$$

where $\Gamma_{\nu\mu}^{\lambda}$ is the usual Riemann connection. For the case of an electron in the field of the Earth the tetrads corresponding to the Schwarzschild metric,

$$ds^{2} = (1 + \tfrac{1}{2}\Phi)^{4}\ (dx^{2}+dy^{2}+dz^{2}) - \left(\frac{1-\frac{1}{2}\Phi}{1+\frac{1}{2}\Phi}\right)^{2} c^{2}dt^{2}, \quad (2.7a)$$

$$\Phi = \frac{GM_{\oplus}}{\rho c^2} \qquad \rho = (x^2+y^2+z^2)^{\frac{1}{2}} \quad , \tag{2.7b}$$

are given in the weak field limit (Φ <<1) by:

$$e^{\mu}_{a}(x) = \delta_{\mu a}(1-\Phi) + 2\delta_{\mu 0}\delta_{a0}\,\Phi \quad , \tag{2.8a}$$

$$e_{b\nu}(x) = \delta_{b\nu}(1+\Phi) - 2\delta_{b0}\delta_{\nu 0}, \tag{2.8b}$$

$$e_{b\nu;\mu}(x) = \frac{1}{c^2}\,[\delta_{\mu\nu}g_b - \delta_{\mu b}g_\nu] \ , \tag{2.8c}$$

$$g_\mu \equiv c^2\,\partial\Phi/\partial x^\mu \quad . \tag{2.8d}$$

Combining Eqs. (2.1) - (2.8) we find after some algebra that the Dirac equation can be written in the form,

$$i\hbar\,\frac{\partial\psi(\vec{r},t)}{\partial t} = \{-i\hbar c(1-2\Phi)\vec{\alpha}\cdot\vec{\partial} - \frac{i\hbar}{2c}\,\vec{\alpha}\cdot\vec{g} + \beta mc^2(1-\Phi)\}\psi(\vec{r},t), \tag{2.9}$$

$$r = \rho(1+\tfrac{1}{2}\Phi)^2 \cong \rho \ ,$$

where $\vec{\alpha}$ and β are the usual Dirac matrices,

$$\vec{\alpha} = \begin{pmatrix} 0 & \vec{\sigma} \\ \vec{\sigma} & 0 \end{pmatrix} \qquad \beta = \begin{pmatrix} \mathbb{1} & 0 \\ 0 & -\mathbb{1} \end{pmatrix} \ . \tag{2.10}$$

The Hamiltonian in Eq. (2.9) is Hermitian when the requisite spatial integrations are carried out using the correct measure, i.e.,

$$\langle H\rangle = \int d^3r\,\sqrt{\hat{g}}\,\psi^\dagger(x)H\psi(x), \tag{2.11}$$

where $\hat{g} = \det g_{ij}$. However, it is more convenient to absorb $\sqrt{\hat{g}}$ into the wavefunction ψ in order that H appear as a Hermitian Hamiltonian when integrated with the respect to the Euclidean coordinates $\int d^3r$. This can be achieved by introducing a new wavefunction $\tilde{\psi} = (1 + \frac{3}{2}\Phi)\psi$ in terms of which the Dirac equation in (2.9) becomes

$$i\hbar \frac{\partial\tilde{\psi}(\vec{r},t)}{\partial t} = \{-\, i\hbar c(1-2\Phi)\vec{\alpha}\cdot\vec{\partial} + \frac{i\hbar}{c}\,\vec{\alpha}\cdot\vec{g}+\beta mc^2(1-\Phi)\}\tilde{\psi}(\vec{r},t)$$

$$= \tilde{H}\tilde{\psi}(\vec{r},t), \qquad (2.12)$$

with a manifestly Hermitian Hamiltonian in the usual sense. We will henceforth drop the "∿" and denote $\tilde{\psi}$ by ψ, and $\tilde{H}$ by H.

The Hamiltonian of Eq. (2.12) is of the general form

$$H = \vec{\alpha}\cdot\vec{A} + \beta B + \beta mc^2 = \mathcal{O} + \mathcal{E} + \beta mc^2 , \qquad (2.13)$$

where the "odd" operator $\mathcal{O}$ couples upper and lower components of the Dirac spinor ψ, whereas the "even" operator $\mathcal{E}$ does not. For later purposes it is desirable to decouple the upper and lower components by means of a Foldy-Wouthuysen (FW) transformation which reduces (2.12) to a 2-component Schrödinger equation. After three iterations of the FW transformation the Hamiltonian in (2.12) assumes the form

$$H = \beta\mu + \mathcal{E} + \frac{\beta}{2\mu}\mathcal{O}^2 - \frac{1}{8\mu^2}[\mathcal{O},[\mathcal{O},\mathcal{E}]] + 0(\frac{1}{\mu^3}) ,$$

$$= \beta\mu(1-\Phi) + \frac{c^2\vec{p}^2\beta}{2\mu} + \frac{3}{2}\beta[-\frac{c^2}{\mu}\Phi\vec{p}^2 + \frac{i\hbar}{\mu}\vec{g}\cdot\vec{p} + \frac{\hbar}{2\mu}\vec{g}\cdot\vec{\sigma}\times\vec{p}] + 0(\frac{1}{\mu^3}) , \qquad (2.14)$$

where $\mu = mc^2$, and $\vec{p} = -i\hbar\vec{\partial}$.

The effective single particle Hamiltonian of Eq. (2.14) corresponds, of course, to the form of the Schwarzschild metric derived in the usual Einstein theory of general relativity. In a more general theory the metric of Eq. (2.7) is replaced in the weak field limit) by

$$-g_{00} = 1-2\Phi + 2\beta\Phi^2, \qquad g_{0i} = 0,\ i = 1,2,3$$

$$g_{ij} = (1+2\gamma\Phi)\delta_{ij}, \qquad (2.15)$$

where the parametrized post Newtonian (PPN) parameters β and γ [3] are defined so that $\beta = \gamma = 1$ gives the conventional general relativity result. The current limits on β and γ for macroscopic systems are discussed in [1].

In order to emphasize the possibility that β and γ may differ at the quantum level from their macroscopic counterparts, we will distinguish the former by primes, so that the coefficient 3/2 in (2.14) gets replaced by $(\gamma' + \frac{1}{2})$.

To obtain the effective interaction for hydrogen we combine the expression for the gravitational interaction of an electron with that for a proton, which is also treated as a Dirac particle. Introducing the center-of-mass (CM) and relative coordinates, $\vec{R}$ and $\vec{r}$ respectively,

$$\vec{R} = \frac{m_e\vec{r}_e + m_p\vec{r}_p}{m_e + m_p} , \qquad \vec{r} = \vec{r}_e - \vec{r}_p , \tag{2.16}$$

we find (neglecting various uninteresting terms)

$$H(e-p) = -(m_e + m_p)\vec{g}\cdot\delta\vec{R} + (\gamma' + \tfrac{1}{2})\frac{\hbar^2 g}{m_R c^2}\,\hat{g}\cdot\{\vec{r}\nabla^2 + \vec{\nabla} - \frac{i}{2}\vec{\sigma}_e\times\vec{\nabla}\}. \tag{2.17}$$

Here m_R is the reduced mass of the electron, $\vec{\nabla} = \partial/\partial\vec{r}$, and $\delta\vec{R}$ is the CM coordinate measured from the surface of the Earth. We see from (2.17) that the first two terms in { }, like the Stark interaction $e\vec{\mathcal{E}}\cdot\vec{r}$, can induce transitions between states of opposite parity such as 2S and 2P. If we multiply and divide the coefficient of { } by e^2, we can rewrite it in the form

$$(\gamma' + \tfrac{1}{2})\frac{\hbar^2 g}{m_R c^2} = (\gamma' + \tfrac{1}{2})\left(\frac{g\hbar}{c}\right)\left(\frac{e^2}{\hbar c}\right)\left(\frac{\hbar^2}{m_R e^2}\right) = (\gamma' + \tfrac{1}{2})\eta\alpha a_o , \tag{2.18}$$

where α is the fine structure constant and a_o is the Bohr radius. We have introduced the constant η

$$\eta = \frac{g\hbar}{c} , \tag{2.19}$$

which sets the energy scale for the gravity-induced transitions, and which expresses in compact form that this term is a relativistic quantum gravitational effect. Table I gives the values of η for various systems of interest. The data for the white dwarfs, entries 3-5, are taken from Ref. [4].

To understand how the gravity-induced admixture of opposite parity states can be detected experimentally we consider, as an

Table 1. Values of $\eta = g\hbar/c$ and $|P_\gamma|$ for various systems. P_γ is calculated assuming that only weak and gravitational interactions are present (see Eq.(2.23) below).

	SYSTEM	η(eV)	$\eta/\eta_\oplus$	$\lvert P_\gamma \rvert$
1.	Earth	2.2×10^{-23}	1	6×10^{-10}
2.	Sun	6.0×10^{-22}	28	2×10^{-8}
3.	40 Eridani B $M = 0.372\ M_\odot$ $R = 0.0152\ R_\odot$	9.7×10^{-19}	4.5×10^4	3×10^{-5}
4.	Sirius B	1.1×10^{-17}	5.3×10^5	3×10^{-4}
5.	WD 2359-43	4.9×10^{-17}	2.3×10^6	1×10^{-3}
6.	Neutron Star $M = 0.4\ M_\odot$ $\rho = 1 \times 10^{15}$ gm cm^{-3}.	3.5×10^{-12}	1.6×10^{11}	≈ 1

example, the decay $2S \to 1S + \gamma$. This decay is highly suppressed since the E1 amplitude is strictly forbidden while the M1 amplitude vanishes to lowest order in α. In the presence of a gravitational field, however, there is admixed into the initial 2S state a 2P component which can decay to 1S + γ via E1. The emitted photon then "remembers" the gravitational field through a correlation between the polarization of the photon and $\vec{g}$.

This correlation can combine with those arising from other sources of 2S - 2P mixing, such as the neutral-current weak interaction or an external electric field, to produce radiation in specific states of linear or circular polarization. The experimental objective is then to measure the polarization of light emitted, say, from a white dwarf and to then separate out the gravitational contribution from all the other sources of polarization. We will return shortly to discuss possible means of carrying out this separation.

To make this discussion more concrete we quote the expressions for the gravity-and Stark- induced nS - nP admixtures: (the weak interaction admixture is discussed in detail in Ref. [5].)

$$\underline{\text{Stark}}: \quad \langle nP_{\frac{1}{2}}m'_J|e\vec{\mathcal{E}}\cdot\vec{r}|nS_{\frac{1}{2}}m_J\rangle = \frac{e\mathcal{E}a_o}{2}\frac{n}{Z}\sqrt{n^2-1}\langle\chi_{\frac{1}{2}}^{m'_J}|\vec{\sigma}_e\cdot\hat{\mathcal{E}}|\chi_{\frac{1}{2}}^{m_J}\rangle \tag{2.20}$$

$$\underline{\text{gravity}}: \quad \langle nP_{\frac{1}{2}}m'_J|(\gamma'+\tfrac{1}{2})\eta\alpha a_o\hat{g}\cdot[\vec{r}\nabla^2+\vec{\nabla}-\frac{i}{2}\vec{\sigma}_e\times\vec{\nabla}]|nS_{\frac{1}{2}}m_J\rangle$$

$$= -(\gamma'+\frac{1}{2})\frac{\alpha\eta}{6}\frac{Z}{n}\sqrt{n^2-1}\langle\chi_{\frac{1}{2}}^{m'_J}|\vec{\sigma}\cdot\hat{g}|\chi_{\frac{1}{2}}^{m_J}\rangle \tag{2.21}$$

In (2.20) and (2.21) the χ's are Pauli spinors with $m_J = \pm\frac{1}{2}$, etc. Using these results, and retaining only the $2P_{\frac{1}{2}}$ intermediate state, it is straightforward to show that the Stark-and gravity-induced amplitudes for $2S \to 1S + \gamma$ are proportional to

$$\mathcal{T} \equiv \langle\chi_{\frac{1}{2}}^{m''_J}|\vec{\varepsilon}^*\cdot\hat{F} + i\vec{\sigma}_e\cdot\vec{\varepsilon}^*\times\hat{F}|\chi_{\frac{1}{2}}^{m_J}\rangle , \tag{2.22}$$

where $\hat{F}$ denotes either $\hat{\mathcal{E}}$ or $\hat{g}$, $\vec{\varepsilon}^*$ is the polarization vector for the emitted photon, and m''_J is the spin of the final 1S electron. An expression such as $\vec{\varepsilon}^*\cdot\hat{g}$ ultimately leads to a correlation between $\hat{g}$ and the polarization of the emitted photon when $|\mathcal{T}|^2$ is computed.

The complete expression for the polarization of the emitted radiation, including the Mi, Stark-, gravity-, and weak-induced amplitudes, is complicated and will be given elsewhere[6]. The polarization will also depend on the extent to which the initial electron is aligned by ambient magnetic fields, such as are known to exist, for example, in white dwarfs[7]. For illustrative purposes we exhibit in Table I an estimate of the circular polarization P_γ assuming that only weak and gravitational interactions are present. We find, taking $\gamma' = 1$ and n=2 (see Eq. (3.11) below),

$$|P_\gamma| \sim \frac{4\pi\sqrt{2}\,\eta\alpha}{G_F\alpha/a_o^3} , \tag{2.23}$$

where G_F is the Fermi constant. We see from Table I that P_γ in (2.23) can be of order 10^{-3} in some white dwarfs which is roughly the present level of sensitivity[7]. One must not conclude from this that gravity-induced opposite parity admixtures are on the verge of being detected in white dwarfs since the polarizations that have already been seen are almost surely due to other mechanisms. These give rise to an enormous background from which the very small gravitational contribution must be extracted. Our hopes for distinguishing between the "signal" (due to gravity) and the background "noise" (due for example to the Stark effect) depend crucially on the fact that the signal has a characteristic "signature". Some components of this "signature" are the following:

1) $P_\gamma \propto |\vec{g}|$: We see from (2.21) that the gravity-induced contribution to P_γ is directly proportional to η and hence to $|\vec{g}|$. This observation can be used to our advantage if we compare $|P_\gamma|$ in various white dwarfs for which $\eta/\eta_\oplus$ varies between $\approx 10^4$-10^6 as we see from Table I.

2) Dependence on Z and n: The most serious backgrounds come from Stark-induced admixtures of opposite-parity levels due to ambient electromagnetic fields. As we see from Eqs. (2.20) and (2.21), the difference in the forms of the Stark and gravity Hamiltonians is manifested in a different dependence of their matrix elements on Z/n:

$$\frac{\langle \text{gravity} \rangle}{\langle \text{Stark} \rangle} \propto \frac{Z^2}{n^2} \,. \qquad (2.24)$$

Thus if we compare hydrogenic systems such as H and He^+ in the same white dwarfs, then the variation of the gravity-induced contribution to P_γ with Z can be used to distinguish it from that due to the Stark effect. The same end can also be achieved by comparing transitions from states with different values of n in hydrogen.

3) $P_\gamma \propto \vec{g}\times\vec{k}$: If the initial 2S electrons have a net polarization, due for example to an ambient magnetic field, then there will be a contribution to P_γ of the form $P_\gamma \propto \vec{s}\cdot\vec{g}\times\vec{k}$ where $\vec{s}$ is the electron spin and $\vec{k}$ is the photon momentum. Such a term can in principle be distinguished from the background contributions by the fact that the latter can be presumed to be "incoherent" over the scale of the star, meaning that they would not exactly mimic the gravitational field $\vec{g}(\vec{r})$. As can be seen from Fig. 1, light reaching the Earth from different parts of the surface of a star (such as the Sun) emanates from regions with different $\vec{g}\times\vec{k}$ and hence different P_γ. Thus the variation of P_γ over the surface of the Sun is another characteristic of the gravity-induced contribution to P_γ. For radiation from distant objects which appear as point sources use can be made of the fact that some of these sources are members of eclipsing binary systems. An example of this is the system Hercules X-1/HZ Herculis[8] which is believed to contain a neutron star and accreted hydrogen orbiting a visible companion. During an eclipse (which occurs ~ every 1.7 days and lasts for several hours) $\vec{g}\times\vec{k}$ varies in a well defined way for light reaching the Earth, and this variation can be used to discriminate between the gravitational and nongravitational contributions to P_γ.

III. COMPARISON OF TERRESTRIAL EXPERIMENTS

We present in this section a heuristic analysis of several terrestrial experiments to test general relativity at the quantum level. In doing this we have several purposes in mind. In the first place we would like to have a rough quantitative estimate of the relative difficulty of different experiments in the hope that one will clearly stand out as the most promising possibility in the foreseeable future. Secondly we would like to illustrate the physical principles that underlie various types of experiments that can be envisioned at present, hoping that this very discussion will stimulate the consideration of other alternatives.

To do this we introduce a "doability quotient" Q which gives a rough measure of the difficulty of a particular terrestrial experiment:

$$Q \equiv \frac{\text{required experimental sensitivity}}{\text{existing experimental sensitivity}}. \quad (3.1)$$

To illustrate the meaning and use of Q we consider the following three experiments.

1) Relativistic COW Experiment: One possibility for testing general relativity at the quantum level is to look for general relativistic effects in the COW experiment[9]. Consider, for example, the relativistic expression for the gravitational potential of a neutron, whose spin we neglect:

$$V(r) = -\frac{GM_{\oplus}(E_n/c^2)}{r}\left\{1 + \gamma' \frac{c^2|\vec{p}_n|^2}{E_n^2}\right\} \quad (3.2)$$

Here $\vec{p}_n$ and E_n are the momentum and energy of the neutron, and γ' is the quantum PPN parameter. In the ultrarelativistic limit $\{\ \} \to (1+\gamma')$ which from (3.2) reproduces the familiar result for massless fields. For the COW experiment we need the nonrelativistic limit of (3.2),

$$V^{NR}(r) \cong -\frac{GM_{\oplus}m_n}{r}\left[1 + (\gamma' + \tfrac{1}{2})\beta_v^2\right], \quad (3.3)$$

where $\beta_v = v/c$. Thus in the COW experiment the relativistic corrections to the Newtonian potential have the effect of simply replacing m_n by an effective mass $m_n^{(e)}$,

$$m_n^{(e)} = m_n\{1+(\gamma' + \tfrac{1}{2})\beta_v^2\}. \quad (3.4)$$

Since $\beta_v = 9.3\times10^{-6}$ for the neutrons in the COW experiment the relativistic correction to be measured in (3.4) is of order $9\times10^{-11}(\gamma'+\frac{1}{2})$ Thus if we take the current sensitivity of the COW experiment to be $\approx 10^{-2}$ [10] then

$$Q = \frac{9\times10^{-11}(\gamma'+\frac{1}{2})}{10^{-2}} = 9\times10^{-9}\ (\gamma'+\tfrac{1}{2}). \tag{3.5}$$

We are thus roughly 8 orders of magnitude away from being able to do such an experiment. In estimating Q we necessarily neglect any consideration of backgrounds since to do otherwise would require a more detailed understanding of a specific experimental setup than we can have at the present stage.

2) Neutron Spin Precession in a Gravitational Field: Dass[11] has suggested measuring the precession of the neutron spin $\vec{s}$ in the Earth's gravitational field as a test of various theories of general relativity. If we restrict our attention to couplings of $\vec{s}$ that conserve parity and time reversal, then the only allowed coupling is that given in Eq. (2.17) which corresponds to a precession

$$\frac{d\vec{s}}{dt} = \vec{\omega}\times\vec{s}\quad ;\vec{\omega} = \frac{(\gamma'+\frac{1}{2})GM_{\oplus}\vec{r}\times\vec{p}_n}{m_n c^2 r^3}, \tag{3.6}$$

where $\vec{p}_n$ is the neutron momentum. (At the surface of the Earth $GM_{\oplus}/r^2 = |\vec{g}| = g$) It follows that the characteristic energy differences which would have to be measured to see such an effect are of order

$$|\langle V^{\sigma}\rangle| \equiv |\langle -\tfrac{1}{2}(\gamma'+\tfrac{1}{2})\eta\vec{\sigma}\cdot\hat{g}\times\vec{p}_n/m_n c\rangle| \cong |\tfrac{1}{2}(\gamma'+\tfrac{1}{2})\eta\beta_v|$$
$$\cong (1.1\times10^{-23}\ \text{eV})\beta_v(\gamma'+\tfrac{1}{2}). \tag{3.7}$$

Present day experimental capabilities correspond to being able to measure a static neutron electric dipole moment μ_E of order $\mu_E \lesssim 1.6\times10^{-24}$ e-cm in an electrostatic field $\mathcal{E} = 10^5$ V/cm. This amounts to measuring an energy difference ΔE,

$$\Delta E = \langle\mu_E\vec{\sigma}\cdot\vec{\mathcal{E}}\rangle \cong 1.6\times10^{-19}\text{eV}, \tag{3.8}$$

and hence for such a system

$$Q \cong \frac{|\langle V^{\sigma}\rangle|}{\Delta E} \cong \frac{(1.1\times10^{-23}\text{eV})\beta_v(\gamma'+\frac{1}{2})}{1.6\times10^{-19}\text{eV}} \tag{3.9}$$

Since $\beta_v \cong 5\times10^{-7}$ in the neutron EDM experiments we get

$$Q \cong 3\times10^{-11}(\gamma'+\tfrac{1}{2}) \tag{3.10}$$

3) Measurement of P_γ in a Terrestrial Experiment: We have discuss previously the possibility of detecting gravity-induced circular polarization of the light emitted from white dwarfs or from hydrogen accreted around neutron stars. While these systems provide the advantages of strong gravitational fields they are not subject to our direct control and hence it is more difficult for us to evaluate the effects of backgrounds than would be the case for a terrestrial experiment. The greater control that we can achieve in terrestrial experiments is, of course, offset by the small size of the expected effects as we now demonstrate. To estimate Q in this case we compare the magnitude of the gravity-induced 2S-2P admixture to that produced by the weak interaction. Since the latter contribution has recently been detected in heavier atoms, and is on the verge of being detected by the Michigan group in hydrogen, the weak 2S-2P admixture can be taken as a measure of the existing experimental sensitivity. We thus have

$$Q \cong \frac{|P_\gamma(\text{earth's gravity})|}{|P_\gamma(\text{weak})|} \cong \frac{|\langle 2P_{\frac{1}{2}}|H(e\text{-}p)|2S_{\frac{1}{2}}\rangle|}{|\langle 2P_{\frac{1}{2}}|H_W|2S_{\frac{1}{2}}\rangle|}$$

$$= \frac{8\pi\sqrt{2}}{3}\;\frac{\eta\alpha(\gamma'+\frac{1}{2})}{G_F\alpha/a_o^3} = 4\times10^{-10}(\gamma'+\tfrac{1}{2}). \tag{3.11}$$

In (3.11) we have identified $\langle H_W\rangle$ with the quantity $\bar{V}$ defined by Dunford, et al. [5] which sets the overall scale of weak effects in hydrogen.

In comparing Eqs. (3.5), (3.10), and (3.11) amongst themselves, or with the results for other proposed experiments, it is probably useful for the time being to regard values of Q which agree to within 2 orders of magnitude or so as representing potential experiments which are "comparably doable".

IV. GRAVITY-INDUCED CP-VIOLATION

We have seen in the previous section that gravity, acting as an external field, can effect transitions between states of opposite parity in hydrogen such as 2S and 2P. In a system such as positronium or charmonium, such a change in parity leads to a change in the eigenvalue of CP for the system as we now discuss. As is well known the charge-conjugation (C) and parity (P) eigenvalues for a fermion-antifermion system $f\text{-}\bar{f}$ are given by

$$C = (-1)^{L+S}, \quad P = (-1)^{L+1}, \quad CP = (-1)^{S+1} . \tag{4.1}$$

Here L and S denote the orbital and spin angular momentum of $f\text{-}\bar{f}$

which is treated in the nonrelativistic limit where L and S are good quantum numbers. Consider now the Hamiltonian $H(f-\bar{f})$, which is the analog for the fermion-antifermion system of $H(e-p)$ in (2.17). Although the detailed expression for $H(f-\bar{f})$ is too long to be presented here, it suffices for our purposes to exhibit the following spin-dependent term denoted by $V^{\sigma}(f-\bar{f})$:

$$H(f-\bar{f}) = V^{\sigma}(f-\bar{f}) + \cdots = -\tfrac{1}{2}(\gamma'+\tfrac{1}{2})\frac{i\hbar\eta}{mc}\,\hat{g}\cdot(\vec{\sigma}_1-\vec{\sigma}_2)\times\frac{\partial}{\partial\vec{r}} + \cdots \quad (4.2)$$

Here $1=f$, $2=\bar{f}$, $\vec{r}=\vec{r}_1-\vec{r}_2$ and m is the mass of f or $\bar{f}$. Since V^{σ} changes both L and S it can change the eigenvalue of CP without at the same time changing the total angular momentum J of the system. V^{σ} can thus admix a CP=-1 $^1P_1$ component into a CP=+1 $^3S_1$ wavefunction. This contrasts with both the Stark Hamiltonian (which does not change CP) and also with the Zeeman Hamiltonian (which changes J as well as CP.) A typical matrix element of V^{σ} in positronium is given by (taking $\gamma'=1$)

$$\langle n=1;\ ^3S_1, S_z=\pm 1\,|V^{\sigma}|\,n=2;\ ^1P_1, L_z=\pm 1\rangle = \mp\frac{8\sqrt{2}}{27}\,Z\alpha\eta. \quad (4.3)$$

We mention in passing that V^{σ} has the property of vanishing when taken between Coulomb wavefunctions with the same principal quantum number n. If we combine the effects of V^{σ} and the weak Hamiltonian H_W then to order $V^{\sigma} H_W$ the gravity-induced parity-violation can be offset by the P- violating and C- violating weak interaction. The net effect is then a C- and CP- violating transition, such as $^1P_1 \leftrightarrow {}^3P_1$.

The fact that V^{σ} leads to a directly calculable CP-violation in fermion-antifermion transitions makes it natural to attempt to connect V^{σ} with the known CP-violating K_L decays. Although there are theoretical and experimental arguments which appear to rule out gravity as a source of CP violation, the relevance of these arguments to the present situation is not at all clear. This is due to the fact that the full expression for $H(f-\bar{f})$ in (4.2), which is rather complicated, contains terms which couple the CM and relative motions of f and $\bar{f}$ to each other, and which at the same time couple each of these to the spin. The dynamics of this CP-violating interaction are presently being studied, motivated in part by the following remarkable numerical relation. Returning to (4.2) we see that the energy scale for $\langle V^{\sigma}\rangle$ is set by the parameter $\eta = 2.15\times10^{-23}$ eV. It is not unreasonable to suppose that if V^{σ} were indeed the source of CP-violation then η would enter in the combination $\eta/c^2\delta m$, where $\delta m = m_L - m_S = 3.56\times10^{-6}\,\mathrm{eV}/c^2$ is the K_L-K_S mass difference which measures the strength of the usual CP-conserving second order weak interaction. If we imagine that the CP-violation was for some reason enhanced by a factor of $(m_K/\delta m)$, where m_K is the kaon mass, then we are led to consider the quantity

$$\frac{\eta}{c^2\delta m}\left(\frac{m_K}{\delta m}\right) = 0.844\times 10^{-3}. \qquad (4.4)$$

Remarkably this gives the experimental value of the CP-violating parameter $\frac{1}{2}\mathrm{Re}\varepsilon$,

$$\begin{aligned} \tfrac{1}{2}\mathrm{Re}\varepsilon &= (0.83\pm 0.03)\times 10^{-3}, \quad \text{from } K^0\to\pi^{\pm}\ell^{\mp}\nu \\ &= (0.80\pm 0.03)\times 10^{-3}, \quad \text{from } K_L\to 2\pi \end{aligned} \qquad (4.5)$$

where ε is given by $|K_L\rangle = [(1+\varepsilon)|K\rangle - (1-\varepsilon)|\bar{K}\rangle]/[2(1+|\varepsilon|^2)]^{\frac{1}{2}}$.

The expression in (4.4) can be recast in the form

$$\frac{\eta}{c^2\delta m}\left(\frac{m_K}{\delta m}\right) = \frac{m_K g(\hbar/c\delta m)}{c^2\delta m} \qquad (4.6)$$

The numerator on the r.h.s. of Eq. (4.6) can be interpreted as the gravitational energy difference of K_L and K_S separated by a vertical distance $(\hbar/c\,\delta m)=5.55$ cm, which is the distance that a relativistic virtual particle would travel in a time $t = \hbar/c^2\delta m$.

V. CONCLUSIONS

We have learned from Dr. Reasenberg's lucid summary at this school[1] that the predictions of general relativity have now been experimentally verified to quite a high precision. His latest result for γ, $\frac{1}{2}(1+\gamma) = 1.000\pm 0.002$, implies that for all practical purposes general relativity in its original form is the correct theory

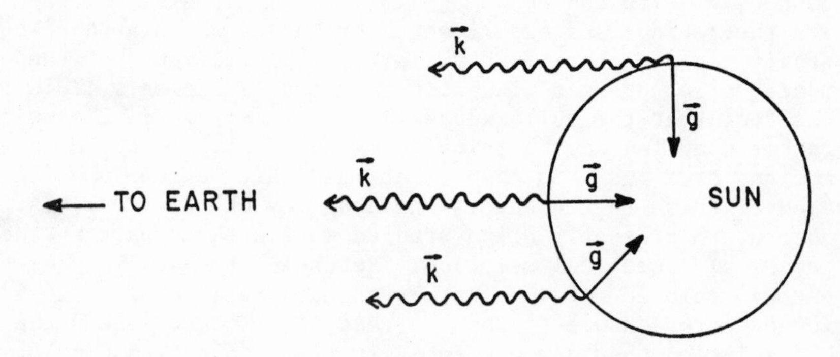

Figure 1. Variation of $P_\gamma \propto \vec{g}\times\vec{k}$ across the surface of the Sun. $\vec{k}$ is the momentum of the photon and $\vec{g}$ is the local acceleration of gravity.

of gravity, at least on the scale of the solar system. We should, of course, devise tests of the other PPN parameters and hopefully such tests will be pursued in the future.

We may now ask where experimental general relativity should go from here. Apart from a search for gravitational radiation and the measurement of the PPN parameters on the solar system scale with greater precision, two other directions naturally present themselves. Professor Bergmann has suggested looking at experiments on a larger (e.g. cosmic) scale, such as by considering experiments which relate our local reference frame to one fixed to the (presumably distant) quasars. We have pointed out the need for looking in the opposite direction, namely at the quantum level, in order to see how the smallest objects in nature (e.g. electrons) see the curvature of space-time.

I am deeply indebted to W.K.Cheng and B.S.Freeman who collaborated with me on various aspects of this problem. I also wish to thank R.R.Lewis, T.J.Moffett, A.W.Overhauser, and H.Primakoff for their help and encouragement, and C.N.Yang for the hospitality of the Institute for Theoretical Physics where part of this work was carried out. Finally I wish to express my appreciation to my colleagues at Stony Brook and Purdue for many numerous discussions. This work was partially supported by the Department of Energy and the National Science Foundation Grant No. PHY-78-11969.

REFERENCES:

1. R. D. Reasenberg, lectures at the "International School of Cosmology and Gravitation," Erice, 1979 (this volume).
2. A. W. Overhauser and R. Colella, Phys. Rev. Lett. 33, 1237 (1974); R. Colella, A. W. Overhauser, and S. A. Werner, Phys. Rev. Lett. 34, 1472 (1975).
3. C.M.Will in "Experimental Gravitation," edited by B. Bertotti, (Academic Press, New York, 1974,) pp. 1-110.
4. H.L.Shipman, Astrophys. J. 228, 240 (1979); T.J.Moffett, T.G. Barnes, and D.S.Evans, Astronom. J. 83, 820 (1978).
5. R. W. Dunford, R.R.Lewis, and W.L.Williams, Phys. Rev. A18, 2421 (1978).
6. E. Fischbach and B.S.Freeman, to be published.
7. J.R.P.Angel, Ann.Rev. Astron. and Astrophys. 16, 487 (1978).
8. D.Gerend and P.E.Boynton, Astrophys. J. 209, 562 (1976).
9. J. Anandan, Phys. Rev. D15, 1448 (1977).
10. A. W. Overhauser, private communication.
11. N.D. Hari Dass, Ann. of Phys.(NY) 107, 337 (1977).

THE MASS-ANGULAR MOMENTUM-DIAGRAM OF ASTRONOMICAL OBJECTS

Peter Brosche

Observatorium Hoher List

D-5568 Daun/Eifel, F.R. Germany

INTRODUCTION

Mass M and angular momentum P are distinguished from other physical parameters by the existence of the respective conservation theorems. They are even distinguished from the energy E by the fact that the latter can be easily radiated away from an astronomical object but not M and P. While M and E can be thought of as a sum of microscopic contributions (gravitational energy can at least be converted into thermal energy or radiation), the large angular momenta of astronomical objects can only be associated with large coordinated motions of large aggregates of mass in large distances. This is because the unit of microscopic angular momentum, Planck's constant $\hbar$, is very small compared with the angular momentum of a celestial body divided by the number of its nucleons. Hence it seems that P is one if not *the* specific quantity defining *macroscopic* objects of astronomical size. If any, then M and P are the quantities which are most probably unchanged since an astronomical object has separated from the rest of the world. Therefore mass-angular momentum-diagrams of such objects should play at least the same important role as a meeting point between theories of the formation and the observations as the Hertzsprung-Russell-diagram plays for the evolution of stars.

If the internal motions V in an object of size R are mainly rotational, the angular momentum is of the order of magnitude $P \approx M\ R\ V$. If, in contrast, the object consists of N subunits in random motion, the r.m.s. expectation value of P is smaller by a factor $\sqrt{N}$. Since nearly all the objects we shall consider are in mechanical equilibrium with their own gravitation, we can use the virial theorem $G\ M^2/R \approx M\ V^2 \approx 2\ E$ for the elimination of one of the parameters M,R,V (G = constant of gravity). In such a way - and furthermore by replacing some parameters by density, angular velocity etc. - one gets a great variety of *equivalent* formulas. They contain always, apart from the finer details not mentioned here, *two* basically independent parameters, e.g. M and P. So there exists no trivial *a priori* expectation for relations between M and P; one is not allowed to 'forget' a second parameter X in one of the equivalent formulas $P = \text{const}\ M^{\alpha}\ X^{\beta}$. Of the many possibilities we mention here

$$P \approx G\ M^2\ /\ V \approx \sqrt{G}\ M^{3/2} \cdot R^{1/2} \tag{1a}$$

$$P \approx 1/\sqrt{N} \quad \text{of the above} \tag{1b}$$

for rotating and random-motion objects respectively. The angular momentum of the rotation of solid bodies is

$$P \approx M\ R^2\ W \tag{2}$$

wherein $M\ R^2$ is the moment of inertia and W the angular velocity; one should bear in mind that for stability reasons equ. (1a) defines an upper limit for the P value of a solid body of given (M,R). Our P values are refered to an axis through the barycenter of the objects. In most cases they are orbital angular momenta and therefore computed with formulas of the type (1a) and (1b). Angular momenta of the subunits could be added vectorially but are usually not significant.

THE ASTRONOMICAL HIERARCHY

For the reasons summarized above the author has investigated the mass-angular momentum-diagram of all classes of astronomical objects as far as the existing data allow their determination (Brosche 1962, 1963). We shall discuss very briefly the situation for the different classes in an ascending order through the hierarchy. As for the hierarchical degree, we make no difference between isolated objects and multiple systems (planets with or without satellites, single or double stars).

Planet-satellite-systems

The total angular momentum P of such a system consists of the part P(rot) due to the planet's rotation plus the orbital part P(orb) of the planet-satellite-motion, if there are satellites; the P(rot) of the latter being negligible. Contrary to what is true for the other classes, the main constituent of P is here P(rot) except in the Earth-Moon-system. But this exception is understandable as the result of tidal interaction (for a recent review, see Brosche and Sündermann, 1978). Likewise, the very slow rotation of Mercury and Venus is not 'primordial' but most probably the consequence of solar tidal torques. All other planet - satellite-systems fulfill a narrow power-law $P \sim M^{\alpha}$ with $\alpha \approx 2$ (Brosche, 1963).

Estimates of masses and diameters of planetoids together with the interpretation of periodical brightness changes as rotation lead to a continuation of the (P,M)-diagram to small masses, but now with a marked deviation from $\alpha \approx 2$ towards $\alpha \approx 5/3$ (Fish, 1967; Hartmann and Larson, 1967).

Stars

We know from observations that the majority of stars are components of double stars and it is very plausible that all other stars have either unresolved companions or planets around themselves (Brosche, 1964). In such a way the characteristic angular momentum would be for all stars of an orbital nature and therefore much larger than that of the rotation of the stars. Otherwise one would encounter here and not in the case of galaxies, where Harrison (1973) fears it, the awkwardness of a bimodal P-distribution (Brosche, 1967a). The relatively most unbiased sample is the one from the solar neighbourhood (Brosche, 1962). It shows a large scatter at a given mass - corresponding to a wide range in semi-major axes of the orbits - and roughly a mean trend $P \sim M^2$.

Stellar clusters

In this case the data are very uncertain since the internal velocities of stellar clusters are at the measurement limit, that is, around 1 km/s. The motions seem to be predominantly random and hence we have to apply equ. (1b). While small V would mean large angular momenta, the random addition of the vectors associated with N = 100 to 10000 stars implies reductions by factors $\sqrt{N}$ = 10 to 100, so that finally we arrive with P/M^2 at the same order of magnitude as in the other cases. Whether the greater N of globular clusters as compared with open clusters really mean a difference in a relative angular momentum scale cannot be decided yet. The famous globular cluster ω Centauri is a counterexample, because of its quite noticeable rotation (Woolley,1964).

Galaxies

The motions in spiral and irregular galaxies are mainly rotational with a small range in V (typically 100-200 km/s). As a consequence, P and M are tightly related with $P \sim M^{\alpha}$ and α not far from 2; least square solutions yield $\alpha \approx 7/4$ (Brosche,1962,1963; Takase and Kinoshita,1967; Aikawa and Hitotuyanagi,1974; Brosche, 1971,1973,1977).

The variation perpendicular to the mean relation can be interpreted as a Maxwellian distribution, whereby the deviation from the mean is correlated with the morphological galaxy type (Brosche,1971,1977). The question whether or not the elliptical galaxies fit into this scheme cannot be answered finally at present since observations of their inner motions are only now published in greater numbers. There are however many indications that these galaxies constitute just the low angular momentum side of the galaxian band in the P-M-diagram (Brosche,1978). The most certain empirical fact not explainable by selection effects etc. is the appearance of an upper limit of rotational velocities of galaxies at about 300 km/s.

Clusters of galaxies

The inner motions in clusters and groups of galaxies and in de Vaucouleurs supercluster are a few 100 km/s, that is, of the same order as within galaxies. Probably the motions are random, but this would not lead to much smaller P since N is only of the order of 100. Much more severe at this highest hierarchical rank is the uncertainty whether or not the clusters are in equilibrium; usually the problem is named the one of the 'missing mass'.

Synopsis

We have three classes with fairly good data: planet-satellite-systems, double stars, galaxies. From these, the two extreme ones show small scatter around a mean P-M-relation. We have uncertain data for stellar clusters and an uncertain situation with regard to the basic equilibrium assumption for clusters of galaxies (but not a contradiction). To make it completely clear, the P-M-diagrams of all classes exhibit a real two-dimensional manifold; there is real variation and not only error dispersion around 'mean' relations.

For each class alone, the P-M-diagrams permit a certain range of possible 'mean' P-M-relations with a variation in the exponent α of a power-law $P \sim M^{\alpha}$. But considering all classes together we encounter the astonishing fact that all can be represented by *one and the same relation*

$$P = \varkappa M^2 \tag{3}$$

Herein, $\varkappa \approx 2.4 \cdot 10^{-15}$ cm^2 g^{-1} s^{-1} must be seen as a logarithmic mean. The exponent $\alpha = 2$ has a very small possible variation, since equ. (3) holds for about 20 powers of ten in the mass, and a deviation of only 1/20 in α would lead to a non-tolerable deviation in P at the well-established planetary or galaxian domain.

OTHER OBJECTS

Elementary particles

There is a strange analogy between equ.(3) and the Regge trajectories of families of elementary particles (Brosche,1969). For the latter, $\Delta P/\Delta M^2$ is about constant and roughly $\hbar/m_p^2$; m_p being the mass of a proton. This ratio divided by the corresponding of the astronomical objects (the constant $\varkappa$) is of the order 10^{35}. So we have one more of the ominous large dimensionless numbers and a well-founded one: it is not based on a 'point' in the microcosm and one in the macrocosm but on relations in both realms. And it is not based on vague extrapolations like the number of nucleons in the universe. It should be noticed that it is not easy to propose only a consistent hypothesis for such numbers, as Gamow's proposal of a changing elementary charge shows (Brosche,1970).

Black holes

We do not want to comment on the question whether there is already empirical evidence for the existence of these objects. According to theory, they can have an angular momentum between zero and an upper limit

$$P = (G/c)\ M^2 \tag{4}$$

(c = velocity of light). So this is again a relation of the type found for astronomical objects, but here the constant is $G/c = 2.2 \cdot 10^{-18}\ cm^2\ g^{-1}\ s^{-1}$, which places the extremely fast rotating black holes about 1000 times below the astronomical objects in a P-M-diagram (Brosche, 1974). If we think that black holes may have evolved from astronomical objects, the black holes possessing maximum P are at the natural transient stage and therefore especially important. A priori, a coherence of the astronomical results with the black hole limit seems not unlikely since the dominating interaction is in both cases the same, gravitation.

The existence of particles with $\hbar/m^2 \approx G/c$ has been proposed by several authors (Motz,1962,1970; Markov,1967; Hokkyo,1968).

FINAL REMARKS

From the very fact that we deal with a hierarchy of objects where each lower-rank object cannot be larger than the higher-rank object of which it is a part, we can derive limits on possible angular momenta. But they are only upper limits for inferior objects and lower limits for superior objects; they do not define the found relation. We have deviations at the lower end of the hierarchy in the nature of the angular momentum (solid-body rotation versus orbital motion elsewhere) and in the power-law; we have perhaps also deviations at the upper end depending on the true interpretation of the missing mass enigma.

At the beginning the existence of many equivalent formulations was mentioned. It should be emphasized however, that there are also 'semi-equivalent' formulations, that is, relations following from a P-M-relation while the reverse is not true. This refers to descriptions using only such quantities in which the organized character of motions does not enter, e.g. to kinetic,potential and total mechanical energy and other measures of the same kind (velocity dispersions,radii). So from a mean P-M-relation $P \sim M^2$ it follows with equ. (1a) $M \sim R$ or furthermore density $\rho \sim R^{-2}$. This parallelism to the Schwarzschild limit $R = 2\ GM/c^2$ was detected by de Vaucouleurs (1970). But from $M \sim R$ we cannot infer the P-M-relation without knowledge of the rotational character of the motions. The statement in terms of P and M is stronger.

The statistical nature of the P-M-diagrams as well as some theories of the formation of galaxies and stars seem to point towards a turbulence picture as basis for the understanding of our results. Assuming constant density and a) the Kolmogorov law $V \sim R^{1/3}$ we arrive at $P \sim M^{13/9}$, b) constant angular velocity leads to $P \sim M^{5/3}$, c) constant V (and N) leads to $P \sim M^2$. But whether or not the observed situation is obtained, the ad hoc assumptions of those 'theories' must be seen as being merely equivalent formulations, not explanations.

Theorists who find it cumbersome to understand the formation of *one* class of objects might not be inclined to consider all at once. But perhaps one should remember the situation sometimes occuring in mathematics that a proof of a more general statement is easier than that of a restricted version of the same theorem. It could be necessary to consider unconventional theoretical possibilities like the one outlined by Treder (1976). And since an empirical connection between two variables can be produced causally via a third, one should also pay attention to other empirical relations as e.g. to the discussion of a dependence between rotation and magnetic field which was recently revived by Sirag(1979). This point was kindly mentioned to the author by Prof. Hehl.

References

Aikawa, T. and Hitotuyanagi, Z., 1974, The Mass-Angular Momentum Relation of Normal Spiral Galaxies, Science Reports Tôhoku, First Series,Vol.LVII:121.

Brosche, P., 1962, Zum Masse-Drehimpuls-Diagramm von Doppel- und Einzelsternen, Astron.Nachr. 286:241.

Brosche, P., 1963, Über das Masse-Drehimpuls-Diagramm von Spiralnebeln und anderen Objekten, Zeitschr. f. Astrophys. 57:143.

Brosche, P., 1964, Die Häufigkeit der Doppelsterne in der Sonnenumgebung, Astr. Nachr. 288:33.

Brosche, P., 1967a, Planetensysteme und das Masse-Drehimpuls-Diagramm, Icarus 6:279.

Brosche, P., 1967b, Die maximalen Rotationsgeschwindigkeiten in Galaxien, Zeitschr. f. Astrophys. 66:161.
Brosche, P., 1969, Eine Analogie zwischen Elementarteilchen und astronomischen Objekten und eine neue große dimenslose Zahl, Naturwiss. 56:85.
Brosche, P., 1970, Konstanz der Elementarladung, Phys. Blätter 26:95.
Brosche, P., 1971, Rotation and Type of Galaxies, Astron. & Astrophys. 13:293.
Brosche, P., 1973, The Manifold of Galaxies, Astron. & Astrophys. 23:259.
Brosche, P., 1974, The mass-angular momentum diagram and the black hole limit, Astrophys. & Space Sci. 29:L7.
Brosche, P., 1977, The Distribution of the Angular Momenta of Galaxies, Astrophys. & Space Sci. 51:401.
Brosche, P., 1978, Hubble sequence,angular momenta and time scales of the early evolution of galaxies, Astrophys. & Space Sci. 57:463.
Brosche, P. and Sündermann, J. (eds.), 1978, "Tidal Friction and the Earth's Rotation", Springer-Verlag, Berlin - Heidelberg - New York.
Fish jr., F.F., 1967, Angular momenta of the planets, Icarus 7:251.
Harrison, E.R., 1973, Galaxy formation and the early universe, in: "Cargèse Lectures in Physics", E. Schatzman, ed., Gordon and Breach, New York - London - Paris.
Hartmann, W.K. and Larson, S.M., 1967, Angular momenta of planetary bodies, Icarus 7:257.
Hokkyo, N., 1968, Quantum Fluctuation of Gravitational Field and Creation of Matter, Progr. Theor. Phys. 39:1078.
Markov, M.A., 1967, Elementary Particles of Maximally Large Masses (Quarks and Maximons), Soviet Phys.-JETP 24:584.
Motz, L., 1962, Gauge Invariance and the Structure of Charged Particles, Nuovo Cimento Serie X, 26:672.
Motz, L., 1973, Gauge Invariance and the Quantization of Mass (of Gravitational Charge), Nuovo Cimento Serie XI, 12B:239.
Sirag, S.-P., 1979, Gravitational magnetism, Nature 278:535.
Treder, H.-J., 1976, Rasshirenie i vrashchatel'nyj moment bolshikh kosmicheskikh mass (russ.), Astrofizika 12:511.
de Vaucouleurs, G., 1970, The case for a hierarchical cosmology, Science 167:1203.
Woolley, R.v.d.R., 1964, Rotation of the Globular Cluster ω Centauri, Nature 203:961.

BIMETRIC GENERAL RELATIVITY THEORY

Nathan Rosen

Department of Physics
Technion-Israel Institute of Technology
Haifa, Israel

I. INTRODUCTION

The Einstein general theory of relativity is the most beautiful structure in all of theoretical physics. It has been remarkably successful in describing gravitational phenomena. It has provided a basis for constructing models of the universe. It has also provided a conceptual framework for discussing large-scale phenomena in general.

If there is room for criticism of the general relativity theory, it is in connection with the question of singularities. A satisfactory theory should be free from singularities, for a singularity implies a breakdown of the physical laws. Now in general relativity one encounters singularities in two situations:

1. Cosmological models based on plausible assumptions expand from an initial singularity, the "big bang". It has been suggested that this singularity would be eliminated if quantum effects were taken into account. However, no one has succeeded up to now in showing how this could be accomplished.

2. The Schwarzschild solution for the field of a particle has a singularity at the location of the particle (in addition to a strange geometry inside the Schwarzschild sphere, where space and time coordinates appear to exchange roles). A sufficiently massive star undergoing gravitational collapse will (according to the generally prevailing opinion) contract until it reaches this singularity at the center.

The purpose of the present work is to consider the possibility of modifying the general relativity theory so as to avoid the above singularities. For this one needs a guiding idea. The present approach is based on the fact that the universe appears to have a fundamental rest frame. This is the reference frame in which the Hubble effect and the black-body background radiation are isotropic. Now it is true that general relativity provides cosmological models characterized by such a rest frame. However, one can try to modify the theory by taking into account the existence of this rest frame — a preferred frame of reference — in the foundations of the theory.

It is proposed to modify the general theory of relativity by introducing a second metric tensor into the theory, so that one gets a bimetric form of general relativity. The second metric is to be associated with the fundamental reference frame of the universe.

II. BIMETRIC GENERAL RELATIVITY

Let us begin by considering the general bimetric formalism. It is assumed that, for a given coordinate system with coordinates x^μ, at each point of space-time there exist two metric tensors, $g_{\mu\nu}$ and $\gamma_{\mu\nu}$, corresponding respectively to the line elements

$$ds^2 = g_{\mu\nu}dx^\mu dx^\nu , \tag{2.1}$$

and

$$d\sigma^2 = \gamma_{\mu\nu}dx^\mu dx^\nu . \tag{2.2}$$

The tensor $g_{\mu\nu}$, the physical or gravitational metric, describes gravitation, and it interacts with matter. With its help one can define the Christoffel symbol $\left\{ {\lambda \atop \mu\nu} \right\}$ and hence covariant differentiation (g-differentiation) denoted by a semicolon (;), and one can form the curvature tensor $R^\lambda{}_{\mu\nu\sigma}$. The tensor $\gamma_{\mu\nu}$, the background metric, can be regarded as, in a certain sense, determining inertial forces. With its help one can define the Christoffel symbol $\Gamma^\lambda_{\mu\nu}$ and corresponding covariant differentiation (γ-differentiation) denoted by a bar ($|$), and one can form the curvature tensor $P^\lambda{}_{\mu\nu\sigma}$.

It is found that there exist two interesting addition theorems. The first one has the form

$$\left\{ {\lambda \atop \mu\nu} \right\} = \Gamma^\lambda_{\mu\nu} + \Delta^\lambda_{\mu\nu} , \tag{2.3}$$

where $\Delta^\lambda_{\mu\nu}$ is a tensor having the same form as $\left\{ {\lambda \atop \mu\nu} \right\}$, but with ordinary partial derivatives replaced by γ-derivatives,

$$\Delta^{\lambda}_{\mu\nu} = \frac{1}{2} g^{\lambda\sigma}(g_{\mu\sigma|\nu} + g_{\nu\sigma|\mu} - g_{\mu\nu|\sigma}) \ . \tag{2.4}$$

The second one can be written

$$R^{\lambda}_{\ \mu\nu\sigma} = P^{\lambda}_{\ \mu\nu\sigma} + K^{\lambda}_{\ \mu\nu\sigma}. \tag{2.5}$$

Here again $K^{\lambda}_{\ \mu\nu\sigma}$ has the same form as $R^{\lambda}_{\ \mu\nu\sigma}$, but with ordinary derivatives replaced by γ-derivatives,

$$K^{\lambda}_{\ \mu\nu\sigma} = -\Delta^{\lambda}_{\mu\nu|\sigma} + \Delta^{\lambda}_{\mu\sigma|\nu} - \Delta^{\lambda}_{\sigma\alpha}\Delta^{\alpha}_{\mu\nu} + \Delta^{\lambda}_{\nu\alpha}\Delta^{\alpha}_{\mu\sigma} \ . \tag{2.6}$$

As a special case, one can have $P^{\lambda}_{\ \mu\nu\sigma} = 0$, so that $\gamma_{\mu\nu}$ describes a flat space-time. This was proposed many years ago [1,2,3] as a modification of the formalism of general relativity, but one that does not change the physical contents of this theory. In this case one has

$$R^{\lambda}_{\ \mu\nu\sigma} = K^{\lambda}_{\ \mu\nu\sigma} \ , \tag{2.7}$$

so that the curvature tensor can be written in two forms, either with ordinary derivatives or with γ-derivatives. Contracting, one gets

$$R_{\mu\nu} = K_{\mu\nu} \ , \tag{2.8}$$

so that the Einstein field equations can also be written in two forms. Working with γ-derivatives has certain advantages. For example, one can obtain the Einstein field equations for empty space from the variational principle

$$\delta\int g^{\mu\nu}(\Delta^{\alpha}_{\mu\beta}\Delta^{\beta}_{\nu\alpha} - \Delta^{\alpha}_{\mu\nu}\Delta^{\beta}_{\alpha\beta})\kappa(-\gamma)^{1/2} d\tau = 0 \ , \tag{2.9}$$

where the scalar κ is given by

$$\kappa = (g/\gamma)^{1/2} \ , \tag{2.10}$$

and $g_{\mu\nu}$ is varied while $\gamma_{\mu\nu}$ is kept fixed. Since the integrand is a scalar density, one can derive an energy-momentum density tensor for the gravitational field in place of the pseudo-tensor that one has in the conventional form of the general relativity theory.

Up to this point we have not considered how the two metric tensors are related. It is natural to assume that, far from matter, where the gravitational field vanishes, $g_{\mu\nu} = \gamma_{\mu\nu}$. However, because of the identities existing among the Einstein field equations, for a given $\gamma_{\mu\nu}$, the solution for $g_{\mu\nu}$

will contain four arbitrary functions. It is therefore desirable to add four equations in order to tie the two tensors together. One possibility is to take

$$\left(\kappa g^{\mu\nu}\right)_{|\nu} = 0 . \tag{2.11}$$

This is a generalization of the De Donder condition often used in general relativity. It should be noted that, while the latter is noncovariant and essentially fixes the coordinate system, (2.11) is covariant and permits general coordinate transformations, under which both tensors transform. However, although Eq. (2.11) seems to be reasonable as a condition on $g_{\mu\nu}$, the fact is that it is arbitrary. Other conditions are possible [3].

The present approach differs from the earlier one in that the background metric is taken as describing, not flat space-time, but rather a space-time related to that of the universe.

Let us begin with some general considerations. In trying to form a picture of the universe, let us take the standpoint that it is something unique in the sense that, although it is based on general physical laws, it involves special conditions. In choosing a background metric $\gamma_{\mu\nu}$ we therefore assume that it describes a space-time which, while not flat, nevertheless has maximum symmetry, i.e., a space-time of constant curvature, so that

$$P_{\lambda\mu\nu\sigma} = \frac{1}{a^2}\left(\gamma_{\lambda\sigma}\gamma_{\mu\nu} - \gamma_{\lambda\nu}\gamma_{\mu\sigma}\right) , \tag{2.12}$$

and hence

$$P_{\mu\nu} = \frac{3}{a^2}\gamma_{\mu\nu} , \tag{2.13}$$

where $\underline{a}$ is a (positive) constant.

In this case, by a suitable choice of coordinates, one can write $d\sigma^2$ in three forms, corresponding to various values of the spatial curvature k,

$$k = 1 : \quad d\sigma^2 = dt^2 - a^2\cosh^2(t/a)(d\chi^2 + \sin^2\chi d\Omega^2) , \tag{2.14}$$

$$k = 0 : \quad d\sigma^2 = dt^2 - e^{2t/a}(dr^2 + r^2 d\Omega^2) , \tag{2.15}$$

$$k = -1 : \quad d\sigma^2 = dt^2 - a^2\sinh^2(t/a)(d\chi^2 + \sinh^2\chi d\Omega^2) , \tag{2.16}$$

where

$$d\Omega^2 = d\theta^2 + \sin^2\theta d\phi^2 . \tag{2.17}$$

These correspond to three different ways of splitting up the space-time into space and time.

In choosing among these possibilities, let us be guided by "the principle of finiteness", according to which one should avoid singularities or infinities in the universe. One would expect the space-time described by $g_{\mu\nu}$ to have the same topology as that described by $\gamma_{\mu\nu}$, in order to avoid singularities. If we consider an isotropic model of the universe, as described by $g_{\mu\nu}$, and look for one containing a finite amount of matter (say, a finite number of baryons), then the model must be spatially closed (k = 1). This then must also be the case for the space described by $\gamma_{\mu\nu}$. Hence we take (2.14), with $k = 1$. The corresponding metric can be regarded as providing a framework, or skeleton, for the universe, so that in the absence of matter $g_{\mu\nu} = \gamma_{\mu\nu}$. With matter present $g_{\mu\nu}$, as determined by the field equations, will be different from $\gamma_{\mu\nu}$.

In view of the fact that now (2.7) and (2.8) do not hold, let us modify the Einstein gravitational field equations by replacing $R_{\mu\nu}$ by $K_{\mu\nu}$. This means that, if we derive the field equations from a variational principle, the latter should contain a term like the integrand in (2.9) to give the left side of the field equations, rather than the corresponding term involving Christoffel symbols that one has in general relativity. We therefore have, as the field equations,

$$K^{\nu}_{\mu} - \frac{1}{2}\,\delta^{\nu}_{\mu}K = -\,8\pi T^{\nu}_{\mu}\,. \tag{2.18}$$

Here T^{ν}_{μ} is the energy-momentum density tensor of the matter or other non-gravitational fields. Indices are raised and lowered with the help of $g_{\mu\nu}$ unless otherwise indicated.

Writing

$$K_{\mu\nu} = R_{\mu\nu} - P_{\mu\nu}\,, \tag{2.19}$$

and taking (2.13) into account, we can put the field equations into the form

$$G^{\nu}_{\mu} = S^{\nu}_{\mu} - 8\pi T^{\nu}_{\mu}\,, \tag{2.20}$$

where G^{ν}_{μ} is the Einstein tensor and

$$S^{\nu}_{\mu} = \frac{3}{a^2}\,\gamma_{\mu\alpha}g^{\alpha\nu} - \frac{1}{2}\,\delta^{\nu}_{\mu}\gamma_{\alpha\beta}g^{\alpha\beta}\,. \tag{2.21}$$

In view of the Bianchi identity,

$$G^{\nu}_{\mu;\nu} \equiv 0\,, \tag{2.22}$$

we have

$$8\pi T^{\nu}_{\mu;\nu} = S^{\nu}_{\mu;\nu} \ . \tag{2.23}$$

In empty space $(T^{\nu}_{\mu} = 0)$,

$$S^{\nu}_{\mu;\nu} = 0 \ . \tag{2.24}$$

It is natural to take this to hold everywhere, so that, from (2.23),

$$T^{\nu}_{\mu;\nu} = 0 \ , \tag{2.25}$$

as in the Einstein theory. Now from (2.21) one finds

$$S^{\nu}_{\mu;\nu} = (3/a^2\kappa)\gamma_{\mu\lambda}(\kappa g^{\lambda\nu})_{|\nu} \ . \tag{2.26}$$

Hence (2.24) gives the condition (2.11) that was considered earlier. However, now it is not an arbitrary choice; it follows from the form of the field equations.

In a given coordinate system, with a given background metric $\gamma_{\mu\nu}$, the gravitational metric $g_{\mu\nu}$ is determined by the field equations (2.20) together with (2.11). We have here a case of over-determination, since there are 14 equations (satisfying the four Bianchi identities) for the ten components $g_{\mu\nu}$. Hence there will be severe restrictions on the initial conditions that can be imposed. In the case of the Einstein field equations one often obtains solutions which do not appear to have physical significance. One can expect that many solutions of this kind will be ruled out by the present overdetermination.

In the rest-frame of the universe, with a suitable choice of coordinates, $\gamma_{\mu\nu}$ is given by (2.14). In any other coordinate system, $\gamma_{\mu\nu}$ is determined by the transformation equations from this coordinate system. While the field equations are covariant, they take on their simplest form in the fundamental rest-frame.

One can expect the scale constant $\underline{a}$ appearing in the field equations to be of the order of $1/H$, where H is the Hubble constant, i.e., of the order of 10^{28} cm. If we are dealing with physical systems having dimensions that are small compared to $\underline{a}$, such as the solar system, then in general S^{ν}_{μ} is negligible in (2.20), and the field equations are equivalent to the Einstein equations. However, in cosmological problems this term should be important.

III. COSMOLOGY

Isotropic Universe

Let us consider a closed isotropic model of the universe (k = 1). Taking a coordinate system in which (2.14) holds, we can write

$$ds^2 = e^{2\phi}dt^2 - a^2\cosh^2(t/a)e^{2\psi}(d\chi^2 + \sin^2\chi d\Omega^2) , \tag{3.1}$$

where ϕ and ψ are functions of t. As usual, let us assume that the universe is filled with matter characterized by a density $\rho(t)$ and a pressure $p(t)$, so that the non-vanishing components of T^{ν}_{μ} are

$$T^0_0 = \rho , \qquad T^1_1 = T^2_2 = T^3_3 = - p . \tag{3.2}$$

It is convenient to introduce the dimensionless variable

$$x = t/a . \tag{3.3}$$

The field equations (2.20) then give two relations

$$e^{-2\phi}(\psi'' + \tanh x)^2 + e^{-2\psi}\text{sech}^2 x = \frac{1}{2}(3e^{-2\psi} - e^{-2\phi}) + \frac{8\pi}{3}a^2\rho , \tag{3.4}$$

$$e^{-2\phi}[2x'' + 3(\psi' + \tanh x)^2 - 2\phi'(\psi' + \tanh x)] + (2e^{-2\phi}+e^{-2\psi})\text{sech}^2 x = \frac{3}{2}(e^{-2\phi}+e^{-2\psi}) - 8\pi a^2 p , \tag{3.5}$$

where a prime denotes a derivative with respect to x.

In addition we have (2.25), which gives the relation

$$\rho' + 3\frac{R'}{R}(\rho + p) = 0 , \tag{3.6}$$

where R is the radius of the universe,

$$R = a\cosh x\, e^{\psi} , \tag{3.7}$$

and also the condition (2.11), which takes the form

$$3\psi' - \phi' = 3\tanh x\,(e^{2\phi-2\psi} - 1) . \tag{3.8}$$

In looking for solutions of the field equations, we make some assumptions about the behavior of the universe, regarded as a unique system with special conditions. We assume that there is a certain moment, which we take as $t = 0$, such that the behavior of the universe is symmetric with respect to it, i.e., that the field variables (ϕ,ψ,ρ) are even functions of t. In that case one can regard the universe as developing from $t = -\infty$ to $t = +\infty$ (or in the reverse direction), or one can think of it as starting from $t = 0$ and developing (in either direction) towards large values of t. From the second point of view, there exists an initial moment $(t = 0)$, and one can specify initial conditions, but if one asks what happened before that, the answer is: the same as what happened after it.

As we shall see, at the initial moment $(t = 0)$ the density has its maximum value, and the universe expands as the time increases. As t goes to infinity, ρ tends to zero. One would expect that $g_{\mu\nu}$ should then approach $\gamma_{\mu\nu}$, as the presence of the matter becomes unimportant, and this will be assumed to be the case. One can say therefore that we have both initial conditions $(t = 0)$ and final conditions $(t = \infty)$, but this refers to the way we go about solving the field equations. One can imagine a "four-dimensional" point of view in which the history of the universe $(-\infty < t < \infty)$ is regarded in its entirety and which is completely determined over its entire length.

Hence we look for a solution of the field equations for $x \gtrsim 0$, taking for $x = 0$,

$$\rho = \rho_o , \qquad \phi = \phi_o , \qquad \psi = \psi_o$$

$$\rho' = \phi' = \psi' = 0$$

with the additional conditions that, as $x \to \infty$, $\phi,\psi \to 0$. For a given value of ρ_o these conditions should determine ϕ_o and ψ_o. It is convenient in the calculation to write

$$z = \psi - \psi_o , \tag{3.9}$$

so that, for $x = 0$, $z = 0$, $z' = 0$; for $x \to \infty$, $z \to -\psi_o$.

What can one say about the initial density? The present standpoint is that this is a fundamental quantity and that its order of magnitude is determined by the fundamental constants c, $\hbar$ and G. From dimensional considerations one can write for the density in conventional units

$$\rho_o = \alpha c^5 \hbar^{-1} G^{-2} \sim 10^{93} \text{ gm/cm}^3 ,$$

and, in the general-relativity units used here,

$$\rho_o = \alpha c^3 \hbar^{-1} G^{-1} \sim 10^{65} \text{ cm}^{-2} , \tag{3.10}$$

where α is a coefficient of the order of unity. We have here an extremely large value, and this leads to large values for $|\phi_o|$ and $|\psi_o|$, as we shall see.

After having found the solution of the field equations, one would like to compare its consequences with observation. Let us consider the Hubble effect. It should first be noticed from (3.1) that the time T as given by a clock at rest in our coordinate system is related to the coordinate time $\underline{t}$ by

$$T = \int_o^t e^{\phi} dt = a \int_o^x e^{\phi} dx . \tag{3.11}$$

The Hubble constant H is given by

$$H = \frac{1}{R}\frac{dR}{dT} , \tag{3.12}$$

with R as in (3.7). It follows that

$$h \equiv Ha = e^{-\phi}(\psi' + \tanh x) . \tag{3.13}$$

The deceleration parameter $\underline{q}$ is defined as

$$q = - R \frac{d^2 R}{dT^2} \bigg/ \left(\frac{dR}{dT}\right)^2 , \tag{3.14}$$

and this can be written

$$q = - 1 + \frac{\phi'}{\psi' + \tanh x} - \frac{\psi'' + \text{sech}^2 x}{(\psi' + \tanh x)^2} . \tag{3.15}$$

In order to be able to apply the solution of the field equations to the description of the present state of the universe, we must determine two quantities, the scale parameter $\underline{a}$ and the value of $\underline{t}$ corresponding to the present time, or the corresponding value of $\underline{x}$. For this purpose we can make use of two pieces of observational data, the mean density of matter in the universe [4], $\rho = 4\times10^{-31}$ gm/cm^3 $= 3\times10^{-59}$ cm^{-2}, and the Hubble constant H, for which we write [5] $H^{-1} = 2\times10^{10}$ ℓt-yr $=$ 2×10^{28} cm.

Dust-Filled Universe

Let us consider the case of a universe filled with dust, so that $p = 0$, this being a fairly good approximation to the present situation in the real universe. In that case (3.6) can be integrated to give

$$\rho = \frac{A}{R^3} = \frac{A}{a^3} e^{-3\psi} \operatorname{sech}^3 x \quad (A = \text{const.}) . \tag{3.16}$$

Let us take

$$y = 2(\phi - \psi) . \tag{3.17}$$

Then (3.8) gives

$$\psi' = \frac{1}{4} y' + \frac{3}{2} \tanh x \, (e^y - 1) . \tag{3.18}$$

Equation (3.5) can now be put into the form

$$y'' - \frac{3}{8} y'^2 + \frac{3}{2} y' \tanh x \, (3e^y + 1) + \frac{1}{2} \tanh^2 x \, (9e^{2y} - 10e^y + 1) + 5(e^y - 1) = 0 , \tag{3.19}$$

while (3.4) can be written

$$\frac{1}{4} y'^2 + y' \tanh x \, (3e^y - 1) + \tanh^2 x \, (9e^{2y} - 10e^y + 1) + 2(1 - e^y) = N , \tag{3.20}$$

where

$$N = \frac{32\pi}{3} a^2 \rho e^{2\phi} , \tag{3.21}$$

so that, by (3.16) and (3.17) ,

$$N e^{\psi - y} \cosh^3 x = 32\pi A / 3a . \tag{3.22}$$

From (3.20) and (3.21) one sees that

$$8\pi a^2 \rho_o = \frac{3}{2} e^{-2\phi_o} (1 - e^{y_o}) , \tag{3.23}$$

so that $y_o < 0$.

Let us first investigate the solution of Eq. (3.19) near $x = 0$ by assuming that y is large and negative so that the exponential functions can be neglected, $e^y \ll 1$. The equation then has the form

$$y'' - \frac{3}{8} y'^2 + \frac{3}{2} y' \tanh x + \frac{1}{2} \tanh^2 x - 5 = 0 . \tag{3.24}$$

The solution satisfying the initial conditions is given by

$$y = y_o - \frac{8}{3} \ln \cos w + 2\ln \cosh x \ , \tag{3.25}$$

with $w = 3x/8^{1/2}$. From (3.17) and (3.18) one then gets (with $\phi_o - \psi_o = \frac{1}{2} y_o$)

$$\phi = \phi_o - 2\ln \cos w \ , \tag{3.26}$$

and

$$\psi = \psi_o - \frac{2}{3} \ln \cos w - \ln \cosh x \ . \tag{3.27}$$

Eq. (3.20) gives

$$N = 2\sec^2 w \ , \tag{3.28}$$

so that, from (3.21) ,

$$8\pi a^2 \rho = \frac{3}{2} e^{-2\phi_o} \cos^2 w \ . \tag{3.29}$$

Comparing this with (3.16), one finds

$$8\pi A/a = \frac{3}{2} e^{\psi_o - y_o} \ . \tag{3.30}$$

It should be noted that, at $x = 0$, $N = 2$ and

$$8\pi a^2 \rho_o = \frac{3}{2} e^{-2\phi_o} \ , \tag{3.31}$$

in agreement with (3.23) in the present approximation.

The clock time T , as given by Eq. (3.11) , is found from (3.26) to be

$$T = \frac{8^{\frac{1}{2}}}{3} a e^{\phi_o} \tan w \ , \tag{3.32}$$

and the radius of the universe, as given by (3.7), now has the form

$$R = a e^{\psi_o} \sec^{2/3} w \ . \tag{3.33}$$

We see that the above solution becomes singular for $x = x^*$, with $3x^*/8^{\frac{1}{2}} = \pi/2$, $x^* = 1.48096 \ldots$ However, in the exact equation (3.19), as x approaches x^* , the exponential functions become large and the character of the solution changes, so that no singularity occurs, although the functions and their derivatives may take on very large values near x^* .

The approximate solution can be valid very close to x^* if $- y_o$ is sufficiently large. If we write

$$x = x^* - \xi \qquad (\xi > 0) , \tag{3.34}$$

then it is found that, for $\xi \ll 1$,

$$y = y_o + \frac{8}{3}\ln\lambda + 2\ln\cosh x^* , \tag{3.35}$$

$$\phi = \phi_o + 2\ln\lambda , \tag{3.36}$$

$$\psi = \psi_o + \frac{2}{3}\ln\lambda - \ln\cosh x^* , \tag{3.37}$$

with $\lambda = 8^{1/2}/3\xi$. If, for example, one has $y_o = -100$, then, for $\xi = 3.89\times10^{-15}$, (3.35) gives $y = -10$, so that the approximation based on $e^y \ll 1$ is still a good one.

However, to get the solution for x larger than x^* one must solve the equations numerically. This was done for Eq. (3.19), for a number of values of y_o , with z obtained by integrating from $x = 0$ the relation

$$z' = \frac{1}{4}y' + \frac{3}{2}\tanh x\,(e^y - 1) . \tag{3.38}$$

Some interesting results were obtained:

The functions are finite near x^* but vary very rapidly with x . In particular, for large $|y_o|$ y has an extremely sharp maximum (> 0) near x^* and then, remaining positive, decreases to zero as x increases. On the other hand, ϕ and ψ , initially negative, increase rapidly near x^* but remain negative in this region. In the range $20 \lesssim -y_o \lesssim 100$ and for $x \gtrsim 1.5$ the functions y, ϕ and ψ are nearly independent of y_o . It is found that ψ remains negative everywhere, tending to zero as x increases, while ϕ reaches zero in the neighborhood of $x = 2.1$, attains a small positive maximum (~ 0.02) near $x = 2.5$ and then decreases to zero as x increases.

The value of y_o determines ϕ_o and ψ_o . In the above range one finds

$$\phi_o = \frac{3}{2}y_o + 4.1 , \qquad \psi_o = y_o + 4.1 , \tag{3.39}$$

and (3.30) then gives

$$8\pi A/a \doteq 91 . \tag{3.40}$$

If we write $D = 8\pi a^2\rho$, then we have approximately

$$D = 91e^{-3\psi}\text{sech}^3 x . \tag{3.41}$$

Going back to (3.13), we see that

$$D/h^2 = 8\pi\rho/H^2 \ , \tag{3.42}$$

depends only on x. For the values of ρ and H mentioned earlier $D/h^2 = 0.3$. From the numerical calculations one finds that for this value $x = 2.6$ approximately. From (3.13) one gets $h = 1.16$, so that $a = 2.3\times10^{28}$ cm. This also gives $Ht = 3.0$. From (3.11) one gets $HT = 1.2$, so that the age of the universe is only slightly larger than $1/H$. For $x = 2.6$, one finds $q = -0.67$. However, it should be recalled that these results are based on the dust-filled model, which cannot provide a good description of the early stages of our universe. Using a more appropriate model for the early history of the universe and then going over to the dust-filled model would probably lead to different values of the above parameters. It should also be remarked that there appears to be considerable uncertainty in the value of q_o as estimated from recent observations [5].

As for the initial conditions, if in (3.10) we take $\alpha = 1$, so that $\rho_o = 4\times10^{65}\ \mathrm{cm}^{-2}$, then from (3.31) and (3.39), with the above value of a, one gets $y_o = -97.6$, so that $\phi_o = -142.3$, $\psi_o = -93.5$. Incidentally, one finds for the initial value of R,

$$R_o = ae^{\psi_o} = 6\times10^{-13}\ \mathrm{cm} \ , \tag{3.43}$$

which is comparable to the size of an elementary particle.

In the region $0 \lesssim x \lesssim x^*$ the values of T/a and R/a are extremely small. Only for $x > x^*$ do they become appreciable. In Fig. 1 R/a as a function of T/a is shown by the curve labelled "DUST".

For large values of x, for which y, ϕ and $|\psi|$ are small compared to 1 and $\tanh x = 1$, Eq. (3.19) can be written

$$y'' + 6y' + 9y = 0 \ , \tag{3.44}$$

the solution of which is given by

$$y = (C_1 + C_2 x)e^{-3x} \ . \tag{3.45}$$

Equation (3.20) becomes

$$N = 2y' + 6y \ , \tag{3.46}$$

and, if (3.45) is substituted into it, one gets

$$N = 2C_2 e^{-3x} \ . \tag{3.47}$$

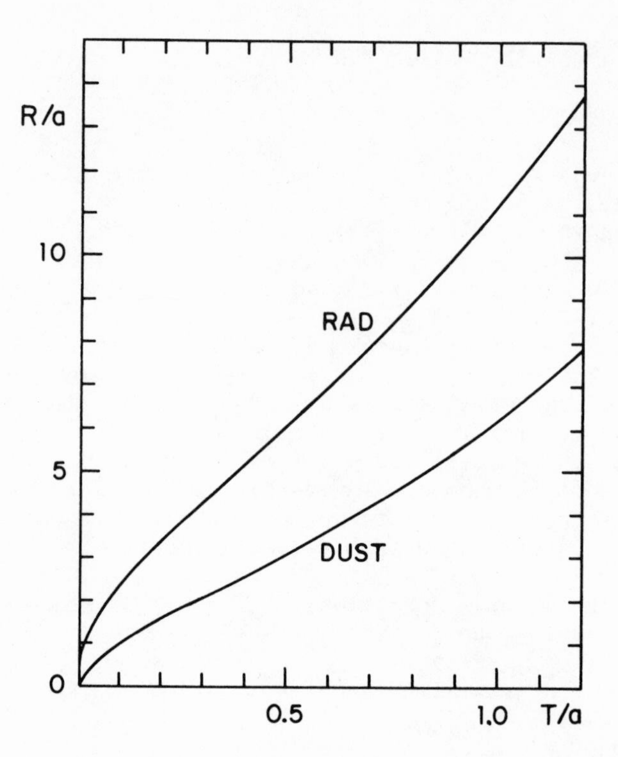

Fig. 1. The radius of the universe R as a function of the clock time T for the dust-filled model (DUST) and the radiation-filled model (RAD), with a the scale parameter.

On the other hand, from (3.21) and (3.16) we have, for large values of x ,

$$N = \frac{256\pi A}{3a} e^{-3x} . \tag{3.48}$$

Using (3.40) we then get

$$C_2 \doteq 485 . \tag{3.49}$$

If one fits y and y' , as obtained from (3.45) , to the values of y and y' obtained numerically for, say, x = 5 , one finds

$$C_1 \doteq - 900 , \tag{3.50}$$

while C_2 is close to the value in (3.49) . A simple calculation then gives, approximately,

$$\phi = (- 305 + 121 x)e^{-3x} , \tag{3.51}$$

$$\psi = (145 - 121\,x)e^{-3x} \qquad (3.52)$$

Radiation-Filled Universe

The dust-filled model considered above is appropriate to the present state of the universe, since the mean pressure of the matter is small. In the distant past, when the density was much greater, the pressure was also large and, as remarked earlier, the assumed model cannot give a good approximation to the situation that existed at that time. During the very early history of the universe, when the density, pressure and temperature were extremely high, there was a period when the universe was dominated by radiation. One can describe this situation by means of a radiation-filled model of the universe. This will be briefly considered here.

For isotropic radiation one has the equation of state

$$p = \frac{1}{3}\rho\,. \qquad (3.53)$$

If this is introduced into Eq. (3.6), one gets

$$\rho = \frac{B}{R^4} = \frac{B}{a^4}\,e^{-4\psi}\,\mathrm{sech}^4\,\frac{t}{a} \qquad (B = \text{const.})\,. \qquad (3.54)$$

Let us make use of the variable y as before. If we add (3.4) and (3.5) taking into account (3.53), (3.17) and (3.18), we get

$$y'' - \frac{1}{4}y'^2 + y'\tanh x\,(6e^y+1) + \tanh^2 x\,(9e^{2y}-10e^y+1) + 4(e^y-1) = 0\,. \qquad (3.55)$$

From (3.4) we have (3.20) with N given by (3.21). However, in light of (3.54) we now have

$$Ne^{2\psi-y}\cosh^4 x = 32\pi B/3a^2\,. \qquad (3.56)$$

If we assume $e^y \ll 1$, Eq. (3.55) takes the form

$$y'' - \frac{1}{4}y'^2 + y'\tanh x + \tanh^2 x - 4 = 0\,. \qquad (3.57)$$

The solution satisfying the initial conditions is then given by

$$y = y_o - 4\ln\cos u + 2\ln\cosh x\,, \qquad (3.58)$$

$$\phi = \phi_o - 3\ln\cos u\,, \qquad (3.59)$$

$$\psi = \psi_o - \ln \cos u - \ln \cosh x , \tag{3.60}$$

with $u = 2^{-\frac{1}{2}} x$.

Eq. (3.20) gives

$$N = 2\sec^2 u , \tag{3.61}$$

so that

$$8\pi a^2 \rho = \frac{3}{4} N e^{-y-2\psi} = \frac{3}{2} e^{-y_o-2\psi_o} \cos^4 u . \tag{3.62}$$

On the other hand, from (3.54) and (3.60)

$$8\pi a^2 \rho = \frac{8\pi B}{a^2} e^{-4\psi_o} \cos^4 u . \tag{3.63}$$

Hence

$$\frac{8\pi B}{a^2} = \frac{3}{2} e^{-y_o+2\psi_o} , \tag{3.64}$$

and

$$8\pi a^2 \rho_o = \frac{3}{2} e^{-y_o-2\psi_o} . \tag{3.65}$$

One also gets

$$T = 2^{-\frac{1}{2}} a e^{\phi_o} \left[\frac{\sin u}{\cos^2 u} + \ln(\sec u + \tan u) \right] , \tag{3.66}$$

and

$$R = a e^{\psi_o} \sec u . \tag{3.67}$$

Obviously the above solution becomes singular for $x = x^*$ with $2^{-\frac{1}{2}} x^* = \pi/2$, $x^* = 2.22144...$ Here again such a singularity does not occur in the solution of the exact equations, although the functions and their derivatives may take on large values for x near to x^* .

Solutions of the equations were obtained numerically. It was found that qualitatively the behavior was somewhat similar to that for the dust-filled model. One must have y_o negative and large. One then finds that y increases very rapidly near x^* reaching a sharp positive maximum and then decreasing to zero as x increases. The functions ϕ and ψ also increase rapidly near x^* , but ψ remains negative while ϕ attains a small positive maximum, both functions then tending to zero as x goes to infinity. For large values of $|y_o|$, say, in the range $60 \lesssim |y_o| \lesssim 160$ one finds that, with $x \gtrsim 2.222$, the functions are nearly independent of the value of y_o . One also finds that in this range of y_o

$$\phi_o = y_o + 3.6 \ , \qquad \psi_o = \frac{1}{2} y_o + 3.6 \ . \qquad (3.68)$$

The relation between R and T is shown in Fig. 1 by the curve labelled "RAD".

It is found that in the range $10^{-50} \lesssim T/a \lesssim 10^{-9}$ one can write in good approximation

$$\frac{R}{a} = 7.15 \left(\frac{T}{a}\right)^{\frac{1}{2}} \ . \qquad (3.69)$$

If one takes $a = 2\times10^{28}$ cm and $\rho_o = 4\times10^{65}$ cm^{-2} , as before, one gets from (3.65) and (3.68)

$$y_o = -\ 145.7 \ , \quad \phi_o = -\ 142.1 \ , \quad \psi_o = -\ 69.25 \ . \qquad (3.70)$$

This gives $R_o = 0.0168$ cm. From these initial values one gets

$$\rho = \frac{3.9\times10^{107}}{R^4} \ , \qquad (3.71)$$

with R in cm and ρ in erg cm^{-3} . On the other hand, if we consider a time T such that muon-pair annihilation has taken place, but the universe is still lepton-dominated, one can write [6]

$$\rho = \frac{9}{2} \sigma\Theta^4 \ , \qquad (3.72)$$

where Θ is the radiation temperature (in °K) and σ is the Stefan-Boltzmann constant (7.59×10^{-15} erg cm^{-3} deg^{-4}). Comparing this with (3.71) one gets

$$\Theta = \frac{1.8\times10^{30}}{R} \ . \qquad (3.73)$$

Now (3.69) can be written

$$R = 1.8\times10^{20}\ T^{\frac{1}{2}} \ , \qquad (3.74)$$

with R in cm and T in sec. Substituting into (3.73) gives

$$\Theta = 10^{10}\ T^{-\frac{1}{2}} \ . \qquad (3.75)$$

The time dependence of the radiation temperature at this stage in the development of the universe is practically the same as in the general relativity theory, and it appears to account for the present cosmic abundance of helium and other light elements [7] .

We see from this cosmological model that, on the basis of the present field equations, one can obtain a satisfactory description of the early stages of the universe without a singular initial state.

IV. THE FIELD OF A PARTICLE

Let us now consider the spherically symmetric gravitational field of a particle at rest in our coordinate system. It was pointed out earlier that, for a small physical system, the present field equations would, in general, agree with the Einstein field equations. However, this may not always be the case; if the field is very intense, the two sets of equations may give different predictions. This could happen, for example, in the case of the field of the particle near the Schwarzschild radius.

One would expect that the field in the vicinity of the particle should not be influenced appreciably by the large-scale curvature of the universe. Hence, in order to simplify the calculations, let us neglect the spatial curvature, and let us take $\gamma_{\mu\nu}$ as given by (2.15), with $k = 0$, instead of (2.14), with $k = 1$. This has the additional effect of changing the time dependence: in place of $\cosh(t/a)$ we now have $\exp(t/a)$. However, for $t/a \gtrsim 1$ the qualitative behavior will be the same.

Even with this simplification we are still faced with a difficult situation: the field equations are partial differential equations (with t and r as independent variables) and are hard to solve. Let us therefore carry out the coordinate transformation

$$t = t' + \frac{1}{2} a \ln (1 - r'^2/a^2) , \tag{4.1}$$

$$r = r' e^{-t'/a} (1 - r'^2/a^2)^{-\frac{1}{2}} . \tag{4.2}$$

Then (2.15) goes over into the static de Sitter line element, which, if we drop the primes, has the form

$$d\sigma^2 = (1-r^2/a^2)dt^2 - (1-r^2/a^2)^{-1}dr^2 - r^2 d\Omega^2 . \tag{4.3}$$

The spherically symmetric field of a particle at the origin is then described by the line element

$$ds^2 = (1-r^2/a^2)e^{\nu}dt^2 - (1-r^2/a^2)^{-1}e^{\lambda}dr^2 - r^2 e^{\mu}d\Omega^2 , \tag{4.4}$$

where λ, μ and ν are functions of r.

Let us take the field equations (2.20) with $T_\mu^\nu = 0$. These are now ordinary differential equations. After some rearrangement one can write them, with a prime now denoting a derivative with respect to r ,

$$(1-r^2/a^2)(\mu''+\frac{3}{4}\mu'^2-\frac{1}{2}\lambda'\mu'+3\mu'/r-\lambda'/r+1/r^2-e^{\lambda-\mu}/r^2)$$

$$=(1/a^2)(r\mu'+\frac{3}{2}e^{\lambda-\nu}-2e^{\lambda-\mu}+\frac{1}{2}) , \quad (4.5)$$

$$(1-r^2/a^2)(\frac{1}{4}\mu'^2+\frac{1}{4}\mu'\nu'+\mu'/r+\nu'/r+1/r^2-e^{\lambda-\mu}/r^2)$$

$$=(1/a^2)(r\mu'-\frac{3}{2}e^{\lambda-\nu}-2e^{\lambda-\mu}+\frac{7}{2}) , \quad (4.6)$$

$$(1-r^2/a^2)(\frac{1}{2}\mu''+\frac{1}{2}\nu''+\frac{1}{4}\mu'^2+\frac{1}{4}\nu'^2-\frac{1}{4}\lambda'\mu'-\frac{1}{4}\lambda'\nu'+$$

$$+\frac{1}{4}\mu'\nu'+\mu'/r-\frac{1}{2}\lambda'/r+\frac{1}{2}\nu'/r)$$

$$=(1/a^2)(r\mu'-\frac{1}{2}r\lambda'+\frac{3}{2}r\nu'-\frac{3}{2}e^{\lambda-\nu}+\frac{3}{2}) . \quad (4.7)$$

We also have the divergence condition (2.11), which can be written

$$(1-r^2/a^2)(2\mu'-\lambda'+\nu'+4/r-4e^{\lambda-\mu}/r) = 2(r/a^2)(1-e^{\lambda-\nu}) . \quad (4.8)$$

If we take $1/a = 0$, the solution of the equations is well known [1] ,

$$ds^2 = \frac{1-m/r}{1+m/r}dt^2 - \frac{1+m/r}{1-m/r}dr^2 - (r+m)^2d\Omega^2 , \quad (4.9)$$

so that

$$e^{\nu} = e^{-\lambda} = (1-m/r)(1+m/r), \quad e^{\mu} = (1+m/r)^2 . \quad (4.10)$$

This solution can be obtained from the usual form of the Schwarzschild solution,

$$ds^2 = (1-2m/r')dt^2 - (1-2m/r')^{-1}dr'^2 - r'^2d\Omega^2 , \quad (4.11)$$

by the relation

$$r' = r + m . \quad (4.12)$$

There is a singularity at $r = m$, the Schwarzschild radius. If we write

$$y = r - m , \quad (4.13)$$

then, for small values of y, the solution can be written

$$e^{\lambda} = 2m/y + 1 , \tag{4.14}$$

$$e^{\mu} = 4 - 4y/m + \ldots , \tag{4.15}$$

$$e^{\nu} = y/2m - y^2/4m^2 + \ldots \tag{4.16}$$

If we assume that $1/a$ is finite but small and substitute the functions of (4.10) into the right sides of the field equations, we see that, near the Schwarzschild radius, the terms involving $\exp(\lambda-\nu)$ behave like $(r-m)^{-2}$. Hence, even for small values of $1/a$, the right sides of the equations become important and cannot be neglected.

To study the behavior of the solution of our field equations near $r = m$, with $m/a \ll 1$, we can write the equations in the simplified form,

$$\mu'' + \frac{3}{4}\mu'^2 - \frac{1}{2}\lambda'\mu' + 3\mu'/r - \lambda'/r + 1/r^2 - e^{\lambda-\mu}/r^2 = \frac{3}{2}e^{\lambda-\nu}/a^2 , \tag{4.17}$$

$$\frac{1}{4}\mu'^2 + \frac{1}{4}\mu'\nu' + \mu'/r + \nu'/r + 1/r^2 - e^{\lambda-\mu}/r^2 = -\frac{3}{2}e^{\lambda-\nu}/a^2 , \tag{4.18}$$

$$\mu'' + \nu'' + \frac{1}{2}\mu'^2 + \frac{1}{2}\nu'^2 - \frac{1}{2}\lambda'\mu' - \frac{1}{2}\lambda'\nu' + \frac{1}{2}\mu'\nu' + 2\mu'/r + \\ + \nu'/r - \lambda'/r = -3e^{\lambda-\nu}/a^2 , \tag{4.19}$$

$$2\mu' - \lambda' + \nu' + 4/r - 4e^{\lambda-\mu}/r = -2(r/a^2)e^{\lambda-\nu} , \tag{4.20}$$

where terms have been neglected which are small compared to other terms having the same behavior near $r = m$.

Let us suppose that the solution of these equations has a singularity at $r = m$. (This m may be different from that characterizing the field of the particle at large distances, although of the same order of magnitude.) We look for a solution near $r = m$ as an expansion in powers of $y = r - m$. One finds

$$e^{\lambda} = \frac{3}{8}b^2(m/y + \frac{11}{8} + \ldots) , \tag{4.21}$$

$$e^{\mu} = b^2(1 - y/2m + \ldots) , \tag{4.22}$$

$$e^{\nu} = \frac{3}{2}(b^2m^2/a^2)(1 - 3y/m + \ldots) , \tag{4.23}$$

where b^2 is the value of e^{μ} for $r = m$.

We see that this solution is different from that given above for $1/a = 0$. While the behavior of e^λ is similar, that of e^ν is quite different. According to (4.16) e^ν vanishes at $r = m$ and is negative for $r < m$, while according to (4.23) e^ν is very small near $r = m$, but remains positive as r decreases below this value. Since, for $1/a = 0$, the signs of e^λ and e^ν both change as one goes into the Schwarzschild sphere, the signature of the metric remains unchanged. Here, with $1/a \neq 0$, only the sign of e^λ changes, so that there is a change in the signature — we get two time-like coordinates —, and the region $r < 2m$ is therefore unphysical.

One can ask what happens to a test particle that falls down to the Schwarzschild sphere, the sphere of radius m. Let us compare the two cases: $1/a = 0$ and $1/a \neq 0$. In the first case, for a particle in radial motion, very close to $y = 0$, we can describe the motion by means of the variational principle

$$\delta \int [(y/2m)\dot{t}^2 - (2m/y)\dot{y}^2]ds = 0 , \tag{4.24}$$

where a dot denotes a derivative with respect to s. Varying t, we have

$$(y/2m)\dot{t} = C \qquad (C = \text{const.}) , \tag{4.25}$$

and from the relation

$$(y/2m)\dot{t}^2 - (2m/y)\dot{y}^2 = 1 , \tag{4.26}$$

we then get

$$\dot{y}^2 = C^2 - \frac{y}{2m} . \tag{4.27}$$

If, for $s = 0$, $y = y_o$ and $\dot{y} = 0$, so that $C^2 = y_o/2m$, the motion is given by

$$y = y_o - s^2/8m . \tag{4.28}$$

As s increases, y decreases and becomes negative for $s^2 > 8my_o$, i.e., the particle enters the Schwarzschild sphere.

Now let us consider the case $1/a \neq 0$. Taking $b^2 = \frac{2}{3}$ for convenience, and assuming that y is very small, we can write the variational principle

$$\delta \int [(m^2/a^2)\dot{t}^2 - \frac{1}{4}(m/y)\dot{y}^2]ds = 0 . \tag{4.29}$$

Varying t gives

$$\dot{t} = (a/m)C \qquad (C = \text{const.}) , \tag{4.30}$$

and putting this into the relation,

$$(m^2/a^2)\dot{t}^2 - \frac{1}{4}(m/y)\dot{y}^2 = 1 , \tag{4.31}$$

gives

$$\dot{y}^2/y = (4/m)(C^2 - 1) = v_o^2/y_o , \tag{4.32}$$

provided we take $\dot{y} = -v_o$, for $y = y_o$. The motion is then given by

$$y = \frac{1}{4}(v_o^2/y_o)s^2 , \tag{4.33}$$

if we take $s = -2y_o/v_o$ (< 0) for $y = y_o$. We see that, as s increases, y decreases until it reaches zero and then increases again. Thus the test particle turns back at the Schwarzschild sphere and does not enter it.

If one wishes to consider the motion of a light ray instead of a particle, one must replace s by some parameter p , and one must replace the unity on the right of (4.26) and (4.31) by zero. One finds that the behavior is qualitatively the same as before: for $1/a = 0$, the ray enters the Schwarzschild sphere, for $1/a \neq 0$, the ray turns back from the surface of this sphere.

We see that, on the basis of the present field equations, the Schwarzschild sphere $(r < m)$ is an impenetrable region, and there is no "black hole".

The coordinates we have been working with are the primed coordinates of (4.1) and (4.2). For the conditions under consideration $(r'/a \ll 1)$ we have

$$t = t' , \qquad r = r'e^{-t'/a} , \tag{4.34}$$

and we see that our conclusions apply to the unprimed coordinates as well.

There are some topics that require investigation. One would like to know more about the field of a particle — not just its behavior near the singularity. It would be good to learn about the field of a rotating body. The question of gravitational collapse is an interesting one.

REFERENCES

1. N. Rosen, Phys. Rev. 57: 147 (1940).

2. A. Papapetrou, Proc. Roy. Irish Acad. 52A: 11 (1948).

3. N. Rosen, Ann. of Phys. 22: 1 (1963).

4. J.R. Gott III and E.L. Turner, Astrophys. J. 209: 1 (1976).

5. A. Sandage, Astrophys. J. 178: 1 (1972).

6. E.R. Harrison, Ann. Rev. of Astron. and Astrophys. 11: 155 (1973).

7. D.W. Sciama, "Modern Cosmology", Cambridge University Press, London (1971), Chaps. 12, 13.

COVARIANCE AND QUANTUM PHYSICS-NEED FOR A NEW FOUNDATION OF QUANTUM THEORY?

Ernst Schmutzer

Sektion Physik, Friedrich-Schiller-Universität

DDR-69 Jena, Max-Wien-Platz 1

ABSTRACT

Starting from the fact that the traditional formulation of quantum mechanics in pictures (Heisenberg picture, Schrödinger picture, etc.) does not fulfil the requirements of covariance, the question of the deeper reasons of this problem is asked. Several arguments for a new foundation of quantum theory on the basis of the formulated "Principle of Fundamental Covariance", combining the General Principle of Relativity (Principle of Coordinate-Covariance in space-time) and the Principle of State-Operator-Covariance (in the space of quantum states), are presented.

The fundamental quantum laws proposed are:

1. laws of motion for the operators, general states and eigenstates on the same level, exhibiting a unified mathematical structure,
2. commutation relations,
3. time-dependent eigenvalue equations.

All these laws fulfil the Principle of Fundamental Covariance (in non-relativistic quantum mechanics with respect to restricted coordinate transformations).

The results of the new foundation of quantum theory are:

1. The theory as a non-relativistic quantum theory is valid for arbitrary frames of reference.
2. The framework of the theory is not limited to the usual quantum theory, but open to a meta-quantum-theory including macroscopic phenomena (thermodynamics, measuring process).
3. The theory can be specialized to the usual quantum mechanics. Therefore all experimental facts on this level are not affected. But the proposed quantum laws do not form a special new picture - they are conceived as the primary laws of quantum theory. No unitary transformation to the Heisenberg picture or to the Schrödinger picture exists.
4. Since the theory is working with time-dependent eigenvalue equations, the whole methodical procedure is changed (we develop a new perturbation theory). Therefore, also questions of interpretation are affected.
5. The fundamentals of quantum field theory are sketched on this basis.

As an application the above theory, especially the new perturbation theory, is applied to a quantum mechanical multilevel system.

EINSTEIN's IDEA OF COVARIANCE

One of mankind's most profound scientific discoveries about the world we live in is that this world does not exist of incoherent facts - but of facts which are (in a very wide sence) causally related by the fundamental laws of nature. These basic laws of nature exhibit - as far as we know till now - the interesting mathematical property of covariance (form-invariance) with respect to arbitrarily continuous coordinate transformations. This discovery was formulated by Einstein in his General Principle of Relativity (also called Principle of General Covariance) in 1915. This principle excellently applies in mechanics, electromagnetism and thermodynamics, but in quantum physics a series of questions arose being on discussion hitherto.

Let us first mention some historical and philosophical aspects and particularily ask the question: Which are the funcamental laws of physics?

The period from Antiquity up to the end of the European Middle Ages accumulated a large store of detailed observational rules and simple laws on motions and processes in nature. It was the ingeneous Galileo who first performed mechanical experiments systematically and evaluated them theoretically - on a rather poor mathematical footing, of course, mostly developed by himself for his own use. The most significant result of his concrete experimental research is his discovery of the laws of projectile motion. But moreover, Galileo was a great theoretician with deep philosophical ideas: Concerning the deep-seated question of the structure of the laws of nature, he presented us - on the basis of his sagacious observations on the anorganic and organic processes on and in a moving ship - with the idea of the Principle of Relativity. For the field of mechanics referring to inertial frames of reference for observers, we nowadays call this principle in its application to Newtonian mechanics the Galilean Principle of Relativity. This special insight into the intrinsic structure of the basic laws of mechanics only became possible, after the famous physicist Newton was able-on the basis of his own mathematical inventions - to formulate the dynamic laws of Newtonian physics (the Newtonian gravitational law has to be considered separately).

Generalizing the Galilean Principle of Relativity to all fields of physics (except gravitation, as was later learned), Poincaré in 1904 and Einstein, recognizing its full extent, in 1905 arrived at the well-known formulation of the Special Principle of Relativity: "In different inertial frames the fundamental laws of physics have the same form".

Using synonymously the term "covariance" for "form-invariance", the Special Principle of Relativity is sometimes also called the "Principle of Lorentz-Covariance", acknowledging Lorentz' invaluable contributions to the Special Theory of Relativity.

Up to this stage of development of the Theory of Relativity, practically all competent physicists agreed in their estimation of the profounder philosophical content of this theory. But what happened after that?

Three obstacles hindered Einstein's progress from the Special Theory of Relativity to the General Theory of Relativity:

1. the restriction to inertial frames of reference,
2. the restriction of the mathematical apparatus to the 4-dimensional Galilean coordinates x, y, z, ct,
3. the problem of gravitation.

Einstein's intensive research between 1908 and 1915 was devoted precisely to these three deep-rooted areas of enquiry.

Which tasks had Einstein to solve in overcoming the three barriers mentioned above?

Overcoming the restriction to inertial frames meant finding a way of formulating the equation of motion for mechanical bodies in such a way that the so-called apparent forces, which in their final form were discovered by Coriolis in about 1830, could be incorporated in a natural manner. In the years 1913/14 Einstein discovered that as a first step this aim could be achieved by the geodesic equation as the mechanical law of motion for a test particle.

At the same time as he was solving this problem, Einstein had to overcome the restriction of the Special Theory of Relativity to Galilean coordinates, i.e. he had to formulate his statements using 4-dimensional curvilinear coordinates. This task led Einstein after a great deal of work to the tensor calculus, i.e. to the so-called Ricci calculus. Fortunately Einstein, who had no special training in this field and who in fact did not even know of the existence of this well established formalism for the description of Riemannian geometry, obtained immense help from Marcel Grossmann. In the years 1913/14 he succeeded in utilizing this mathematical apparatus which soon proved to be the adequate tool for describing the phenomenon of gravitation. It took another two years, until 1915, for Einstein to formulate the General Principle of Relativity as a generalization of the Special Principle of Relativity:
"In different frames of reference, arbitrarily moving, the fundamental laws of physics have the same form".

This principle has been and is a point of issue to the present day. Long discussions have taken place for more than half a century. In his controversary against Infeld and others, Fock came to the conclusion that the General Principle of Relativity is physically empty. Therefore, he even rejected the use of the name "General Theory of Relativity".

It is not our intention to go further into the details of this controversary here /1/. In a few words, our opionion is as follows: The General Principle of Relativity plays a basic role as regards the logical construction of the General Theory of Relativity. It has a deep-rooted physical-philosophical content, i.e. it reveals a deep insight into the intrinsic structure of the laws of nature and it exhibits a better understanding of the unity of nature. The fact that the laws of nature are of the same form for arbitrarily moving observers teaches us that Einstein's idea of the General Principle of Relativity (of course, this historical name can be misleadingly interpreted) is of great scientific value with respect to the objective character of the laws of nature.

In our opinion-taking, according to Einstein, the General Principle of Relativity in the above interpretation as the basis of the General Theory of Relativity - Einstein achieved two important things in 1915: first, he developed the General Theory of Relativity, and second, within this framework he elaborated his theory of gravitation. His original paper of 1916 states these two things explicitly in two different sections. Following this line, we must clearly distinguish from the conceptual point of view between the General Theory of Relativity and Einstein's Theory of Gravitation as a part of it.

Einstein's equation of mechanical motion and Maxwell's equations of electromagnetism have a mathematical structure such that they fulfil Einstein's General Principle of Relativity, i.e. they keep the same form if we pass from one observer to another one by the 4-dimensional spatial-temporal coordinate transformations

$$x^{i'} = x^{i'}(x^m) \qquad (1)$$

(Latin indices run from 1 to 4, Greek indices from 1 to 3). But which situation do we meet in quantum physics?

ON THE NEED OF COVARIANCE FOR QUANTUM MECHANICS

Let us start our consideration on quantum theory with the following basic question: Does quantum theory obey Einstein's line of covariant thinking? To simplify the situation, let us first restrict our investigations to quantum mechanics, i.e. let us first exclude quantum field theory from our considerations. Of course, we cannot apply Einstein's General Principle of Relativity to non-relativistic quantum mechanics, since general 4-dimensional coordinate transformations are not appropriate. As we estimate the situation, it is not the 4-dimensional coordinate transformations which are the rational essence of Einstein's idea, but the philosophical line of covariance of the laws of nature; and in this respect it is legitimate for us to pose the question of covariance with respect to quantum mechanics But to answer this question, we first have to state: which are the fundamental laws of quantum mechanics?

Probably most of us agree in the statement that Dirac's method of description of the quantum phenomena by operators and states, using the bras and kets, is the most appropriate mathematical apparatus to reflect these microcosmic phenomena. But in contrast to mechanics, electromagnetism, gravitation etc., where we have a general consensus on the question of the fundamental laws, the situation in quantum mechanics is quite different. A clever student being asked this question in his examinations will probably reply:
"I think the best mathematical tool to describe quantum mechanics is Dirac's bra-ket-space in which the quantum mechanical operators act. I also have no doubt that the equation of motion for the operators, the equation of motion for the states and Heisenberg's commutation relations for the observables are the fundamental laws of quantum mechanics. But to write down the concrete form of this set of laws, please allow me to ask: In which picture shall I work?"

The fact that our hypothetical student answers the question with another question shows you the deeper essence of this problem: The fundamental laws of quantum mechanics, as we use them up to now, do not follow the idea of covariance with respect to the corresponding transformations of this field of physics, namely unitary transformations. The form of the fundamental laws of quantum mechanics depends on the picture used; and as we know, an infinite variety of equivalent pictures exists.

In this stage of the development of our ideas we have to answer a question of principle, all further considerations being dependent on it:

Alternative I: Are we theoretically fully satisfied by a final description of the quantum phenomena on the basis of mean values only? If so, then all further endeavours can be abandoned, because at the level of mean values the quantum mechanical equations are covariant with respect to unitary transformations. But is this point of view not a purely pragmatic one, a barrier to our deeper reasoning on this subject? Since I myself was deeply impressed by Einstein's covariant thinking on nature during my studies in physics, I never was fully content with this lack of covariance of conventional quantum mechanics. For nearly three decades, especially in periods of lecturing on quantum mechanics, I was thinking intensively about this problem.

Alternative II: Microcosmic phenomena correspond to the description of nature on the level of operators and states. This version means that the mathematical apparatus of operators and states is the adequate and therefore the primary apparatus of microphysics. Adopting this position, we start from a level which is deeper than the level of mean values. In this case it is legitimate to pose again the question of covariance to the fundamental laws of quantum mechanics. This question occupied me for about 20 years; and it became clear to me that - following this line - the foundation of quantum mechanics had to be changed in some very essential points. Quite naturally, the outcome could not be a new picture, because a new picture would be equivalent to the pictures already known, and would therefore not lead to covariance.

Of course, I leave it to the reader whether he is willing or not to follow my line of a picturefree covariant operator/state quantum mechanics. Nevertheless, I believe that such basic questions of physics should be reasoned out again and again to gain full theoretical transparency.

Our new foundation of quantum theory was published in various articles /2/, particularily also in a short version in the proceedings of the IV course of the International School of Cosmology and Gravitation at Erice in 1975 /3/. Therefore we can avoid to repeat the presentation of our theory and rather use the opportunity of the VI course of this school to give some more arguments for our point of view and to inform on several new results of application.

CHANGE OF THE TRADITIONAL BASIS OF QUANTUM THEORY

The traditional formulation of quantum mechanics is based on the Heisenberg picture

$$\text{a)}\quad \frac{dF^{(H)}}{dt} = \frac{\partial F^{(H)}}{\partial t} + \frac{i}{\hbar}\left[H^{(H)}, F^{(H)}\right], \quad \text{b)}\quad \frac{d|\Psi\rangle^{(H)}}{dt} = 0 \qquad (2)$$

or on the Schrödinger picture

$$a)\ \frac{dF^{(S)}}{dt} = \frac{\partial F^{(S)}}{\partial t}, \quad b)\ \frac{d\,|\Psi\rangle^{(S)}}{dt} = \frac{1}{i\hbar} H^{(S)}\,|\Psi\rangle^{(S)}, \quad (3)$$

which are connected by a unitary transformation (F general operator, $|\Psi\rangle$ general state). All other pictures obtained by unitary transformations are equivalent.

Usually the Heisenberg picture or the Schrödinger picture are related to inertial frames of reference. If quantum mechanical phenomena are investigated by a non-inertial observer, the above quantum mechanical equations must be transformed by a unitary transformation in a special way. The construction of the corresponding unitary operator (transition from an inertial frame to a non-inertial frame) sometimes leads to considerable difficulties concerning the physical uniqueness. In such dubious cases it is convenient to start with the general-covariant Dirac equation and to pass over from this quantum field equation to quantum mechanics /4/. But this procedure properly does not satisfy the need for a quantum mechanics in a covariant form, which has to be valid in any frame of reference ab initio. If such a picture-free formulation of the basic laws of quantum mechanics were to exist, according to our opinion it would then have to be cousidered as the proper primary basis of quantum physics.

In the literature quoted above we could show that such a picture-free quantum mechanics can be constructe in a consistent manner. Our new formulation starts with the generalized vector of quantum state

$$|\Psi\rangle = |\Psi(Q_K(t),\ P_K(t),\ t)\rangle, \quad (4)$$

surpassing the conception of Hilbert space vector ($Q_K(t)$ position operator, $P_K(t)$ momentum operator). As the basic laws of quantum mechanics we take: Heisenberg's commutation rules:

$$a)\ [Q_K,\ Q_L] = 0\ , \quad b)\ [P_K, P_L] = 0, \quad c)\ [Q_K, P_L] = i\hbar\,\delta_{KL}; \quad (5)$$

Heisenberg's equation of motion of an operator:

$$\frac{dF}{dt} = \frac{\partial F}{\partial t} + \frac{i}{\hbar}\ [H,\ F]\ , \quad (6)$$

where $F = F(Q_K(t),\ P_K(t), t)$ is a general operator and $H(Q_K(t),\ P_K(t), t)$ is the Hamiltonian of the quantum mechanical system;

equation of motion of a general state:

$$\frac{d\,|\Psi\rangle}{dt} = \frac{\partial|\Psi\rangle}{\partial t} + \frac{i}{\hbar}\;H\,|\Psi\rangle \;; \tag{7}$$

equation of motion of an eigenstate:

$$\frac{d\,|f_\sigma\rangle}{dt} = \frac{\partial|f_\sigma\rangle}{\partial t} + \frac{i}{\hbar}\;H|f_\sigma\rangle \;; \tag{8}$$

eigenvalue equation:

$$F|f_\sigma\rangle = f_\sigma\;|f_\sigma\rangle \tag{9}$$

which in contrast to usual quantum mechanics is time-dependent if the operator F depends on time.

One should mention that our propounding of the conception of the state vector according to (4) allows us to distinguish between the total and the partial derivative of the state. The explicit time dependence corresponds to external temporal influences.

The derivatives of operators and states with respect to operators are defined by the relations:

$$\text{a)}\quad \frac{\partial F}{\partial Q_K} = \frac{1}{i\hbar}\left[F,P_K\right], \qquad \text{b)}\quad \frac{\partial F}{\partial P_K} = -\frac{1}{i\hbar}\left[F,Q_K\right],$$

$$\text{c)}\quad \frac{\partial|\Psi\rangle}{\partial Q_K} = -\frac{1}{i\hbar}\,P_K|\Psi\rangle\,, \qquad \text{d)}\quad \frac{\partial|\Psi\rangle}{\partial P_K} = \frac{1}{i\hbar}\,Q_K|\Psi\rangle\,. \tag{10}$$

For the statistical interpretation of quantum mechanics we need the following notions:
norm of a state:

$$\langle\Psi|\Psi\rangle = 1 \;; \tag{11}$$

mean value of an operator:

$$f = \langle\Psi|\,F|\Psi\rangle\,; \tag{12}$$

expectation value of an operator:

$$\bar{f} = \langle\check{\Psi}|\,F|\,\check{\Psi}\rangle, \tag{13}$$

where the transformed state $|\check{\Psi}\rangle$ is given by

$$|\check{\Psi}\rangle = N\,|\Psi\rangle \qquad (\text{"static aspect"}) \tag{14}$$

with

$$i\hbar \frac{\partial N}{\partial t} = HN. \tag{15}$$

This quantum theoretical scheme (4) to (15) is fully covariant with respect to an arbitrarily time-dependent unitary transformation described by a unitary operator U:

$$\text{a)}\quad |\bar{\Psi}\rangle = U\,|\Psi\rangle\ , \qquad \text{b)}\quad \frac{\partial|\bar{\Psi}\rangle}{\partial t} = U\,\frac{\partial|\Psi\rangle}{\partial t}\ , \tag{16}$$

$$\text{a)}\quad |\bar{f}_\sigma\rangle = U|f_\sigma\rangle\ , \qquad \text{b)}\quad \frac{\partial|\bar{f}_\sigma\rangle}{\partial t} = U\,\frac{\partial|f_\sigma\rangle}{\partial t}\ , \tag{17}$$

$$\text{a)}\quad \bar{Q}_K = U\,Q_K\,U^+\ , \qquad \text{b)}\quad \bar{P}_K = U\,P_K\,U^+\ , \tag{18}$$

$$\text{a)}\quad \bar{F}=F(\bar{Q}_K,\bar{P}_K,t)=UFU^+, \qquad \text{b)}\quad \frac{\partial\bar{F}}{\partial t} = U\,\frac{\partial F}{\partial t}\,U^+\ , \tag{19}$$

$$\bar{H} = U\,H\,U^+ - i\hbar\,\frac{dU}{dt}\,U^+\ . \tag{20}$$

Our new formulation of quantum mechanics has various advantages. We mention some of them:

1. Because of the covariance the basic equations have the same form in any arbitrarily moving frame of reference. The situation is analogous to that of classical canonical mechanics.
2. The asymmetry with respect to time of the equations (2) and (3) vanishes.
3. In contrast to traditional quantum mechanics the general state and the eigenstate obey the same law of motion.
4. In contrast to traditional quantum mechanics there do not exist different transformation laws of the Hamiltonian, depending on the picture used. We got the unique transformation law (20). This fact is very important because we don't believe that such a fundamental quantity as the Hamiltonian does not obey a unique transformation law. One should again compare this uniqueness with the situation in classical canonical mechanics.

From these considerations as well as several others - including philosophical ones - we were led to the following generalization of Einstein's idea of the General Principle of Relativity:

Principle of Fundamental-Covariance

Fundamental physical laws are form-invariant:

1. with respect to coordinate transformations in space-time (General Principle of Relativity)
   → Principle of Coordinate-Covariance,

2. with respect to operator-state-transformations in the space of quantum states
   → Principle of State-Operator-Covariance.

In other words:
"Fundamental physical laws have the same form for two observers of different states of motion, independently of the coordinates or operators and states being used."

It is convenient to sketch the network of our picture-free covariant quantum mechanics and the traditional quantum mechanical pictures in the following logical scheme:

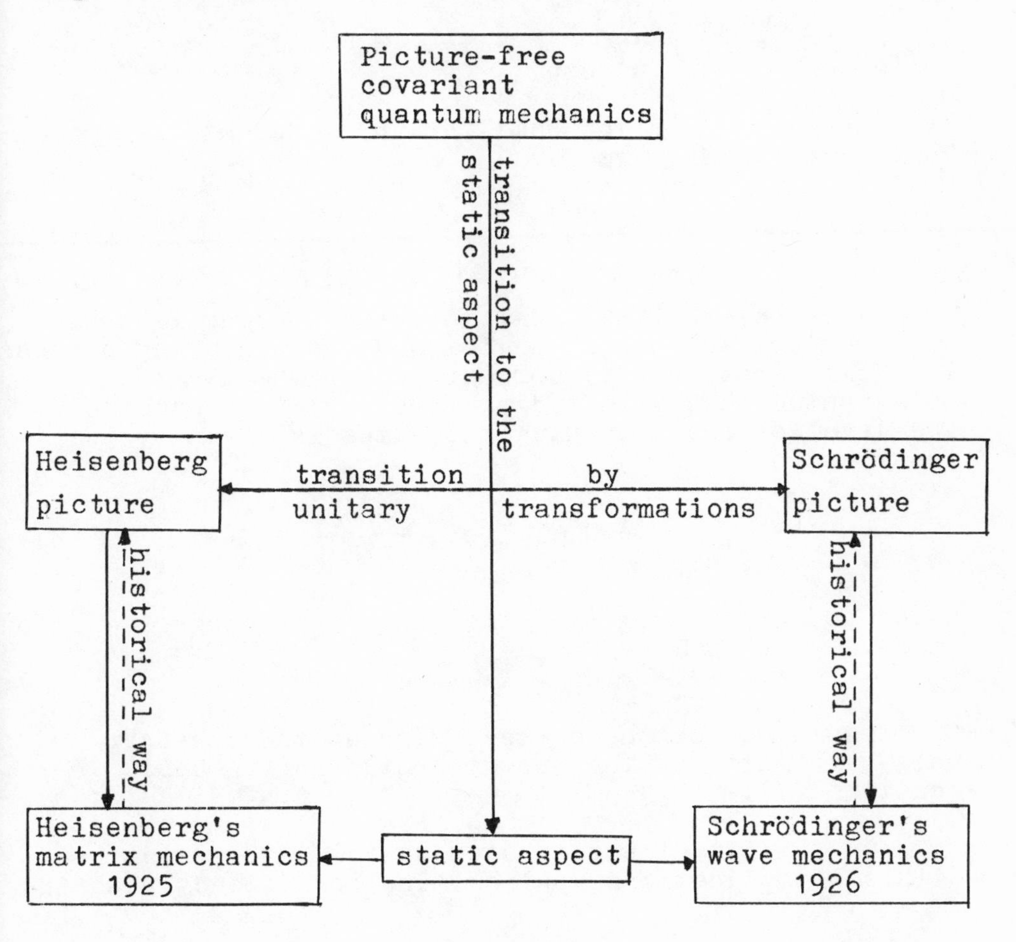

SCHRÖDINGER REPRESENTATION

Using the scalar products

$$\text{a)}\ \Psi(q,t) = \langle q|\check{\Psi}\rangle \ , \quad \text{b)}\ \check{\Psi}(q,t) = \langle q|\Psi\rangle \ , \tag{21}$$

$$\text{a)}\ \varphi_\nu(q,t) = \langle q|\check{h}_\nu\rangle \ , \quad \text{b)}\ \phi_\nu(q,t) = \langle q|h_\nu\rangle \ , \tag{22}$$

where the eigenvalue equation

$$H|h_\nu\rangle = h_\nu|h_\nu\rangle \tag{23}$$

for the Hamiltonian and according to (14) the transformation

$$|\check{h}_\nu\rangle = N|h_\nu\rangle \tag{24}$$

hold, we find from (7) the generalized Schrödinger wave equation

$$\frac{\partial\Psi}{\partial t} = \left(\frac{\partial\Psi}{\partial t}\right)_{expl} + \frac{1}{i\hbar} H_D \Psi \ . \tag{25}$$

The quantity H_D is the Hamiltonian differential operator Furthermore, the term

$$\left(\frac{\partial\Psi}{\partial t}\right)_{expl} = \langle q| \frac{\partial|\check{\Psi}\rangle}{\partial t} \tag{26}$$

describes external temporal influences being important with respect to thermodynamics and the measuring process The full equation (25) corresponds to the level of our metaquantum theory including irreversible processes. The level of usual quantum mechanics is given by the restriction

$$\left(\frac{\partial\Psi}{\partial t}\right)_{expl} = 0 \ , \quad \text{i.e.} \quad \frac{\partial|\Psi\rangle}{\partial t} = 0 \ . \tag{27}$$

Hence we obtain from (25) the usual Schrödinger equation

$$i\hbar\frac{\partial\Psi}{\partial t} = H_D\Psi \ . \tag{28}$$

One should notice the correct sign at the imaginary unity, in spite of the opposite sign in the basic equation (7).

On this level Born's statistical interpretation with all implications applies. The consistency with the experimental facts is guaranteed. Our further investigations will dwell on this level.

The eigenvalue equation in Schrödinger representation, corresponding to (23), reads

$$H_D \phi_\nu = h_\nu \phi_\nu \ . \tag{29}$$

NEW PERTURBATION METHOD FOR A TIME-DEPENDENT HAMILTONIAN AND APPLICATION TO A MULTI-LEVEL SYSTEM

The time-dependent Hamiltinian of a system is decomposed in the usual way into a time-independent and a time-dependent part

$$H_D = H_D^o + H_D^t (t). \tag{30}$$

We use the time-independent eigenvalue equation

$$H_D^o \ \chi_\sigma = E_\sigma \chi_\sigma \tag{31}$$

and expand

$$\phi_\nu (q,t) = \sum_\lambda c_{\nu\lambda} (t) \ \chi_\lambda (q). \tag{32}$$

Inserting into (29) we are left with the linear system

$$\sum_\lambda c_{\nu\lambda} \left[\delta_{\sigma\lambda} \ (E_\lambda \ -E_\nu \ )+H_{\sigma\lambda} -h_\nu^{\ t} \delta_{\sigma\lambda}\right]= 0 \tag{33}$$

for the coefficients $c_{\nu\lambda}$, where

$$\text{a) } H_{\sigma\lambda}(t) = \int \chi_\sigma^* H_D^t \chi_\lambda \, dq, \quad \text{b) } h_\nu^{\ t} = h_\nu - E_\nu . \tag{34}$$

The secular equation corresponding to (33) reads

$$\det \left| H_{\sigma\lambda} + \delta_{\sigma\lambda}(E_\lambda - E_\nu - h_\nu^{\ t}) \right| = 0. \tag{35}$$

Hence the quanties $h_\nu^{\ t}$ can be calculated.

In contrast to Dirac's perturbation method we expand with respect to the time-dependent eigenfunctions:

$$\Psi = \sum_\sigma S_\sigma(t) \ \phi_\sigma(q,t). \tag{36}$$

In the usual way we find the system of linear differential equations

$$\dot{S}_\mu - \frac{1}{i\hbar} h_\mu S_\mu + \sum_\sigma S_\sigma \varphi_{\mu\sigma} = 0. \tag{37}$$

with

$$\varphi_{\mu\sigma} = \int \phi_\mu^* \ \frac{\partial \phi_\sigma}{\partial t} \ dq = - \varphi_{\sigma\mu}^* . \tag{38}$$

The physical background of Dirac's perturbation theory is the stationary system which is affected by a (relatively weak) time-dependent perturbation during a time interval. The above perturbation theory /5/ is based on a time-dependent background, i.e. if the background problem can be solved exactly (adiabatic time-parametric problems etc.), the use of our perturbation theory is advantageous.

The transition probability for the transition from an initial state χ_i before the perturbation to a final state χ_λ after the perturbation reads:

$$w_{\lambda(i)} = \sum_{\sigma,\varkappa} S^*_{\sigma(i)} S_{\varkappa(i)} c^*_{\sigma\lambda} \, c_{\varkappa\lambda} \quad . \tag{39}$$

According to our new interpretational aspects concerning the time-dependent eigenstates the probability of presence of a system, which started from the (time-dependent) initial state ϕ_i, in the (time-dependent) final state ϕ_λ is defined by

$$W_{\lambda(i)} = \left| S_{\lambda(i)} \right|^2 \quad . \tag{40}$$

To prove our perturbation theory at a practical example we investigated an N-level system. This problem is very important in non-linear optics. In the case of a time-dependent Hamiltonian describing the system (including radiation), the eigenvalues are calculated from the secular equation (35) being an algebraic equation of degree N for the quantities $h_\nu{}^t$. The resulting time-dependent shift of the levels is a direct consequence of the basic ideas concerning the time-dependent behaviour of quantum systems.

In the paper cited above /5/ we did the calculations in all details for a 2-level system with temporally periodic interactions by electromagnetic radiation.

IDEAS ON GENERAL-RELATIVISTIC QUANTUM FIELD THEORY IN A METRIC FIELD

The deeper reason of our proposal of a covariant picture-free quantum theory is of course to find a new approach to a fully covariant general-relativistic quantum field theory. The difficulties in this field of research are well known. According to our former considerations, in our opionion the basis of such an approach should be the above formulated Principle of

Fundamental Covariance. Since this theory has not been worked out in detail till now, we can only sketch the basic ideas using the symbolism of our book /6/.

The general state will get the structure

$$|\phi\rangle = |\phi\,(U_\Omega\,,\,U_{\Omega,\alpha}\,,\,\Pi^\Omega,\,g_{mn}(x^i))\rangle \tag{41}$$

(U_Ω field operators; Π^Ω momentum operators; g_{mn} metric; α = 1,2,3; m,n,i = 1,2,3,4), while an operator density will be of the structure

$$f = f(U_\Omega,\,U_{\Omega,\alpha}\,,\,\Pi^\Omega,\,g_{mn}(x^i)). \tag{42}$$

The generalization of the fundamental laws from above leads us to the simultaneous system of equations (P_i 4-momentum operator):

$$\frac{\partial f}{\partial x^i} = \left(\frac{\partial f}{\partial x^i}\right)_{expl} + \frac{1}{i\hbar}\left[P_i, f\right], \tag{43}$$

$$\frac{\partial|\phi\rangle}{\partial x^i} = \left(\frac{\partial|\phi\rangle}{\partial x^i}\right)_{expl} + \frac{1}{i\hbar}\,P_i\,|\phi\rangle, \tag{44}$$

$$P_i\,|p_{i\mu}\rangle = p_{i\mu}\,|p_{i\mu}\rangle\;, \tag{45}$$

$$\frac{\partial|f_\mu\rangle}{\partial x^i} = \left(\frac{\partial|f_\mu\rangle}{\partial x^i}\right)_{expl} + \frac{1}{i\hbar}\,P_i\,|f_\mu\rangle\;. \tag{46}$$

This set of equations should be the basis of further research.

REFERENCES

1. V.A.Fock, "The Theory of Space, Time, and Gravitation", Pergamon, New York (1959).
E.Schmutzer, "Relativistische Physik", Teubner, Leipzig (1968).
2. E.Schmutzer, Nova Acta Leopoldina (Suppl.No.8)44:79 (1976); Exp.Tech.Phys. 24:131 (1976); Gen.Rel.Grav. 8:699 (1977).
3. E.Schmutzer, General Relativity and Quantum Theory, in: "Topics in Theoretical and Experimental Gravitation Physics", V.De Sabatta and J.Weber, ed., Plenum Press, New York (1977).

4. E.Schmutzer and J.Plebanski, Fortschr.Phys.25:37(197
5. E.Schmutzer, Acta Phys.Polon. B8:569 (1977).
6. E.Schmutzer,"Symmetries and Conservation Laws of Phys
Pergamon,Oxford (1972).

RELATIVISTIC EQUATIONS OF MOTION OF "SPIN PARTICLES"

Bronisław Średniawa

Institute of Physics
Jagellonian University
Cracow, Poland

The problem of the structure of physical systems in microphysics can be treated either from the point of view of quantum theories (through non-local field theories) or from the standpoint of classical theories. We shall be concerned here with the classical aspect of the problem trying to show how the structure of physical systems was derived starting from relativistic theories. We shall call such structures "spin particles" not maintaining, however, that they are real models of elementary particles. (Such structures are also called "relativistic tops"). One can consider classical equations of motion of spin particles in different ways, starting from different branches of physics. In this report we shall mainly treat the methods of deriving these equations which were worked out by M. Mathisson, J. Weyssenhoff and their collaborators, discussing also shortly some results of other authors. It is not possible to give in a short report the complete review of the investigation about relativistic spin particles, which are continued up to now. We shall be concerned only with free spin particles, and shall not consider here the numerous papers dealing with the behaviour of spin particles in the presence of electromagnetic and meson fields. We sketch the derivation of the equations of motion of spin particles moving slower than light and moving with the velocity of light and discuss the solutions of the equations of motion in both cases. (The work of "Cracow school" of Mathisson, Weyssenhoff and collaborators leads to practically soluble equations). We shall see, that starting from different theories we shall obtain structures of similar kind.

2. THE FREE SPIN PARTICLE MOVING SLOWER THAN LIGHT

a) The derivation of the equations of the spin particle from the equations of the gravitational field

M. Mathisson [1,2] developed Einstein ad Grommer's idea of deriving the equations of motion of free particles from the equations of the gravitational field as the singularities in this field. Mathisson derived a variational principle for the linearized field taking as background the gravitational field of general relativity.

We shall formulate this principle in the simplified form([2], [4])taking Minkowski space as the background.

Let us consider the linearized gravitational field on the background of the Minkowski metrics*

$$g_{\alpha\beta} = \overset{\circ}{g}_{\alpha\beta} + \gamma_{\alpha,\beta} , \qquad \gamma_{\alpha\beta} \text{ small},$$

where

$$\overset{\circ}{g}_{11} = \overset{\circ}{g}_{22} = \overset{\circ}{g}_{33} = 1 , \ \overset{\circ}{g}_{00} = -1, \ \overset{\circ}{g}_{\alpha\beta} = 0 \text{ for } \alpha \neq \beta.$$

We enclose the singularities of the gravitational field with a tube in four-dimensional space and we continue the field (1) to the interior of the tube obtaining the gravitational field without singularities.

Now we consider the gravitational energy-momentum tensor $T^{\alpha\beta}$ satisfying the conservation law

$$\partial_\beta T^{\alpha\beta} = 0 , \quad T^{\alpha\beta} = T^{\beta\alpha} . \qquad (2)$$

Let us choose a four-dimensional domain $D^{(4)}$ and an arbitrary vector field ξ_α vanishing on its boundary with its derivatives $p_{\alpha\beta} = \partial_\beta \xi^\alpha$ and define the volume integral over $D^{(4)}$

$$\int_{D^{(4)}} \partial_\beta \left(\xi_\alpha T^{\alpha\beta} \right) d^4x = \int_{D^{(4)}} T^{\alpha\beta} \partial_\beta \xi_\alpha \, d^4x + \int_{D^{(4)}} \xi_\alpha \partial_\beta T^{\alpha\beta} d^4x .$$

Transforming the left-hand side integral into the surface integral, using the boundary condition for the

---

* $\alpha, \beta = 0, 1, 2, 3; \quad i, k = 1.2.3.$

field ξ^α and the conservation law (2) we get

$$\int_{D^{(4)}} T^{\alpha\beta} \partial_\beta \xi_\alpha = 0 \qquad \text{or} \qquad \int_{D^{(4)}} T^{\alpha\beta} p_{\alpha\beta}\, d^4x = 0. \tag{3}$$

We consider the domain $D^{(4)}$ as a tube formed by time-like lines, closed with areas perpendicular to these lines. We assume also that $T^{\alpha\beta} = 0$ outside $D^{(4)}$. Inside $D^{(4)}$ we choose a time-like line L (which shall be regarded as the world-line of the particle

$$x^\alpha = x^\alpha(\tau) \quad ; \qquad \tau_1 \leq \tau \leq \tau_2 \tag{4}$$

(τ is the proper time). Then we take perpendicular sections $\Sigma(\tau)$ in each point of the line and we expand $p_{\alpha\beta}$ into a Taylor series on $\Sigma(\tau)$ around the point of the line. In this manner we can transform the volume integral into a line integral ([1],[5],[2]), obtaining thus Mathisson's variational principle

$$\int_{D^{(4)}} T^{\alpha\beta} p_{\alpha\beta}\, d^4x = \int_{L\ \tau_1}^{\tau_2} \{ m^{\alpha\beta} p_{\alpha\beta} + d^{\lambda\alpha\beta} \partial_\lambda p_{\alpha\beta} + \cdots \} d\tau = 0, \tag{5}$$

where $m^{\alpha\beta}$, $d^{\lambda\alpha\beta}$ (being tensors and functions of τ) are the successive gravitational multipole moments, having in the proper system of the particle ($u^i = 0$) the form

$$m^{\alpha\beta} = \int_{\Sigma(\tau)} T^{\alpha\beta} d^3x \tag{6}$$

$$d^{i\alpha\beta} = \int_{\Sigma(\tau)} x^i T^{\alpha\beta} d^3x, \qquad d^{4\alpha\beta} = 0, \tag{7}$$

where $\Sigma(\tau)$ is the section of the tube, perpendicular to the world line. From (7) it follows that $d^{\lambda\alpha\beta} u_\lambda = 0$.
Equations of motion of the unipole particle. If we consider the first term of the right-hand side of (5) only, we can easily show that

$$m^{\alpha\beta} = m_o u^\alpha u^\beta, \quad m_o = \text{const}. \tag{8}$$

Defining the time-like vector

$$G^\alpha = m_o u^\alpha \tag{9}$$

as the momentum of the particle ($m_o \neq 0$ is the rest-mass of the particle) we get from the variational principle following equations of motion ($\cdot = d/d\tau$)

$$\dot{G}^\alpha = 0 \tag{10}$$

being the equations of geodetics.

<u>Equations of dipole particle [1], [2], [6].</u> Taking into consideration the two first terms in the variational principle one can show that $d^{\lambda\alpha\beta}$ has the form

$$d^{\lambda\alpha\beta} = \frac{1}{2}(s^{\alpha\lambda} u^\beta + s^{\lambda\beta} u^\alpha) + n^\lambda u^\alpha u^\beta, \tag{11}$$

where

$$s^{\alpha\beta} = -s^{\beta\alpha}, \quad s^{\alpha\beta} u_\beta = 0, \quad n^\lambda u_\lambda = 0. \tag{12}$$

We define the momentum G^α of the particle as the vector

$$G^\alpha = (m_o + \dot{n}^\lambda u_\lambda) u^\alpha + s^{\alpha\beta} \dot{u}_\beta + \dot{n}^\alpha, \tag{13}$$

which we assume to be time-like. (It can be easily shown that $m_o = m^{\alpha\beta} u_\alpha u_\beta$ is a constant). Thus we get from the variational principle the equations of motion of the "spin particle"

$$\dot{G}^\alpha = 0 \tag{14}$$

$$\frac{d}{d\tau}(s^{\alpha\beta} + n^\alpha u^\beta - n^\beta u^\alpha) = G^\alpha u^\beta - G^\beta u^\alpha. \tag{15}$$

Eliminating G^α from (13) and (14) we get the differential equation of the third order in respect to x^α. Now there arise two simple possibilities:

1. Putting $n^\alpha = 0$ we get the momentum of the "particle"

$$G^\alpha = m_o u^\alpha + \frac{1}{c^2} s^{\alpha\beta} \dot{u}_\beta \tag{16}$$

and its equations of motion

$$\dot{G}^{\alpha} = 0 \quad , \tag{17}$$

$$\dot{s}^{\alpha\beta} = G^{\alpha}u^{\beta} - G^{\beta}u^{\alpha} \quad , \tag{17'}$$

which are Mathisson's equations [1] in the form given by Weyssenhoff and Raabe [7]. These equations were solved in nonrelativistic approximation by Mathisson [8] and in the general case by Weyssenhoff and Raabe [7] who obtained in a suitably chosen coordinate system C (in which $G^{i} = 0$) a uniform circular motion, the radius of the circle K being arbitrary and the velocity v constant. In C , which is the rest system of a circle as a whole the role of the proper angular momentum is played by the quantity

$$s_{c} = s_{\alpha\beta}\, s^{\alpha\beta} \left(1 - \frac{v^2}{c^2}\right)^{-\frac{1}{2}} = \mathrm{const.}$$

The bivector $s^{\alpha\beta}$ may be called the spin bivector.
2. Putting $s^{\alpha\beta} =$ we get [6]

$$G^{\alpha} = (m_{0} + \dot{n}^{\lambda} u_{\lambda}) u^{\alpha} + \dot{n}^{\alpha} \tag{19}$$

and the equations of motion

$$\dot{G}^{\alpha} = 0 , \tag{20}$$

$$n^{\alpha} = k \dot{u}^{\alpha} \quad , \tag{20'}$$

where k is a scalar depending on τ .
Now we postulate that n^{α} has to be a quantity characterizing the particle, and the simplest assumption is that

$$n^{\alpha} n_{\alpha} = \mathrm{const}, \tag{20''}$$

which implies that k = const. Then the solution of (20) - (20") in the coordinate system where $G^{i} = 0$ is also the uniform circular motion; so the existence of $n^{\alpha} \neq 0$ introduces also spin properties. The particle, characterized by n^{α} was considered primarily by Hönl and Papapetrou [9]. They stated that the uni-

form circular motion satisfies their equations.
Remark 1. The solutions of the equations of motion resulting from consideration of three first terms in (5) (giving the quadrupole particle) turned to give also uniform circular motion in the appropriate system of reference [6].
Remark 2. Another method to derive the equations of motion of pin particle was applied by J. Lubański [10] who expanded gravitational retarded potentials into the series of multipole terms and obtained Mathisson's equations.
Remark 3. Let the dimensions of the particle characterized by m_0 and n^α tend to zero. In order to preserve $n^\alpha \neq 0$ in this limit, we must ([7],[11]) assume the existence of the negative mass; a particle of this kind is called a unipol-dipol particle. The assumption $n^\alpha = 0$, $s^{\alpha\beta} \neq 0$ allows to consider particles possessing only positive masses.
Remark 4. Equations (16) - (17') were obtained from the equations of gravitational field also by Nagy [11], who used the method of Infeld [12] and assumed pseudoeuclidian metrics as the background.

b) Derivation of the equations of motion of the spin particle from the theory of spin-fluid

Weyssenhoff and Raabe [7] assumed as starting point the equations of an incoherent fluid (in which the pressure $p = 0$) not exposed to external forces, in special relativity. These equations can be written in the form

$$\partial_\alpha T^{\alpha\beta} = 0, \tag{21}$$

where

$$T^{\alpha\beta} = \rho_0 u^\alpha u^\beta \tag{22}$$

is the symmetrical energy-momentum tensor.

Weyssenhoff and Raabe generalized these equation to the case of spin-fluid in the following way:
1° They accepted the unsymmetrical energy-momentum tensor

$$T'^{\alpha\beta} = g^\alpha u^\beta , \tag{23}$$

where g^α isn't necessarily parallel to u^α . Introducing the relativistic generalization of the substantial derivative

$$\frac{Df}{D\tau} = \partial_\nu (f u^\nu) \tag{24}$$

one can write the equation $\partial_\alpha T'^{\alpha\beta} = 0$ for the tensor (23) in the form

$$\frac{D g^\alpha}{D\tau} = 0 . \tag{25}$$

2^0 They introduced the density of proper angular momentum as a bivector $\sigma^{\alpha\beta}$, having vanishing components σ^{14}, σ^{24}, σ^{34} in rest-frame of the small fluid volume (what implies $\sigma^{\alpha\beta} u_\beta = 0$). Therefore the law of conservation of angular momentum is completed to

$$\frac{D}{D\tau} (x^\alpha g^\beta - x^\beta g^\alpha) + \frac{D}{D\tau} \sigma^{\alpha\beta} = 0 . \tag{26}$$

Thanks to (24) this can be written

$$\frac{D}{D\tau} \sigma^{\alpha\beta} = g^\alpha u^\beta - g^\beta u^\alpha . \tag{26'}$$

Integrating (24) and (26) over a small volume, in which u^α and $\dot{u}^\alpha$ are constant, and defining

$$G^\alpha = \int g^\alpha dV_0 ; \quad s^{\alpha\beta} = \int s^{\alpha\beta} dV_0 ; \rho_0 = \frac{1}{c^2} g^\alpha u_\alpha ; m_0 = \int \rho_0 dV , \tag{27}$$

where dV_0 is the proper volume one gets the equations (16) - (17').

c) Derivation from a straightforward generalization of classical mechanics

In classical mechanics of material points the momentum $\vec{G}$ is parallel to the velocity. Let us abandon this assumption but accept that for the free particle

$$\frac{d\vec{G}}{dt} = 0 \tag{28}$$

and calculate the derivative of the angular momentum

$$\vec{l} = \vec{r} \times \vec{G}$$

$$\frac{d\vec{l}}{dt} = \frac{d}{dt}(\vec{r} \times \vec{G}) = \vec{r} \times \frac{d\vec{G}}{dt} + \vec{v} \times \vec{G} = \vec{v} \times \vec{G}.$$

Angular momentum is not a constant of motion. In order to preserve the law of conservation of total angular momentum we introduce the "intrinsic" (spin) angular momentum $\vec{s}$ such that

$$\frac{d\vec{s}}{dt} = \vec{G} \times \vec{v} . \tag{29}$$

Thus

$$\frac{d}{dt}(\vec{l} + \vec{s}) = 0. \tag{30}$$

Generalizing these equations to special relativity and defining the bivector $s^{\alpha\beta} = (\vec{s}, \vec{q})$; $(\vec{q} = \frac{1}{c}\vec{v} \times \vec{s})$ we get $s^{\alpha\beta} u_\beta = 0$ and equations (28), (29), (30) will go over again to the equations (16) - (17′).

The free particle moving with the velocity of light.

In order to fix the radius of the circle K and to obtain some kind of spin particle, Weyssenhoff and Raabe [14] passed with the velocity v of the particle to the velocity of light. In order to do it they took an arbitrary parameter π on the world-line (increasing with time t) instead of the proper time τ . (We denote: $' = d/d\pi$, $w^\alpha = dx^\alpha/d\pi$) and performed the limiting procedure $v \to c$. Then the equations of motion (12), (16) - (17′) pass into

$$G'^{\alpha} = 0 , \tag{31}$$

$$s'^{\alpha\beta} = G^\alpha w^\beta - G^\beta w^\alpha , \tag{32}$$

$$s^{\alpha\beta} w_\beta = 0 , \tag{33}$$

$$w^\alpha w_\alpha = 0 . \tag{34}$$

Taking the time t as the parameter π , denoting $v^\alpha = dx^\alpha/dt$ (v^α is not, of course, a vector), and choosing coordinate system C in which $G^i = 0$ we obtain as the solution of equations (31) - (34) the uniform circular motion. The radius of the circle is:

$$r_c = \frac{s_c}{c\, M_c} \tag{35}$$

and for angular velocity we obtain:

$$\omega_c = \frac{M_c\, c^2}{s_c} \tag{36}$$

where

$$s_c^2 = \frac{1}{2} s_{ik}\, s^{ik} \quad , \quad M_c = \frac{1}{c}\, G^4 \tag{37}$$

are in constants and. $M_c \neq 0$.

J. Weyssenhoff [16] regarded this particle as the classical model of Dirac's electron, because putting $M>0$, and equal to the electron mass, and $s = \hbar/2$ he obtained for this particle the following similarities with the Dirac's electron:

1. The momentum $\vec{G}$ is in general not parallel to the velocity $\vec{v}$
2. The velocity v is equal to the velocity of light
3. Angular velocity ω_c and the radius r_c are of the order of the angular velocity and amplitude of Schrödinger's Zitterbewegung
4. Moreover, it turns out that the stability condition for the circle K in a variable magnetic field is the same as the condition for not creating of positon-negaton pairs in this field.

3. DERIVATION OF EQUATIONS OF MOTION OF THE FREE-SPIN PARTICLE FROM THE VARIATIONAL PRINCIPLE WITH HIGHER DERIVATIVES

Since the systems of equation which we have considered, can be reduced to the higher order differential equations with respect to x^α, the problem arises of deriving them from the Hamilton variational principle with higher derivatives. This problem was first considered by Bopp [15]. This idea was resumed by J. Weyssenhoff [16] who formulated the homogeneous canonical formalism with higher derivatives. He considered the lagrangian $\mathcal{L}(x'^\alpha, x''^\alpha)$ subject to the conditions:

1. $\mathcal{L}$ is lorentz-invariant,
2. $\int_P^{P_2} \mathcal{L}\, d\pi$ does not depend on the parametrization,

3. $\mathcal{L}$ remains finite when $(ww) \equiv w^{\alpha} w_{\alpha} \rightarrow 0$.
The simplest lagrangian statisfying these conditions is

$$\mathcal{L} = -2\gamma \left\{ \sqrt{-ww} + \sqrt{l_0} \sqrt[4]{(w'w') - \frac{(ww')^2}{(ww)}} \right\}, \quad (38)$$

where γ and l_0 are constants.
(Let us note that the structure constant l_0 of dimension of length, has appeared). Variational principle based on this lagrangian implies for $(ww) \rightarrow 0$, i.e. for the particle moving with the velocity of light, the equations of motion (31) - (35), in which

$$s^{\alpha\beta} = w^{\alpha} l^{\beta} - w^{\beta} l^{\alpha}, \quad \text{where} \quad l_{\alpha} = \frac{\partial \mathcal{L}}{\partial w'^{\alpha}}. \quad (39)$$

Then it turns out that in this case the lagrangian (38) is a constant of motion. If for the motion with the velocity lower than that of light we postulate with Borelowski [17] that the lagrangian should be also a constant, we obtain also the equations (16) - (17´). Therefore we obtain from the homogeneous Weyssenhoff´s variational principle (derived from the postulates of the general character) with Borelowski´s condition, the same equations as before.

4. BI-POINT MODEL OF SPIN PARTICLE

Let us consider [20] (in connection with the bi-local field theory [18] (see also [19]) the system of two points with the world-lines given by $x^{\alpha}_{(1)}(\pi)$, $x^{\alpha}_{(2)}(\pi)$, $\tau_1 = \tau_1(\pi)$, $\tau_2 = \tau_2(\pi)$). We define

$$x^{\alpha} = \frac{1}{2}(x^{\alpha}_{(1)} + x^{\alpha}_{(2)}), \quad r^{\alpha} = x^{\alpha}_{(2)} - x^{\alpha}_{(1)} \quad (40)$$

and assume that the equations of these lines follow from the variational Hamilton principle with the lagrangian $L(r^{\alpha}, r'^{\alpha}, x'^{\alpha})$, satisfying the postulates

(1) - (3) of Weyssenhoff (see § 3). We then obtain

$$L = -2\gamma\left\{\sqrt{-(x'x') + \frac{(r x')^2}{(r r)}} + \sqrt{(r' r') - \frac{(\tau r')^2}{(r r)}}\right\}, \quad (41)$$

where γ is a constant.

If we add the supplementary assumption

$$(r r) = \mathrm{const} \equiv l_0 \neq 0, \quad (42)$$

then from equations of motion (41) and (42), with solution in the center of mass Σ_0 system of both points, we obtain an uniform circular motion of both points around the center of mass (in Σ_0 there is $\tau_1 = \tau_2 = \tau$).

5. CONCLUSIONS AND FINAL REMARKS

Our considerations showed that starting from different theories, namely from general relativity with symmetrical energy-momentum tensor, from hydrodynamics with non-symmetrical energy momentum tensor, from the Hamiltonian formalism with higher derivatives and even from a classical model of a particle intimately connected with bilocal field theory, we obtain an extended flat structure. As far as to the sixties this structure was by many physicists regarded as the classical model of an elementary particle and there were made attempts to quantize this model in order to get the mass-spectrum of elementary particles. Today we can only say that general relativity as well as compulsory generalizations of relativistic mechanics lead to the appearance of some structures having with flat character, but we do not connect them more intimately with the structures of elementary particles.

Finally, let us remark that the interest in problems of structures following from classical theories and from the variational principle with higher derivatives did non stop in sixties and early seventies. As berg [21], Hanson and Regge [22] and Musicki [23]. The paper of Hanson and Regge, which appeared in 1974, dealt with the generalization of theories we have discussed just above. They consider a singularity (which they call symmetrical top) which is a world-line with a tetrapod attributed to every point of it. Starting from very general assumptions the authors derived a differential equation of this world-line, without solving it, but pointing out the problems which are still waiting for solution.

REFERENCES

[1] M. Mathisson, Acta Phys. Pol. 6, 163 (1937)
[2] M. Mathisson, Proc. Cambr. Phil. Soc. 38, 331 (1940)
[3] A. Einstein, J. Grommer, Sitzungsber. d. preuss. Ak. d. Wiss., phys.-mat. Klasse, 2 (1927)
[4] B. Średniawa, Thesis 1947 (unpublished)
[5] A. Bielecki, M. Mathisson, J. Weyssenhoff, Bull. de l'Acad. Pol. de Sc. et de Lettres, Classe de Sc. Math. at Not. 22 (1937)
[6] B. Średniawa, Acta Phys. Pol. 9, 99 (1947)
[7] J. Weyssenhoff, A. Raabe, Acta Phys. Pol. 9, 7 (1947)
[8] M. Mathisson, Acta Phys. Pol. 6, 218 (1937)
[9] H. Hönl, A. Papapetrou, Zf. f. Phys. 112, 512 (1940)
[10] J. Lubański, Acta Phys. Pol. 6, 356 (1937)
[11] K. Nagy, Acta Phys. Ac. Sc. Hung. 7, 325 (1957)
[12] L. Infeld, Bull. Ac. Pol. CLIII, 213 (1955)
[13] J. Weyssenhoff, Acta Phys. Pol. 9, 26 (1947)
[14] J. Weyssenhoff, A. Raabe, Acta Phys. Pol. 9, 19 (1947)
[15] F. Bopp, Zf. f. Naturf. 1, 197 (1948), 3a, 537 (1948)
[16] J. Weyssenhoff, Acta Phys. Pol. 11, 49 (1951)
[17] Z. Borelowski, Acta Phys. Pol. 20, 619 (1961)
[18] M. Born, Rev. Mod. Phys. 21, 465 (1949)
[19] J. Rayski, Helv. Phys. Acta 36, 1081 (1963)
[20] Z. Borelowski, B. Średniawa, Acta Phys. Pol. 25, 609 (1964)
[21] P. Havas, J.N. Goldberg, Phys. Rev. 128, 398 (1972)
[22] A.J. Hanson, T. Regge, Ann. of Phys. 87, 498 (1974)
[23] D. Mušicki, Journ. of Phys. A (Great Britain) 11, 39 (1978).

ANGULAR MOMENTUM OF ISOLATED SYSTEMS IN GENERAL RELATIVITY

Abhay Ashtekar

Universite'de Clermont-Fd., 63170 Aubiere, France
and
Max Planck Institut, Föhringer Ring 6, 8 München 4o, FRG

1. INTRODUCTION

The purpose of this article is to outline the present status of the issue of introducing a meaningful notion of angular momentum of isolated gravitating systems. More precisely, the aim is three-folds: i) to point out the conceptual difficulties which, in the framework of general relativity,make the introduction of angular momentum more difficult than than that of, say, energy-momentum; ii) to summarize ways in which these difficulties have been overcome to a certain extent; and, iii) to discuss theorems which unify the resulting definitions. The article is addressed to non-experts. The emphasis is therefore on the basic ideas involved; detailed proofs of various assertions can be found in the original papers which are referred to at appropriate places in the text. Also, the exposition is intended to be complementary to winicour's review {1} of the subject. In particular, some of the issues discussed by Winicour in detail are skipped entirely or discussed only briefly and the developments that have taken place since the completion of {1} are stressed.

The material is organized as follows. Sec.2 recalls definitions of angular momentum which become available in the presence of isometries. Sec.3 summarizes the conceptual difficulties which arise when one attempts to introduce similar definitions in absence of isometries. Sec.4 reviews the situation at null infinity and Sec.5 at spatial infinity. Theorems relating various definitions are discussed in Sec. 6.

2. DEFINITIONS IN PRESENCE OF ISOMETRIES.

A. The Komar Scalar.

Fix a space-time (M, g_{ab}), i.e., a smooth, 4-dimensic al manifold M equipped with a metric g_{ab} of signature $(- + + +)$. Let us suppose that M is topologically R^4 and g_{ab} satisfies Einstein's vacuum equation outside a spatially compact world tube. Then, given any Killing fie K^a, the Komar {2} integral

$$Q_K := \int_{S^2} \varepsilon_{abcd} \nabla^c K^d \, dS^{ab} \tag{1}$$

is independent of the particular choice of the 2-sphere S^2, surrounding the sources, made in its evaluation. Indee using the fact that every Killing field K^a must satisfy $\nabla_a \nabla^a K_c = - R_{ac} K^a = 0$ if Einstein's equation is satisfied, it follows that $F_{ab} := \nabla_a K_b$ satisfies source-fr Maxwell's equations in the region in which R_{ab} vanishes and (1) is precisely the charge integral for this F_{ab}. *If K^a happens to be a rotational Killing field, i.e., if K^a is spac like with closed orbits, one can interpret Q_K as the " Z-component" of angular momentum.* Due to surface independence of (1), Q_K is, in the terminology of Newman and Penrose {3} , an abs lutely conserved quantitiy. In particular, if (M, g_{ab}) happens to be asymptotically flat at null infinity, one c evaluate the integral in (1) on a 2-sphere cross-section of $\mathcal{I}$; irrespective of how much radiation pours out of $\mathcal{I}$ or pours in to $\mathcal{I}^-$, the value of Q_K is independent of t cross-section . The Komar scalar Q_K provides what is p haps the simplest and the most unambiguous notion of angu lar momentum in general relativity. Indeed, a viability c terion for possible definitions in more general contexts that they should reduce to this notion in the axi-symmetr context.

Remarks: (i) In order for Q_K to represent an obse vable on the space of all gravitational fields, we must have a well-defined way of normalizing the Killing vector K^a. (Given a Killing field K^a, rK^a is also a Killing field for any real number r.) In the axi-symmetric conte this is achieved by demanding that along any integral cur of K^a, the affine paramenter should run from 0 to 2π. (ii) It is easy to check that if M is spatially compact Q_K must vanish for all Killing fields. If the region in which R_{ab} vanishes admits several (equivalence classes of) 2-spheres which can not be continuously deformed in each other - i.e., if the dimension of the second homoto group exceeds one- the same Killing field can give rise to several distinct conserved quantities. However, they obey an obvious addition law. (iii) If K^a is a Killing field also in the source region, one can construct a conserved quantity $\underline{Q}_K$ in an obvious way: $\underline{Q}_K := \int_\Sigma T_{ab} K^b dS^a$ where Σ is a Cauchy surface. Q_K can also be expressed a as an integral over a Cauchy surface : $Q_K = \int_\Sigma R_{ab} K^b dS^a$ Thus, in general, $Q_K \neq \underline{Q}_K$! This inequality has certain puzzling aspects. Consider the case when K^a is a stationary Killing field. In this case, Q_K represents the *total* mass associated with (M, g_{ab}) -including the contributions of the gravitational field itself- while $\underline{Q}_K$ is often interpreted as the *inertial* mass associated with sources alone. One is therefore tempted to interpret the difference $Q_K - \underline{Q}_K$ as contribution to mass from the gravitational field. However, the interpretation is not so clear cut: if K^a is a rotational Killing field, Q_K equals $\underline{Q}_K$,and one is forced to conclude that although the gravitational field can contribute to mass, it can not contribute to angular momentum !

B. The Hansen Dipole Moment.

The next notion of angular momentum arises from the definition of multipole moments available in the case of stationary isolated systems. Consider a space-time (M, g_{ab}) in which the metric satisfies the vacuum equation outside some world tube. Let it admit a Killing field t^a which is timelike outside a spatially bounded region and has, furthermore the property that the three dimensional manifold S of its orbits is asymptotically Euclidean. (For detailed definitions, see {4} and {5}.) The manifold S inherits three basic fields from the 4-geometry: a metric h_{ab} which is positive definite outside some compact region, a scalar field $\lambda \equiv -t.t$, the norm of the Killing field, which is positive outside this compact set and a vector field $\omega_a = \varepsilon_{abcd} t^b \nabla^c t^d$, the twist of the Killing field. Furthermore, the asymptotic conditions on (S, h_{ab}) imply that there exists a manifold $\tilde{S}$ with a metric $\tilde{h}_{ab}$ such that, as a point set, $\tilde{S} = S \cup \Lambda$, Λ being a single point, with $\tilde{h}_{ab} = \Omega^2 h_{ab}$ on S. Furthermore, Ω admits an extension to $\tilde{S}$ and vanishes at Λ sufficiently rapidly to enable one to interpret Λ as"the point at infinity of S".

Hansen{4} has defined a set of multipole moments using the asymptotic behavior of h_{ab}, λ , and ω_a . It is conjectured that these moments might suffice to characterize the 4-geometry completely outside a suitable world-tube. (For a partial result, see {6}) There are two sets of moments : one associated with mass and the other with angular momentum. Each of these is represented by a trace-free totally symmetric tensor at the point Λ at spatial infinity, which has a certain transformation property under conformal rescalings of $\tilde{h}_{ab}$. These rescalings can be regarded as "shifts of origin about which multipole moments are defined" and the transformation properties of multipole moments are compatible with this interpretation.

It is the angular momentum dipole moment -represente by a vector S_a at Λ - that is of interest to the present analysis. This S_a is defined by: $S_a := \lim_{\to \Lambda} \delta_a \tilde{\Omega}^{-1/2} \lambda \cdot \omega$ where ω is a potential for ω_a , $\omega_a = \delta_a \omega$, and where δ denotes the gradient on S. (Existence of ω is ensured by Einstein's vaccum equation which holds near Λ .) The vect S_a is interpreted as the *total spin vector* of the gravit ting system described by (M, g_{ab}).

Remarks:(i) As in the case of the rotational Killing field, in order to compare spins associated with differen space-times, it is essential to normalize the stationary Killing vector in a 'universal' way. The usual convention is to require that t^a be future-directed and unit at sp tial infinity, i.e., that $\lim \lambda = 1$. (ii) Let (M, g_{ab}) admit an additional Killing field R^a which is rotational Then one would expect the spin vector S_a to point along the axis of R^a and its magnitude to equal the Komar scal associated with R^a. In the framework of {4} however,thi issue is still open. (iii)The angular momentum monopole moment m^* is defined by $m^* := \lim \tilde{\Omega}^{-1/2} \lambda \cdot \omega$. Although one expects m^* to vanish in any physically reasonable space-time, for the class of space-times considered in{4} there exists no proof to this effect. If an example is actuall found in which asymptotic conditions of {4} are satisfied but in which m^* fails to vanish, one would be forced to conclude that these conditions fail to capture the notion of asymptotic flatness to a satiafactory extent. Indeed, if $m^* \neq 0$, S_a fails to be conformally invariant and can not therefore be regarded as a spin vector.

3. EXTENSION TO ASYMPTOTICALLY FLAT SPACE-TIMES: DIFFICULTIES

In the axi-symmetric space-times, angular momentum arises as a conserved scalar quantitiy while in the stationary case, it arises as a 3-vector at the point Λ at spatial infinity. We now consider space-times which are asymptotically flat in some suitable sense but which do not, in general, admit any isometries. What should the nature of angular momentum be in this case ? Let us begin by recalling the situation in special relativity. In this case, the notion of angular momentum is intertwined with the structure of the Poincaré group. More precisely, angular momentum emerges as a skew tensor field M_{ab} in Minkowski space with the familiar transformation property:

$$M_{ab}(x') = M_{ab}(x) + P_{[a} T_{b]} \qquad (2)$$

for all points x and x' in Minkowski space where P_a is the 4-momentum of the system and T_a the vector connecting x and x'. $M_{ab}(x)$ is interpreted as the angular momentum of the system about the origin x. Although M_{ab} is a tensor field, because of (2), it has only 6 rather than 6^{∞} independent components. A simple transformation law such as (2) is possible only because the Poincaré group has precisely a four paramenter family of Lorenetz subgroups, i.e., only beacuse just as any two points of Minkowski space are related by a translation, so are any two Lorentz subgroups of the Poincaré group.

One might expect that if one restricts oneself to asymptotically flat space-times, one would again obtain Poincaré group as the group of asymptotic symmetries enabling one to introduce a 'tensorial' angular momentum also for isolated gravitating systems. Unfortunately, the issue is not so simple. To see this, consider a typical Killing field K^a in Minkowski space:

$$K^a = \underline{F}^{ab} X_b + \underline{V}^a \qquad (3)$$

where $\underline{F}^{ab}$ is a fixed skew tensor and $\underline{V}^a$, a fixed vector, and where X_b is the position vector of any point relative to a fixed origin. In asymptotically flat spacetimes, we wish to consider vector fields analogous to this K^a. Since, by assumption, the curvature goes to zero as one moves away from sources, the ambiguities introduced by curvature in what is meant by 'analogous' will, presumably, become smaller as one moves away. However, in Minkowski space, the part of K^a which is linear in X^a (the Lorentz part) dominates the part independent of X^a(the translation part) in this limit. Hence, apriori, it is quite possible for the asymptotic symmetrty group to become much larger than the Poincaré group even though in its strucutre, it continues to mimic the Poincaré group: given a vector field S^a which resembles K^a of (3) at large distances from sources, if $\underline{F}^{ab} \neq 0$, S^a+B^a will also resemble K^a provided B^a remains bounded. That is, if $\underline{F}^{ab} \neq 0$, there is a 'competition' between the fall-off of curvature and the unbounded growth of asymptotic Killing fields analogous to K^a. The precise nature of the asymptotic symmetry group therefore depends quite sensitively on the detailed fall-off properties of curvature.

Indeed, a typical situation {7 - 11} is the following: if one imposes only those boundary conditions which ensure that energy-momentum is well defined, the asymptotic symmetry group G emerges as a semi-direct product of an infinite dimensional Abelian group S with the Lorentz group L . Thus, although G resembles the Poincaré grou in its structure, the translation subgroup of the Poincar group is replaced by the *infinite* dimensional S. Elements of S are called supertranslations.S may admit a preferr 4-dimensional subgroup T which reduces to the usual tra slation subgroup in Minkowski space. Futhermore, T may be a normal subgroup of G . Yet, G need not admit *any* preferred Poincaré subgroup: as the intuitive argument given above suggests, it is the elements of G which resemble K^a of (3), with $\underline{F}^{ab} \neq 0$, which inherit supertranslation ambiguities in the transition from Minkowski space-times asymptotically flat ones.The symmetry group G now has infinite -rather than just four- parameter family of Lore tz subgroups. Cosequently, a canonical tensorial angular momentum with just 6 independent components can simply nc be introduced in principle! Using any *one* Poincaré subgroup of G, one *can* hope to introduce a tensorial angula momentum satisfying (2). However, angular momentum corresponding to different Poincaré sub-groups will fail to be related in a simple way. Hence, the resulting notion of angular momentum will be *qualitatively* different from the one in special relativity.

On the other hand, if the asymptotic curvature does fall off sufficiently fast to enable one to remove ambiguities discussed above, then the situation simplifies. In this case, a preferred Poincaré subgroup of G can be selected in a natural way and the angular momentum can, a least in principle, recover its special relativistic stat

4. NULL INFINITY

Fix a space-time (M, g_{ab}). Let us suppose that it is asymptotically flat in the sense of {12}. Then, one can 'attach' to M a null boundary $\mathcal{I}$ and introduce a smmoth met $\hat{g}_{ab}$ on $\hat{M} \equiv M \cup \mathcal{I}$ such that , on M , $\hat{g}_{ab} = \Omega^2 g_{ab}$. As c approaches $\mathcal{I}$, Ω goes to zero precisely at such a rate that one can interpret $\mathcal{I}$ as 'the null cone at infinity'. The group of asymptotic symmetries for this class of spac times is the BMS group B {7 - 9} which belongs to the family of groups G considered in Sec.3. Thus, B admits ar infinite dimensional Abelian normal subgroup BS -isomorpł to the group of smooth functions on a 2-sphere under addi tion- of BMS supertranslations and B/BS is isomorphic t the Lorentz group L . Finally, although B admits a prefer red 4-dimensional subgroup BT of BMS translations, it does not admit a preferred Lorentz subgroup, nor a Poincaré subgroup. Consequently, irrespective of the details of the mathematical expressions involved, the notion of angular momentum is forced to be qualitatively different from that in special relativity.

Fix a Poincaré subgroup P of B. Then, at least at the conceptual level, the situation becomes similar to that in special relativity: Given a cross-section s of $\mathcal{I}$, one can try to introduce a tensorial angular momentum -analogous to M_{ab} - using generators of this P and various physical fields on s. The cross-section s represents a 'retarded instant' of time. Hence by comparing values of angular momentum evaluated on two different cross-sections, one can compute the angular momentum carried away by gravitational waves between the two retarded instants. There have been several attempts at obtaining expressions for angular momentum in this broad framework{13 - 17}, each of which exploits a particular, well understood aspect of angular momentum of simpler physical systems. Thus, for example, in {13} one generalizes the expression of the Komar integral using 'asymptotic Killing fields' in place of exact ones; in {14} one considers Einstein's theory as a gauge theory and computes conserved quantities associated with local Lorentz rotations; and, in {17} one introduces a phase-space strucutre on the space of fields on $\mathcal{I}$ representing radiation and obtains expressions for fluxes of angular momentum as Hamiltonians generating suitable canonical transformations on this phase space. These investigations have led to two *distinct* expressions: approaches {13,16,17} lead to one and {14,15} to another. (For details, see {1}.) The difference between the two expressions (involves only the asymptotic shear and) disappears both in the weak field as well as the stationary limit. However, there does exist a situation which can be used to distinguish between the two: in axi-symmetric space-times *with* gravitational radiation, the expression in {13, 16, 17} is always compatible with the Komar integral while that in {14,15} need not be. Irrespective of these details, it is fair to say that the status of angular momentum at null infinity is still far from being satisfactory: in none of the available approaches can one see a simple relation between the angular momentum tensors associated with *diffrent* Poincaré subgroups of B. This lack of information is often referred to as the 'supertranslation ambuity'.

There does exist a class of space-times, however, in which these ambiguities disappear: space-times in which Bondi news {7,8,9} vanishes identically. In this case, one can introduce an additional structure on $\mathcal{I}$ - a 4-parameter family of shear-free cross sections- and select a canonical Poincaré subgroup of B consisting of elements of B which preserve this new structure. The 4-parameter family of these cross sections naturally inherits the structure of Minkwski space. Angular momentum now emerges as a tensor field $M_{\alpha\beta}$ on this Minkowski space satisfying the transformation property (2). We have:

$$M_{\alpha\beta}(s)\ F^{\alpha\beta}(s) \ := \ \int_s {}^*\hat{K}_{abcd}\ \hat{l}^a \hat{X}^b_{F(s)}\ d\hat{S}^{cd} \qquad (4)$$

for all shear-free cross sections s of I. Here, ${}^*\hat{K}_{abcd} = \mathrm{Lim}\ \Omega^{-1}\,{}^*\hat{C}_{abcd}$ is the dual of the asymptotic Weyl curvature; l^a is the null vector field, orthogonal to the given cross-section s satisfying $l^a\nabla_a\Omega = -1$ and $X^a{}_F$ is the BMS vector field on I corresponding to the Lorentz rotation about s generated by $F_{\alpha\beta}$ in the Minkowski space of shear-free cuts. (Greek indices refer to this Minkowski space and the latin ones to M. For details concerning (4), see, e.g., {18}.) Bondi news vanishes if and only if the asymptotic curvature $\hat{K}_{abcd}$ satisfies $\hat{K}_{abcd}\,\hat{n}^a\,\hat{n}^c = g\,\hat{n}_b\,\hat{n}_d$ on I where $\hat{n}_a = \hat{\nabla}_a\Omega$. In terms of the physical Wely curvature, this condition implies that '$1/r$ and $1/r$ parts' of C_{abcd} vanish at null infinity. Thus, as remarked in Sec.3, the reduction of the symmetry group is intertwined with the rate at which the curvature dies off as one moves away from sources.

Remarks: Consider a situation in which the Bondi news has compact support on I. In this case, one can select a Poincaré subgroup of the BMS group using shear-free cross sections in the past of this compact region as well as in the future. However, in general, the two Poincaré subgroups will be *distinct* and a meaningful comparison between the associated angular momenta -i.e. the evaluation of the angular momentum radiated away- will not be possible !

5. SPATIAL INFINITY

Since it is the presence of radiation that introduces ambiguities in the notion of angular momentum at null infinity and since radiation emitted at finite times can never reach spatial infinity, one might expect the notion to become unambiguous in the spatial asymptotic regime. To a certain extent, this expectation turns out to be correct.

Fix a space-time (M, g_{ab}). Let us suppose that it is asymptotically flat at spatial infinity in the sense of {11} or {19}. Then, one can attach to M a point i^0 and introduce a metric $\hat{g}_{ab}$ on the completed space-time such that *every* point of M is space-like related to i^0, and, on

M, $\hat{g}_{ab} = \Omega^2 g_{ab}$. The asymptotic properties of Ω enable one to interpret i^0 as 'the point at spatial infinity', or, equivalently, as 'the vertex of the null cone at infinity'. In general, the Weyl curvature of $\hat{g}_{ab}$ is singular at i^0, the 'strenght of the singularity' being a measure of the total mass associated with the space-time in pretty much the same way as the 'strength of the Coulomb singularity' of the Maxwell field is a measure of the electric charge. However, the differentiability requirement on $\hat{g}_{ab}$ is such that $\Omega^{1/2}\hat{C}_{abcd}$ admits a limit $\underline{C}_{abcd}(\eta)$ at i^0 depending on the tangent vector η, at i^0, to the curve of approach in a smooth way. Consider the hyperboloid Ψ of unit space-like vectors in the tangent space at i^0. $\underline{C}_{abcd}$ is completely characterized by two smooth fields $E_{ab} := \underline{C}_{ambn}(\eta)\,\eta^m\eta^n$ and $B_{ab} := {}^*\underline{C}_{ambn}(\eta)\,\eta^m\eta^n$ on this Ψ. These two fields *- called the electric and the magentic parts of the asymptotic Weyl curvature-* carry information about the '$1/r^3$'-part of the physical Weyl tensor. E_{ab} has a direct physical interpretation: it governs the radial geodesic deviations for time like geodesics in the far field region. B_{ab}, on the other hand, has no simple interpretation; we shall see in Sec.6 that for a large class of physically interesting systems, B_{ab} vanishes identically. The definitions of asymptotic flatness itself includes only those conditions which ensure the finiteness of the ADM energy-momentum and does not imply $B_{ab} = 0$.

The group of asymptotic symmetries is called the Spi group. (Spi stands for spatial infinity and rhymes with Scri.) This group will be denoted by SG. Again, SG belongs to the class of groups G discussed in Sec.3. The only difference between the BMS group B and SG is in the 'size' of the supertranslation subgroup: the group SS of Spi supertranslations is isomrphic to the additive group of smooth functions on the hyperboloid Ψ, rather than on a 2-sphere. Hence, in the introduction of angular momentum one seems to be faced with the same conceptual problems as those at null infinity.

At the technical level, the problems are even worse at this stage: One simply does not have physical fields on Ψ which can carry information about angular momentum! That is, even if one is willing to accept the supertranslation ambiguities as in the case of null infinity, one not write down a viable formula for angular momentum. This is because the only non-zero conserved quantitiy that one can construct from E_{ab} is the (ADM) energy-momentum while all such quantities constructed from B_{ab} vanish identically! (Note, incidently, that, in particular, the magnetic analogue of energy-momentum -and hence

also the angular momentum momopole- vanishes identically
This is in sharp contrast with the situation in Maxwell
theory where the satisfaction of Maxwell's equations nea
infinity does not imply that the magnetic charge must va
sh. Also, compare with the situation in Sec 2.B.) In any
case, from physical considerations, one would expect the
information about angular momentum to reside in the '$1/r$
part' of the physical Weyl curvature and both E_{ab} and B
refer only to the '$1/r^3$ -part'.

It turns out, however, that both conceptual and tec
cal difficulties can be resolved at one stroke. Demand t
B_{ab} should vanish on Ψ . (Again a restriction on the fa
off properties of curvature!) Then, one can select a can
nical Poincaré subgroup of SG. Futhermore, since $B_{ab} = 0$
one can consider the 'next order' field $\hat{\beta}_{ab} := \lim_{\to} {}^*\hat{C}_{amb}$
$\eta^m \eta^n$. This $\hat{\beta}_{ab}$ contains information about the[10] $1/r^4$-
part' of the physical Weyl curvature. Define a tensor $\hat{M}$
at i^0 by:

$$\hat{M}_{ab} F^{ab} := 1/2 \int_{S^2} \hat{\beta}_{ab} X^a_{*F} dS^b \quad (5)$$

where F^{ab} is any skew tensor at i^0, X^a_{*F} is the Lorent
rotation in the tangent space of i^0 generated by ${}^*F_{ab}$ a
S^2 is any 2-sphere cross-section of the hyperboloid Ψ a
i^0. . The integral on the right can be shown to be indepe
dent of the particualr choice of the 2-sphere made in it
evaluation. Conformal rescalings of $\hat{g}_{ab}$ which respect
various asymptotic conditions -including $B_{ab} = 0$- giv
rise to Spi-translations. Under these rescalings, $\hat{\beta}_{ab}$
transforms precisely in the way required for $\hat{M}_{ab}$ to sa
fy the analogue of equation (2). (For details, See {10}.
Thus, for the class of space-times in which the '$1/r^3$-pa
of the Weyl tensor is completely determined at spatial i
finity be E_{ab}, equation (5) provides a definition of ang
lar momentum which is free of supertranslation ambiguiti

Remarks: (i) Although the condition $B_{ab} = 0$ is qui
simple when expressed in terms of the Weyl tensor C_{abcd}
of the 4-metric g_{ab}, its expression in terms of the initial data on any given Cauchy surface is quite involved.
Hence, in retrospect, it not surprising that the reducti
of the asymptotic symmetry group to the Poincaré could n
be achieved in a simple way in the approaches based on t
initial value formulation. Indeed, in these approaches,
structure of the asymptotic symmetry group itself was qu
ambiguous: the supertranslation subgroup was varyingly
claimed to be the additive group of smooth functions on
2-sphere, the group of pairs of such functions, a 4-dime
sional group and the trivial group consisting of just th

identity element; the 3+1 splitting made the structure quite opaque ! (ii) Note that in both the spatial and the null regimes, the reduction of the asymptotic symmetry groups requires stronger conditions than the ones which ensure finiteness of energy-momentum. This situation is not surprising: already for linear fields in Minkowski space, the fall-off required for angular momentum to be well-defined is stronger than the one required for energy-momentum to be well-defined.

6. RELATION BETWEEN VARIOUS DEFINITIONS

We have seen various contexts in which the notion of angular momentum of isolated systems is meaningful. The obvious question now arises: What is the relation between all these definitions? Fortunately, a number of theorems on this issue are now available and together they provide a unified and 'compact' picture of the situation.

Let us begin with space-times which are asymptotically flat at spatial infinity in the sense of {11} and {19}.For angular momentum to be well-defined, B_{ab} must vanish on Ψ. Fortunately, we have {19} :

Theorem: If the space-time is, in addition, stationary axi-symmetric, $B_{ab} = 0$ on Ψ .

We can therefore compare the resulting† tensorial angular momentum $\hat{M}_{ab}$ at i^0 with the Hansen dipole moment in the stationary case and the Komar scalar in the axi-symmetric case. The results are the following {19}:

Theorem: Let (M, g_{ab}, t^a) be a stationary space-time which is asymptotically flat in the sense of {4} as well as {19} Then the natural projection from M on to the manifold S of orbits of t^a sends the spin vector $\hat{S}_a := \hat{\varepsilon}_{abcd}\, \hat{M}^{cd}\, \hat{p}^b$ at i^0 to the Hansen dipole moment S_a to the point Λ at infinity of S, where $\hat{\varepsilon}_{abcd}$ is the alternating tensor i^0 compatible with $\hat{g}_{ab}$ and where $\hat{p}^a$ is the ADM energy-momentum vector at i^0.

Theorem: Let (M, g_{ab}, R^a) be axi-symmetric and asymptotically flat at spatial infinity in the sense of{19}. Then, $Q_R = \hat{M}_{ab}\, \hat{F}^{ab}$, where Q_R is the Komar scalar and $\hat{F}^{ab}$ is the tensor at i^0 defined by $\hat{F}^{ab} = \hat{\nabla}^{[a} R^{b]}$.

These theorems say that the definition introduced in

---

†Note that $B_{ab} = 0$ does not imply that $\hat{\beta}_{ab}$ exists:the manetic part of the asymptotic Weyl curvature may fall off for example, as $(\ln r) r^{-4}$. In what follows, we assume that $\hat{\beta}_{ab}$ exists.

Sec.5 reducesto those of Sec.2 in presence of isometries. Note that this agreement is non-trivial since each defini tion involves a distinct geometrical object: the Komar sc lar refers to the rotatioanl Killing vector, the Hansen c pole moment to the stationary Killing vector and the angu lar momentum tensor at i^0 ,to the asymptotic properties c Weyl curvature. That these distinct definitions agree whe they become available simultaneously, is one of the many deep and subtle aspects of Einstein's vacuum equation. Th two theorems together imply that in axi-symmetric station ary space-times satisfying conditions in {4} and {19},the Hansen dipole moment does point along the axis of the rot tional Killing field, its magnitude being equal to the Komar scalar associated with R^a, as one intuitively expe cts.

Next, let us consider space-times which are stationa and asymptotically flat at null and spatial infinity in t sense of {11} or {17}. Then, in addition to the Hansen di pole moment and the tensorial angular momentum at i^0, the is also available a tensorial angular momentum $M_{\alpha\beta}$ in th Minkowski space of shear-free cross-sections of . Set $S_\alpha = \varepsilon_{\alpha\beta\lambda\omega}P^\beta M^{\lambda\omega}$. Since $M_{\alpha\beta}$ has the transformation prope ty analogous to (2), S_α is a constant vector field in th asymptotic Minkowski space: Physically, it represents the spin vector of the isolated system at null infinity.One c now show the following result {17}:

Theorem: There exists a natural isomorphism $\hat{\Psi}$ between the space of constant vector fields in the Minkowski space of shear-free cross sections and the tangent space of i^0 .Fur thermore, $\hat{\Psi}(S^\alpha) = \hat{S}^a$, where $\hat{S}^a$ is the spin vector at i^0.

Thus, once again, all definitions agree.

Finally, consider axi-symmetric space-times which ar asymptotically flat at null infinity in the sense of {]2} As remarked in Sec.4, inspite of the presence of gravita- tional radiation, the component of the angular momentum tensor {13,16,17} at null infinity selected by the rota- tional Killing field,equals the Komar scalar defined by this Killing field. (For details, see {13} and {1})

7. DISCUSSION

It appears at this stage that the major problems whi are still open in this field are pretty much concentrated around the definitions at null infinity in presence of gr vitational radiation.Indeed, in the spatial regime, the only obvious open issue is the relation of the definition at i^0 with that proposed by Sommers{20} in terms of a projective completion of the space-time, an issue which should not be difficult to resolve once a precise definition of asymptotic flatness, encompassing gravitational fields which are not necessarily weak, is found in terms of the projective completion. In the null regime, perhaps the most outstanding issue is that of supertranslation ambiguities. The general consensus seems to be that these ambiguities are genuine and one just has to 'live with them'. That is, the general feeling seems to be that the qualitative change suffered by the notion of angular momentum in presence of gravitational radiation is an unaviodable feature associated with the transition from special to general relativity analogous to the change suffered by the notion of energy in the transition from Newtonian mechanics -where it is an absolute notion- to speical relativity -where it becomes observer-dependent . However, to put this viewpoint on a sound footing, one must understand the operational significance of supertranslation ambiguities. Such an investigation is, to our knowledge, not yet undertaken.

Even if one accepts these ambiguities, there still remains the issue of finding the 'correct' formula. The fact that two distinct candidates are already avaialable suggests that the situation is likely to be complicated. Using the agreement with the Komar integral as a viability criterion one may eliminate one of the two available choices. However, it is not clear that yet other alternatives do not exist. Perhaps what is needed is a set of strong viability conditions against which one can test various candidates. Alternatively, one might prove general theorems -analogous to the result, in the energy-momentum case, that the Bondi mass can never increase in time on $\mathcal{I}^+$ - which can provide strong support in favor of one of these candidates.

Finally, there is a set of problems associated with the relation between null and spatial regime in non-stationary contexts. The condition $B_{ab}=0$ selects a preferred Poincaré subgroup of the Spi group. Does it also select a preferred Poincaré subgroup of the BMS group ? If so, what is the relation of this subgroup with the one selected via the Newman-Penrose {21} prescription?? Can one extend the last theorem of Sec.6 to obtain a relation between the past limit of angular momentum on $\mathcal{I}^+$ and angular momentum at i^0 in the radiating case ?

Acknowledgments : I thank Robert Geroch, Richard Hansen, Anne Magnon-Ashtekar, Roger Penrose, Michael Streubel and Jeff Winicour for disscussions and correspondance on the issues discussed here.

References:

1. Winicour, J. to appear in the *Einstein Centennial Volume*, ed. P.G. Bergmann, J.N. Goldberg & A. Held, (Plenum, New-York)
2. Komar, A. (1959) Phys. Rev. 113, 934
3. Newman, E.T. and Penrose, R. (1968) Proc. Roy. Soc. (London) A305, 175
4. Hansen, R.O. (1974) J. Math. Phys. 15, 46
5. Geroch, R. (1970) J. Math. Phys. 11, 2580
6. Xanthopoulos, B.C. (1978) Pre-print.
7. Bondi, H., Metzner, A.W.K. and Van der Berg, M.J.G. (1962) Proc. Roy. Soc. (London) A269, 21.
8. Sachs, R.K. (1962) Proc. Roy. Soc. (London) A270, 103
9. Penrose, R. (1974), in *Group Theory in Non-linear Physics* ed. A.O. Barut (D. Reidel, Dordrecht)
10. Ashtekar, A. and Hansen, R.O. (1978) J. Math. phys. 19, 1542
11. Ashtekar, A. to appear in the *Einstein Centennial Volume, ed.* P.G. Bergmann, J.N. Goldberg, & A. Held, (Plenum, New-York
12. Geroch, R. and Horowitz, G.T. (1978) Phys. Rev. Lett. 40, 20
13. Tamburino, L. and Winicour, J. (1966) Phys. Rev. 150, 1039; Winicour, J. (1968) J. Math. Phys. 9, 860
14. Bramson, B.D. (1975) Proc. Roy. Soc. (London) A341, 463
15. Streubel, M. (1978) Gen. Rel. & Grav. 9, 551
16. Prior, K.R. (1977) Proc. Roy. Soc. (London) A354, 379
17. Ashtekar, A. and Streubel, M. (1979) Pre-print.
18. Ashtekar, A. and Streubel, M. (1979) J. Math. phys. (to appe
19. Ashtekar, A. and Magnon-Ashtekar, A. (1979) J. Math. Phys. 20, 793
20. Sommers, P.D. (1978) J. Math. Phys. 19, 549
21. Newman, E.T. and Penrose, R. (1966) J. Math. Phys. 7, 863

ISOMETRIES AND GENERAL SOLUTIONS OF NON-LINEAR EQUATIONS

F.J. Chinea

Departamento de Métodos Matemáticos, Facultad de Ciencias Físicas, Universidad de Madrid, Madrid-3 and Departamento de Física Fundamental, U.N.E.D., Madrid-3, Spain

1. INTRODUCTION

There has been much interest in recent years in the physical applications of certain non-linear partial differential equations, such as the Sine-Gordon and Korteweg-de Vries equations. It is our purpose here to discuss some differential geometric ideas that are common in General Relativity and which may prove of interest in their study.

We shall consider equations in one time and one space variable. The possible relevance of geometric ideas is based on the fact that a number of such equations may be associated in a rather natural way with spaces of constant curvature[1].As is well known, a space of constant curvature is homogeneous under a (pseudo-)orthogonal group of isometries, and there exists an induced action on the solutions of the corresponding equation. This fact provides a method for generating new solutions from known ones, and even finding the general solution in some specific cases.

A similar group-invariance property appears in the case of stationary, axially symmetric solutions of the Einstein equations[2,3], and in the study of the so-called chiral fields (exemplified by fields constrained to take values on a n-dimensional sphere)[4].

The geometric setting will be described briefly in the next section. In section 3 we illustrate the use of isometries for finding general solutions of non-linear equations by treating explicitly the case of the Liouville equation. Finally, the group-theoretic aspects of the problem are dealt with in the last section. The problem of finding the general solution of a certain second order equation is seen to be equivalent to one of finding the solution of a set of first order equations, or, alternatively, constructing all realizations of a three dimensional semisimple Lie algebra as vectorfields on a two-dimensional manifold, with the generators being restricted by a set of two constraints.

2. GEOMETRIC SETTING

Consider the two-dimensional metric

$$ds^2 = \sin^2 \frac{1}{2}\, \phi(x,t)dt^2 + \cos^2 \frac{1}{2}\, \phi(x,t)dx^2 \qquad (2.1)$$

depending on a function $\phi(x,t)$. It is easily seen[5] that the surface described by (2.1) is of constant curvature if and only if $\phi(x,t)$ satisfies the Sine-Gordon equation

$$\phi_{tt} - \phi_{xx} = k \sin\phi \qquad (2.2)$$

with the scalar curvature given by

$$R = -4k \qquad (2.3)$$

Other equations may be cast in this form. For instance, the Liouville equation

$$\phi_{uv} = ke^{\phi} \qquad (2.4)$$

corresponds to a metric

$$ds^2 = e^{\phi(u,v)}dudv \qquad (2.5)$$

In general, one has a metric given by

$$ds^2 = g_{ij}(\phi)dx^i dx^j \qquad (i,j = 1,2) \qquad (2.6)$$

where g_{ij} depends on ϕ and (possibly) on its derivatives, and the corresponding wave equation is simply

$$R(\phi) = \text{constant} \qquad (2.7)$$

If g_{ij} depends only on ϕ and not on higher derivatives of the field, the resulting partial differential equation will clearly be of second order.

In the case of constant negative curvature, there exists a remarkable set of transformations which map a given surface into a new surface of equal constant curvature. These transformations, developed by Bianchi and Bäcklund[5], are obtained by means of a finite geometric construction (as opposed to the usual infinitesimal transformations which are typically encountered in differential geometric problems). Such mappings, referred to as Bäcklund transformations in the literature, may be put in the form of a pair of differential equations. For the Sine-Gordon case, they may be written as

$$\begin{aligned} \phi_u + \psi_u &= 2\lambda \sin\frac{\phi-\psi}{2} \\ \phi_v - \psi_v &= \frac{2}{\lambda} \sin\frac{\phi+\psi}{2} \end{aligned} \qquad (2.8)$$

$$\left(u = \frac{1}{2}(t+x),\ v = \frac{1}{2}(t-x)\right)$$

where ϕ and ψ are the fields corresponding to, respectively, the first and second surfaces, and λ a constant. Once equations (2.8) are written down, one may ignore their geometric meaning and simply enquire about their differential consequences. It is easy to see that, remarkably enough, the integrability conditions of (2.8) are precisely the Sine-Gordon equations for ϕ and ψ. Similarly, one may write down the Bäcklund transformation for the surfaces associated with the Liouville equation,

$$\begin{aligned} \phi_u + \psi_u &= 2\sqrt{2\lambda}\ \mathrm{sh}\ \frac{\psi-\phi}{2} \\ \phi_v - \psi_v &= -\frac{\sqrt{2}}{\lambda}\exp\left(\frac{\psi+\phi}{2}\right) \end{aligned} \qquad (2.9)$$

Bäcklund transformations have being used extensively to derive soliton solutions of the corresponding equations, obtain conservation laws, etc. Generalizations of such transformations to a higher number of independent variables have been proposed. However, it is clear that in higher dimensions the number of fields connected by a given set of such transformations should be greater than two, for otherwise the integrability conditions force the fields to satisfy additional equations besides the desired ones.

Due to their representing surfaces of constant curvature, the equations we are considering admit a group of transformations corresponding to the isometry group of the surface. A given solution will be mapped into a new solution whose functional form with respect to the new variables coincides with that of the former with respect to the old variables. For a surface of constant negative curvature the corresponding group is SO(2,1). This fact will be used in the next section to describe a method for constructing the general solutions of such equations.

3. ISOMETRIES AND GENERAL SOLUTIONS

Suppose one has a surface associated with a certain non-linear equation, described by $g_{ij}(\phi)$. As we have seen g_{ij} admits a transitive isometry group. Ordinarily, one knows g_{ij} explicitly and computes its Killing fields, from which the corresponding group structure may be inferred:

$$g_{ij} \to \xi_{(\mu)(i;j)} = 0 \to \left[\xi_{(\mu)}, \xi_{(\nu)}\right] = c_{\mu\nu}{}^{\rho}\, \xi_{(\rho)} \to c_{\mu\nu}{}^{\rho}$$

(the greek index labels the three independent Killing fields). In the present case, however, we know in advance that the structure coefficients $c_{\mu\nu}{}^{\rho}$ are precisely those of the so(2,1) algebra. It is then possible to proceed backwards, so to speak:

$$c_{\mu\nu}{}^{\rho} \to \xi^{i}_{(\mu),j}\, \xi^{j}_{(\nu)} - \xi^{i}_{(\nu),j}\, \xi^{j}_{(\mu)} = c_{\mu\nu}{}^{\rho}\, \xi^{i}_{(\rho)}$$

$$\to g_{ij,k}\, \xi^{k}_{(\mu)} + g_{ik}\, \xi^{k}_{(\mu),j} + g_{kj}\, \xi^{k}_{(\mu),i} = 0 \to \phi = \phi(g_{ij})$$

If the general solution of

$$\left[\xi_{(\mu)}, \xi_{(\nu)}\right] = c_{\mu\nu}{}^{\rho}\, \xi_{(\rho)} \tag{3.1}$$

is found, then the general solution of the corresponding wave equation may be obtained. Notice that this method depends on the fact that equation (3.1) is covariant when expressed in terms of ordinary derivatives due to its antisymmetry. This allows for its solution before the metric is known.

As an example, we may find the general solution of the Liouville equation[6]. Two of the Killing equations for the metric (2.4) are

$$\xi^{1}_{(\mu),v} = 0 \qquad \xi^{2}_{(\mu),u} = 0 \tag{3.2}$$

Using this fact, equations (3.1) may be solved easily. One finds

$$\underset{(1)}{\xi} = \left(\frac{\cos\theta}{\theta_u}, \frac{\cos\zeta}{\zeta_v}\right) \qquad \underset{(2)}{\xi} = \left(\frac{\sin\theta}{\theta_u}, \frac{\cos\zeta}{\zeta_v}\right)$$

$$\underset{(3)}{\xi} = \left(\frac{1}{\theta_u}, \frac{1}{\zeta_v}\right) \tag{3.3}$$

for arbitrary functions $\theta = \theta(u)$, $\zeta = \zeta(v)$. The remaining Killing equation may be used to find ϕ_u, ϕ_v ("the remaing equation" is really three equations, due to the presence of three independent Killing fields)

$$\begin{aligned} \phi_u &= \frac{\theta_{uu}}{\theta_u} - \frac{\theta_u}{\sin(\theta-\zeta)}\left[1+\cos(\theta-\zeta)\right] \\ \phi_v &= \frac{\zeta_{vv}}{\zeta_v} + \frac{\zeta_v}{\sin(\theta-\zeta)}\left[1+\cos(\theta-\zeta)\right] \end{aligned} \tag{3.4}$$

From (3.4) we get the general solution

$$\phi(u,v) = \ln\frac{\lambda}{2}\ \frac{\theta_u\,\zeta_v}{\sin^2\frac{1}{2}(\theta-\zeta)} \qquad (\lambda = -1/k) \tag{3.5}$$

4. GROUP THEORETIC ASPECTS

The procedure used in the previous section may be cast into a form which brings out in a more transparent way the underlying group properties. First, notice that the metric of a space of constant curvature may be written as

$$g^{ij} = \sigma K^{\mu\nu}\,\underset{(\mu)}{\xi^i}\,\underset{(\nu)}{\xi^j} \tag{4.1}$$

where σ is a constant, $\underset{(\mu)}{\xi^i}$ the Killing fields, and $K^{\mu\nu}$ the Killing-Cartan form of the corresponding isometry group. (This may be easily seen by writing down the metric for a specific space of constant curvature and then using the fact that any two spaces of the same curvature may be mapped isometrically onto each other). The Killing equations are now redundant, in the sense that (4.1) is automatically invariant.

Introducing the three-vectors

$$p_\mu = \underset{(\mu)}{\xi^1} \qquad q_\mu = \underset{(\mu)}{\xi^2} \tag{4.2}$$

the metric(4.1) may be expressed as

$$g^{11} = \sigma p^2 \qquad g^{12} = \sigma p.q \qquad g^{22} = \sigma q^2 \tag{4.3}$$

where the scalar product in (4.3) is taken with respect to the constant Killing-Cartan metric (a(++-) metric in the so(2,1)case).

The Lie algebra equations (3.1) may be written as

$$\begin{aligned} p &= p_u \times p + p_v \times q \\ q &= q_v \times q + q_u \times p \end{aligned} \tag{4.4}$$

It is clear that solving for the field ϕ in $g_{ij}(\phi)$ yields two relations among the three components of the metric, and consequently on the p,q vectors. Such constraints are to be imposed in solving (4.4). Turning back to our previous example, we see that the Liouville case is characterized by the constraints

$$p^2 = 0 \qquad q^2 = 0 \tag{4.5}$$

(these equations replace eqs. (3.2)). Using (4.5) and (4.4) one obtains p and q in a straightforward way. Finally, the general solution ϕ may be obtained without integrations by writing it as

$$\phi = - \ln \frac{\sigma}{2} (p.q) \tag{4.6}$$

In general, we see that the constraints imposed by part of the Killing equations on the solutions of (3.1) may be substituted by functional constraints among the g^{ij}. When the associated partial differential equation is of the second order, such equations are finite (in general algebraic). For instance, the Sine-Gordon metric (2.1) is characterized by the algebraic constraints

$$\begin{aligned} & p.q = 0 \\ & \frac{1}{p^2} + \frac{1}{q^2} = 0 \end{aligned} \tag{4.7}$$

The problem of finding the general solutions of the given wave equations reduces then to finding the general solution of (4.4) with two constraints on p,q. This is equivalent to finding all realizations of so(2,1) in terms of homogeneous differential operators of the first order acting on a two-dimensional space, with the Casimir-like element g^{ij} satisfying a pair of finite constraints.

Acknowledgements

The author wishes to thank L. Abellanas, G.G. Alcaine, L. Martínez Alonso and F. Guil for conversations, and J. Burzlaff for bringing Ref.1 to his attention.

References

1. R. Sasaki, Niels Bohr Institute Preprint NBI-IIE-79-3 (1979).
2. R. Geroch, J. Math. Phys. 13, 394 (1972).
3. W. Kinnersley, D.M. Chitre, J. Math. Phys. 19, 1926 (1978).
4. K. Pohlmeyer, Comm. Math. Phys. 46, 207 (1976).
5. L.P. Eisenhart, A Treatise on the Differential Geometry of curves and surfaces, Dover, N.Y., 1960.
6. F.J. Chinea, Phys. Lett. 72A, 281 (1979).

ON THE VISUAL GEOMETRY OF SPINORS AND TWISTORS

Hans Hellsten

Institute of Theoretical Physics

University of Stockholm

1. INTRODUCTION

In the present paper a number of results will be presented, indicating that the usual component formalism of spinor and twistor algebra can be cast into a visually geometric form. Approaching the subject in a somewhat unorthodox manner, we succeed in constructing a faithful model of linear, complex 2-space in terms of Euclidean 3-space and a 2-plane. This model enables us to interpret various algebraic operations on spinors in a visually geometric way. For instance, spinor addition will be brought into correspondence with ordinary addition of 3-vectors. Interpreting twistors as pairs of spinors in the model, we find the interpretation of twistor norm and twistor translation to be of particular interest. In fact, translation of null twistors, when interpreted, ties the Euclidean 3-space and the 2-plane to Minkowski space in a natural manner; such a tie was not supposed to exist from the outset. Our spinor interpretation should be contrasted to, for example, the null flag interpretation[1], the latter being essentially of a topological, non-linear nature. It would thus be very difficult to try to form the sum of two spinors, or decompose a spinor into base components, resting the argument on null flags. In our interpretation no such difficulties arise.

A detailed presentation of our visualization scheme is beyond the scope of this paper. More details, and also some background material, can be found in earlier works by the author [2,3,4]. The possibility of visualizing twistors was noted only recently, however, and has not been described elsewhere. We will focus our attention on those properties of the visualization scheme, which are needed in order to interpret twistor translation and twistor norm. These concepts remain meaningful (in contrast to e.g. Lorentz transformations) in the sub-class of twistors, for which the π-spinors, of the corresponding spinor pairs (π, ω), are normed (see (2.4)).

In order to simplify the discussion as much as possible, we restrict ourselves to this class of twistors, though the interpretation of twistor translation and twistor norm in the general case should not prove much more complex. We shall not discuss Lorentz or conformal transformations of twistors, the details of the interpretation of these transformations have as yet (june -79) not been looked into.

2. THE SPINOR SPACE MODEL

Regard Euclidean 3-space E and a 2-plane C contained in E. Pick out a point $P \in C$ and form the vector spaces E_P, C_P consisting of all affine vectors PQ, P being fixed whereas Q varies in E or C. For the time being we do not think of E as in any particular way related to space-time. Rather, we shall rely on the intrinsic properties of E_P and C_P in order to construct out of them a space isomorphic to $\underline{C}^2$ (linear, complex 2-space). For this purpose we assume that there is a preferred unit vector in C_P, which we denote by 1_P. Given this vector we may identify C_P with a linear, complex 1-space in the usual way. Thus the complex numbers $\underline{C}$ will act as scalars, an element $\rho e^{i\alpha} \in \underline{C}$ causing in C_P a rotation through the angle α and a re-scaling by the factor ρ. Denote an arbitrary element in C_P by z_P and scalar action upon this element by $z' z_P$ (where $z' \in \underline{C}$). We also adopt exponential notation in C_P, defining $e_P^{i\alpha} \equiv e^{i\alpha} 1_P$. We now make our main definition, i.e. we introduce a (rather peculiar!) form of bilinear product $E_P \otimes C_P$ as follows: It may be checked that the following three axioms are mutually consistent*

A) $\rho_1(x_P \otimes z_P) + \rho_2(y_P \otimes z_P) = (\rho_1 x_P + \rho_2 y_P) \otimes z_P$,

B) $z_1(x_P \otimes z_P) + z_2(x_P \otimes z'_P) = x_P \otimes (z_1 z_P + z_2 z'_P)$,

C) $z_P \otimes (z z'_P) = (\bar{z} z_P) \otimes z'_P$.

Here $x_P, y_P \in E_P$; $z_P, z'_P \in C_P$; $\rho_1, \rho_2 \in \underline{R}$ (the real numbers) and z, z_1, $z_2 \in \underline{C}$. Bar denotes complex conjugation.

Geometrically, these axioms have the following interpretation. From axioms A and B it follows that $x_P \otimes \rho z_P = (\rho x_P) \otimes z_P$ for $\rho \in \underline{R}$. Thus the expression $x_P \otimes z_P$ determines a vector pair $(\rho x_P, z_P/\rho)$ up to an arbitrary choice of the scale factor $\rho \in \underline{R}$. Axiom C completes A and B in stating that in the degenerate case when both factors of $x_P \otimes z_P$ lie in C_P, we may not only change their relative length, but we are also free to rotate x_P and z_P in C_P, keeping the angle between them constant.

* Note that there is a slight difference between these axioms and those of ref. 4 (in that paper a 3-dimensional half-space W_P was used instead of E_P). The two sets of axioms are equivalent, though for our purpose the present set appears to be more efficient.

The space $E_P \otimes C_P$ is fully isomorphic to $\underline{C}^2$ as we now will show. We denote by $a_P \in E_P$ the unit normal to the plane C_P. An arbitrary element of E_P may be decomposed into a vector along a_P and a vector in C_P, the element thus written $\rho a_P + z 1_P$ for some unique pair of numbers $\rho \in \underline{R}$, $z \in \underline{C}$. In a corresponding manner, we write for an arbitrary element of $E_P \otimes C_P$ (making use of the rules A and C in turn)

$$(\rho a_P + z 1_P) \otimes z' 1_P = (\rho a_P) \otimes z' 1_P + (z 1_P) \otimes z' 1_P = \\ = a_P \otimes (\rho z') 1_P + 1_P \otimes (\bar{z} z') 1_P \ . \tag{2.1}$$

Proceeding inversely, we start with an arbitrary pair of complex numbers z_1, $z_2 \in \underline{C}$, to which we may assign a unique element of $E_P \otimes C_P$ as follows:

$$a_P \otimes z_1 1_P + 1_P \otimes z_2 1_P = a_P \otimes z_1 1_P + \\ + (\bar{z}_2 \bar{z}_1^{-1} 1_P) \otimes z_1 1_P = (a_P + \bar{z}_2 \bar{z}_1^{-1} 1_P) \otimes z_1 1_P \tag{2.2}$$

(here $z_1 \neq 0$; in the case $z_1 = 0$ the corresponding element of $E_P \otimes C_P$ is just $1_P \otimes z_2 1_P$). Evidently (2.1) constitutes a unique decomposition of the elements of $E_P \otimes C_P$. Defining

$$a_P \otimes z 1_P \cong \begin{pmatrix} z \\ 0 \end{pmatrix}, \quad 1_P \otimes z 1_P \cong \begin{pmatrix} 0 \\ z \end{pmatrix}, \tag{2.3}$$

we obtain a one-to-one mapping between $E_P \otimes C_P$ and $\underline{C}^2$, transforming components into components. According to rule B the transformation of components is linear, thus the mapping is an isomorphism. In view of this isomorphism we shall refer to the elements of $E_P \otimes C_P$ as spinors.

Let us regard a normed spinor, that is a spinor for which the components $(z_1, z_2) \in \underline{C}^2$ are restricted according to

$$|z_1|^2 + |z_2|^2 = 1 \ . \tag{2.4}$$

It is an elementary exercise indeed to prove that such a spinor can be represented as $x_P^N \otimes e_P^{i\alpha} \in E_P \otimes C_P$, where x_P^N is some unit vector in E_P. Spinors of this form will be of importance to us in the next paragraph.

3. THE TILTING OF THE SPACE $E_P \otimes C_P$

Henceforward, S^2 will stand for the fixed 2-sphere of unit diameter, located in E in such a way that $P \in S^2$ and C_P is a tangent plane to S^2 (see Figure 1). By means of this 2-sphere, it will become possible to re-express the elements of $E_P \otimes C_P$ in terms of any other space $E_Q \otimes C'_Q$, the plane C'_Q being tilted with respect to C_P. Let $1'_Q$ be any tangent vector to S^2 of unit length, having $Q \in S^2$ as its foot-point (thus $1'_P$ means some arbitrary direction at P, in general different from 1_P). We define C'_Q as the tangent plane at Q, containing $1'_Q$ as its preferred unit vector.

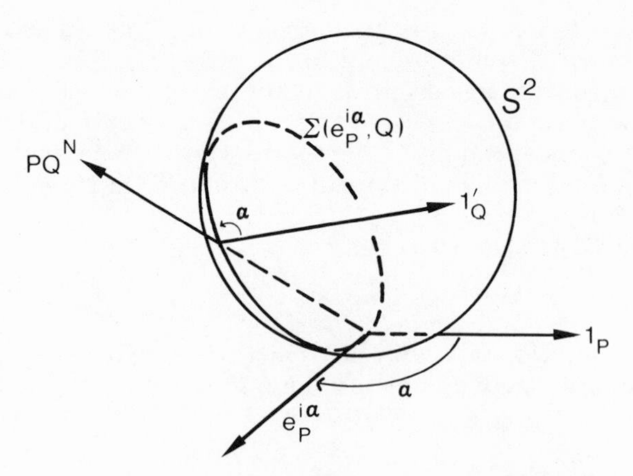

Fig. 1. The tangent vector $1'_Q$, associated to the normed spinor $PQ^N \otimes e_P^{i\alpha}$.

Given C'_Q, the space $E_Q \otimes C'_Q$ can be constructed in exactly the same manner as $E_P \otimes C_P$. The point in introducing a class of spaces in this way is of course to make the formalism symmetric: It will be found that each normed spinor $x_P^N \otimes e_P^{i\alpha} \in E_P \otimes C_P$ uniquely determines a new space $E_Q \otimes C'_Q$; $Q \in S^2$ and an isomorphism

$$E_P \otimes C_P \sim E_Q \otimes C'_Q \quad , \tag{3.1}$$

with the property that $x_P^N \otimes e_P^{i\alpha}$ is taken into the "symmetrical" spinor $1'_Q \otimes 1'_Q \in E_Q \otimes C'_Q$ by the isomorphism, i.e.

$$x_P^N \otimes e_P^{i\alpha} \sim 1'_Q \otimes 1'_Q \quad . \tag{3.2}$$

We will now describe exactly how (3.1) is constructed.

We need to regard the set of oriented circles on S^2, passing through the fixed point P. Each such circle is uniquely characterized by a pair $e_P^{i\alpha}$, Q, where $e_P^{i\alpha} \in C_P$ is tangent to the circle at P, and $Q \in S^2$ is another point through which the circle passes. Alternatively, we may characterize the circle by a triple of points U, Q, R $\in S^2$, through which it passes in turn. The point U will always be chosen to coincide with P. In the first case we denote the circle $\Sigma(e_P^{i\alpha}, Q)$, in the second case we denote the circle $\Sigma(P,Q,R)$. Given the tangent vector $1'_Q$, the circle $\Sigma(e_P^{i\alpha}, Q)$ may also be denoted $\Sigma((e')_Q^{i\beta}, P)$, where $(e')_Q^{i\beta} \equiv e^{i\beta} 1'_Q$ and β is the angle between $1'_Q$ and $\Sigma(e_P^{i\alpha}, Q)$.

For the normed spinor $x_P^N \otimes e_P^{i\alpha}$, let us write $x_P^N = PQ^N$, where Q is the point of intersection between x_P^N and S^2 (in the degenerate case when x_P^N is a tangent vector to S^2 we regard $\lim_{P \to Q} PQ^N \otimes e_P^{i\alpha}$). The following definition is just a geometrical re-phrasing of the ordinary mapping of spinors on null flags, restricted to the particular case of normed spinors[4]: To the normed spinor $PQ^N \otimes e_P^{i\alpha}$ we relate the unit tangent vector $1'_Q$ at Q, the direction of $1'_Q$ being such that

$$\Sigma(e_P^{i\alpha}, Q) = \Sigma((e')_Q^{i\alpha}, P)$$

(see Figure 1). Given a normed spinor $PQ^N \otimes e_P^{i\alpha}$, a space $E_Q \otimes C'_Q$ is thus determined. The isomorphism (3.1) is now defined as

$$PR^N \otimes \rho e_P^{i(\gamma+\delta)} \sim QR^N \otimes \rho (e')_Q^{i\gamma} \quad . \tag{3.3}$$

Here, $PR^N \otimes \rho e_P^{i(\gamma+\delta)}$ denotes an element of $E_P \otimes C_P$, whereas $QR^N \otimes \rho (e')_Q^{i\gamma}$ denotes the corresponding element of $E_Q \otimes C'_Q$. The angle δ depends on the point R and is defined as the angle between the circles $\Sigma(P,Q,R)$ and $\Sigma(e_P^{i\alpha}, Q)$, measured at P, i.e.

$$\Sigma(P,Q,R) = \Sigma(e_P^{i(\alpha+\delta)}, Q) \ . \tag{3.4}$$

It is not difficult to prove that the expression (3.3), in the limit $R \to Q$, yields the expression (3.2), thus (3.3) is indeed a transformation taking the spinor $PQ^N \otimes e_P^{i\alpha}$ into "symmetrical" form. Moreover, (3.3) fulfils a number of "natural properties"[4]. For instance, we have the following property. If

$$PQ^N \otimes e_P^{i\alpha} = z_1 (a_P \otimes 1_P) + z_2 (1_P \otimes 1_P) \tag{3.5}$$

($z_1, z_2 \in \underline{C}$), then (3.3) corresponds to a SU_2 matrix

$$r = \begin{pmatrix} z_2 & -z_1 \\ \bar{z}_1 & \bar{z}_2 \end{pmatrix} \tag{3.6}$$

transforming components of spinors in $E_P \otimes C_P$, with respect to the base $a_P \otimes 1_P$, $1_P \otimes 1_P$, into components of spinors in $E_Q \otimes C'_Q$, with respect to the base $a_Q \otimes 1'_Q$, $1'_Q \otimes 1'_Q$ (a'_Q being the unit normal of C'_Q).

4. RELATION TO TWISTORS AND MINKOWSKI SPACE

In contrast to the customary representation of twistors[1]

$$Z^\alpha \longleftrightarrow \begin{pmatrix} \pi_{A'} \\ \omega^A \end{pmatrix} \tag{4.1}$$

($\pi_{A'}$, ω^A are complex 2-vectors; the indices A, A' being "abstract"), we shall represent them in the form

$$\begin{pmatrix} \bar{\pi}^{A} \\ \omega^{A} \end{pmatrix} . \tag{4.2}$$

Here, $\bar{\pi}^{A}$ is obtained from $\pi_{A'}$ by complex conjugation and raising of the index, by the anti-symmetric inner product. Care will, of course, be taken to preserve the ordinary transformation properties of twistors. Thus, translations of twistors will take place according to

$$\begin{aligned} \bar{\pi}^{A} &\rightarrow \bar{\pi}^{A} , \\ \omega^{A} &\rightarrow \omega^{A} + i\, x^{AA'} \pi_{A'} \end{aligned} \tag{4.3}$$

($x^{AA'}$ is the usual Minkowski Hermitean matrix).

We shall explore some of the consequences of defining twistors in terms of our geometrical formalism. Let the label A refer to spinor space $E_P \otimes C_P$. We define

$$\begin{pmatrix} \bar{\pi}^{A} \\ \omega^{A} \end{pmatrix} \cong \begin{pmatrix} x_P^N \otimes e_P^{i\alpha} \\ y_P \otimes z_P \end{pmatrix} \tag{4.4}$$

(as was pointed out in the introduction, we restrict ourselves to the case that $\bar{\pi}^{A}$ is normed). Here, if

$$\bar{\pi}^{A} = \begin{pmatrix} z_1 \\ z_2 \end{pmatrix} , \quad \omega^{A} = \begin{pmatrix} z_3 \\ z_4 \end{pmatrix} \tag{4.5}$$

then

$$\begin{aligned} x_P^N \otimes e_P^{i\alpha} &= z_1 (a_P \otimes 1_P) + z_2 (1_P \otimes 1_P) , \\ y_P \otimes z_P &= z_3 (a_P \otimes 1_P) + z_4 (1_P \otimes 1_P) . \end{aligned} \tag{4.6}$$

We may re-represent the pair of spinors of $E_P \otimes C_P$ in terms of any other space $E_Q \otimes C_Q'$, according to the formula (3.3):

$$\begin{pmatrix} x_P^N \otimes e_P^{i\alpha} \\ y_P \otimes z_P \end{pmatrix} \sim \begin{pmatrix} u_Q^N \otimes (e')_Q^{i\beta} \\ v_Q \otimes z_Q' \end{pmatrix} . \tag{4.7}$$

At the same time we must transform the components of the spinors by means of a transformation

$$\begin{pmatrix} r_A^B & 0 \\ 0 & r_A^B \end{pmatrix} \begin{pmatrix} \bar{\pi}^{A} \\ \omega^{A} \end{pmatrix} = \begin{pmatrix} \bar{\pi}'^{B} \\ \omega'^{B} \end{pmatrix} . \tag{4.8}$$

Here r_A^B is the 2 x 2 unitary matrix constructed in the way (3.6). The label B refers to the space $E_Q \otimes C_Q'$. There will be a correspondence

$$\begin{pmatrix} \bar{\pi}'^B \\ \omega'^B \end{pmatrix} \cong \begin{pmatrix} u_Q^N \otimes (e')_Q^{i\beta} \\ v_Q \otimes z'_Q \end{pmatrix} ,$$

exactly analogous to (4.4), as it was defined by (4.5), (4.6).

Let us examine the concept of twistor norm in view of the correspondence (4.4). The twistor norm is defined as

$$Z^\alpha \bar{Z}_\alpha = 2\,\mathrm{Re}\,(z_1 z_4 - z_2 z_3) \tag{4.9}$$

(cf. (4.5)). Thus, for a twistor of the form

$$\begin{pmatrix} \bar{\pi}^A \\ \omega^A \end{pmatrix} \cong \begin{pmatrix} 1_P \otimes 1_P \\ (a_P + z\,1_P) \otimes z'\,1_P \end{pmatrix} , \tag{4.10}$$

the twistor norm is $2\,\mathrm{Re}\,z'$, as is seen by comparison with (2.2). For an arbitrary twistor (4.4), we pass to the space $E_Q \otimes C'_Q$ for which $x_P^N \otimes e_P^{i\alpha} \sim 1'_Q \otimes 1'_Q$. In this space

$$y_P \otimes z_P \sim (a_Q + z\,1'_Q) \otimes z'\,1'_Q , \tag{4.11}$$

where the right-hand side is determined according to (3.3). From (3.6), (4.5) and (4.8) we obtain that the twistor norm of (4.4) is $2\,\mathrm{Re}\,z'$, the number z' being given by (4.11). We have thus arrived at a form of geometrical interpretation of the twistor norm. An immediate application: A null twistor (i.e. a twistor with a vanishing norm) can always be represented in the form

$$\begin{pmatrix} 1'_Q \otimes 1'_Q \\ v_Q \otimes i'_Q \end{pmatrix} , \tag{4.12}$$

where $i'_Q \equiv i\,1'_Q$.

We wish to interpret twistor translations in terms of (4.4). We find that this can be done according to the following scheme. Assume that our 3-space E is the spacelike hypersurface of (real) Minkowski space $\underline{R}^{1,3}$, corresponding to some fixed value of the time coordinate x_o. Regard the 2-sphere S^2 as the intersection with E of the light cone, focusing at the origin O of $\underline{R}^{1,3}$. For every space E_Q; $Q \in S^2$, there is a natural projection of $\underline{R}^{1,3}$ on E_Q, taking place along null rays parallel to the null vector OQ (see Figure 2). If we represent an arbitrary element of $\underline{R}^{1,3}$ in the form $\tau\,OU$, where $U \in E$, OU is an affine vector and τ is some real factor of proportionality, the projection reads

$$\tau\,OU \in \underline{R}^{1,3} \rightarrow \tau\,QU \in E_Q . \tag{4.13}$$

Clearly, the Hermitean matrices $x^{AA'}$ can be brought into a one-to-one

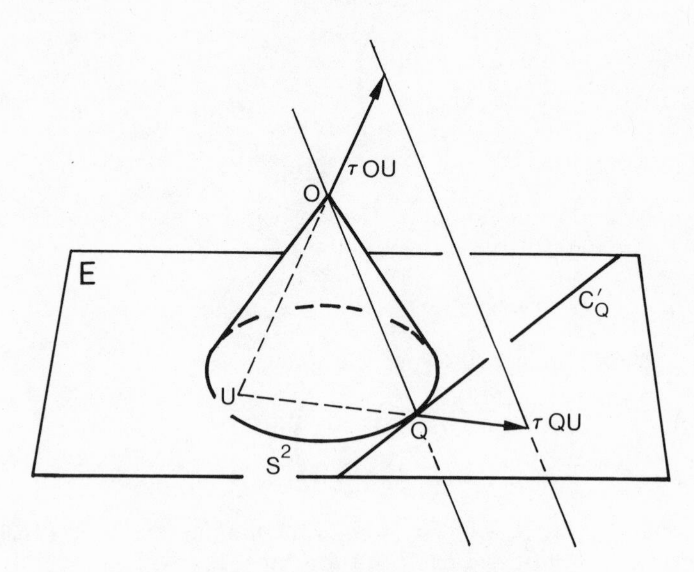

Fig. 2. The projection $\underline{R}^{1,3} \to E_Q$

relation to the vectors τ O U. This relation, which we assume to be given in some fixed way, allows the following interpretation of the spinors $i x^{AA'} \pi_{A'}$ of (4.3). In the particular case that $\bar{\pi}^A \cong 1_P \otimes 1_P$ it is not difficult to see that one may write

$$i x^{AA'} \pi_{A'} \cong (\tau PU) \otimes i_P \ .$$

Moreover, if $\bar{\pi}^A \cong PQ^N \otimes e_P^{i\alpha}$ and $i x^{AA'} \pi_{A'} \cong x_P \otimes z_P$ one may prove that

$$i x^{AA'} \pi_{A'} \cong x_P \otimes z_P \sim (\tau Q U) \otimes i'_Q \ ,$$

where the right-hand side spinor belongs to the space $E_Q \otimes C'_Q$, for which $PQ^N \otimes e_P^{i\alpha} \sim 1'_Q \otimes 1'_Q$; <u>the point U, the factor τ and the matrix $x^{AA'}$ are the same in both formulas</u>.

We have arrived at a completely geometrical characterization of twistor translations. In order to translate the twistor (4.4) we pass to the space $E_Q \otimes C'_Q$, where $\bar{\pi}^A$ obtains its "symmetrical" form. For a translation in $\underline{R}^{1,3}$ along the 4-vector τ O U; U $\in$ E, the translation of the twistor becomes

$$\begin{pmatrix} 1'_Q \otimes 1'_Q \\ v_Q \otimes z'_Q \end{pmatrix} \quad \to \quad \begin{pmatrix} 1'_Q \otimes 1'_Q \\ v_Q \otimes z'_Q + (\tau Q U) \otimes i'_Q \end{pmatrix} \tag{4.14}$$

Here, as long as the twistor is non-null, we cannot directly perform the addition on the right-hand side. We must either decompose each term of the sum according to (2.1), and add the terms componentwise, or use the geometrical procedure of reference 4. However, for a null twistor (4.12), the sum can be written down directly as

$$v_Q \otimes i'_Q + (\tau Q U) \otimes i'_Q = (v_Q + \tau Q U) \otimes i'_Q \, . \tag{4.15}$$

Formula (4.15) expresses the consistency of the interpretation of null twistors as describing null rays in Minkowski space. The null twistor (4.12) corresponds to the null ray, which is mapped in the single element $v_Q \in E_Q$ by the projection (4.13). The formulas (4.14), (4.15) state that this correspondence is translation invariant.

ACKNOWLEDGEMENTS

The author would like to express his thanks to A. Bette, B. Laurent, U. Lindström and B. Sten. These persons have in different ways contributed to the ideas behind the present work.

REFERENCES

1. R. Penrose, "Twistor Algebra", J. Math. Phys., 8 (1967), 345 - 366.

2. H. Hellsten, "A visual description of 2-component spinor calculus", USIP-report 75-16, Institute of Theoretical Physics, University of Stockholm, Stockholm (1975).

3. H. Hellsten, "A materially founded description of spinor algebra and its connection with Minkowski geometry", USIP-report 77-08, Institute of Theoretical Physics, University of Stockholm, Stockholm (1977).

4. H. Hellsten, "Visual geometry and the algebraic properties of spinors", to appear in J. Math. Phys.

GRAVITON PHOTOPRODUCTION IN STATIC ELECTROMAGNETIC FIELDS AND SOME ASTROPHYSICAL APPLICATIONS

Sree Ram Valluri*

University of Regina, Regina, Canada

ABSTRACT

A review of the photoproduction processes that give rise to gravitons is presented. Some astrophysical applications are also considered.

1. INTRODUCTION

Einstein's theory predicts that gravitational radiation (GR) is produced in extremely small quantities in ordinary atomic processes. The probability that a transition between two atomic states will proceed by emission of GR rather than electromagnetic radiation (EMR) is $GE^2/e^2 \sim 3 \times 10^{-54}$ (E = 1 e.v.). The interaction of photons with gravitons and other elementary particles is a topic of great interest for at least a few reasons enumerated below:

1) On astrophysical grounds because of the possible insight into astrophysics and perhaps into cosmology and various applications and consequences on both domains.

2) It is possible in a not too distant future to produce very intense beams and test therefore the theoretical predictions. Also the Universe has several interesting strong sources of EMR like quasars, pulsars etc. and gravitons could be generated by the interaction of the emitted EMR with the fields around the objects themselves. Such studies will also help in the establishment of a "Gravitational Wave" (GW) astronomy.

---

* From July 1st at the Institute of Physics, Georgia Institute of Technology, Atlanta, GA. 30332 U.S.A.

3) A complete theory of Maxwell-Dirac that would take care of the logarithmic infinities would be possible by including gravitons as has been done by Salam[1].

The weak field approach to GR is a good approximation since observable GR is likely to be of low intensity and a precise meaning to the concept of a graviton can only be given for a weak field solution of Einstein's equations

$$R_{\mu\nu} - \frac{1}{2} g_{\mu\nu} R = \kappa T_{\mu\nu} \quad (\kappa = \sqrt{16\pi G}), \tag{1.1}$$

$$g_{\mu\nu} = \eta_{\mu\nu} + \kappa h_{\mu\nu}, \tag{1.2}$$

$h_{\mu\nu}$ is the gravitational field and $\eta_{\nu\mu}$ is the Minkowski tensor. We follow the Lagrangian formulation of Gupta[2] in connection with the quantization of a linearized gravitational field.

We give a short review of the processes of graviton photoproduction in this paper and discuss some astrophysical application.

The processes considered are mainly linear. Most of the interactions have an interest in themselves and serve to ascertain in a clearer way, the significance and role of gravitation in physics: units h = c = 1 are used unless otherwise indicated.

2. PROCESSES OF PHOTOPRODUCTION

Gertsenshtein [3] used the linearized theory to consider the resonance of an e.m.w and a g.w. and calculate the conversion efficiency. The problem of the e.m. response of a capacitor to an incident GW has been investigated by Lupanov [4].

Photoproduction processes of the type $\gamma + e \rightarrow e + g$, shown in Figs. 1 have been first studied by Vladimirov [5] using the formalism of Gupta. Their work shows that a measurement of the $\sin^4\theta$ effect (which occurs in the diff. cross-section) and of the Doppler red shift would make possible a complete estimate of velocity and direction of motion of astrophysical objects.

Boccaletti et al. [6] have considered the creation and absorption of gravitons in the presence of the catalytic action of the Earth $\mathcal{M}$. The process

$$\gamma + \mathcal{M} \rightarrow \mathcal{M} + \gamma + g \quad , \tag{2.1}$$

shown in Fig. 2, is with an incident laser photon flux $\sim 10^{28}$ photons (cm^2), $\mathcal{M}$ is the mass of the Earth. The cross-section $\sigma \sim 10^{-37}$ cm^2. The interaction Lagrangian of Gupta

$$L_I = -\kappa\, h_{\mu\nu}\, T_{\mu\nu} \quad , \tag{2.2}$$

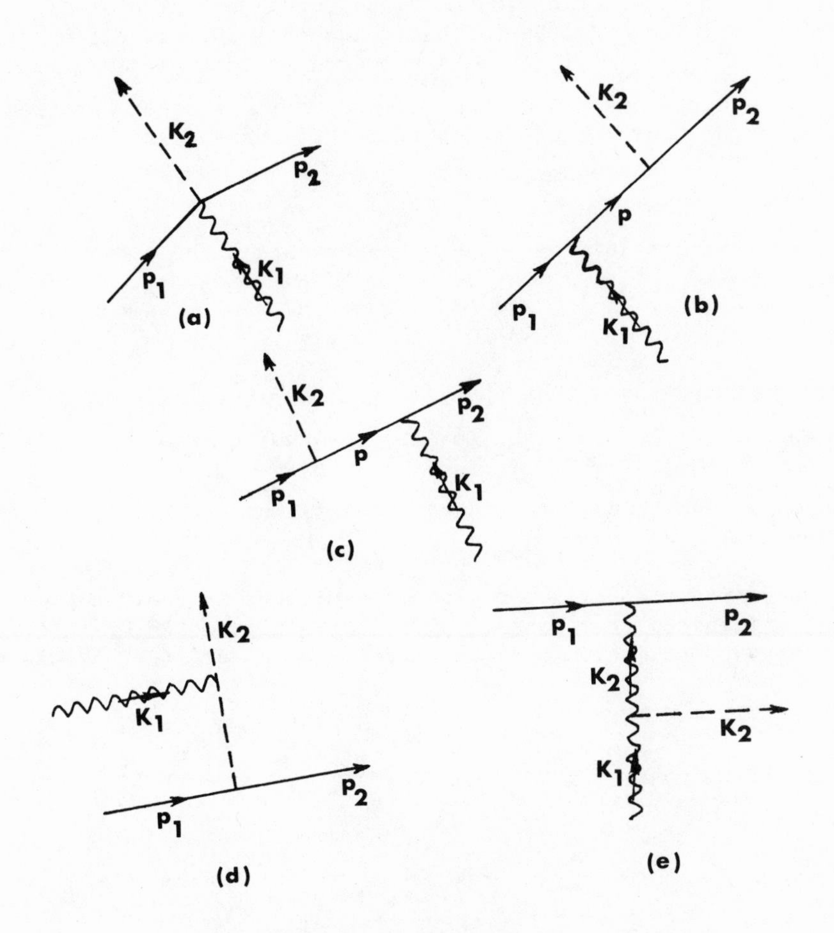

Figures 1(a-e): The process $\gamma + e \rightarrow e + g$

Electron line

Photon line

Graviton line

($T_{\mu\nu}$ is the e.m. tensor for energy momentum) is used in the calculations.

De Sabbata et al. [7] have also suggested laboratory experiments utilizing intense laser beams $\sim 10^3$ W with a magnetic field of 10^5 gauss for an interaction length $\sim 10^4$ cm. They calculate a conversion efficiency $\sim 10^{-25}$. This means that 10^{-3} gravitons/(cm^2 sec) with $E_{graviton} \sim 1$ e.v. are produced in this highly effective process. Weber and Hinds [8] studied the processes of creation of gravitons by Coulomb and magnetostatic fields and estimate

$$\sigma \;\; \approx 8\pi^2 \;\frac{GWR}{c^4} \tag{2.3}$$

For a galaxy $R \sim 10^{22}$ cm, $B \sim 10^{-6}$ gauss and W the total field energy $\sim B^2R^3$

$$\sigma \sim 10^{28} \;\text{cm}^2 \quad . \tag{2.4}$$

Their work indicates that

1) Such interactions might lead to observable effects over a long period of time;

2) The red shift of light cannot be explained as a "tired light" phenomenon.

Gauge invariance is not clearly exhibited in their approach which originated from a Hamiltonian formulation of GTR. The Lagrangian formulation used by Papini and Valluri (P.V.)[9] is re-

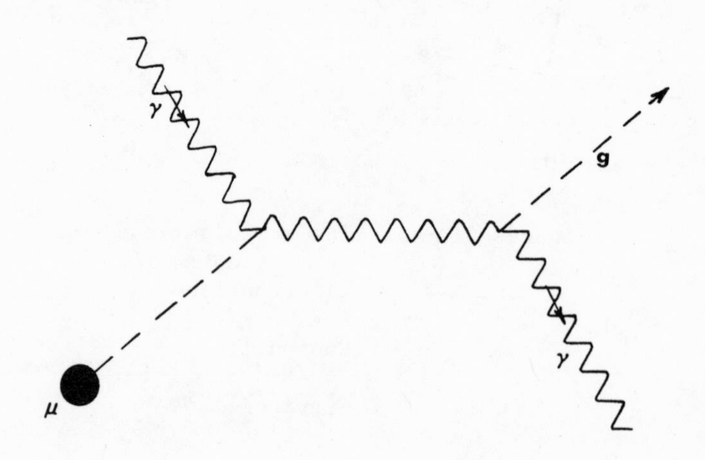

Figure 2: Graviton photoproduction in a static external field.

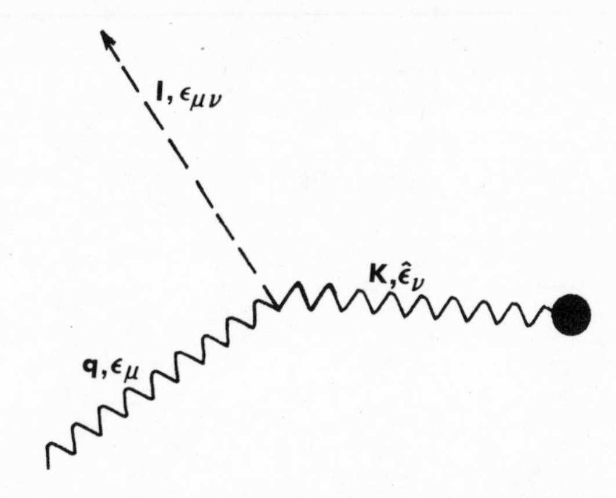

Figure 3: First order Feynman Diagram for Photon conversion into a Graviton. ● Stands for the external field.

assuring in this respect. For a dipole magnetic field in the presence of currents (vector potential case)[10)]

$$\sigma_d^{P.V.} = 4\pi^3 \; \frac{61}{15} \; \frac{G\mu^2\nu^2}{c^6} \; [\, 1- \frac{49}{61} \cos^2\beta] \qquad (2.5)$$

is obtained. μ is the dipole momenta and β is the angle between the dipole orientation and the direction of the incident photon. The diagrams for the first order and the radiative corrections are shown in Figs. 3, 4 and 5, respectively. Eq. (2.5) differs by a factor of 2π from the result of DeLogi and Mickelson[11)] who used a Feynman perturbative approach. Ginzburg and Tsytovich[12)] used the techniques of transition scattering to get for zero currents (scalar potential case)

$$\sigma_s^{GT} = \frac{112\pi^3}{30} \; \frac{G\mu^2\nu^2}{c^6} \; [1 - \frac{1}{7} \cos^2 \beta] \quad . \qquad (2.6)$$

For the Coulomb field, the differential cross-section has a characteristic Rutherford peak[11)] at $\theta = 0$ which can be avoided either by Debye shielding for scattering in a plasma or by assuming that the incident GW front has a width D. For astrophysical sources, recoil is negligible and

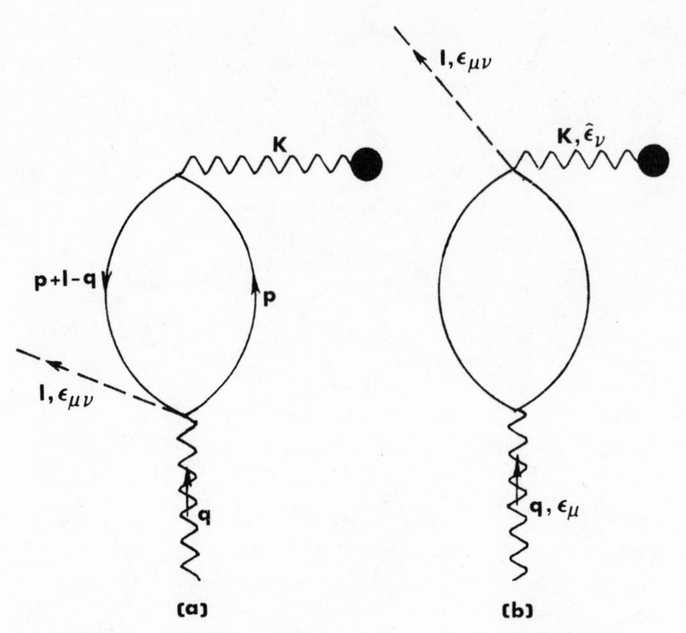

Figure 4: The two vertex radiative correction.

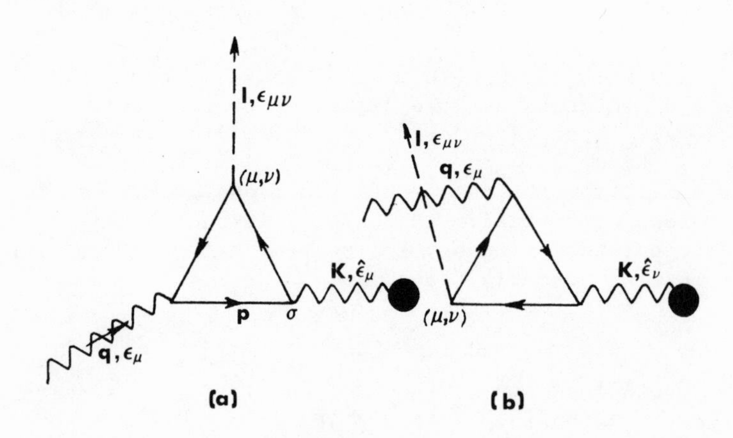

Figure 5: The 3 vertex radiative correction.

$$\frac{\partial T^{\mu\nu}}{\partial x_\nu} \approx 0 \tag{2.7}$$

instead of $\frac{\partial T^{\mu\nu}}{\partial x_\nu} \neq 0$ (exact condition).

An appropriate gauge like the transverse traceless (TT) simplifies work.

For the radiative corrections [10] the e.m. energy momentum tensor is modified as follows:

$$T_{\mu\nu} \rightarrow T'_{\mu\nu} = [1 + \frac{5\alpha}{18\pi} + k^2\ \pi_f(k)]\ T_{\mu\nu} \tag{2.8}$$

[$\alpha = \frac{e^2}{4\pi}$ and $\pi_f(k)$ is related to the vacuum polarization]. σ is independent of h since the interaction has a classical limit.

We present estimates of the gravitational power for some astrophysical objects below.

Table I: Estimates from astrophysical objects

Object	Spectral region	Emission power = L_g (erg/s)	Flux at Earth
The Sun	Infra-red	5×10^{12} (photoproduction Papini and Valluri)	$1.7\text{x}10^{-2}/(\text{cm}^2\text{sec})$
		6×10^{14} (Weinberg, [13])	$1.8\text{x}10^{0}/(\text{cm}^2\text{sec})$
		5×10^{15} (bremsstrahlung, Carmeli [14]) Complete spectrum including the tail)	$1.7\text{x}10/(\text{cm}^2\text{sec})$

The power due to classical quadrupole radiation for the Jupiter-Sun system is 7.6×10^{11} erg/s in an entirely different frequency range. Bremsstrahlung and photoproduction constitute the strongest source of GR in our planetary system.

Object	Region	L_g (erg/s)	Flux at Earth
Quasar 3c273	Infra-red	10^{32}	$\times 10^{-25}$ erg/(cm^{2}sec)
	Optical	10^{29}	$\times 10^{-29}$ "
	Radio	10^{23}	10^{-35} "

The fluxes are << 10^{-12} erg/cm^{2}s due to broad band bursts produced by large explosions in quasar and galactic nuclei as conjectured by Press and Thorne [15].

Object	Region	L_g (erg/s)	Flux
Seyfert galaxy (NGC 1068)	Infra-red	3×10^{25}	10^{-27} erg/cm^{2}s
Galactic center	"	8×10^{22}	"

The flux is << than 10^{38} erg/s for a gravitational luminosity of a galaxy due to the motion of its double stars.

Object	Region	L_g (erg/s)	Flux at Earth (erg/cm^{2}s)
Neutron star	Soft x-ray	$\sim 10^{34}$	$\sim 10^{-9}$

Therefore $L_g^{neutron\ star} > L_g^{sun}$ and $L_{e.m.}^{sun}$.

Object	Region	L_g (erg/s)	Flux at Earth (erg/cm^{2}s)
Pulsar NP0532	Radio	$\sim 10^{20}$	$\sim 10^{-23}$
	Optical	$\sim 10^{21}$	$\sim 10^{-22}$
	Infra-red	$\sim 10^{25}$	$\sim 10^{-18}$
	x-ray	$\sim 10^{32}$-10^{34}	$\sim (10^{-11}$-$10^{9})$
	γ-ray	$\sim 10^{31}$	$\sim 10^{-12}$

Power emitted via photoproduction can be > quadrupole contribution for pulsars of ellipticity < 10^{-3} and could play a role in the cooling of the star. There can be GR emission when dipole moment axis and rotation axis coincide unlike the classical case.

3. CONCLUSIONS

1) The presence in the Universe of a background of GR distributed over the complete spectrum due to the overall population of astrophysical objects is expected. Boccaletti et al. [16] estimated the graviton energy density as 10^{-19} erg/cm^3 << the e.m. cosmologica energy density of 10^{-14} erg/cm^3 which must be modified in the light of current astrophysical knowledge.

2) The efficiency of some quantum processes can be at least as high as that of the classical ones.

3) Due to the extreme weakness of their interaction, gravitons are almost unabsorbable and could hence convey to us information about the various evolutionary stages of the Universe.

It is heartening to note that Joseph Taylor and co-workers [17] have indirect evidence for the observations of "GWS" from observations of the slowing down of the orbital motion of the pulsar PSR 1913 + 16. There is a promise of a new window with the advent of "GW" astronomy.

ACKNOWLEDGMENTS

I would like to thank Professors V. De Sabbata and P. Bergmann for enabling me to come to Erice and giving me the opportunity of presenting this work. I would also like to thank Professor Abdus Salam for kind hospitality at the International Centre for Theoretical Physics, Trieste, where the manuscript was prepared.

REFERENCES

1) A. Salam, The Physicist's Conception of Nature, Ed.J.Mehra (1973) (Reidel Publishing Company, Dordrecht, The Netherlands), pp. 430-446.
2) S. N. Gupta, Proc. Phys. Soc. (London) 65A (1952) 161, 608.
3) M. E. Gertsenshtein, Sov. Phys. JEPT 14 (1962) 84.
4) G. A. Lupanov, Sov. Phys. JETP 25 (1967) 76.
5) Y. S. Vladimirov, Sov. Phys. JETP 16 (1963) 65 and 18 (1964) 176.
6) D. Boccaletti, V. De Sabbata, C. Gualdi and P. Fortini, Nuovo Cimento 48 (1967) 58.
7) a) V. De Sabbata, D. Boccaletti and C. Gualdi, Sov. J. Nucl. Phys. 8 (1969) 537; Yad. Fiz. 8 (1968) 924; b) V. De Sabbata, in "Topics in Theoretical and Experimental Gravitation Physics" edited by V. De Sabbata and J. Weber (1977) (Plenum Press, New York), pp. 69-81.
8) J. Weber and H. Hinds, Phys. Rev. 128 (1963) 2414.
9) G. Papini and S. R. Valluri, Can. J. Phys. 56 (1978) 801.
10) G. Papini and S. R. Valluri, Can. J. Phys. 53 (1975) 2306, 2313, 2315; 54 (1976) 76.
11) W. K. Delogi and A. R. Mickelson, Phys. Rev. D16 (1977) 2915.
12) V. L. Ginzburg and V. N. Tsytovich, Sov. Phys. Radiophys. Quantum Electronics 18 (1975) 25.
13) S. Weinberg, Phys. Rev. 140 (1965) 516.
14) M. Carmeli, Phys. Rev. 158 (1967) 1248.

15) W. H. Press and K. S. Thorne, Ann. Rev. Astron. Astrophys. 10 (1972) 344.
16) D. Boccaletti, V. De Sabbata, C. Gualdi and P. Fortini, Nuovo Cimento B54 (1968) 34.
17) J. Taylor, L. A. Fowler and P. M. McCulloch, Nature 277 (1979) 437.

INVARIANT DEDUCTION OF THE GRAVITATIONAL EQUATIONS FROM THE PRINCIPLE OF HAMILTON

by Attilio Palatini

Rendiconti del Circolo Matematico di Palermo

10 August 1919, 43, 203-212

[Translation by Roberto Hojman and Chandrasekher Mukku]

TRANSLATOR'S NOTE

In this translation of Palatini's article, we have tried to adhere as closely as possible to the original, not only as regards the original text, but also the choice of english equivalents for technical expressions. It should be noted that Palatini not only uses superscripts for contravariant indices but appends round brackets to them. This is not to be confused with the modern use of round brackets - denoting symmetrization. $\{^{jk}_{\ i}\}$ is the historical form of the Christoffel symbols. In keeping with the summation convention, they are nowadays written as $\{^{\ i}_{jk}\}$. We retain the historical form. To avoid sources of confusion, we have introduced extra labelling of equations. These are with greek indices. We would like to take this opportunity to thank Professors P.G. Bergmann and V. De Sabbata for their kind hospitality at Erice.

INTRODUCTION

It is already well known that in the general theory of relativity, physical space is characterized by a quadratic differential form (that mixes space and time)

$$ds^2 = \sum_{ij} g_{ij}\, dx^i\, dx^j \qquad (1)$$

in the differentials of the four co-ordinate variables $x_0 = t, x_1, x_2, x_3$, whose coefficients g_{ij} are the gravitational potentials of Einstein. The discriminant of (1) - essentially negative - will be denoted by (g).

The mutual interdependence of all physical phenomena and the geometric nature of the space is completely determined by the ten gravitational equations

$$G_{ik} - \left(\frac{1}{2} G + \lambda\right) g_{ik} = -\kappa T_{ik} \quad , \tag{2}$$

where $G_{ik} = \sum_{0h}^{3} \{ih,hk\}$ is the Riemann curvature tensor; $G = \sum_{0ik}^{3} G_{ik} g^{(ik)}$ is the mean curvature of the four-dimensional space (1); T_{ik} is the energetic tensor that is determined from all the elements - stresses, quantity of motion, energy density and flux - that characterize the physical phenomena; κ and λ are two universal constants.

After these gravitational equations were discovered by Einstein efforts were made to derive them from a variational principle just as one derives the equations of Lagrange from Hamiltonian's principl in ordinary mechanics.

This goal was reached by Einstein himself, establishing a new Hamiltonian principle that was made precise by Hilbert and Weyl [1)].

However the procedures followed by these authors do not conform to the spirit of the absolute differential calculus, because in deriving the invariant equations, one must use non-invariant formulae.

My aim is to reach the same goal, while preserving the invariance of all the formulae at every step. In doing this, I will take advantage of the results obtained in my note: "On the foundations of the absolute differential calculus" (see the preceding note in this volume; Rend. Circ. Mat. Palermo, Vol.43, 1919); Hereafter referred to as N.

1. FUNDAMENTAL POSTULATE

We begin by introducing with Hilbert, the following fundamental postulate: The laws of physics depend on a unique, universal function H having the following properties:

(a) It is invariant with respect to general co-ordinate transformations;

(b) It depends on the gravitational potentials $g^{(ik)}$ and on the corresponding Christoffel and Riemann symbols and

(c) it depends on the elements that characterize the physical phenomena.

However, we have no a priori knowledge of the explicit form of the universal function H and must therefore introduce some hypotheses.

From the point of view of the synthesis of all physical phenomena, it is convenient to suppose that

$$H = G + L + 2\lambda \quad ,$$

where λ is a universal constant, G (the mean curvature of the four-dimensional space) is a term that contains the information and characterizes the influence of the space-time on the behaviour of the phenomena, and L is a term that includes all the manifestations of physical origin except those that are intimately related to the structure of space-time itself.

2. STRUCTURE OF THE FUNCTION L. REDUCED MECHANICAL SCHEME

From the speculative point of view, it seems desirable to attribute to all these manifestations (direct or indirect) an electromagnetic origin (as should be the case for the luminous and thermodynamic phenomena). The expression for L should depend in a complicated way on the parameters fixing the electromagnetic state of the system, and the gravitational equations should not be isolated from those governing the behaviour of all the other phenomena.

Having in mind the possibility of adopting the study to concrete cases, it is convenient to limit oneself to the consideration of the gravitational field by itself and to collect everything that arises from the set of physical phenomena (excluding gravitation), into a specific function of position and time, precisely in an energetic tensor T_{ik}.

A similar situation is found in ordinary mechanics when wishing for instance, to study the motion in a conservative field, of a material point on a frictional surface, the energetic analysis of the phenomena (that might lead one to consider the thermal aspects of the problem when taking into account the heat dissipated due to friction) is replaced by introducing a position dependent non-conservative frictional force.

In the usual mechanical approach, taking into account a whole set of circumstances (giving rise to loss of kinetic energy) would be impossible, or at least quite laborious to analyse with profit. Instead, one is led to consider as given, forces that are not derivable from a potential. In the same way, in the Einsteinian scheme, it is satisfactory to study, instead of L (whose precise expression should depend on a whole on a set of phenomena, making its explicit study impossible or undesirable) the tensor T_{ik}.

By using T_{ik}, an appropriate matter Lagrangian *)

$$L = \kappa \sum_{0}^{3}{}_{ik} \; T_{ik} \, g^{(ik)} \tag{3}$$

can be constructed so as to lead us to the gravitational equations. κ denotes a universal constant of homogeneity. The given elements of the tensor T_{ik} are not to be considered as independent of the $g^{(ik)}$, instead, one should take the products $\sqrt{g}\, T_{ik} = \mathfrak{T}_{ik}$ (which constitute the so-called tensor of volume associated to the tensor T_{ik}).

3. PRINCIPLE OF HAMILTON

With these assumptions, and taking the form of the universal function to be

$$H = G + L + 2\lambda$$

with

$$L = \kappa \sum_{0}^{3}{}_{ik} \; T_{ik} \, g^{(ik)} ,$$

we want to show that the gravitational equations follow from the variational principle

$$\delta \int_S H \, dS = 0 \quad . \tag{4}$$

*) For such a purpose one can again invoke the mentioned analogy with classical mechanics, by noting that from Hamilton's variational principle $\delta \int (T + U)\, dt = 0$ [T kinetic energy, U potential] is valid for the case of conservative forces, one can go to the generalized principle, valid for any force with components X_i $(i = 1,2,3)$ by substituting for U the linear expression $\sum_{1}^{3}{}_i X_i x_i$ and assuming $\delta X_i = 0$. The expression (3) for L is in a sense the analogue of $\sum_{1}^{3}{}_i X_i x_i$.

Here S denotes an arbitrary region of the four-dimensional space-time and δ denotes a variation with respect to the potentials $g^{(ik)}$ with the condition that $\delta g^{(ik)}$ (and their first and second derivatives) vanish on the boundary of S.

Before proceeding with the proof of our proposition, it is necessary to establish some preliminary formulae.

4. PRELIMINARY FORMULAE

Variation of the Christoffel symbols: Let us begin with the identities *)

$$\frac{\partial g_{nk}}{\partial x_j} - \sum_p \left[\left\{ {nj \atop p} \right\} g_{pk} + \left\{ {kj \atop p} \right\} g_{np} \right] = 0 \quad , \qquad (5)$$

essentially expressing the well-known lemma of Ricci. They can be easily verified by using the expressions for the Christoffel symbols of the second kind.

With the above definition of δ, we write $\delta g_{nk} = e_{nk}$ and applying δ to (5) one gets

$$\frac{\partial e_{nk}}{\partial x_j} - \sum_p \left(\left\{ {nj \atop p} \right\} e_{pk} + \left\{ {kj \atop p} \right\} e_{np} \right) - \sum_p \left(g_{pk} \delta \left\{ {nj \atop p} \right\} + g_{np} \delta \left\{ {kj \atop p} \right\} \right) = 0.$$

The first two terms constitute the covariant derivative of the system e_{nk} (cf. formula (14) of N for the particular case $m = 2$), therefore

$$e_{nk|j} = \sum_p \left(g_{pk} \delta \left\{ {nj \atop p} \right\} + g_{np} \delta \left\{ {kj \atop p} \right\} \right) \quad . \qquad (\alpha)$$

Permuting k with j and then h with j, and summing up the two equations thus obtained and then subtracting (α) one gets

$$\eta_{hkj} = \sum_p g_{pj} \delta \left\{ {hk \atop p} \right\} \quad , \qquad (6)$$

where

$$\eta_{nkj} = \frac{1}{2} \left(e_{nj|k} + e_{kj|n} - e_{nk|j} \right) \quad .$$

*) In this section, summation indices have no limits, so that all considerations here, will be valid not only for the four-dimensional ds^2 of Einstein, but for any ds^2.

Multiplying (6) with $g^{(ij)}$ and summing over j, we immediately obtain

$$\delta\{{}^{hk}_{i}\} = \eta^{(i)}_{hk} , \tag{7}$$

where

$$\eta^{(i)}_{hk} := \sum_j g^{(ij)} \eta_{hkj} . \tag{8}$$

Variation of Riemann symbols and explicit expression for G: Eq.(8) immediately reveals $\eta^{(j)}_{hk}$ to be a mixed system, twice covariant and once contravariant.

From the fundamental formula that defines the covariant derivative of a mixed system (cf. formula (13) of N) we get

$$\eta^{(i)}_{hk|j} = \frac{\partial \eta^{(i)}_{hk}}{\partial x_j} - \sum_\ell \left(\{{}^{hj}_{\ell}\}\eta^{(i)}_{\ell k} + \{{}^{kj}_{\ell}\}\eta^{(i)}_{h\ell} - \{{}^{\ell j}_{i}\}\eta^{(\ell)}_{hk} \right) . \tag{9}$$

Let us now consider the Riemann symbols of the second kind

$$\{hi,kj\} = \frac{\partial}{\partial x_j}\{{}^{hk}_{i}\} - \frac{\partial}{\partial x_k}\{{}^{hj}_{i}\} + \sum_\ell \left(\{{}^{hk}_{\ell}\}\{{}^{\ell j}_{i}\} - \{{}^{hj}_{\ell}\}\{{}^{\ell k}_{i}\} \right)$$

and act on them with the symbol δ. Having in mind Eq.(7) one finds

$$\delta\{hi,kj\} = \frac{\partial \eta^{(i)}_{hk}}{\partial x_j} - \frac{\partial \eta_{hj}}{\partial x_k}$$

$$+ \sum_\ell \left(\eta^{(\ell)}_{hk}\{{}^{\ell j}_{i}\} + \eta^{(i)}_{\ell j}\{{}^{hk}_{\ell}\} - \eta^{(\ell)}_{hj}\{{}^{\ell k}_{i}\} - \eta^{(i)}_{\ell k}\{{}^{hj}_{\ell}\} \right) .$$

Applying(9) [adding and subtracting $\sum_\ell \eta^{(i)}_{\ell h} \{{}^{kj}_{\ell}\}$ from the right-hand sides] we get

$$\delta\{hi,kj\} = \eta^{(i)}_{hk|j} - \eta^{(i)}_{hj|k} .$$

Since $G_{hj} = \sum_k \{hk,kj\}$, it follows that

$$\delta G_{hj} = \sum_k \left(\eta^{(k)}_{hk|j} - \eta^{(k)}_{hj|k} \right).$$

Therefore, for the variation of the mean curvature

$$G = \sum_{ik} G_{ik}\, g^{(ik)}$$

one has

$$\delta G = \sum_{ik} G_{ik}\, \delta g^{(ik)} + \sum_{ihk} g^{(ik)} \left(\eta^{(h)}_{ih|k} - \eta^{(h)}_{ik|h} \right) \quad . \qquad (10)$$

Defining

$$i^{(k)} := \sum_{ih} \left(g^{(ik)} \eta^{(h)}_{ih} - g^{(ih)} \eta^{(k)}_{ih} \right) \quad ,$$

it can be immediately verified that

$$\sum_{ihk} g^{(ik)} \left(\eta^{(h)}_{ih|k} - \eta^{(h)}_{ik|h} \right) = \sum_{k} i^{(k)}_{|k}$$

then by virtue of formula (17) of N, Eq.(10) can be written as

$$\delta G = \sum_{ik} G_{ik}\, \delta g^{(ik)} + \frac{1}{\sqrt{g}} \sum_{k} \frac{\partial(\sqrt{g}\, i^{(k)})}{\partial x_k} \quad . \qquad (11)$$

5. DEDUCTION OF THE GRAVITATIONAL EQUATIONS

Defining $d\omega := dx_0\, dx_1\, dx_2\, dx_3$, one has

$$ds = \sqrt{g}\, d\omega$$

and (4) can be written as

$$\delta \int_S \left\{ (G + 2\lambda) \sqrt{g} + \kappa \sum_{ik=0}^{3} \mathfrak{T}_{ik}\, g^{(ik)} \right\} d\omega = 0 \quad ,$$

or, remembering that $\mathfrak{T}_{ik}$ should be regarded as being independent of $g^{(ik)}$

$$\int_S \left\{ \delta G \sqrt{g} + (G + 2\lambda)\, \delta\sqrt{g} + \kappa \sum_{ik=0}^{3} \mathfrak{T}_{ik}\, \delta g^{(ik)} \right\} d\omega = 0 \quad . \qquad (12)$$

Now

$$\delta \sqrt{g} = \sum_{ik} \frac{\partial \sqrt{g}}{\partial g^{(ik)}}\, \delta g^{(ik)} \quad ;$$

but as is well known

$$\frac{\partial \sqrt{g}}{\partial g^{(ik)}} = -\frac{1}{2} \sqrt{g}\; g_{ik} \quad .$$

Therefore

$$\delta \sqrt{g} = -\frac{1}{2} \sqrt{g} \sum_{ik} g_{ik}\, \delta g^{(ik)} \quad .$$

Let us now substitute this into Eq.(12), and use (11) for δG. Also, writing

$$\int_S \sum_{0k}^{3} \frac{\partial(\sqrt{g}\, i^{(k)})}{\partial x_k}\, d\omega = \sum_{0k}^{3} \int_S \frac{\partial(\sqrt{g}\, i^{(k)})}{\partial x_k}\, d\omega$$

allows one to use Green's lemma to convert the volume integral into a surface integral. Consequently, the integral vanishes by virtue of the expression for $i^{(k)}$ and the assumption that the variation of the potentials and their derivatives vanish on the boundary of S.

So one is left with

$$\int_S \sum_{0\,ik}^{3} \left\{ G_{ik} - \left(\frac{1}{2} G + \lambda \right) g_{ik} + \kappa\, T_{ik} \right\} \delta g^{(ik)}\, dS = 0 \quad .$$

Given the arbitrariness of S and $\delta g^{(ik)}$, the usual prescription gives

$$G_{ik} - \left(\frac{1}{2} G + \lambda \right) g_{ik} = -\kappa\, T_{ik} \quad . \qquad (\beta)$$

Thus, the gravitational equations have been derived from the variational principle while keeping the calculations invariant throughout.

6. DIFFERENTIAL CONDITIONS FROM CONSERVATION PRINCIPLES

Let us recall that the elements of the energetic tensor T_{ik} are open to a simple physical interpretation; stresses, density and energy flux [2], and we should not forget that such a tensor is constructed from all physical phenomena except gravitation. It then follows that the so-called conservation theorems must hold, that is to say for each material system considered, and for each of its elementary portions, the components of the external force applied to the system and the power density (rate of energy transferred to the system from external sources) must vanish. In other words, the T_{ik} components constitute a double system with vanishing divergence. This can be expressed, in the notation of the absolute differential calculus, as

$$\sum_{0}^{3}{}_{k} \; T_{ik}^{(k)} = 0 \quad .$$

If we now denote by A_{ik}, the left-hand side of Eqs.(β), then we must have

$$\sum_{0}^{3}{}_{k} \; A_{ik}^{(k)} = 0 \quad . \tag{13}$$

One might be led to imagine that these relations between the g_{ij}'s, impose a restriction on the possible forms of ds^2, characterizing the Einstein manifold.

However, it is easy to prove that Eqs.(13) are satisfied identically. In order to prove this, one may use the same methods that allowed us to deduce the gravitational equations, following a criterion already indicated by Weyl [3].

Under a change of variables, the parameters x_0, x_1, x_2, x_3 are substituted by new ones related to the old ones by

$$x_i' = x_i + \xi^{(i)}, \quad i = 0,1,2,3 \quad , \tag{14}$$

where $\xi^{(i)}$ denote four arbitrary infinitesimal functions of x_0, x_1, x_2, x_3 and constitute a simple contravariant system.

Let us now determine the variations δg_{ik} suffered by the coefficients of the fundamental form

$$ds^2 = \sum_{0}^{3}{}_{k} \; g_{ik} \, dx_i dx_k$$

under the transformations (14).

Subjecting ds^2 to variation, it is found that

$$\delta ds^2 = \sum_{0}^{3}{}_{ikj} \left\{ \frac{\partial g_{ik}}{\partial x_i} \xi^{(j)} + 2g_{ij} \frac{\partial \xi^{(j)}}{\partial x_k} \right\} dx_i dx_k \quad ,$$

or, by defining ξ_i to be the reciprocal elements of the elements $\xi^{(i)}$, i.e.

$$\xi^{(j)} = \sum_0^3 g^{(ij)} \xi_i ,$$

$$\delta ds^2 = 2 \sum_{0\ ik}^3 \left\{ \frac{\partial \xi_i}{\partial x_k} - \sum_{0\ j}^3 \left\{{ik \atop j}\right\} \xi_j \right\} dx_i\, dx_k \quad .$$

The term in parenthesis is immediately recognized as the covariant derivative of the system ξ_i and therefore we can rewrite δds^2 as

$$\delta ds^2 = 2 \sum_{0\ ik}^3 \xi_{i|k}\, dx_i\, dx_k \quad ,$$

or, using symmetry,

$$\delta ds^2 = \sum_{0\ ik}^3 (\xi_{i|k} + \xi_{k|i})\, dx_i\, dx_k \; .$$

Therefore for the variations δg_{ik}, one gets

$$\delta g_{ik} = \xi_{i|k} + \xi_{k|j} \quad . \qquad (\gamma)$$

To get the variation of the reciprocal elements $g^{(ik)}$, one uses the following identity

$$\sum_{0\ p}^3 g^{(ip)} g_{qp} = \varepsilon_{iq} \quad .$$

Applying the symbol δ to this identity, we find

$$\sum_{0\ p}^3 \delta g^{(ip)} g_{qp} + \sum_{0\ p}^3 g^{(ip)} \delta g_{qp} = 0 \quad ,$$

or, multiplying by $g^{(kq)}$ and summing over the q index

$$\delta g^{(ik)} = - \sum_{0\ pq}^3 g^{(ip)} g^{(kq)} \delta g_{pq}$$

and finally (γ) gives

$$\delta g^{(ik)} = - \sum_{0\ pq}^{3} g^{(ip)} g^{(kq)} (\xi_{p|q} + \xi_{q|p}) \quad . \qquad (15)$$

We now consider the expression $I = \int_S (G + \lambda)\, dS$ (where G and λ are defined above). I is an invariant under any change of variables, in particular under the transformation (14).

One then deduces that the variation δI that I suffers under the transformation (14) must vanish, i.e.

$$\delta I = \delta \int_S (G + \lambda)\, dS = 0.$$

Proceeding as in Sec.5 one gets

$$\int_s \sum_{0\ ik}^{3} \left\{ G_{ik} - (\frac{1}{2} G + \lambda) g_{ik} \right\} \delta g^{(ik)}\, dS$$

$$= \int_S \sum_{0\ ik}^{3} A_{ik}\, \delta g^{(ik)}\, dS = 0 \quad .$$

Substituting the expression for $\delta g^{(ik)}$ from Eq.(15), and noting that A_{ik} is a symmetric system, one obtains

$$\int_S \sum_{0\ ikpq}^{3} A_{ik}\, g^{(ip)} g^{(kq)} \xi_{p|q}\, dS = 0 \quad .$$

Integrating by parts and using the formula (23) established in N

$$\int_S \sum_{0\ ikpq}^{3} A_{ik|q}\, g^{(ip)} g^{(kq)} \xi_p dS = \int_S \sum_{0\ ik}^{3} A_{ik}^{(k)} \xi^{(i)}\, dS = 0 \ .$$

Since the region of integration S and the functions $\xi^{(i)}$ are arbitrary, one concludes that

$$\sum_{0\ k}^{3} A_{ik}^{(k)} = 0 \quad .$$

Padova, August 1919.

REFERENCES

1) D. Hilbert, Die Grundlagen der Physik (Erste Mitteilung [Gottingen Nachrichten, Sitzung, 20 November 1915]; H. Weyl, Zur Gravitationstheorie [Annalen der Physik, Bd. LIV (1917), pp.117-145], or, Raum. Zeit und Materie (Berlin, Springer, second edition 1919).

2) Cf. T. Levi-Civita, Sulla espressione analitica spettante al tensore gravitazionale nella teoria di Einstein [Rendiconti della R. Accademia dei Lincei, Serie V, Vol. XXVI, 1 semestre 1917, pp.381-391], p.383 and 384.

3) See Ref.1 from p.121.

ON A GENERALIZATION OF THE NOTION OF RIEMANN CURVATURE AND SPACES WITH TORSION

M. Elie Cartan

Comptes Rendus 27 Feb. 1922, *174*, 593-595

(Translation by G. D. Kerlick)

In a recent note (Cartan, 1922)[(+)], I indicated how, in an Einsteinian universe with a given ds^2, one may define geometrically the stress-energy tensor attached to each volume element of that universe. It is that tensor which, when set equal to zero, yields the law of gravitation for regions devoid of matter. The definition which I gave has the curvature of space to intervene by means of a certain rotation associated with every infinitesimal closed contour; that rotation which is introduced also gives the strength to Levi-Civita's notion of parallel transport.

That notion of parallel transport was well presented by its author by means of geometrical considerations, but it is very difficult to define in a precise manner without calculation. But it is possible, it seems to me, to show the profound significance of generalizing that same notion of space; this leads us at the same time to geometrical images of a material universe richer than our own, at least as we usually think of it. It also shows the true reason for the fundamental laws obeyed by the energy tensor (a law of symmetry and a law of conservation).

Let us limit ourselves to the case of three dimensions, the generalization to four being easy. Imagine a space which in the immediate neighborhood of each point has all the characteristics of Euclidean space. The inhabitants of this space know, for example, how to relate the points infinitesimally nearby to a given point A by means of an orthogonal triad with origin A; we further suppose

---

(+) M. E. Cartan, Comptes Rendus, *174* (1922) 437.

that the space has a law which permits one to relate to the triad whose origin is at A, all triads having origins A' infinitesimally nearby to A; in particular we suppose that it makes sense to say that two directions, one issuing from A and the other from A' are parallel. By definition, *such a space will be defined by the law of mutual reference* (Euclidean in nature) *between two triads infinitesimally nearby.*

A space like that of the preceding section is *not completely defined by its* ds^2; one has left undetermined part of the operation which permits the passing from one triad whose origin is A to the triad infinitesimally nearby whose origin is A', that is the *translation* AA'. But there is still a rotation which, the ds^2 having been given, may still be defined by an arbitrary law.

Having posed this, when one describes an infinitesimal closed contour starting from a point A and returning to it, the difference between the space under consideration and a Euclidean one is made manifest in the following fashion: attach a triad of reference to each point M of the contour; to go from the triad attached at M to one attached at M' infinitesimally nearby, one must effect an infinitesimal translation and rotation, of which one knows the components with respect to the moving frame attached at M. Imagine, then, that this set of infinitesimal displacements be effected *in a Euclidean space*, starting from an arbitrarily chosen initial triad at A. When the moving triad at point M of the non-euclidean space departs from A and returns, having traversed the infinitesimal closed contour, *one does not recover the initial triad in the Euclidean space*, but rather, in order to recover it, one must effect a complementary displacement, whose components will be well defined with respect to the initial triad at A. This complementary displacement is independent of the law by which one attaches a triad at each point M of the contour.

Thus, to each infinitesimal closed contour of the space are associated an infinitesimal translation and an infinitesimal rotation (of the same order of magnitude as the area bounded by the contour), which manifests the difference between this space and the Euclidean one. The rotation may be represented by a vector whose origin is at A, and the translation by a couple. One may thus demonstrate the following *conservation law:* If one considers an infinitesimal volume, *the vectors and couples associated to the surface elements which bound the volume are in equilibrium.*

Thus one has a geometrical image of a continuous medium which is in equilibrium under the action of elastic forces, but in that case where these forces are manifested on the surface elements of the volume not only by a single force (tension or compression), but also by a couple (torsion).

Return now to the case where one simply gives a ds^2. An easy calculation shows that among all the laws of mutual reference between two triads infinitesimally nearby and compatible with the given ds^2, *there is one and only one such law whereby the translation associated with every closed contour vanishes.* This is the law which leads to Levi-Civita's notion of parallel displacement. If one adopts this law, the couple (considered above) disappears, *and this is the reason why the elastic stress tensor satisfies a law of symmetry.*

In the general case where there is a translation associated with every infinitesimal contour, one may say that the given space differs from Euclidean space in two ways: first, by its Riemannian *curvature,* which gives a rotation, and second, by its *torsion,* which gives a translation.

In a space with curvature and torsion, the method of the moving triad permits, just as in Euclidean space, the construction of a theory of the curvature of lines and surfaces. A *straight line* will be characterized by the property of having at all points a zero (relative) curvature, that is to say it conserves step by step the same direction. *The straight line is therefore not necessarily the shortest distance from one point to another;* these lines coincide only in spaces devoid of torsion and in certain exceptional spaces with torsion.

A very simple example of that last case is the following: imagine a space F which corresponds point by point with a Euclidean space E, the correspondence preserving distances. The difference between the two spaces is the following: two orthogonal triads issuing from two points A and A' infinitesimally nearby in F will be parallel when the corresponding triads in E may be deduced one from the other by a given helicoidal displacement (of right-handed sense, for example), having as its axis the line joining the origins. The straight lines in F thus corresponds to the straight lines in E: They are the geodesics. The space F thus defined admits a six-parameter group of transformations; it would be our ordinary space as viewed by observers whose perceptions have been twisted. Mechanically, it corresponds to a medium having constant pressure and constant internal torque.

I would like to add that the preceding considerations, which, from the point of view of continuum mechanics, appeared in the beautiful work of MM.E. and F. Cosserat for the Euclidean case, likewise appear in Weyl's theory of generalized spaces, and may be generalized in similar fashion.

COMMENTS ON THE PAPER BY ELIE CARTAN: SUR UNE GENERALISATION DE LA NOTION DE COURBURE DE RIEMANN ET LES ESPACES A TORSION

Andrzej Trautman

Instytut Fizyki Teoretycznej
Hoza 69
00-681 Warszawa, Poland

This paper is a real gem, written in s style characteristic of Elie Cartan: it contains important new ideas, but no precise definitions, theorems or equations.

The author generalizes the notion of parallel transport of vectors, introduced by Levi Civita. This generalization is motivated by physical considerations: Cartan refers to his earlier paper on the stress-energy tnesor in Einstein's theory and to the work of the brothers E. and F. Cosserat on continuous media with an intrinsic angular momentum.

The geometry considered by Cartan is that of a three-dimensional manifold with a metric tensor g and a linear connection ω which is Euclidean - or metric - i.e. compatible with g. The condition of compatibility may be written as

$$Dg_{\mu\nu} = 0 \quad , \tag{1}$$

where D is the exterior covariant derivative, $Dg_{\mu\nu} = dg_{\mu\nu} - \omega^{\rho}{}_{\mu}g_{\rho\nu} - \omega^{\rho}{}_{\nu}g_{\rho\mu}$, and $\mu,\nu,\rho = 1,2,3$.

Cartan emphasizes that eq.(1) does not completely define the connection form $\omega^{\mu}{}_{\nu}$, namely

$$\omega^{\mu}{}_{\nu} = \gamma^{\mu}{}_{\nu} + \kappa^{\mu}{}_{\nu} \quad , \tag{2}$$

where γ is the Levi Civita connection and the tensor-valued form $\kappa_{\mu\nu} = g_{\mu\rho}\kappa^{\rho}{}_{\nu}$ is skew in the pair (μ,ν) , but otherwise arbitrary. He interprets formula (2) by saying that infinitesimal parallel transport defined by ω consists of a translation, i.e. parallel transport defined by the Levi Civita connection, and of a rotation given by κ .

To understand the rest of the paper, it is convenient to introduce, following Cartan, a moving frame, i.e. a field (θ^μ) of three linearly independent 1-forms (triad). In terms of these basis forms, the curvature and torsion 2-forms may be written as

$$\Omega^\mu{}_\nu = d\omega^\mu{}_\nu + \omega^\mu{}_\rho \wedge \omega^\rho{}_\nu = \frac{1}{2} R^\mu{}_{\nu\rho\sigma} \theta^\rho \wedge \theta^\sigma \quad ,$$

$$\Theta^\mu = d\theta^\mu + \omega^\mu{}_\nu \wedge \theta^\nu = \frac{1}{2} Q^\mu{}_{\rho\sigma} \theta^\rho \wedge \theta^\sigma \quad .$$

Cartan considers next a field of triads, defined by parallel transport along a closed curve (loop), and a radius vector along the loop. A vector field (u^μ) is parallel if it is covariantly constant,

$$Du^\mu = 0 \quad . \tag{3}$$

The integrability condition of (3) is

$$\Omega^\mu{}_\nu u^\nu = 0 \quad .$$

In a curved space, in general there are no parallel vector fields, but eq.(3) can always be integrated along a curve. When this is done along a small loop, the vector u changes, approximately, by

$$\Omega^\mu{}_\nu u^\nu$$

times the surface element spanned by the loop. Similarly, a radius (position) vector field (r^μ) satisfies

$$\nabla_\mu r^\nu = \delta^\nu_\mu \quad \text{or} \quad Dr^\mu = \theta^\mu \quad , \tag{4}$$

and the integrability condition

$$\Omega^\mu{}_\nu r^\nu - \Theta^\mu = 0$$

determines the change in r when the position vector field is built along a loop by integration of eq.(4). Cartan points out that the position vector is not only <u>rotated</u> - as is a vector undergoing parallel transport - but suffers also a shift or <u>translation</u> proportional to Θ^μ times the surface element spanned by the loop.

Introducing the completely skew tensor $(\eta_{\mu\nu\rho})$, $\eta_{123} = \sqrt{\det(g_{\mu\nu})}$, one can represent the density of rotation by the vector (-valued 2-form)

$$t_\mu = \frac{1}{2} \eta_{\mu\nu\rho} \wedge \Omega^{\nu\rho} \tag{5}$$

and the density of translation by the "couple"

$$s_{\mu\nu} = -\eta_{\mu\nu\rho} \wedge \theta^\rho . \tag{6}$$

The Bianchi identities

$$D\Omega^\mu{}_\nu = 0 \qquad \text{and} \qquad D\theta^\mu = \Omega^\mu{}_\nu \wedge \theta^\nu$$

together with eq.(1), imply the conservation laws italicized by Cartan in the sixth paragraph of the paper

$$Dt_\mu = 0 \quad , \tag{7}$$

$$Ds_{\mu\nu} = \theta_\nu \wedge t_\mu - \theta_\mu \wedge t_\nu \quad . \tag{8}$$

These equations are interpreted as conditions of equilibrium of a continuous medium under the action of elastic forces with a non-vanishing (if $s_{\mu\nu} \neq 0$) density of moments, i.e. under the action of torsion-inducing forces (seventh paragraph).

Vanishing of torsion is equivalent to $s_{\mu\nu} = 0$ and $\kappa^\mu{}_\nu = 0$, and, by virtue of eq.(6), implies the symmetry of the stress tensor $t^{\mu\nu}$, defined by

$$t^\mu = \frac{1}{2} t^{\mu\nu} \eta_{\nu\rho\sigma} \wedge \theta^\rho \wedge \theta^\sigma \quad .$$

Incidentally, all of these considerations generalize to higher-dimensional spaces, and in particular, to the four-dimensional spacetime. In the latter case, $\eta_{\mu\nu\rho}$ is replaced by $\eta_{\mu\nu\rho\sigma}\theta^\sigma$, where $(\eta_{\mu\nu\rho\sigma})$ is the completely skew tensor in four dimensions. An essential change occurs in eq.(7), which is replaced by

$$Dt_\mu = \frac{1}{2} \eta_{\mu\nu\rho\sigma} \theta^\sigma \wedge \Omega^{\nu\rho} \quad . \tag{9}$$

In a subsequent paper [1], Cartan develops a theory of space, time and gravitation based on the geometry of a four-dimensional space with a metric compatible with its linear connection, which need not be symmetric. In that theory, $t_{\mu\nu}$ is interpreted as the stress-energy tensor. Led by analogies with special and general relativity, Cartan imposes on it a conservation law of the form (7); together with the identity (9) this results in a highly restrictive algebraic constraint. Presumably, the constraint discouraged Cartan, who later never returned to his theory of 1923. According to our present views, conservation laws in any geometric theory of gravitation result from Bianchi identities. Therefore, eq.(9) should be accepted as a differential (local) conservation law. It has been shown [2] that from an isometry in spacetime, and eqs.(8) and (9), there follows a global conservation law, similar to the law known in the Riemannian case.

Let us now return to Cartan's short note in the Comptes Rendus. In its tenth paragraph Cartan defines straight (autoparallel) lines and points out that, in general, they are different from Riemannian geodesics (shortest curves). These two sets of lines coincide if and only if the torsion tensor $Q_{\mu\nu\rho} = g_{\mu\sigma} Q^{\sigma}_{\ \nu\rho}$ is completely skew. Such is the case of R^3 with a linear connection defined by helicoidal displacements. If $\theta^1 = dx$, $\theta^2 = dy$ and $\theta^3 = dz$, then the coefficients of the connection are $\omega_{\mu\nu} = \alpha\eta_{\mu\nu\rho}\theta^{\rho}$, where α is the pitch of the helicoidal motion. Both curvature and torsion are constant. This is a 'parity-violating' space: the torsion tensor defines a preferred orientation.

Acknowledgments

These comments have been written, at the suggestion of Friedrich Hehl, during my stay at the International Centre for Theoretical Physics, Trieste. I am grateful to the Centre for hospitality.

References

1. E. Cartan, "Sur les variétés à connexion affine et la théorie de la relativité généralisée", Ann. Ec. Norm. 40:325 (1923); 41:1 (1924).
2. A. Trautman, The Einstein-Cartan theory of gravitation, in "Ondes et Radiations Gravitationnelles", Colloque Intern. du CNRS, No.220, CNRS, Paris (1973).

INDEX

Abelian generators, 182
Affine motions, groups of, 80-83
AGCT Lie algebra, 213, see also
Lie algebra
d'Alembert operator, 74
Algebras
Clifford, 128, 182, 241
Fiber, 213
Grassman, 228
Poincaré, 228, 244, 249
Stony Brook, 221, 224
superconformal, 179-182,
see also Superalgebras
Almost complex structure,
defined, 283
ALSEP packages, on Moon, 329
Angular momentum
of black holes, 380
definitions in, 445-446
of galaxies, 378
of galaxy clusters, 378
of isolated systems, 435-438
vs. mass, 375
of plant-satellite systems,
377
of stars, 377
of stellar clusters, 377-378
Anholonomic connection
antisymmetry of, 26
torsion tensor and, 27
Anholonomic coordinate system,
72
Anholonomic curvature,,
P-transformation of, 22
Anholonomic frames, spinor
dynamics of, 64
Anholonomic Lorentz index, 76
Anholonomic tetrad frames, 8
Anholonomity, torsion and, 27
Anti-self-duality, 267, 270
Anti-self-dual vacuum space-
time, 312
Apollo 11 optical retro-
reflector, 331
Arecibo Ionospheric
Observatory, 320
Arecibo planetary radar,
characteristics of, 321
Argand plane, Riemann sphere
and, 289, 304
Asthenodynamics, Weinberg-Salam
model of, 197
Astrometric satellite, proposed,
353
Astronomical hierarchy, angular
momenta in, 376-379
Astronomical objects, mass-
angular momentum
diagram of, 375-381
Asymptotic Weyl curvature, 443
Axial form factor, 145
Axial vector, of tetrad field,
83-84
Axial weak current, gravity
and, 146
Axisymmetric space-time, 438
Axisymmetric stationary field,
Dirac test particle in,
86-90

Bargmann-Wigner vector, 134

Baud length, 320
Bays theorem, in probability estimation, 334-335
Bianchi identities, 44, 214, 249, 254
  contracted, 54, 105-108, 111-112, 114-116
"Big bang" model, 383
Bimetric formalism, 384-387
Bimetric general relativity theory, 383-404
Binary system, gravity research in, 318
Black holes
  angular momentum of, 380
  Schwarzchild sphere and, 404
Bondi news, space-time and, 441-442
Bosons
  dimensions of, 229-230
  massless, 233
  spin of, 229
Bra-ket space, 412
Breitenlohner equations, 255
BRS equations, 210

Cabibbo angle, 161-162
Cartan curvature, comments on, 493-496
Cartan-Maurer equations, 200-204, 209-210
Cauchy data, symmetry group of mappings of, 174-176
Cauchy problem, local Lorentzinvariance and, 117-120
Charge conjugation matrix, 241
Charge conjugation-parity violation, gravity-induced, 370-372
Charles Stark Draper Laboratory, 317
Christoffel symbol, 27, 238
  defined, 384
  historical form of, 477,479
  variation of, 481
Classical mechanics, spin particles and, 429-430
Clifford algebra, 128, 182, 241
Coboundary concept, 306
Cochain, defined, 306
Cocycle, defined, 306
Cohomology, sheaf, 303-310
Cohomology group, 306
Commutation relations, of global P-transformations, 15-16
Compactified complex Minkowski space, 297, see also Minkowski space
  Poincare invariance in, 298-300
Complexification, in complex vector space, 276-278
Complex linear vector space, 275-276
Complex manifolds, 275-285
  deformation of, 309-311
  extension of, 281-285
  large family of, 281-282
Complex numbers, in twistor theory, 289
Complex scalars, multiplication of, 276
Complex structure
  "almost," 283
  tangent-space, 282
Complex vector space
  complexification in, 276
  complex structure and, 276-277
Conservation law
  equilibrium and, 490
  spin particles and, 424-425
Conservation principles, differential conditions for, 484-487
Conservation theorem, gravitational equations and, 484-487
Contour integrals, in twistor theory, 303-309
Contracted Binachi identities, 111-112, see also Bianchi identities
  defined, 105
  Noether forms of, 114-116
  notation for, 106-108
Cosmology, in bimetric general relativity theory, 389-400

Covariance
  Einstein's idea of, 408-411
  principles of, 407, 417
  quantum physics and, 407-421
COW experiment, relativistic, 368-369
Curvature-square term, rotons and, 56
Curvature tensor, 385

Dali group manifold, see Soft (Dali group manifold)
De Donder condition, 386
Deformation tensors, 19
De Sitter algebra, 247-248
De Sitter gravity, 218
Diffeomorphisms, 122-123
  commutator of, 175
Differential geometry, Lie groups and, 197-203
Differential Very Long Baseline Interferometry, 338
Diff(M), see Diffeomorphisms
Dipole moment
  electric, 369
  Hansen, 437-438
Dirac electron
  motion equations for, 36 n.
  in rotational potential measurement, 36
Dirac equation
  Dirac matrices and, 362
  general-covariant, 414
  in non-Riemannian space-time, 127-131
  quasiclassical limit of, 98-103
Dirac field, Lagrangian of, 56-57, 142
Dirac Lagrangian
  of Dirac field, 56-57, 142
  local Lorentzinvariance and, 119
Dirac large numbers hypothesis, 167-169, 338-339
Dirac mappings, subset of , 175-176
Dirac mass operator, 99
Dirac matrices, constant, 361
Dirac-matter test particle, 134-136
Dirac-Maxwell theory, 57
Dirac particle, 86-90
  dynamics of, 98-99
  in Riemann-Cartan space, 141
  "Zitterbewegung" of, 100
Dirac perturbation theory, 419-420
Dirac spinor, hadron and, 152
Dirac theory, spin motion and, 98
Dirac-Weyl neutrino equation, 303
Divergence relations, integration of, 131-134
DLBI, see Differential Very Long Baseline Interferometry
Doability quotient, 368

Earth, motion of around Sun, 337-338
ECSK (Einstein-Cartan-Sciama-Kibble) theory, 44, 52-54
Einstein-Cartan theory, 2, 143, 158
"Einstein choice," 79-80
Einstein equations, see also Einstein field equations
  curved twistor space and, 311-315
  holomorphic, 283
Einstein equivalence principle, 24, see also Equivalence principle
Einstein field equations, 385-387
Einstein general relativity theory, 63-64, see also General relativity
Einstein gravitation theory (1915), 2
Einstein Lagrangian, locally Lorentz invariant, 76
Einstein linearized field equation, 75

Einstein tensor, 79, 239-240, 387
Electric dipole moment, 369
Electromagnetic field, of hadron, 163-166
Electron field, gravitational moments of, 57
Elementary particle, see also Particle; Particle physics
  angular motion of, 379
  spin density of, 168-169
  torsion and strong gravity in, 139-169
Energy-momentum density tensor, 385-387
Energy-momentum tensor
  divergence of, 126
  of gravitational field, 42 n.
  macroscopic, 35
  in special relativity, 128
Energy-momentum theorem, 71
Equivalence principle
  Einsteinian, 24
  in general relativity theory, 94
  local, 25
  lunar laser ranging data and, 338
  minimal and non-minimal coupling in, 34
  in Riemann-Cartan geometry, 93-103
  strong and weak, 93-96
Experimental gravitation, measurement of, 317-355, see also Gravitation

Fermi-Bose symmetry, 227-229
Fermion-antifermion system, 371
Fermions
  massless, 233
  spin of, 229
Feynman diagram, for gravitons, 471
"f" gravity, 145
Fiber algebra, 213
Field, holonomic index of, 9
Field equations
  Euler-Lagrange equations as, 108
  in Poincaré gauge field theory, 44
Field Lagrangian
  Lorentz invariance of, 72 n.
  weak field approximation and choice of, 73-80
Field momentum, in Poincare gauge field theory, 44
Field of a particle, 400-404
Field theory, classical, 9, see also Poincaré gauge field theory
Fierz rearrangements, 240, 242-243
Finiteness principle, 387
Foldy-Wouthuysen transformation, 363
Fourier transform model, of Martin surface, 323-325
Free particle, at maximum velocity, 430
Free spin particle, motion equations for, 431-432
Function L, structure of, 479-480
Fundamental covariance, principle of, 407, 417, 420-421

Galaxies, angular momentum of, 378
Galaxy clusters, angular momentum of, 378
Galilean Principle of Relativity, 409
Gauge bosons, propagating, 49
Gauge field equations, 30-44
  Reimann-Cartan space-time and, 5
Gauge field Lagrangian, 37-40
  plausible structure of, 56
  selection of, 45-49
Gauge field momentum current, 42
Gauge field spin current, 42
Gauge field strength, 44
Gauge field theory, see Poincaré gauge field theory

Gauge formalism, coordinate and Lorentz invariance in, 20
Gauge group
  multisymplectic bundle, 112-114
  Poincaré group as, 120-122
Gauge potentials, 17-20, 44
General field equations, differential identities and, 123-124
  see also Field equations
General relativity, see also General Principle of Relativity; Relativity
  angular momentum of isolated systems in, 435-448
  bimetric, 383-404
  covariance and, 408-411
  equations of motion in, 125-126
  equivalence principle in, 336-337
  Hamiltonian formulation of, 174
  Palatini formulation of, 233
  quantization of, 257
  self-dual fields and, 257
  self-duality principle and, 267-273
  terrestrial experiments in, 368-370
  tests of at quantum level, 359-373
General relativity theory, 63, 416-417, see also General relativity
  energy-momentum theorem of, 71
  equivalence principle in, 94
Ghost fields, 210
Global interial frames, 10-11
Global P-transformation, 16, see also P-transformation
  Minkowski space-time and, 29
Global supersymmetry, algebra of, 252-253
Gordon decomposition argument, 56-59
GR, see General relativity; General relativity theory
Graded Jacobi identity, 179
Graded Lie algebras, 179
Grassman algebra, 205, 228
Grassman ring, 183-186
Gravitation, see also Gravity
  experimental, 315-355
  gauge symmetries and, 117-124
  geometric theories of, 1-4
  metric structure in measurement of, 1-2
  Newtonian theories of, 2
  Nordström theory of, 3
  Poincaré gauge field theory of, 5-59
  Riemann-Cartan theories and, 2
Gravitational equations, invariant deduction of from Hamilton principle, 477-487
Gravitational field
  energy-momentum tensor of, 42 n.
  neutron spin precession in, 369
Gravitational Lagrangian
  curvature and, 110
  general field equations for, 123-124
Gravitational radiation, from atomic processes, 467
Gravitation measurements
  parameter estimation in, 334
  by planetary landers, 328-329
  radar in, 319-325
  radio interferometry in, 329-331
  from spacecraft, 326-329
Gravitation theory, development of, 3
Graviton photoproduction, in static electromagnetic fields, 467-475
Gravitons
  defined, 49
  photon conversion to, 471
  in Poincaré gauge field theory, 5, 48-50
  propagating, 54-56
Gravity, see also Gravitation
  de Sitter, 218
  local gauge theory of, 231
  macroscopic, 27
  particle physics and, 7

Gravity (continued)
  Poincaré group and, 11-12
Gravity-induced charge conjugation-parity violations, 370-372
Gravity wave detectors, 318 n.
Group manifold
  rigid, see (Rigid) group manifold
  spontaneous fibration of, 213

Hadron
  Dirac spinor and, 152
  electromagnetic field of, 163-166
Hadronic characteristic radius, 165
Hadronic current, 153
Hadronic matter density, 159
Hadronic spin density, 167
Hadronic weak current, 145
Hamiltonian, time-dependent, 419-420
Hamiltonian constraints, Lie algebra and, 175
Hamilton variational principle, 480-481
Hansen dipole moment, 437-438
Hausdorff space, 305
Heisenberg theory, in quantum mechanics, 413-414
Helicity operator, in twistor quantization, 300
Hermiticity, of complex matrix, 293
Hilbert space, 288
Hilbert space vector, 414
Holonomic index, 9
Hydrogen atom, in gravitational field, 361-367

Induction tensor density concept, 37 n.
Inertial frames
  "dragging" of, 88
  local kinematic, 23-25
Inertial mass, weak equivalence principle and, 94
Inertial spin, Lamor frequency and, 94
Infinitesimal generator, 44
Infinitesimal Lorentzrotation, 121
Infinity twistors, 298
Interferometry, see also POINTS
  in gravity measurements, 329-331
  Very Long Baseline, 329, 338

Jacobi identity, 69, 211, 244
Jupiter, first-order light bending by, 354
Jupiter-Sun system, quadrupole radiation from, 473

Kahler manifold, 285
Killing-Carton metric, 453-454
Killing equations, 452-454
Killing field
  asymptotic, 441
  rotational, 436, 438
Killing metric, 186, 218
Killing vector, 436
Killing vector fields, 208, 211-212
Klein-Gordon equation, 148
Klein-Gordon field, 280
Komar scalar, 436-437
Korteweg-de Vries equations, 44

Lagrangian (function)
  differential forms for, 55 n. 109
  of Dirac field, 56-57, 142
  scalar-field, 35
  gauge-field, 32, 37-40
  matter-field, 110, 122
  pseudo-closed, 216
  semileptonic, 149-152
  supergravity, 220-221
Lagrangian field theory, super Lie algebra and, 246
Lagrangian multiplier, 51
  potential variation and, 70 n.
Lagrangian quadratic, for WRSS groups, 215-218
Lamor frequency, inertial spin and, 94
Large numbers hypothesis, 167-169, 338-339

Laser ranging system, in gravitational measurements, 331-333
Leptonic processes, equation for, 144
Lepton-nucleon interaction, 144
Levi-Civita connection, 97, 127, 283-284, 493
Levi-Civita tensor, 141
Lick Observatory, 331
Lie algebra
  commutators and anticommutators in, 244
  vector field and, 231-232
  world points in, 175
  Yang-Mills potential and, 262
Lie group manifold, 177
Lie groups, differential geometry and, 197-203
Lie subalgebra, 181
  grade zero, 193
Lie superalgebras, 178-179
  classification of, 186-187
Light, red shift of, 470
Liouville equation, 450-451
  general solution of, 452
LIS, see Local inertial system of reference
LNH, see Large numbers hypothesis
Local equivalence principle, 25, see also Equivalence principle
Local interial system of reference, 93
Local kinematical inertial frames, 23-25
Local Lorentzinvariance, Cauchy problem and, 117-120
Local orthonormal tetrad field, 67
Local P-transformations, see also P-transformations
  closure of, 11, 21-23
  of gauge potentials, 17-20
  infinitesimal functions and, 67-68
  unchanged frames and coordinates in, 64
Local P-transformed field, 121
Lorentz bundle, 219
Lorentz covariance, 409
Lorentz double-matrix, 218
Lorentz invariance, 65
  strong vs. weak, 20 n.
Lorentz rotation, infinitesimal, 121
Lunar laser ranging system, 331-333
Lunar retroreflectors, 331-333

McDonald Observatory, 332
Macroscopic energy-momentum tensor, 35, see also Energy-momentum tensor
Macroscopic gravity, 27
Macroscopic matter, degenerate case of, 34-40
Majorana fields, 230
Majorana spinors, 205
  symmetry properties for, 242
Mariner 9 spacecraft
  radiation pressure from, 328
  time-delay observations from, 344-345
Mark I VLBI system, 330
Mars
  empirical surface correlation function for, 325
  gravitational potential of, 328
  Mariner 9 time-delay observations on, 344-345
  perihelion advance on, 340
  precession rate for, 346
  retroreflectors for, 333
  Viking mission to, 344
Martian surface, Fourier transform model of, 323-325
Mass, vs. angular momentum, 375
Mass-angular momentum diagram, for astronomical objects, 375-381
Mathisson equation, 133
Mathisson force, on spinning particle, 33
Mathisson variational principle, 425-427
Matter
  action function of, 10-11

Matter (continued)
coupling of space-time to, 30-44
gauge fields of, 8
macroscopic, 34-40
Poincaré behavior of, 8
scalar, 64-73
spin aspect of, 64
Matter distribution, equivalence property and, 64
Matter field distribution, 14-15, 18
Matter field Lagrangian, 110, 122
Matter fields, P-transformations of, 12-15
Maxwell equations, self-dual solution of, 257
Mercury, perihelion advance in, 340
Metric affine theory, 43
Metric compatible connection, 13
Metric field, post-Newtonian expansion of, 83-84
Metric postulate, of space-time physics, 27
Metric structure, compatibility of, 2
Metric tensor, 1-2
Microphysics, physical systems structure in, 423
Minimal coupling principle, 140-145
equivalence principle and, 34
Minkowskian world, gauge frames and, 24
Minkowski metrics, 8, 424
notation for, 107
Minkowski space, 291, 424, 442
compactified complex, 297-300
complexified, 296
twistors and, 461-465
Momentum conservation law, volume-force densities and, 33
Momentum current, of gauge fields, 42
Moon
ALSEP packages on, 329
motion of around Sun, 337-338
retroreflectors on, 331-333

Multisymplectic bundle, gauge group action and, 112-114

Neutron spin precession, in gravitational field, 36
Newtonian theories, gravitation and, 2
Noether forms, contracted Bianc identities and, 114-116
Noether identities
of general gauge field Lagrangian, 37-40
of matter Lagrangian, 31-34
Nonlinear equations
group theoretic aspects of, 453-455
isometries and general solutions of, 449-455
Non-Riemannian space-times, test particle motion in 125-136
Nordström theory, 3
Null infinity, angular momentum and, 440-442
Null twistor, 295

Orientation, local standard of, 13
Orthonormal tetrad field, 67
Orthnormal tetrad frames, 13-14
Orthosymplectic groups, superunitary, 196
Orthosymplectic sequences, 188-191

Palatini formulation, 233, 477-487
Palatini variation, 40 n.
Papapetrou equation, 133, 142
Parameter estimation, in gravitation measurement, 334
Parametrized post-Newtonian parameters, 319, 363
Parity states, gravity-induced admixture of, 364-366
Particle, translational motion of, 96
Particle field, 400-404

Particle physics, 7
Pauli-Lubanski spin vector, 100, 292
Pauli spin matrices, 259
Penrose twistor approach, 174
PEP, see Planetary Ephemeris Program
Perihelion advance, of inner planets, 340-341
PG, see Poincaré gauge
Photons, massless, 7
Photoproduction processes, graviton production in, 465-475
Physical systems structure, in microphysics, 423
P-invariance, local, 17
Planck length, 54, 72, 124
Planet
  light reflection in reradiation from, 328
  perihelion advance of, 340-341
  radar scattering from, 320-321
Planetary Ephemeris Program, 319, 323
Planetary lander, 328-329
Planet-satellite systems, angular momentum of, 377
Poincaré algebra, 228
  extension of, 244
  graded, 249
Poincaré-Birkhoff-Witt theorem, 178
Poincaré gauge, 8
Poincaré gauge field theory, 5-59
  kinematic content of, 94
  macroscopic limit of, 63-90
  scalar matter distributions in, 66-73
Poincaré group, 203-207
  as gauge group, 120-122
  gravity and, 11-12
  in special relativity, 5
Poincaré invariance
  in compactified complex Minkowski space, 298-300
  world points and, 173
Poincaré subalgebra, 182
Poincaré subgroup, angular momentum of, 441, 447
Poincaré tensors, gauge fields as, 68
Poincaré transformation, see P-transformation
Points, existence of in space-time, 287
POINTS (Precision Optical Interferometry in Space), 317, 352-355
PPN, see Parametrized post-Newtonian parameters
Precision Optical Interferometry in Space, 317, 352-355
Principal bundle, 177, 197, 212, 219
  defined, 208
  gauge theory and, 207-210
Principle of Coordinate Covariance, 407, 417
Principle of Equivalence, see Equivalence principle
Principle of Fundamental Covariance, 407, 417, 420-421
Principle of General Covariance, 408
Principle of Lorentz Covaraince, 409
Principle of State-Operator-Covariance, 407
Probability estimation, Bayes' theory in, 334-335
Projective twistor sphere, 265-266, 292
Pseudo-closed Lagrangian, 216
Pseudocurvature, 216
Pseudo-Riemannian metric, 105
Pseudo-Riemannian real space, 287
P-tensor, 21
P-transformations
  of anholonomic curvature, 22
  commutation relations of, 15-16
  global, 16, 29
  local, see Local P-transformations

P-transformations (continued)
  of matter fields, 12
Pulsars, gravity research on, 318

Quantum field theory, 420-421
Quantum laws, and Principle of Fundamental Covariance, 407-408
Quantum mechanics
  Dirac method for, 412
  general relativity and, 411
  local validity of, 7-8
Quantum physics, covariance and, 407-421
Quantum systems, gravity effects in, 360
Quantum theory, change of traditional basis of, 413-417
Quasi-linearity hypothesis, 45-48

Radar scattering, from planets, 320-321
Radar signals, in gravity measurements, 319-325
Radio interferometry, in gravitation measurements, 329-331
Red shift, 470
Relative orientation, local standard of, 67
Relativistic COW experiment, 368-369
Relativistic quantum gravity effects, 360
Relativistic theory, defined, 1
Relativity
  Galilean principle of, 409
  special theory of, 409-410
Relativity theory
  Einsteinian vs. bimetric, 383
  metric parameter β in, 340-341
Retroreflectors, on lunar surface, 331-333
Ricci calculus, 410
Ricci rotation coefficients, 95
Ricci tensor, 151
Riemann-Cartan geometry, equivalence principle in,
Riemann-Cartan space-time
  anholonomic and holonomic views of, 25-29
  Dirac particle in, 141
  freely falling non-rotating elevator in, 24
  gauge field equations and, 5
  limiting cases in, 28
  propagation equation for test spin in, 86-87
  P-transformations from scalar field in, 66-67
Riemann-Cartan theories, gravitation and, 2
Riemann curvature, generalization of, 489-491, 493-496
Riemannian geometry
  Einstein theory and, 2-3
  teleparallelism geometry and, 102-103
Riemannian metric, complex, 283
Riemannian normal coordinate system, 95-96
Riemann sphere, Argand plane and, 289, 304
Riemann symbols, variation of, 482
Riemann tensor, 108
  cyclic identity in, 272
  decomposition of, 267
  tetrad components of, 268
(Rigid) group manifold, 177, 21
  defined, 197
RNCS, see Riemannian normal coordinate system
Rotational potential, momentum law and, 36
Rotons
  curvature-square term and, 55-56
  defined, 49-50
  in Poincaré gauge field theory, 5, 48-50
  propagating, 54-56
  suppression of, 50-54
RQG effects, see Relativistic quantum gravity effects

Scalar field Lagrangian, 35

Scalar matter, dynamics of, 64-65
Scalar matter distributions, Poincaré gauge kinematics and, 66-73
Schrödinger equation, 363, 413-414, 418-419
Schrödinger theory, in quantum mechanics, 413-414
Schwarzschild metric, 361, 363
Schwarzschild radius, 166, 401-402
Schwarzschild solution, for particle field, 383, 401-402
Schwarzschild sphere, 403-404
Self-dual fields, 257-273
Self-duality
  general relativity and, 267-273
  twistor theory and, 288-289
Semileptonic Lagrangian, 149-152
Semileptonic processes, equations for, 144
Semileptonic weak interactions, 147-148
Shapiro time-delay effect, 343
Sheaf cohomology theory, 303-310
Short baseline interferometry, 342, see also Very short baseline interferometry
Sine-Gordon equations, 449, 451
Soft (Dali) group manifold, 177, 197, 210, 212
Solar potential, 317
Solar probe, 317
Solar probe, 317, 349-351
Solar system
  complexity of, 317, 335
  model of, 335
  relativity testing in, 317
Solar system bodies, ephemerides of, 319
Solar system dynamics, study of, 323-325
Space(s)
  law of mutual reference and, 490
  with torsion, 489-491
Spacecraft
  Doppler tracking of, 326
Spacecraft (continued)
  gravitation measurements from, 326-329
  location of, 327
  time-delay normal points for, 328
Space $E_p \otimes C_p$, tilting of, 459-461
Space-time
  angular momentum definitions in, 445-446
  asymptotically flat, 438-440
  axisymmetric, 438
  coupling of matter to, 30-44
  curved twistor space and, 311
  as manifold, 173
  Riemann-Cartan, see Riemann-Cartan space-time
Space-time geometry, 12-30
Space-time physics, metric postulate in, 27
Spatial infinity, angular momentum and, 442-445
Special Principle of Relativity, 409
Special relativity
  Dirac field energy-momentum tensor in, 128
  local validity of, 7-8
Special Theory of Relativity, 409
Spherical target, Doppler resolution of, 322-323
Spin bivector, 427
Spin connection, field equation for, 237
Spin current, of gauge fields, 42
Spin density
  of elementary particle, 168
  of universe, 168
Spin-fluid theory, spin particle motion equations and, 428-429
Spinless particles, in Riemann geometry, 4
Spin motion, in Riemann-Cartan space-time, 86-87
Spinning matter, Weyssenhoff model of, 160

Spinning particle, Mathisson force on, 33
Spinor connection coefficient, transformation law for, 270
Spinor dynamics, anholonomic frames and, 64
Spinor formalism, 258-262
Spinor matter, teleparallelism theory and, 75 n.
Spinor parity, 182
Spinors
  defined, 258
existence of, 64
  visual geometry of, 457-465
Spinor space model, 458-460
Spin particles
  bi-point model of, 432-433
  defined, 423
  derivation of equations for, 424-428
  relativistic equations of motion for, 423-434
Spin-torsion coupling constant, 140-145
Spin vector, total, 438
Spontaneous fibration, of group manifold, 213, 218
Star
  angular momentum of, 377
  apparent position of, 354
Stark effect, 361, 367
Stark Hamiltonian, 371
Starlight, solar deflection of, 341-342
Stefan-Boltzmann constant, 399
Stellar clusters, angular momentum of, 377-378
Stony Brook algebra, 221, 224
Straight line, as "shortest distance," 491
Strong coupling constant, 66 n., 156-157
Strong equivalence principle, 95
Strong gravity, 139-140
  coupling constant in, 156-157
  defined, 49
  particle electromagnetic field and, 162-169
Strong gravity (continued)
  torsion and, 158-159
  weak interaction and, 145-162
Subalgebra, Lie, 181, 193
Subradar point, 320
Sun-Earth-Moon system, 337-338
"Sun Grazer" study, 350
Superalgebras, 178-179
  Cartan-type, 193
  classical, 187-191
  exceptional, 191-193
  hyperexceptional, 191-193
  Lie, 178-179, 186-187
  nonclassical, 193
  supertraceless, 187-188
Superconformal algebras, 179-18
Super de Sitter group, 219
Superdeterminant, 185
Supergauges, 178
Supergravity, 219-224
  defined, 228-229
  diffeomorphisms and, 174
  extended geometric, 224
  formulation of, 231-232
  gauge action invariance in, 236-240
  as mathematical theory, 227
  quantum gravity and, 197
  superspace and, 253-255
Supergravity Lagrangian, 220-22
Supergroups, 194-196
Super Lie algebra, Lagrangian field theory and, 246, see also Lie superalgebras
Superlinear sequence, 187
Supermanifolds, 183-186
Supermatrix, 185
Super-Poincaré algebras, 179-182, 244
Super-Poincaré groups, 177, 203-207, 219
Superspace, 249-255
  defined, 250
  structure of, 250-251
  supergravity and, 253-255
Supersymmetry, 174, 178, 191
  global, 252

Supersymmetry transformations, 231
Supertranslational ambiguity, 441, 448
Supertranslational subgroups, 443
Superunimodular supergroups, 196
Superunitary supergroups, 196
Symmetry laws, 491
Symplectic structure, of two-dimensional complex linear vector space, 258

Teleparallelism theory, 50-52, 54, 93
  Riemannian geometry of, 102-103
  spinor matter and, 75 n.
Test particles, motion of in non-Riemannian space-time, 125-136, see also Dirac particle
Tetrad field
  axial vector of, 83
  energy-momentum tensor density of, 71
  post-Newtonian expansion of, 83-86
Tetrad field equation, affine motion groups and exact solutions of, 80-83
Thirring-Lense effect, 88
Thomas precession, law of, 97
Time-delay normal points, for orbiting spacecraft, 328
"Tired light" phenomenon, 470
Tordions, 49 n.
Torsion
  strong gravity and, 158-159
  weak interaction and, 50 n.
Torsion tensor
  axial vector of, 84
  equations for, 108
  holonomic components of, 27
Twistor(s)
  defined, 291-294
  Minkowski space and, 461-465
  visual geometry of, 457-465
Twistor norm, 463
Twistor quantization, 300-303
Twistor spaces, curvature of, 311-315
Twistor theory, 287-315
  complex field in, 288
  contour integrals and sheaf cohomology in, 303-307
  "success" of, 288
  twistor description in, 291-294
  twistor functions in, 305-308
Twistor wave-function, 300

Unitary supergroups, 196
Universal spin precession, 97
Universe
  dust-filled, 391-397
  isotropic, 389-391
  radiation-filled, 397-400
  rest-frame of, 388
  spin density in, 168-169
  total spin of, 168

Variational principle
  Hamiltonian, 480-481
  spin particle motion equations and, 431-433
Vector meson dominance hypothesis, 162
Venus Orbiting Imaging Radar, 317, 341, 351-352
Venus radar mapping mission, 317
Very Long Baseline Interferometry 329
Viking planetary landers, 328-329,
  time-delay effects for, 347
Viking Relativity Experiment, 344
VLBI, see Very Long Baseline Interferometry
VOIR, see Venus Orbiting Imaging Radar

Weak equivalence principle, in Riemann-Cartan geometry, 93-103
Weak field approximation, Lagrangian choice and, 73-80
Weak gravity, Einsteinian, 49
Weak interaction, strong gravity and, 145-162

Weakly reducible symmetric groups, 213
Weitzenböck space, 120
Weyl curvature, asymptotic, 443
Weyl curvature tensor, 311
Weyssenhoff fluid model, 160
Weyssenhoff variational principle, 432
World lines, 287
World points, 173-176
WRS group, 213
WRSS groups, 213-218

Yang-Mills fields, 262-267, 271, 288
Yang-Mills gauge potential, 209
Yang-Mills gravity, 49
Yang-Mills Lagrangian, 210
Yang-Mills potential, 262
Yang-Mills tetrad field equation, 63
Yang-Mills theory, 42 n., 234, 248

Zeeman effect, 361
Zeeman Hamiltonian, 371